CHEMISTRY RESEARCH AND APPLICATIONS

APPLIED SURFACE CHEMISTRY OF NANOMATERIALS

Chemistry Research and Applications

Additional books in this series can be found on Nova's website
under the Series tab.

Additional e-books in this series can be found on Nova's website
under the e-book tab.

Nanotechnology Science and Technology

Additional books in this series can be found on Nova's website
under the Series tab.

Additional e-books in this series can be found on Nova's website
under the e-book tab.

CHEMISTRY RESEARCH AND APPLICATIONS

APPLIED SURFACE CHEMISTRY OF NANOMATERIALS

MOHAMED M. CHEHIMI
AND
JEAN PINSON
EDITORS

nova publishers
New York

For permission to use material from this book please contact us:
Telephone 631-231-7269; Fax 631-231-8175
Web Site: http://www.novapublishers.com

Library of Congress Cataloging-in-Publication Data

Applied surface chemistry of nanomaterials / [edited by] Mohamed M. Chehimi and Jean Pinson (University Paris Diderot, Paris, France, and others).
pages cm
Includes bibliographical references and index.
ISBN 978-1-62808-351-4 (hardcover)
1. Nanostructured materials--Surfaces. 2. Surface chemistry. I. Chehimi, Mohamed Mehdi, editor of compilation. II. Pinson, Jean, editor of compilation.
TA418.9.N35A676 2013
620.1'15--dc23

2013022833

Published by Nova Science Publishers, Inc. † New York

CONTENTS

Preface **vii**

DIAMOND **1**

Chapter 1 Diamond Nanoparticles: A Review of Selected Surface Modification Methods for Bioconjugation, Biology and Medicine **3**
Sabine Szunerits and Rabah Boukherroub

Chapter 2 Diamond Nanostructures: Technologies, Properties and Electrochemical Applications **33**
Nianjun Yang, Waldemar Smirnov and Christoph E. Nebel

CSP2 **55**

Chapter 3 Chemistry on Graphene **57**
Cécilia Ménard-Moyon

Chapter 4 Carbon Nanotubes: Surface Modification and Applications **95**
Mohamed M. Chehimi, Jean Pinson and Zakaria Salmi

Chapter 5 Property-Oriented Surface Engineering of Carbon Nanotubes for Polymer Nanocomposites **145**
Jin-Kai Yuan, Sheng-Hong Yao, De-Long He and Jinbo Bai

METALS **171**

Chapter 6 Atomic Sized Metallic Point Contacts: Fabrication Methods and Characteristics of Quantized Conductance **173**
Jalal Ghilane and Jean Christophe Lacroix

Chapter 7 Preparation, Surface Modification and Functionalization of Anisotropic Ferromagnetic Metal Nanoparticles: Towards Multifunctional Materials **205**
Jean-Yves Piquemal, Kahina Aït Atmane, Guillaume Viau, Mona Ibrahim, Philippe Decorse, Aazdine Lamouri and Mohamed M. Chehimi

Chapter 8	Gold Nanoparticles at Polymeric Interfaces *Samia Mahouche-Chergui, Sandrine Lépinay, Mourad Lakhlifi, Mohamed Guerrouache, Mohamed M. Chehimi and Benjamin Carbonnier*	**241**
Polypyrrole Nanostructures		**255**
Chapter 9	Nanostructured Polypyrrole Materials: Focus on Templateless Synthetic Methods and on Some Applications *Catherine Debiemme-Chouvy*	**257**
Chapter 10	Carbon Nanotubes/Polypyrrole Hybrids *Matej Mičušík and Mária Omastová*	**279**
Insulating-conductive Composites and Catalytic Materials		**301**
Chapter 11	Surface Coating of Soft Materials with Conducting Polymer-Metal Nanocomposites *Syuji Fujii and Yoshinobu Nakamura*	**303**
Chapter 12	Carbon Fiber and Nano/Micro Hybrid Structures and their Interface in Advanced Composites *Delong He, Jing Zhang, Johan Saba, Jinkai Yuan, Fabien Miomandre and Jinbo Bai*	**319**
Index		**347**

PREFACE

Nanoscience and nanotechnology are blooming in scientific publications and patents, one finds 96000 papers in the Web of Science database and 21200 patents in the FPO database under the heading "nano". These numbers give an idea of the present impetus of the research on these topics.

At the same time nanotechnology is becoming part of our life. Microelectronics that is in the heart of our computers and smartphones was probably the first industry to jump in the domain below 100 nm; the 90 nm technology appeared in 2004 and the last generation of transistors and their connections are manufactured with the 22 nm technology. Polymer-carbon nanotubes (with diameters well below 100 nm) are now commercially available, for example one can now buy tennis racquets and bicycles made of carbon nanotube composites. There are also several companies that produce materials for nanomedicine such as gold nanoparticles for drug delivery or radiation therapy enhancement, silver nanoparticles for fighting infections and healing wounds. Catalytic nanoparticles are now marketed as catalysts for fuel cells.

These achievements have been possible thanks to the development of analysis tools such a transmission electronic microscope (TEM), scanning tunneling microscope (STM) that permit to observe nanoparticles, nanotubes,… down to a few nm, but also very sensitive IR, Raman, XPS, ellipsometry machines that can analyze down to monomolecular films (monolayers).

Many of these nanotechnologies are based on nanoobjects, some of them have been known for a long time, but could not be properly characterized. Gold nanoparticles have been manufactured by A. Cassius since 1679 for colouring glasses with purple colors. Silver *nanoparticles* have a long history in medicine. On the contrary some nanoobjects at the heart of nanotechnologies have been discovered recently: fullerene in 1985 (Harold Kroto, Robert Curl and Richard Smalley), nanotubes in 1991 (Sumio Iijima), graphene in 2004 (Andre Geim and Konstantin Novoselov). There are now a large number of nanoobjects described in the literature: nanoparticles (made of many materials: gold, iron oxides, diamonds,…); semiconducting quantum dots where the electrons are confined and quantized; nanowires (made of metals, semiconductors or dielectrics, but also organic nanowires).

The common property of all these nanoobjects is the increasing importance of their surface compared to their volume as the size decreases and single wall nanotubes or graphene can be considered as pure surfaces. Therefore most of their properties are expressed through their surface, this situation leads to some important drawbacks, one of them is their tendency to agglomerate and therefore lose their nano character. But conversely the importance of their surfaces opens new perspectives in their potential use: if one is able to confer new properties to their surface, new applications can be expected. It is the object of this book to describe the large number of surface modifications of these nanoobjects and the numerous applications of these surface functionalizations.

The first part of the book deals with carbon that can be hybridized sp3 or sp2. Diamond nanoobjects are constituted of sp3 carbons, while nanotubes and graphene are from sp2 carbons.

Chapter 1 by Sabine Szunerits and Rabah Boukherroub deals with nanodiamonds (ND); they first describe their preparation and the preparation of surface functionalized ND. Then they describe the covalent modification of these species but also their post functionnalization that involves reacting the first formed chemical function with other reactive molecules thus opening a variety chemical structures on the surface of ND that open new applications by binding to therapeutic materials.

Chapter 2 by Nianjun Yang, Waldemar Smirnov and Christoph E. Nebel describes diamond nanostructures such as nanopillars and nanowires; they first report on the different methods that permit to create these structures and then on their properties (interfacial properties, electrochemical activity and their functionalization. The last part of the chapter describes the applications of these nanostructures for the detection of DNA hybridization and the direct electrochemistry of cytochrome c.

Chapter 3 by Cécilia Ménard-Moyon concerns graphene, the last discovered of these carbon nanoobjects. The introduction describes the structure of graphene and the interest of its functionalization, the second part focuses on covalent functionalization where a chemical bond is formed between the surface and the modifying group; the third one presents the non-covalent functionalization such as π-stacking and interactions with surfactants and polymers that does not disrupt the extended π-conjugation of graphene.

Chapter 4 Mohamed M. Chehimi, Jean Pinson and Zakaria Salmi describe the surface Mmodification of carbon nanotubes (CNT) and their applications. The introduction deals with the different exceptional properties of CNT, but also with the difficulties in using these species. The second part concerns the many mathods that are now available for the covalent surface modification of CNT, while the third one is concerned with non covalent functionalization. Not surprisingly, the chemical methods that permit the functionnalization of grapheme and nanotubes are somewhat similar; among the covalent modifications, one finds oxidations leading to grapheme oxide (GO) and oxidized CNT; radical reactions starting from diazoniums salts; cycloaddition reactions,....Among the non-covalent modifications, adsorption of pyrene, DNA or surfactants are used with both graphene and CNT. Finally some applications of these surface modifications are described, for example the debundling of CNT and the formation of composites.

Chapter 5 by Jin-Kai Yuan, Sheng-Hong Yao, De-Long He, and Jinbo Bai describes the surface engineering of carbon nanotubes for polymer nanocomposites. Carbon nanotubes (CNT) exhibit unprecedented mechanical, electrical, and thermal properties as well as nanometer scale diameter and high aspect ratio, which make them an ideal reinforcing agent for high-performance multifunctional polymer nanocomposites. However, their potential has not been fully realized in composites so far due to the inherent agglomeration tendency and weakness at the interface. Various surface engineering techniques aim at solving such problems. The chapter also emphasizes the mechanical, electrical but also dielectric properties of CNT/polymer composites. This Chapter constitutes a follow up to Chapter 4 as it links surface chemistry of CNT to the mechanical, electrical and dielectric properties of the corresponding CNT/polymer composites.

Chapters 6,7,8 are concerned with metallic nanoobjects.

Chapter 6 by Jalal Ghilane and Jean Christophe Lacroix describes atomic sized metallic point contacts. When the size of a metallic conductor reaches the diameter of a few atoms, the current becomes quantized. The different methods mechanical, chemical or electrochemical that permit the formation of these atomic wires are described. Quantized conductance atomic switch are discussed as well as deviation from integer values of G_0 in quantized point contact; finally protected (modified) point contacts are described.

Chapter 7 by Jean-Yves Piquemal, Kahina Aït Atmane, Guillaume Viau, Mona Ibrahim, Philippe Decorse, Aazdine Lamouri, and Mohamed Mehdi Chehimi concerns ferromagnetic metal nanoparticles and their surface modification. The first part of this review, based on selected examples from the most recent literature, presents the preparation of anisotropic metal ferromagnetic nanoparticles, mainly based on iron, cobalt and nickel, by vapor-phase and liquid-phase processes. In the second part, the different strategies reported for the surface modification and the functionalization of these particles are described. Surface modification is realized during particle synthesis and allows, in some favorable cases, to drive the anisotropic growth. Surface functionalization is generally performed after the preparation of the anisotropic particles, *i.e.* by a two-step procedure. Using these approaches, oxidation-protected anisotropic magnetic metal nanomaterials can be obtained as well as multifunctional materials exhibiting ferromagnetism and at least one other interesting physicochemical property.

Chapter 8 by Samia Mahouche-Chergui, Sandrine Lépinay, Mourad Lakhlifi, Mohamed Guerrouache, Mohamed M. Chehimi and Benjamin Carbonnier describes gold nanoparticles at polymeric interfaces. It first reports on designing nano-assemblies made of gold nanoparticles (GNP) at polymeric interface. Their different applications are then described: i) for separation (chromatography, by incorporating GNPs into polymer-based monoliths), ii) for sensing applications involving, for example, the combined use of ultrathin molecularly imprinted polymer and enhanced transmission surface plasmon resonance spectroscopy and iii) catalytic applications with assemblies such as CNT/polymer brushes/GNP.

Chapter 9 and 10 are concerned with organic nanowires.

Chapter 9 by Catherine Debiemme-Chouvy describes nanostructured polypyrrole materials; it focuses on their templateless synthetic methods and on some of their applications. First the synthetic methods chemical (for example by two phase oxidation of pyrrole) or electrochemical (in aqueous, organic or ionic liquid solution) that permit to obtain these nanoobjects are reported. Then different applications are detailed i) for (bio)sensors: detection and determination of analytes in solution (for example spermidine) or in gaseous phase such (H_2) and ii) to energy storage (for example for supercapacitors with capacitance and pseudocapacitance).

Chapter 10 by Matej Mičušík and Mária Omastová is concerned with carbon nanotubes/polypyrrole hybrids. Their preparation is first described by polymerization of pyrrole in the presence of CNT; one obtains CNT where polypyrrole films or fibrils are grown on the surface, they can be further post functionnalized. Their different applications to supercapacitors, sensors, actuators, electromagnetic shielding are then reported..

The final two chapters describe the different composites that can be obtained with nanoobjects.

Chapter 11. by Syuji Fujii and Yoshinobu Nakamura concerns the surface coating of soft material with conducting polymer-metal nanocomposite. This contribution reviews a facile and new chemical approach developed in order to enable one-step nanoprecise coating of conducting polymer-noble metal nanocomposite on soft substrates. It describes one-step facile coating of polymer particles or fibers with conducting polymer-noble metal nanocomposites by chemical oxidative polymerization in environmentally benign aqueous media. Characterization methods with respect to morphology, colloidal stability, surface/bulk chemical composition and conductivity by scanning and transmission electron microscopies, laser diffraction particle size analysis, X-ray photoelectron spectroscopy, thermogravimetric analysis, elemental analysis and four-point probe technique are presented. The nanocomposite-coated materials as recyclable catalyst for aqueous organic reactions are also described.

Chapter 12 by Delong He, Johan Saba, Jing Zhang, Fabien Miomandre and Jinbo Bai describes carbon fiber and nano/micro hybrid structures and their interface in advanced composites. These composites are particularly important as they are part of modern airplanes. The methods for the fabrication of carbon fibers, their inclusion in composites are reported. Nano/micro hybrids are described; they include, for example, CNT grown on carbon fibers. The surface modification of carbon fibers is also reported. This final chapter insists on the mechanical properties of these composites.

As a conclusion, this book covers the surface modification of nanoobjects starting with the numerous chemical reactions that permit to tailor mechanical, electrical, magnetic, medical properties, among others, of the nanoobjects and their composites. We have enlightened the key importance of these modifications for the future academic and industrial development of these nanoobjects.

The authors and editors of this book wish you a fruitful reading.

DIAMOND

In: Applied Surface Chemistry of Nanomaterials
Editors: Mohamed M. Chehimi and Jean Pinson

ISBN: 978-1-62808-351-4

Chapter 1

DIAMOND NANOPARTICLES: A REVIEW OF SELECTED SURFACE MODIFICATION METHODS FOR BIOCONJUGATION, BIOLOGY AND MEDICINE

Sabine Szunerits* and Rabah Boukherroub†
Institut de Recherche Interdisciplinaire (IRI),
Villeneuve d'Ascq, France

1. INTRODUCTION

Integrating biological components with nanomaterials for applications in drug delivery or medical diagnostics has been widely explored using a spectrum of approaches. Recent studies have shown that nanomaterials enabling drug delivery, such as cancer therapeutics, could revolutionize the field of pharmacology and help tackle diseases at a molecular level. [1-3] The use of nanomaterials in biomedical applications is of great interest since their size scale is similar to biological molecules and structures.

Among the nanomaterials being considered, nanoscale diamond particles also referred to as nanodiamonds (NDs) are very promising. [4] While their applications in surface coatings and in nanocomposites [5] were highlighted in the literature, their excellent chemical stability and biocompatibility in biological systems make NDs promising for biomedical applications. [6] Functional NDs are expected to rival their inorganic (*e.g.,* SiO_2, TiO_2) and metal counterparts (*e.g.,* Au, Ag) as drug-delivery vehicles, as templates for biological sensing, or as probes in bioimaging. It is anticipated that the attractive properties of NDs will be exploited in a similar manner to other carbon nanoparticles, quantum dots, and metallic nanoparticles, for the development of therapeutic agents for diagnostic probes, delivery vehicles, gene therapy, anti-viral and anti-bacterial treatments, tissue scaffolds, and the development of novel medical devices. [6, 7]

* E-mail: sabine.szunerits@iri.univ-lille1.fr.
† E-mail: rabah.boukherroub@iri.univ-lille1.fr.

The first records of the production of diamond nanoparticles date back to the 1960s, when a group of Soviet scientists discovered single crystals of cubic diamond particles in soot produced by detonating an oxygen-deficient TNT/hexogen composition in inert media without using any extra carbon source. [8] Production of NDs in the US started only in 1988, [9] and today there is a multitude of commercially available methods allowing the production of large quantities of NDs at rather low-cost. [10]

NDs have been rapidly gaining popularity in biological imaging and drug delivery applications since colloidal suspensions of individual diamond particles with diameter of 4-10 nm are commercially available. An additional appealing feature of NDs is their intrinsic fluorescence (Figure 1). Diamond crystallites with a nominal size of 100 nm are capable of producing stable fluorescence from color centers after surface treatment with strong oxidative acids. [11] These results were not completely unanticipated as the fluorescence of NDs originates from point defects embedded in the crystal lattice, the most noteworthy being the negatively charged nitrogen-vacancy center $(N\text{-}V)^-$, which is the dominant end product of thermally annealing or irradiation-damaged type Ib diamond. [12] These fluorescent NDs are considered non-toxic alternatives to semiconducting quantum dots for biomedical imaging. [12-15] Concerns about the toxicity of other carbon-based nanomaterials (carbon nanotubes, fullerenes, carbon nanoparticles, etc.) brought the safety of NDs into questions. However, a wide variety of studies examining the behavior NDs in cultured cells and complete organisms have concluded that the particles are highly biocompatible. [2, 11, 16, 17] In comparison to other carbon-based nanostructures, NDs were indeed found to be less toxic. [6, 17] In addition to their essentially non-toxic character to cells and living animals and their stable intrinsic fluorescence, the nitrogen vacancy defect allowed fluorescent NDs to be also detected by magneto-optical techniques. [18]

One important aspect concerns the study of the surface structures of nanodiamond. [19] In the absence of passivants, the surface of NDs is decorated by under-coordinated "dangling bonds", which are energetically unfavorable and highly reactive. The surface of ND is for this reason modified with functional moieties, including carboxyl, carbonyl, lactone, ketone and hydroxyl groups or reconstructed through graphitization (Figure 1). [20, 21] This inhomogeneity hampers direct functionalization of NDs due to uncontrolled surface chemistry. It is thus imperative to have a complete control of the surface chemistry of these particles. Depending upon the groups present on the NDs surface, different immobilization strategies can be sought after. Over the years, a variety of gas phase and chemical-based wet techniques have been developed to modify the surface of NDs. [22] Some of these approaches also helped in the reduction of cluster formation (aggregation), one of the major problems still encountered using NDs. [23]

Small molecules and biomolecules can be immobilized on NDs surface through a variety of mechanisms such as physisorption, electrostatic binding, specific recognition and covalent coupling. Physisorption involving van der Waals or electrostatic interactions has been widely used to prepare stable biomolecule-NDs conjugates. This strategy has been successfully applied for non covalent immobilization of bovine insulin, [24] horse cytochrome *c*, [25] lysozyme, [26-29] and apoobelin. [30] In most cases, the proteins retain their activity upon adsorption on the NDs surface.

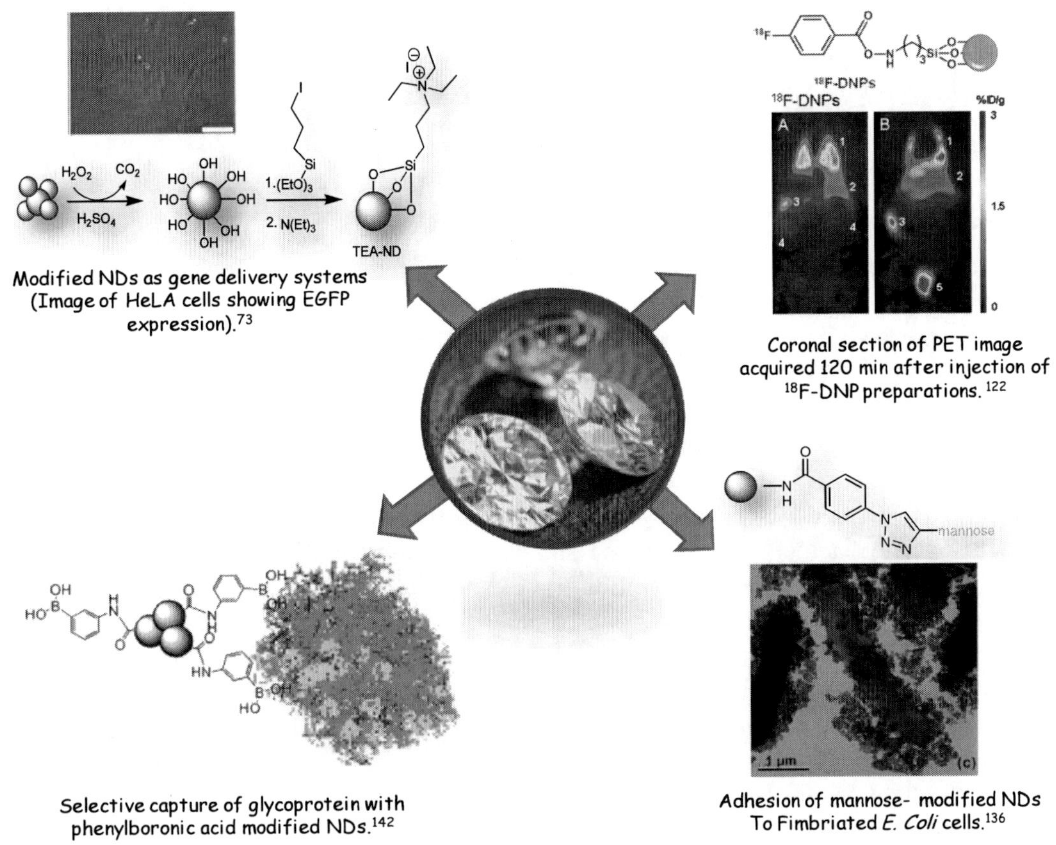

Figure1. Interest of nanoscale diamond particles, called nanodiamonds (confocal scanning image and fluorescence spectra.

Drug immobilization through physisorption on NDs surface has also been demonstrated. NDs functionalized with doxorubicin hydrochloride (DOX), [31, 32] and 10-hydroxycamptothecin (HCPT) [33] have been successfully used to deliver such drugs in cells while preserving their efficacy.

Biomolecules chemisorption with covalent bond formation on the NDs surface is a more controlled way for obtaining a stable biomolecule-NDs conjugate, but requires a surface pre-activation by grafting specific reactive group. This chapter discusses the currently available methods for reliable and well-controlled surface modification of the NDs surface and highlights some applications of NDs.

2. Production Methods of NDs

To better understand the origin of the complex surface chemistry of NDs (Figure 1), one has to consider the conditions used for their production. A variety of methods are currently available and used for the fabrication of non-fluorescent or fluorescent NDs. Most industrial methods for the production of non-fluorescent NDs are based on detonation techniques, [8, 34] Mypolex, ball milling of synthetic or natural diamond microcrystals, [35] and laser ablation. [36] Other approaches based on plasma-assisted chemical vapor deposition (CVD),

[37] irradiation of graphite, [38] chlorination of carbides [39] or autoclave synthesis from supercritical fluids [40] have also been reported, but their production scale is rather low. The potentially industrial methods for the fabrication of NV fluorescent nanodiamonds (FNDs) are based on high pressure high temperature (HPHT) diamond. [41, 42]

Detonation NDs is still one of the most widely used NDs source. The production of NDs from explosives such as trinitrotoluene (TNT) and hexogen (60/40 wt.%) in a closed-metallic chamber results in diamond-containing soot (Figure 2). This man-made diamond did not attract much attention at the beginning due to a continued failure in disintegrating ND particles into primary particles. [43] While the particle size of its primary crystallites is ~5 nm with a very narrow size distribution, [43, 44] due to harsh conditions in the reaction chamber, detonated ND particles exist mainly in the form of strongly bound agglomerates. [45] The particles are not only linked by the usual electrostatic interactions by also *via* covalent bonds between surface functional groups as well as by soot structures surrounding each primary particle. [46, 47] The detonation soot can be purified using oxidizing mineral acids (HNO_3, mixtures of H_2SO_4 and HNO_3, $K_2Cr_2O_7$ in H_2SO_4, KOH/KNO_3, HNO_3/H_2O_2 under pressure, etc), [48] as the reactivity of disordered sp^2 carbon is higher than that of diamond, thus removing non-diamond impurities. During the cleaning step, the surface of nanodiamond is covered with a variety of functional groups such as hydroxyl, carbonyl, carboxyl, anhydrides and lactones (Figure 2). These purified ND particles show still a strong tendency to agglomerate, which can withstand ultrasonic treatment. To overcome this limitation, de-aggregation process in suspension by milling detonated NDs with ceramic microbeads (ZrO_2, SiO_2) or by microbead-assisted ultrasonic disintegration are applied to nanodiamond. [19, 49, 50] Stable colloidal solutions of individual ND particles of 4-5 nm in diameter are obtained. However, contamination of NDs with difficult-to-remove zirconia, the high cost of zirconia microbeads, and NDs amorphization (or even graphitization) in the course of milling are major drawbacks of this method. Retreatment of the de-agglomerated particles with oxidizing chemicals is therefore needed, which results in a partial re-aggregation of the primary particles. [50, 51] Other attempts have been undertaken to overcome the agglomeration problem. ND aggregates disintegration using mechanochemical treatment, that is, milling in the presence of chemicals such as surfactants and electrolytes (e.g., sodium oleate) was investigated. [52, 53] Mechanical action was applied in the form of grinding or sonication in the presence of sodium oleate and resulted in a reduction of the aggregate size of the treated NDs in water from 2 μm down to 40-60 nm. However, no particles smaller than 20 nm were observed.

Stable and bright fluorescent NV-rich NDs have been prepared by electron irradiation (typically 2 MeV) and annealing at high temperature (800-900°C) in vacuum of type Ib diamond microcrystals. [54-56] A high yield method for the production of isolated ultrasmall diamond nanoparticles was reported by Boudou et al. [41] The technique relies on electron irradiation (10 MeV) and annealing at high temperature (800°C) in vacuum of type Ib high pressure high temperature (HPHT) diamond microcrystals. The powder obtained by high energy ball milling of fluorescent HPHT diamond microcrystals was converted into an aqueous colloidal dispersion of highly crystalline ultrasmall nanoparticles with a mean size less than or equal to 10 nm. The overall yield of this method is about 15% (w/w) of the microdiamond mass converted into $\leq$10 nm diamond nanoparticles. [41] The group of Chang has developed a practical method to scale up the production of fluorescent nanodiamonds (FNDs) from synthetic type Ib diamond powders using either a home-built 40 KeV He^+ beam or a research grade 3 MeV H^+ beam, followed by thermal annealing at 800°C. [14, 57] Deagglomeration of the NDs was achieved by strong acid treatment (H_2SO_4-HNO_3). The FNDs have been successfully used as photoluminescent probes for tracking intracellular processes. [14, 57, 58]

More recently, dry salt or sucrose assisted milling was successfully used as simple, inexpensive, and efficient alternative method for the deaggregation of nanodiamonds. [59] The technique uses non toxic and noncontaminating compounds such as sodium chloride and sucrose, which can be easily removed at the end of the milling process. Using the dry media assisted milling with subsequent pH adjustment, it is possible to produce stable aqueous nanodiamond colloidal solutions with particles <10 nm in diameter.

High-temperature hydrogen treatment resulted, moreover, in stable single-particle aqueous nanodiamond suspension of 4-5 nm. [23] Independent of the method chosen, one concern that needs to be addressed in any case is the possibility of re-aggregation over time, during drying and/or after surface functionalization due to capillary forces or attractive van der Waals forces, pulling the particles together.

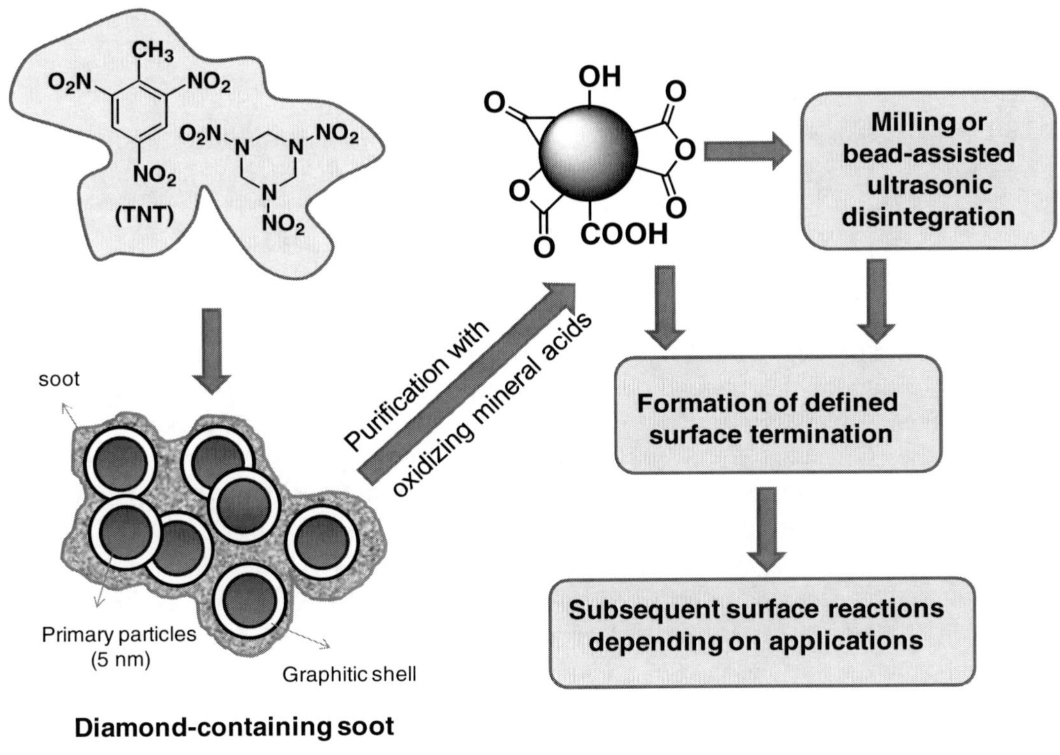

Figure 2. Formation of detonation NDs with well-defined surface terminations: A mixture of TNT (60 wt%) and hexogen (40 wt %) are detonated in a closed metallic chamber in an atmosphere of N_2, CO_2 and liquid or solid water forming a diamond-containing soot. The soot is purified with oxidizing mineral acids forming oxidized NDs. The particles can be either used directly or treated with different methods to homogenize the surface functions. These particles can be used for subsequent surface reactions.

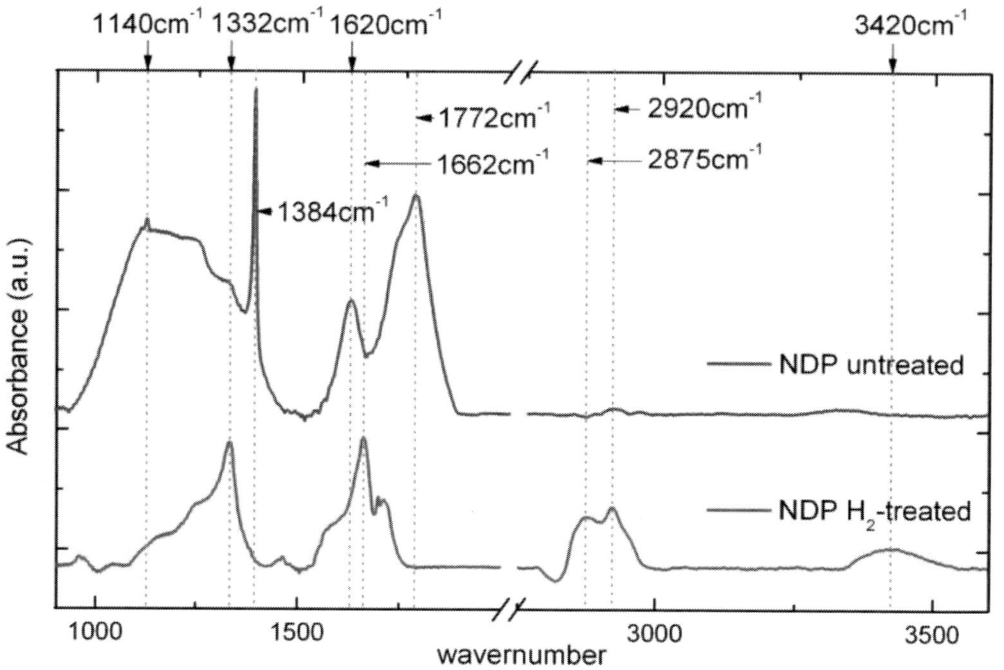

Figure 3. FTIR spectrum of untreated ND particles (reprint with permission from Ref. [23]).

3. Preparation of ND with Well Defined Surface Terminations

As outlined above, a distinct feature of ND particles compared to carbon nanotubes or carbon nanoparticles is that many different functional groups are present on the ND surface (Figure 2). Fourier-transform infrared spectroscopy (FTIR) is well suited to characterize the different surface functional groups present on ND powders. Figure 3 shows a representative FTIR spectrum of as-prepared diamond particles produced *via* detonation synthesis. [23]

The broad band starting from 1000 cm^{-1} with a maximum at 1140 cm^{-1} is due to the superposition of bands associated with nitrogen and ethereal -C-O-C- groups. [60] The presence of carboxyl groups is indicated by the band at 1384 cm^{-1} due to the OH bending vibration in COOH. The band at 1772 cm^{-1} is unambiguous evidence of C=O stretching modes and thus for the presence of carbonyl groups. The OH bending modes of surface hydroxyl groups are seen at 1620 cm^{-1}. Rather weak bands at 2875-2920 cm^{-1} due to stretching modes of CH_2 and CH_3 are observed, suggesting the presence of alkyl groups. Additionally, in some cases a band at 1332 cm^{-1}, assigned to C-C stretching vibrations of the diamond lattice with nitrogenous impurities, can be seen. [61] The highly inhomogeneous, but rich surface chemistry of NDs makes it a challenge for direct functionalization but opens the scope for multifunctional approaches. It is obvious that for a high reproducibility of further reactions, the NDs surface should be as homogeneous as possible. The initial control of the surface termination will eventually decide on the quality of the final functionalized material and its applicability.

Different strategies are used to homogenize the surface functions in order to confer nanodiamond well defined surface terminations prior to surface derivatization. Several attempts can be listed:

(i) Strong acid treatment.
(ii) Oxidation under controlled atmosphere (air, air/ozone, ozone), resulting in COOH/OH modified particles. [47, 62]
(iii) Fenton reaction treatment forming OH modified particles.
(iv) Chemical reduction with borane forming OH modified particles. [63]
(v) Annealing under controlled atmosphere forming fluorinated and chlorinated particles.
(vi) Surface reduction in hydrogen atmosphere. [23, 64-66]

Surface functionalization is often applied to customize nanomaterials for the sought after application. In addition to the different strategies investigated to prepare homogenous surface functions on nanodiamond, follow up reactions using the relevant surface function have been developed to form covalent bonds between the nanoparticles' surface and the immobilized unit.

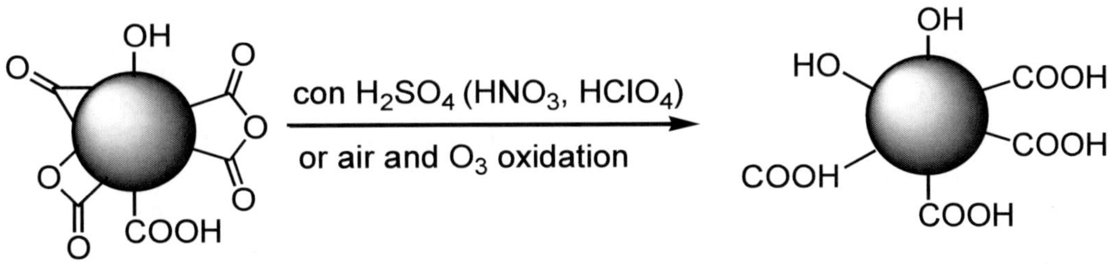

(A)

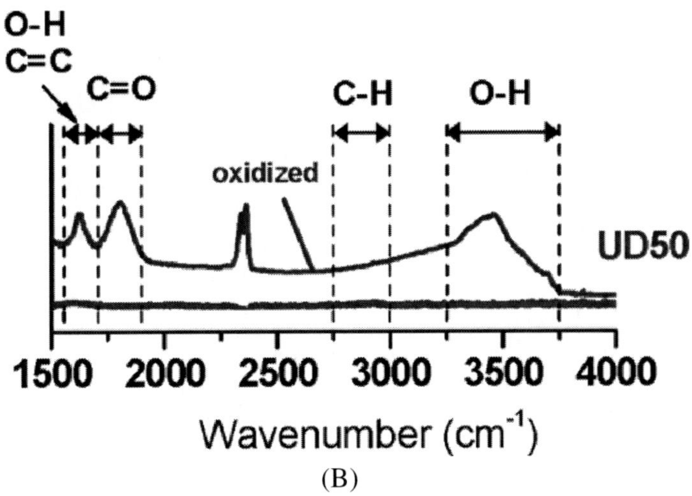

(B)

Figure 4. (A) Formation of hydrophilic ND particles carrying hydroxyl and/or carboxylic acid groups; (B) FTIR spectrum of ND after air oxidation for 5 h at 425°C (reprint with permission from Ref. [47]).

3.1. Preparation of Carboxylated NDs

Nanodiamonds modified with polar groups such as carboxylic acid and hydroxyl functions are convenient starting materials due to the hydrophilic character of the surface groups and the possibility to covalently link functional molecules via amination, silanization or esterification reactions. These particles can be formed from detonated diamond by air [47] or air/ozone [62] purification or by a strong oxidative acid treatment (Figure 4A). [48, 67] Acid treated NDs exhibit a surface density of 7% of carboxyl groups, [68] being sufficient for the covalent immobilization of biopolymers containing amino groups (e.g. polylysine). [69, 70] Other oxidants such H_2O_2 [71] or H_2O_2/H_2SO_4 (piranha) [72] were also used to form carboxyl functions on the NDs surface while removing the non-diamond carbon.

Air oxidation of NDs soot for 5 h at 425°C is of particular interest as it results in a complete removal of CH_2 and CH_3 groups and a significant increase in the amount of OH, carboxylic acid, anhydrides and/or cyclic ketones as seen in the FTIR analysis (Figure 4B). [47] Furthermore, the sp^3/sp^2 ratio increased from 0.3 to 19, indicating the destruction of sp^2 carbon in addition to homogenization of the surface functions on the NDs surface. This purification approach uses non toxic and non-corrosive chemicals making the process thus highly appealing.

3.2. Preparation of Hydroxylated NDs

A different approach for surface function harmonization has been described by the group of Martin and is nowadays widely used. [73, 74] Aggressive Fenton treatment using H_2O_2 and Fe_2SO_4 under strong acidic conditions (H_2SO_4) can serve to remove significant amounts of undesirable soot matter, while at the same time to introduce a large density of surface hydroxyl groups that can be subsequently used as sites for the covalent anchoring of alkyl chains or aromatic rings (Figure 5A). The Fenton reaction is based on the production of highly aggressive $OH^\bullet$ radicals able to partially oxidize the exposed surface carbons. The types of oxygen functions introduced by the Fenton treatment were analyzed by FTIR spectroscopy (Figure 5B). Depending on the conditions of the Fenton treatment, the intensity of the hydroxyl band at 3400 cm^{-1} is the most intense band observed. A concomitant appearance of C-O single bond at 1200-1100 cm^{-1} was also observed. The C=O and C=C groups previously present in the raw NDs are maintained during the oxidation process as the intensity of the 1750 and 1630 cm^{-1} bands remained unaltered.

A different route to hydroxylated NDs is described by Krueger et al. [46, 75] It is based on the reduction of carbonyl groups by reaction with various reducing agents such as lithium aluminum hydride ($LiAlH_4$) or borane (BH_3). The use of borane is preferred as a considerable amount of $Al(OH)_3$ residues is formed during the reduction with $LiAlH_4$, which require NaOH [76] or HCl [77] treatment at 90°C overnight to be eliminated. In case of borane reduction, the boronic acid produced upon hydrolysis of boronic ester is easily removed by washing/centrifugation cycles. This treatment resulted also in a significant smaller particles' aggregation and is currently one way to overcome agglomeration problems.

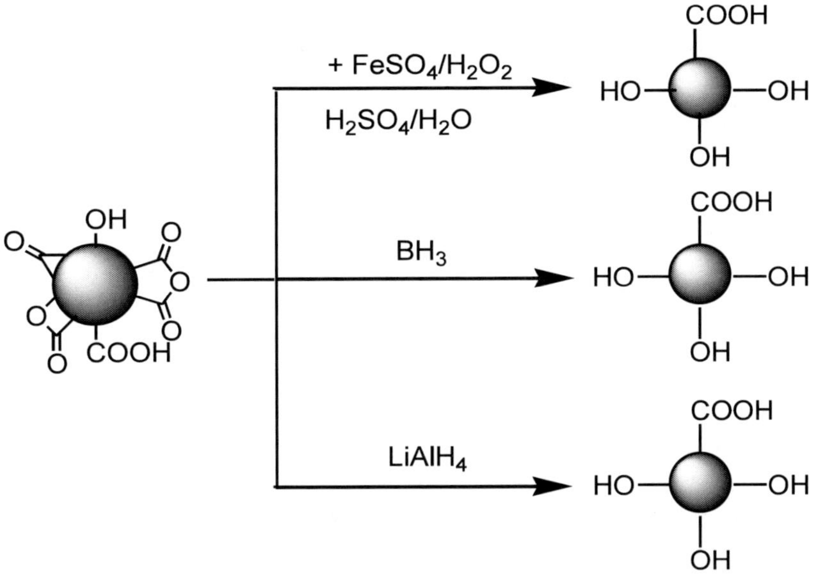

(A)

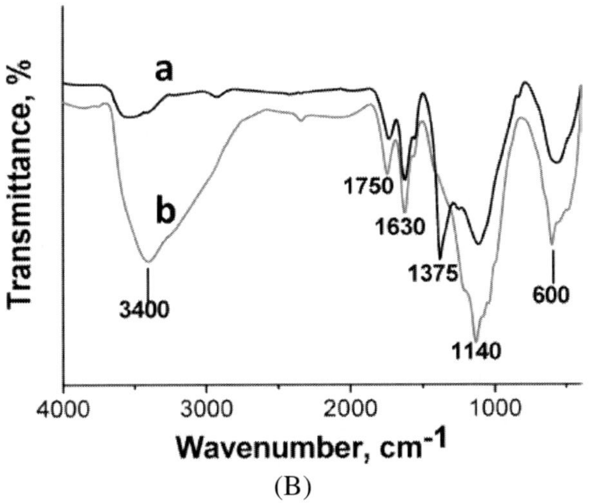

(B)

Figure 5. (A) Formation of hydroxyl-terminated NDs. (B) FTIR spectra of NDs before (a), and after Fenton treatment (b) (Reprint with permission from Ref. [73]).

The resulting hydroxylated nanodiamonds carried up to 0.5 mmol g^{-1} of OH groups. These OH groups could then be used for subsequent silylation or acylation reactions. However depending on the nature of the initial surface groups on the NDs, the borane reduction is not always the optimal way to remove all kinds of carbonyl functions. Namely esters or lactones cannot be easily reduced by boranes (and can still be identified after the borane reduction, e.g., in FTIR spectra). [75]

Girard et al. used photochemical irradiation of hydrogenated nanodiamond to introduce hydroxyl functionality on the surface. [66] Direct UV irradiation at λ=254 nm of hydrogenated NDs in a quartz pipe filled up with water vapors for 6 h led to the formation of fully hydroxylated NDs surface. FTIR analysis showed bands at 1050 and 3300 cm^{-1}, characteristic of C-OH and OH groups. The absence of bands between 1700 and 1800 cm^{-1} suggests that the oxidation process occurs without COOH formation.

3.3. Preparation of Hydrogenated NDs

Homogenization of oxidized diamond by surface hydrogenation turned out to be rather difficult as most reactions using reducing agents did not lead to the complete removal of the oxygen-containing groups. Early reports on high temperature treatment in hydrogen, plasma-ionized hydrogen, nitrogen and *in vacuo* of ultradispersed detonation diamond (UDD) powder suggested that the generally acknowledged hydrogenation reduction reaction was not the main reaction occurring in the process of treating UDD in hydrogen. In fact, the multiple effects of thermal decomposition, decarbonylation and/or decarboxylation and structural changes in ultradispersed diamond lattice at elevated temperatures were the cause of the changes in UDD modified by hydrogen, nitrogen and *in vacuo.* [61] The formation of C-H bonds on diamond powder was obtained by using hydrogen gas at elevated temperatures. High temperature (400-850°C) treatment of detonation nanodiamonds has led to the formation of C-H, but also an increase of C-OH bonds on the surface. [23, 78, 79] Ando et al. showed that thermal treatment of oxidized diamond powder in H_2 environment leads to C-H termination for temperatures below 900°C. Above 900°C, the amount of C-H bonding decreased and the formation of graphitic carbon phase was observed. [80-83]

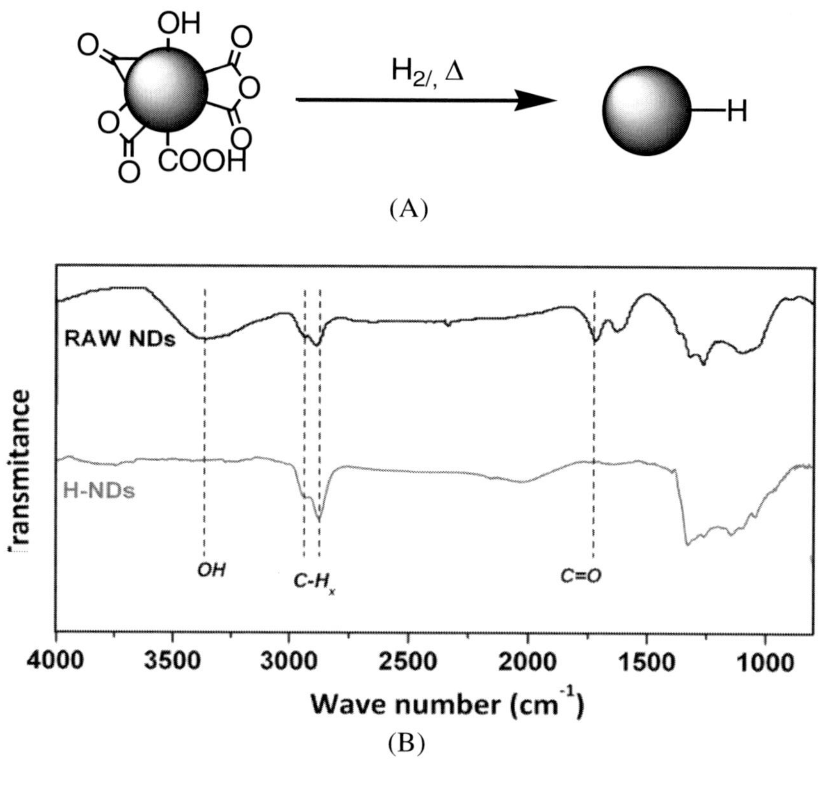

Figure 6. (A) Formation of hydrogenated NDs; (B) FTIR spectrum of raw (a) and surface hydrogenated (b) detonation NDs (reprint with permission from Ref. [64]).

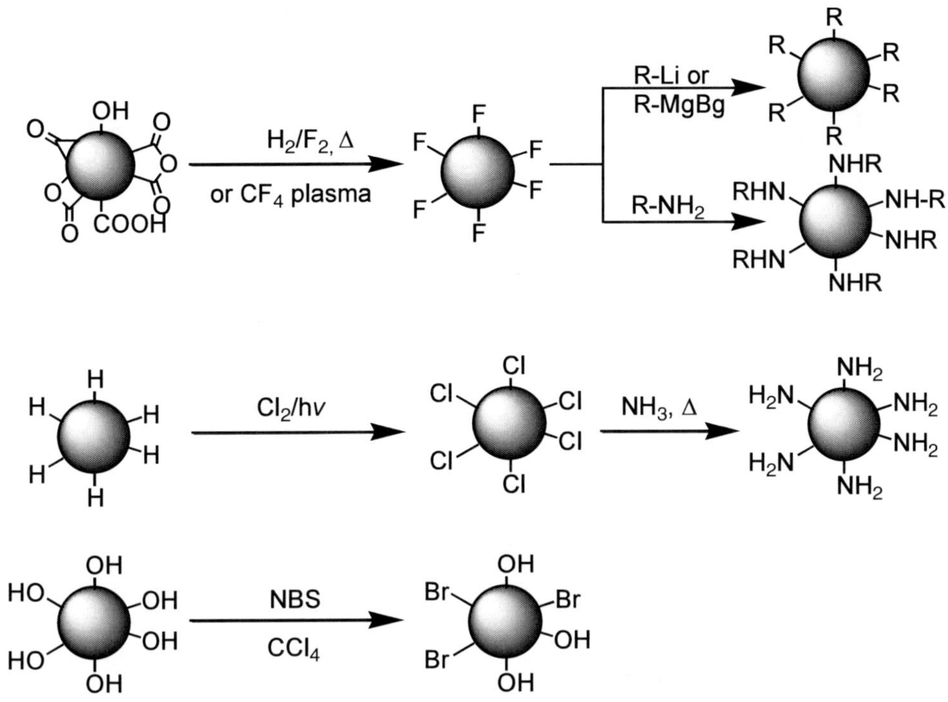

Figure 7. Halogenation and post-functionalization strategies of NDs.

The use of high temperatures for homogeneous hydrogenation of small particles is not suitable, as it can lead to the formation of amorphous and sp^2 carbon shells around the sp^3 core. Approaches relying on low temperature treatment or using other reduction mechanisms will be more appropriate. In this context, hydrogen plasma treatments are an alternative. Loh et al. used microwave hydrogen plasma to partially hydrogenate acid-treated NDs. The treatment was repeated twice to achieve sufficient amounts of C-H bonds and homogeneous functionalization of the surface. [84] Recently, Girard and co-workers showed that homogenous hydrogenation can be achieved in a microwave plasma reactor where the nanodiamonds were placed in a rotating quartz tube to ensure the uniformity of the hydrogenation process (Figure 6A). [64, 66, 85] After hydrogenation, FTIR characterization clearly emphasized the oxygen removal with the disappearance of the C=O and OH stretching bands located at 1720 cm^{-1} and 3400 cm^{-1}, respectively (Figure 6B). [64] Apart from the "fingerprint" area between 900-1400 cm^{-1}, the H-terminated NDs only exhibited signals located around 2900 cm^{-1} assigned to $C\text{-}H_x$ stretching bands.

3.4. Preparation of Halogenated NDs

Halogenated nanodiamonds can be considered as a suitable starting material for grafting more complex moieties on the NDs surface. One of the first attempts to halogenate nanodiamonds was based on the fluorination. [20] The conditions of fluorination are harsh and consist in annealing under controlled atmosphere with an elemental fluorine/hydrogen mixture at 470°C (Figure 7). [20] These fluorinated nanodiamond surfaces were used as a precursor for the preparation of a series of functional NDs by subsequent reaction of the C-F surface species with alkyllithium reagents, diamines and amino acids in nucleophiliic substitution reactions. [20]

Ray et al. applied atmospheric pressure plasma with CF_4 for the grafting of fluorine atoms at the surface of nanocrystalline diamond films. [86] XPS and FTIR analysis revealed a high concentration of fluorine on the surface (nearly 50 at. %), consisting mainly of $-CF_2-$ and $-CF_3$ functional groups.

Besides NDs fluorination, the chlorination and bromination of the NDs surface were also investigated, although their introduction proved to be more difficult. The most successful method for the introduction of chlorine atoms onto NDs is currently based on the photochemical reaction of molecular chlorine in the gas phase [87] or in the liquid phase [88] with hydrogenated NDs. The chlorinated NDs surface is hydrolytically stable in air for several days and for a few hours in 0.05 M NaOH aqueous solution. [88] Saito et al. described a radical reaction using sulfuryl chloride (SO_2Cl_2) and azobisisobutyronitrile (AIBN) in chloroform at 50°C to chlorinate hydrogenated diamond powder. [89] Thermal reaction of hydrogenated diamond powder with molecular chlorine is another approach used to chlorinate the NDs surface. [82] The reaction is performed at temperatures ranging from room temperature to 500°C. The weight change of the diamond powder after treatment with chlorine at different temperatures was measured. It was found that by increasing the treatment temperature, the weight of the diamond increases. A maximum value corresponding to weight gain equivalent to 52-60% was reached at 250°C, suggesting that all hydrogen atoms on the diamond surface have been substituted with chlorine atoms. The stability of the chlorinated diamond was evaluated by exposing to water vapor at different temperatures. The results indicated that water vapor reacts with the chlorinated surface even at room temperature to give OH groups on the surface. A higher temperature treatment led to a more pronounced oxidation of the chlorinated diamond surface and the surface termination changed from C-OH to C=O bonds. [82] Other methods include the thermal chlorination using Cl_2 or CCl_4 vapor. [78]

The reaction of Fenton treated ND particles (oxidized NDs) with *N*-bromosuccinimide (NBS) in CCl_4 was used for partial bromination of the ND surface. Chemical and XPS analyses of the resulting surfaces indicated the presence of 3.53% bromine (0.44 mmol g^{-1}), corresponding at least to one bromine atom per 170 carbon atoms. [74] However, this reaction does not lead to full halogenation of the surface and some hydroxyl functions are still present after the NBS treatment.

3.5. Preparation of Aminated NDs

Amine termination is a well-established route to link a large variety of functional molecules such as biomolecules, polymers, etc…onto surfaces through amide bond formation, reductive amination, nucleophilic attack or direct condensation reaction. Amination of NDs represents thus an interesting strategy for building complex and functional surfaces for various applications. Although, direct amination of hydrogen-terminated polycrystalline and ultracrystalline diamond films under photochemical [90] or using NH_3 plasma [91-93] has been described in several reports, amination of NDs was generally conducted in a two-step process. Indeed, under controlled atmosphere, chlorinated diamond powder (0.5-3 μm in size) easily reacts with ammonia under thermal [82, 94] or photochemical conditions [95] to yield NH_2 termination. The reaction of fluorinated NDs with anhydrous ethylenediamine in the presence of pyridine at about 130°C for 24 h under nitrogen atmosphere allowed to incorporate $-NH_2$ terminal groups onto the surface (Figure 7). [20]

A multi-step process was used to prepare aminomethyl-terminated ND starting from fully reduced NDs-OH. The NDs-OH particles were prepared from ND particles, produced *via* detonation, by $LiAlH_4$ reduction. The OH termination was transformed to *p*-toluenesulfonyl or tosyl (OTS) termination by subsequent reaction with *p*-toluenesulfonyl chloride or tosyl chloride (OTS). OTS displacement using potassium cyanide (KCN) at 95°C in dimethylsulfoxide (DMSO) allowed the introduction of –CN functional groups onto the NDs surface. Finally, $LiAlH_4$ reduction of NDs-CN gave the corresponding NDs-NH_2. The NDs-NH_2 particles were successfully used for conjugating to *N*-hydroxysuccinimide derivatives of TAMRA and biotin. [96] Other strategies were also investigated to establish the highly versatile amino groups on nanodiamonds, including the use of aminated silanes [46, 75] and aminated aromatic moieties. [84]

3.6. Graphitization of NDs

While most of surface homogenization routes consist in generating one particular functional group on the NDs surface, removal of the surface functional groups was also exploited for the homogenization of NDs. In this transformation, partial or full removal of the surface functional groups was achieved by thermal treatment [97-110] or irradiation with high energy electrons [111-113] and leads to the formation of graphitic shell (bucky diamonds) or carbon onions (multishell fullerene-like objects). The graphitization process upon thermal annealing depends on several parameters such as temperature, duration, environment, and the prevailing vacuum. For example, a graphitization onset was observed at around 800°C when the annealing was performed in argon ambient. Above this temperature, a phase transition process from cubic diamond to graphite occured and the size of the ND crystallites decreased from ~50 Å to ~25 Å. [97] Annealing in low vacuum (1 Pa) or in an argon atmosphere at 1400°C transformed all the ND particles into onion-like carbon. [102, 109]

The electrochemical properties of ND particles subjected to different annealing treatments were investigated using $Fe(CN)_6^{3-/4-}$ as a redox probe. Annealing in vacuum at 900-1100°C produced a continuous fullerenic shell wrapped around a diamond core with a high conductivity and electrochemical activity. In contrast, when annealing was performed below 850°C, the background current became lower in the supporting electrolyte and the charge transfer of the $Fe(CN)_6^{3-/4-}$ redox couple decreased. The reduced background current was attributed, in large part, to a significant decrease of electroactive surface carbon-oxygen functionalities upon thermal annealing. [101]

4. Covalent Surface Modifications

Although the nanodiamond core is physically and chemically strong, its rich surface chemistry is easily modified with a variety of methods.

4.1. Nucleophilic Substitution of C-X Bonds

As previously discussed, the chemical reactivity of carbon-halogen bonds on NDs has been exploited for the introduction of amino groups onto the NDs surface. [20, 82, 94, 95] Nucleophilic displacement of the carbon-chlorine bond of chlorinated NDs with CN^- [88] or organolithium reagents [88, 89] has been successfully used for the preparation of nitryl-, and alkyl- or phenyl-terminated NDs surfaces. For example, the direct reaction of chlorinated NDs with *n*-, *sec*-, and *tert*-butyllithium in tetrahydrofuran at 30°C for 4 h led to the incorporation of alkyl moieties onto the surface. The fraction of the alkylated surface carbons to the total surface carbons was calculated based on pyrolysis products obtained at 1040°C and found to be 0.055 for *n*-butyl, 0.040 for *sec*-butyl and 0.047 for *tert*-butyl modified NDs surfaces. [89] Even though C-F bonds are not easily displaced by organic nucleophilic reagents, the reaction of fluorinated NDs with alkyllithium reagents and amino acids led to the formation of alkyl- and amino acid ND derivatives with improved solubility in polar organic solvents and a reduced particle agglomeration. [20]

4.2. Functionalization of Carboxylated NDs through Amide Bond Formation

One of the most reliable grafting methods for the immobilization of large proteins and enzymes is the coupling of amino groups from their side chain onto carboxyl-terminated NDs through the formation of amide bonds (Figure 8A). The reaction of amines with COOH groups involves the activation of the COOH group using water-soluble *N*-hydroxysuccinimide (NHS) in the presence of a coupling agent like *N*-ethyl-N'-(3-dimethylaminopropyl) carbodiimide (EDC). The resulting ester reacts with amine-containing molecules to form an amide bond. Various small molecules and biomolecules, including poly-Lysine, [12] polyarginine, [70] transferrin, [114] trypsin, [115] have been covalently immobilized through this process.

On the other hand, acid chloride formed by activation of the COOH groups with thionyl chloride also leads to amide bond formation when reacted with amines (Figure 8A). Gogotsi et al. produced in this way hydrophobic NDs by using dodecylamine as amine-reactive linker. [13, 116] Incorporation of long hydrocarbon alkyl chains onto the NDs surface improved its dispersibility in hydrophobic solvents such as benzene, toluene, chloroform, dichloromethane, etc.

Interestingly, the NDs modified with dodecylamine (NDs-ODA) showed blue fluorescence upon excitation at 410 nm (Figure 8B). Although the origin of this fluorescence is not clear, its high stability over time (2 years of storage) suggests that it does not result from an interaction with solvent and is not specific for solvated NDs. [13] Similarly, acyl chloride-terminated NDs was reacted with imidazolium bromide bearing propylamino groups to yield NDs functionalized with ionic liquid. The resulting adduct can be easily dispersed in ionic liquids with uniform size distribution. [117] Similar chemical route was employed earlier by Ushizawa et al. for covalent immobilization of aminated DNA onto acyl-terminated diamond powder (1-2 μm). [48]

Amide bond formation on NDs can be also achieved by the reaction of aminated NDs with molecules bearing a carboxylic acid functional group. This route was successfully applied to graft azide containing groups onto NDs surface. [118]

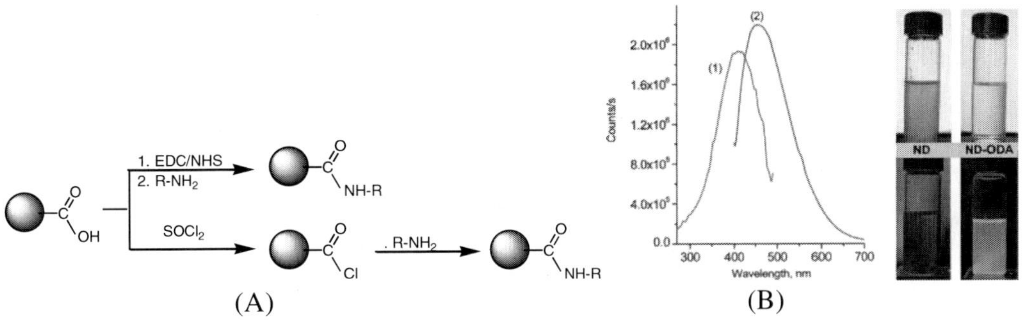

Figure 8. (A) Schematic reaction of carboxylated nanodiamonds with amines; (B) Excitation monitored at 450 nm emission (1) and emission at 410 nm excitation (2) spectra of NDs-ODA dispersion in dichloromethane; photographs of NDs and NDs-ODA (0.004% wt) dispersions in dichloromethane with visible (upper row) and UV (365 nm, lower row) illumination (reprint with permission from Ref. [13]).

4.3. Functionalization of Hydroxylated NDs

The use of hydroxylated nanodiamonds allows a great variety of chemical reactions to be carried out involving C-O bond rupture, replacement by other types of carbon bonds (bromination and arylation) as well as approaches that are based on the cleavage of the O-H bond and replacement by other O-X bonds (e.g. silylation) (Figure 9A). One widely used approach is the esterification reaction of the hydroxyl groups of NDs with carboxylic acid derivatives. [74, 119] Reaction of NDs-OH with carboxylic acid chlorides allowed introducing long alkyl chain onto the ND particles surface and improved their dispersibility in organic solvents. [119] The effect of chain length of alkyl-terminated NDs on the agglomeration properties of the particles was evaluated by TEM (Figure 9B). [74] These images present many similarities in the particle size with an important decrease in particle agglomeration for long alkyl chains. We have shown for example that the use of 4-azidobenzoic acid enables in an easy manner the introduction of azide function ($-N_3$) onto the surface of NDs, which can be further used for linking of alkynyl-terminated molecules in a "click" chemistry approach (see section 5). [120]

One of the most typical reactions of alcohols is arylation under conventional Friedel-Crafts conditions. [74] Methoxybenzene (anisole) and *N,N*-dimethylaniline have been linked to HO-terminated NDs in the presence of a catalytic amount of anhydrous $AlCl_3$ (Figure 9A).

The other widely used strategy is based on silanization reaction between the hydroxyl groups on NDs and trialkoxysilane molecules (Figure 9A). Depending on the choice of the alkoxy groups, the surface loading can be tuned between 0.3-1.5 mmol g^{-1}. [46] Some concerns of this reaction are related to possible strong cross-linking of the nanoparticles and intramolecular condensation reactions, leading to increased agglomerate size and a large organic shell surrounding the NDs core. An efficient way to counterbalance this and to obtain fully deagglomerated particles is the use of monoalkoxysilanes, which are less prone to cross-linking side reactions and by using bead assisted sonic disintegration during the silanization reaction. [121] The use of 3-aminopropyltrimethoxysilane is particularly appealing as the terminal amino group can be subsequently reacted with the carboxylic terminus of various molecules. [63, 75, 122-124] Biotinylated NDs were thus formed with a surface loading of biotin of about 1.45 mmol g^{-1}. [63]

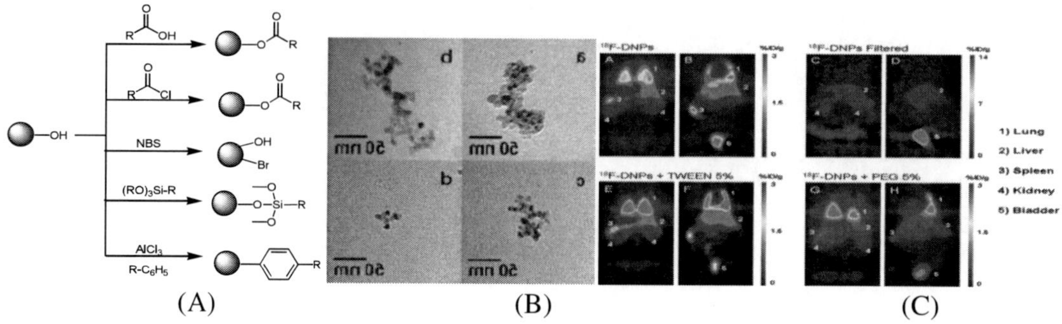

Figure 9. (A) Representation of a selection of grafting reactions using hydroxylated NDs; (B) TEM images of alkyl-functionalized NDs formed by the reaction of HO-terminated NDs with CH_3COCl (a), $CH_3(CH_2)_6COCl$ (b), $CH_3(CH_2)_{10}COCl$ (c), and $CH_3(CH_2)_{14}COCl$ (d) (reprint with permission from Ref. [74]); (C) Coronal sections of positron emission tomography images acquired 120 min after injections of the ^{18}F-labled NDs (reprint with permission from Ref. [122]).

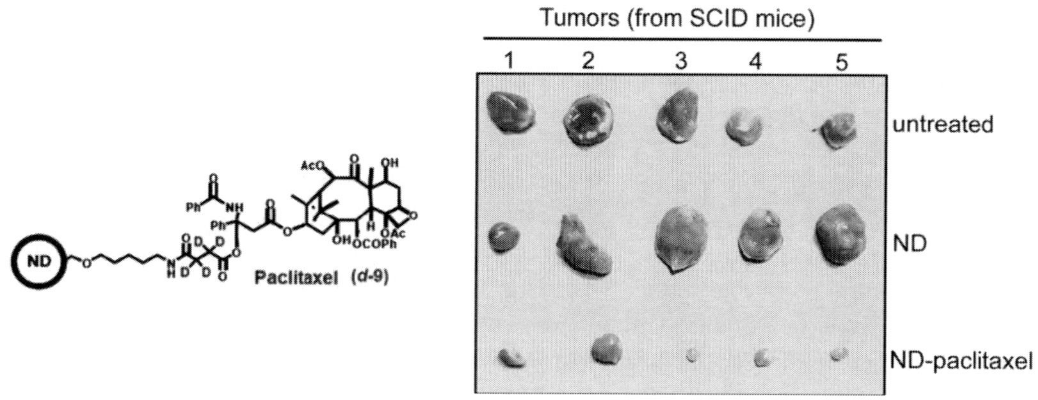

Figure 10. Effect of Paclitaxel-modified NDs (10 µg/mL for 48 h) on tumorigenesis of lung carcinoma tumor xenograft in SCID mice (reprint with permission from Ref. [125]).

Isocyanate-terminated silane was used by Martin et al. to link thionine through a subsequent nucleophilic addition reaction. [73] The thionine modified NDs were able to enter the cell nuclei. Furthermore, triethylammonium functionalized NDs using silane chemistry were successfully used for gene delivery. [73] Rojas et al. used this approach to label ND particles with ^{18}F by reacting the amine-terminated NDs with *N*-succinimidyl 4-[^{18}F] fluorobenzoate. [122] The radioactive labeled particles were used to study their *in vivo* biodistribution by positron emission tomography (Figure 9 C). It was found that radiolabeled diamond particles accumulate mainly in the lung, spleen, and liver and were excreted into the urinary tract.

Covalent linking of cancer drugs such as paclitaxel can additionally be achieved on $LiAlH_4$ reduced carboxylated NDs in a multi-step process. [125] The synthetic procedure for the conjugation of NDs and paclitaxel consists in: i) carboxylation, ii) $LiAlH_4$ reduction of the carboxyl groups to hydroxyl functions, iii) coupling of 6-(chlorohexyloxy)-tetrahydropyran, iv) deprotection of the THP groups to yield a hydroxyl termination, v) transformation of OH groups to methanesulfonate termination by the reaction with methanesulfonyl chloride, vi) reaction with NH_4OH to yield NH_2 termination, vii) coupling of paclitaxel-2'-succinate in the presence of (2-ethoxy-1-ethoxycarbonyl-1,2-dihydroquinoline (EEDQ). Paclitaxel-modified NDs, obtained via this approach, markedly blocked the tumor growth and formation of lung cancer cells in xenograft SCID mice in comparison to unmodified NDs and to non treated samples as seen in Figure 10. [125]

More recently, the interest of dopamine and its derivatives for the direct functionalization of NDs has been shown (Figure 11). Dopamine, chemically known as 4-(2-aminoethyl)-benzene-1,2-diol, has sparked great interest as an anchor for the functionalization of metal oxide surfaces because of the stability and strength of the resultant five-membered metallocyle chelate. Hydroxylated NDs were modified via the hydroxyl groups with dopamine bearing azide function or thermally responsive poly-N-isopropylacrylamide (PNIPAM). [126]

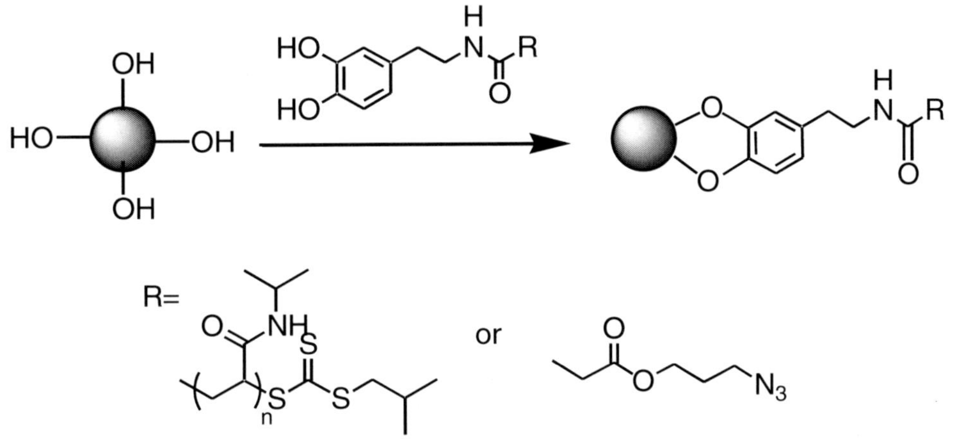

Figure 11. Direct functionalization of hydroxylated NDs with dopamine derivatives carrying azide or poly-*N*-isopropylacrylamide function (reprint with permission from Ref [126]).

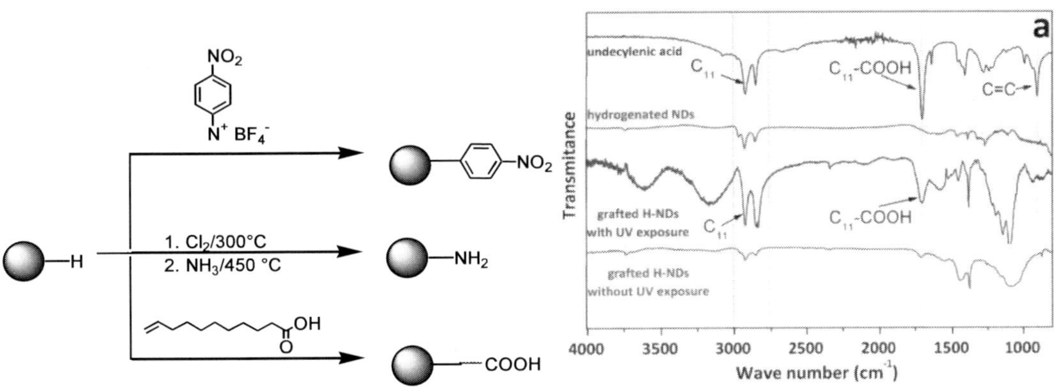

(A) (B)

Figure 12. (A) Reactions of hydrogenated NDs; (B) FTIR spectra of hydrogenated NDs before and after reaction with undecylenic acid with and without UV exposure (reprint with permission from Ref. [66]).

4.4. Functionalization of Hydrogenated NDs

While hydrogenation of NDs was a challenging task, once the right conditions were found [64, 66, 85] the hydrogenated NDs are accessible to a variety of reactions (Figure 12A). The most widely investigated on diamond surfaces is the reaction of functional alkenes with H-terminated interfaces in a photochemical process. [127] Functional alkenes such undecylenic acid reacts with hydrogenated NDs under 254 nm UV irradiation. The FTIR spectrum of H-terminated NDs after reaction with undecylenic acid shows clearly the presence of the bands at 1700 cm^{-1} and 2800 cm^{-1}, characteristic of the acid. [66] Moreover, the absence of a band at 900 cm^{-1} related to vinyl function suggests that the grafting process occurs via C-C bond formation (Figure 12B). [66]

The other widely used approach for a robust and well controlled grafting process of functional groups on NDs is based on the use of diazonium salts, either formed *in situ* or directly applied as a salt. [66, 84, 85, 128, 129] In this reaction, the reduction of the diazonium group is not supported electrochemically and the charge exchange is based on chemical reactions occurring at the interface forming C-C bonds. [130] It is believed, that the covalent grafting occurs through the creation of a phenyl radical by means of an electron transfer from the NDs to the aryldiazonium salt. Dahoumane et al. showed, recently, that new bioactive fluorescent NDs can be obtained via diazonium salt chemistry and atom transfer radical polymerization (ATRP). [128] The NDs were first modified by the ATRP initiator using the electroless reduction of BF_4^-, $^+N_2$-C_6H_4-$CH(CH_3)$-Br, which initiates the ATRP of *tert*-butyl methacrylate at the surface of the nanodiamonds (Figure 13). The grafted chain thickness increased linearly with respect to time, reaching a plateau at ≈2 nm. Bovine serum albumin (BSA) was linked to the NDs by first hydrolyzing the new nanohybrid structures into poly(methacrylic acid) followed by covalent coupling through EDC/NHS activation procedure. [128]

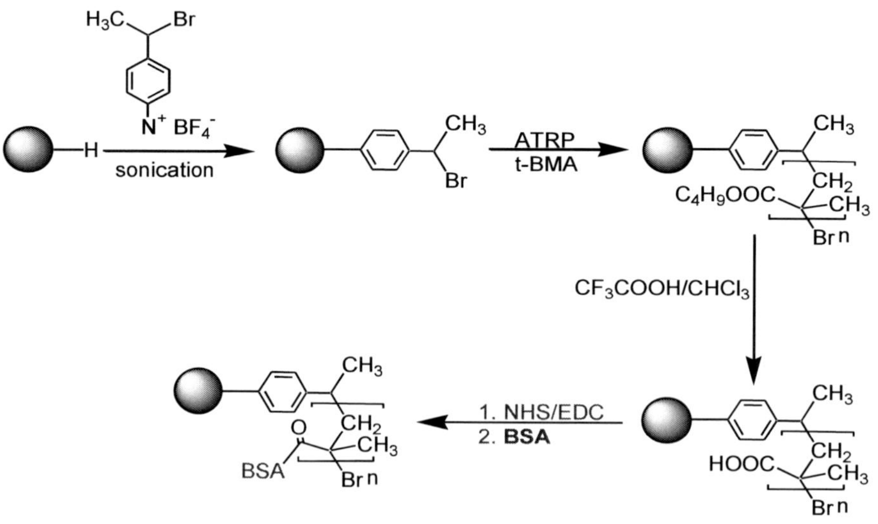

Figure 13. Surface modification of fluorescent NDs with BSA through electroless grafting of bromoethyl aryl diazonium salt, followed by surface-confined ATRP with *tert*-butyl methacrylate, hydrolysis to poly(methacrylic acid) and covalent linking of BSA (reprint with permission from Ref. [128]).

The photochemical reactivity of the C-H bonds terminating hydrogenated diamond powder was also exploited for the incorporation of thio and thiocarbonyl functional groups onto the NDs surface. This has been achieved by the direct irradiation of the hydrogenated NDs surface in the presence of elemental sulfur. [131] The sulfur-modified diamond was linked to gold nanoparticles and film through sulfur-containing moieties.

4.5. Functionalization of Annealed NDs

Diels-Alder reaction was also performed on thermally annealed NDs (partially graphitized) for grafting organic moieties through stable C-C bonds. [132] In this process, the partially graphitized NDs act as the dienophile and undergo a [4+2] cycloaddition with *in situ* generated *ortho*-quinodimethanes. This reagent has been chosen since the formation of an aromatic system in the course of the reaction ensures the irreversibility of the grafting. Conventional dienes turned out to yield only low surface loadings due to the favored *retro*-Diels-Alder reaction. Subsequent electrophilic reactions on the aromatic ring enabled the introduction of various functional groups on the NDs surface. [132] Thermally annealed ND particles at 1200°C for 1-3 h were successfully functionalized in a free radical addition process by sonication in a solution containing monomers and a radical initiator, benzoyl proxide. [133] It is to be noted that a short microwave irradiation (10-15 s) is required after benzoyl peroxide addition to ensure decomposition of the radical initiator and subsequent chain propagation of polymerization within a short time. Using this strategy, oligomers with various functionalities such as $–C(=O)OCH_3$, -COOH, $-NH_2$ or aliphatic moieties were grafted on the NDs surface. The chemically modified NDs can be dispersed in solvents for many days without apparent precipitation.

Arylation of thermally annealed NDs using aryl diazonium salts has been recently described. [104] The arylated ND particles showed an improved solubility in different solvents. For instance, the carboxylated sample displayed high solubility of 0.63 mg/mL in water and 0.24 mg/mL in PBS buffer. Furthermore, the agglomerate size of the original NDs was substantially reduced to ~20-50 nm by this chemical procedure without using ultrasonication or milling. [104] Reduction of graphenic edges of annealed NDs by sodium in liquid ammonia leads to NDs salt. This material can be reacted with alkyl or aryl halides to yield a radical anion that dissociates spontaneously into a radical and halide. The free radical adds readily to the aromatic rings of the annealed NDs. [107] While the annealed NDs exhibit no solubility in water, surface functionalization with 5-bromovaleric acid increased significantly the solubility of the corresponding sodium salt modified NDs in water to 160 mg/L. [107]

5. Post-Functionalization

Basically all kinds of organic reactions can be carried out on suitable surface terminated nanodiamonds. Most of the current problems are not connected to the chemical reaction itself, but is mostly related to the yield of the transformation and the possibility of occurring side reactions on the surface of NDs. Subsequent reactions can also affect some basic properties of NDs such colloidal stability, particle size, zeta-potential, etc.

5.1. Through "Click" Chemistry

One class of reaction that is very attractive for the quick, simple and efficient grafting of organic moieties is the so-called "click" chemistry. As already discussed for diamond thin films, the copper (I) catalyzed cycloaddition of azides to alkynes is the most versatile method for the conjugation of complex organic moieties with nanoscopic objects such as carbon nanotubes and quantum dots. [134]

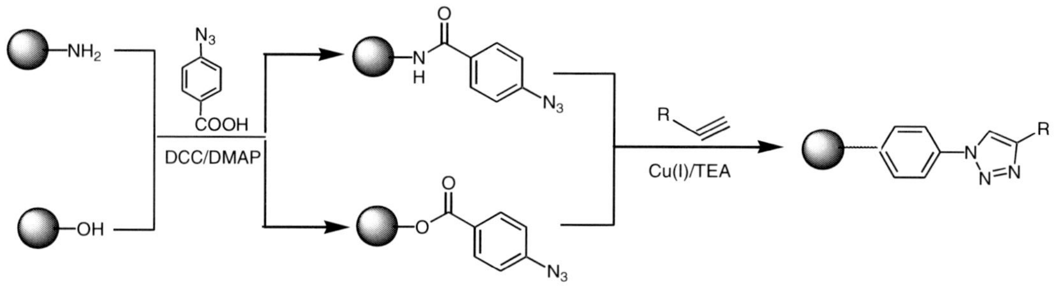

(A)

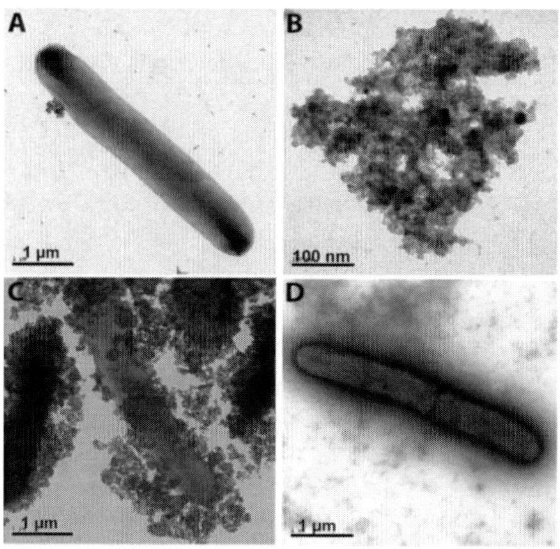

(B)

Figure 14. (A) "Click" chemistry on amine- or hydroxyl-terminated NDs; (B) TEM images of glyco-NDs with either fimbriated and or non-fimbriated *E. coli*. Cells were incubated during 40 min in PBS in the presence of different NDs and cells were washed twice in PBS and re-suspended in TRUMP buffer for fixation and TEM observation. Images clearly show an increase in NDs-mannose adhesion to cell surface when type 1 fimbriae are present. A. Fimbriated bacteria in the presence of NDs-galactose; B. Images of NDs-mannose alone; C. Fimbriated cells with NDs-mannose; D. Non fimbriated *E. coli* cells with NDs-mannose. In each panel, NDs were used at a final concentration of 10 μg/mL (reprint with permission from Ref. [136]).

This technique was recently adopted to "click" a variety of different compounds bearing a terminal acetylene group (1-decyne, electroactive ethynylferrocene, fluorescent N-propargyl-1-pyrenecarboxamide, pyrene, glycans, zonyl polymer) to azide-terminated NDs (Figure 14A). [118, 126, 135] The efficient covalent grafting of NDs with various carbohydrate analogs using a "click" chemistry strategy was recently demonstrated. [136] These newly developed "glyco-NDs", when presenting the appropriate sugar analog on their surface are able to selectively and very efficiently inhibit an *E. coli* K-12 derivative strain, constitutively expressing type 1 fimbriae, to adhere to various eukaryotic cells. In addition the glyco-NDs are able to inhibit bacterial biofilm formation. The dramatic affinity of the mannose-modified NDs compared with that of alternate mannose functionalized nanoparticles reported, such as those derived from gold (RIP of 200) or fullerene (RIP of 3) [137, 138] was put down to their efficient harnessing of a multivalency effect, which might be put down in part to the intrinsic physiochemical properties of the NDs core and its dimensions as well as to the fact that mannose moieties are linked to this core *via* an aromatic triazole tether motif expected to be a much better ligand for FimH than mannose alone.

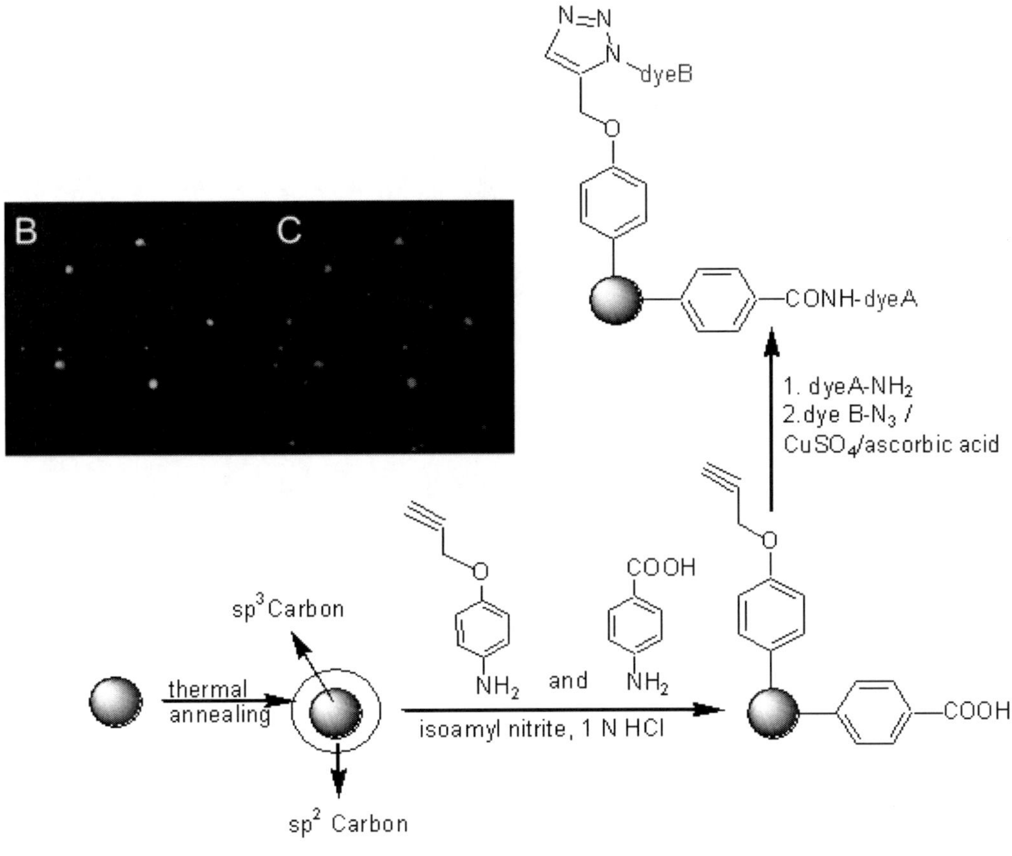

Figure 15. Orthogonal surface functionalization of NDs using different diazonium salts. The subsequent formation of triazole and amide bonds enables the independent grafting of two different fluorescent dyes Oregon Green (by amide bond) and TAMRA (by triazole formation) (reprint with permission from Ref [139]).

Transmission electron microscopy (TEM) images, performed on samples where bacteria had been treated with sugar-functionalized NDs, were used to visualize how mannose interferes with type 1 fimbriae-mediated adhesion (Figure 14B). While galactose-modified NDs did not show any binding to fimbriated cells, mannose-modified NDs were clearly aggregated around fimbriated bacteria whereas, no or scarcely any binding was observed for mannose-modified NDs on afimbriated cells, establishing that the adhesion was FimH specific.

The use of "click" reaction, tolerating a large variety of functional groups, allows also achieving orthogonal functionalization of NDs. [139] Krueger et al. prepared ND particles displaying different reactive moieties (alkynyl and carboxylated ligands) on their surface. This way, it is possible to address both reactive sites in such ways that the respective other functionality is not affected and remains available for the selective grafting of a second moiety. [139] Two different dyes, TAMRA and Oregon Green were successfully coupled to the NDs surface through "click" chemistry and amide bond formation (Figure 15).

More recently, the interest of dopamine and its derivatives for the direct functionalization of NDs has been shown (Figure 15A). Dopamine, chemically known as 4-(2-aminoethyl)-benzene-1,2-diol, has sparked great interest as an anchor for the functionalization of metal oxide surfaces because of the stability and strength of the resultant five-membered metallocyle chelate. Hydroxylated NDs were modified via the hydroxyl groups with dopamine bearing azide function or thermally responsive poly-*N*-isopropylacrylamide (PNIPAM).[126] Similarly, the NDs functionalized through the direct reaction of hydroxylated NDs with dopamine derivative bearing an azide group (section 4.3) were further derivatized with alkynyl-pyrene via copper(I)-catalyzed Huisgen 1,3-dipolar cycloaddition. The presence of pyrene grafted onto NDs was confirmed by UV/Vis spectroscopy by absorption features at 276 and 342 nm (Figure 16). The surface loading with pyrene was estimated to be 0.27 mmol g^{-1}. [126]

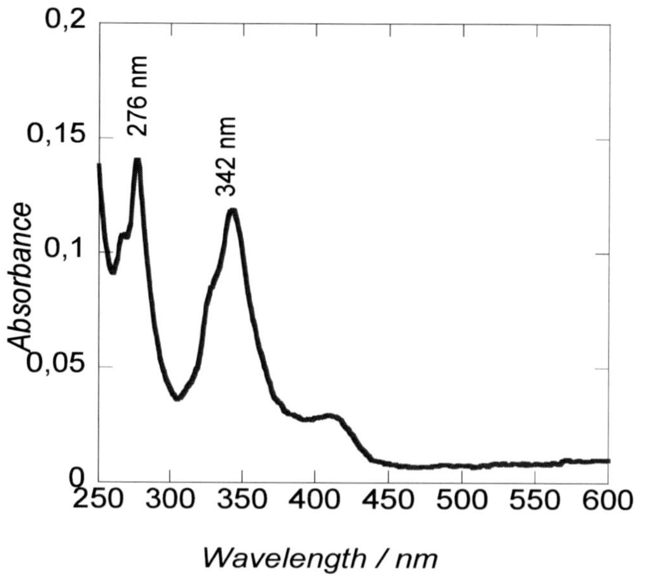

Figure 16. UV-vis spectrum of NDs-Py particles in CH_3CN (reprint with permission from Ref [126]).

5.2. Others

Introduction of easy functionalizable primary alcohol groups (oxyhexanol) on hydroxylated ultradispersed NDs was used as a starting material for covalent attachment of drug molecules and chiral ligands. [76] L-proline-functionalized NDs were investigated as heterogeneous enantioselective catalysts in the asymmetric aldol reaction. Depending on the experimental conditions, the reaction of benzaldehyde derivatives with cyclohexanone yielded 5-10% of the product with 41-87% ee. [76] Independently, NDs covalently modified with alcohol dehydrogenase were successfully used as a biocatalyst for enantioselective reduction of acetophenone to phenoxyethanol with ee > 99% and a reaction efficiency of 49%. [140] Benzoquinone modified hydroxylated NDs were used as efficient mean for simultaneous covalent immobilization of two different proteins (BSA, IgG) in a relatively straightforward way. [141] The functionalized NDs are able to specifically bind to the target antigen (mouse IgG). The results pave the way for possible applications of NDs as carriers to address deliveries of bioactive substances (i.e., drugs) to various biological targets *in vivo*. [141] Covalent modification of NDs with aminophenylboronic acid was applied for selective capture of glycoproteins from unfractionated protein mixtures. [142] A loading capacity of ~350 mg of glycoprotein/g of NDs could be attained on NDs that have been silanized with an alkyl linker chain prior to linking the phenylboronic acid. The role of the alkyl spacer chain is to form an exclusion shell, which suppresses nonspecific binding of nonglycated proteins. [142]

Conclusion and Perspectives

A large amount of effort has been put into the understanding and controlling the surface chemistry of nanodiamonds over the last decade. Hydrogenated, hydroxylated, carboxylated, halogenated and aminated ND nanoparticles can be formed by physical and chemical approaches. To date, NDs can be functionalized either covalently or electrostatically with a vast array of therapeutic molecules, including DNA, RNA, hormones, proteins, vitamines, and small-molecules. [63, 84, 123, 143-149] However, many challenges remain such as the production of nanodiamond particles with sizes smaller than 4 nm while preventing their aggregation. The wide range of potential applications of nanodiamonds in the biomedical area will continue to drive reaearch in this field forward. In the case of the use of nanodiamonds for drug delivery, the initial studies are extremely promising. A number of key studies, including the evaluation of the safety and therapeutic efficiency of nanodiamond conjugates not only in mice but also in human related models, will have to be performed in the nearby future so that the translation to patients could be made possible. Additonally, the fate of nanodiamonds once in the body has to be evaluated. Although initial data suggesting that NDs are eliminated through the bladder, presumably as urine, are encouraging, the precise mechanism of nanodiamonds elimination from complete organs has to be the subject of future investigations. Next to these rather applied questions, there is still a long list of fundamental questions which need to be addressed. Although the surface chemistry approaches developed for detonation nanodiamonds can be easily transferred to fluorescent diamond formed from HTHP diamond microcrystals, it is not clear how surface termination will influence the brightness of the fluorescence from NV centers? Indeed, a recent report highlighted the sensitivity of the NV centers to the chemical and electrical environment. [150] In the future, even stricter requirements on the flexibility of the functional groups built on the nanoscale diamond crystals will be demanded due to the use for complex and sophisticated applications of NDs in biomedial or high technological materials. The future on the developement of highly flexible and versatile surface chemistry on NDs is thus bright.

Acknowledgment

The Centre National de la Recherche Scientifique (CNRS), the Université Lille1, the Nord-Pas de Calais Région, and the Institut Universitaire de France (IUF) are gratefully acknowledged for financial support.

References

[1] Davis, M. E.; Zuckerman, J. E.; Coi, C. H. J. *Nature* 2010, *464*, 1067.
[2] Huang, H.; Pierstorff, E.; Osawa, E.; Ho, D. *Nano. Lett.* 2007, *7*, 3305.
[3] Liu, Z.; Fan, A. C.; Rakhra, K.; Sherlock, S.; Goodwin, A.; Chen, X.; Yang, Q.; Felsher, D. W.; Dai, H. *Angew. Chem. Int. Ed.* 2009, *48*, 7668.
[4] Krueger, A. *J. Mater. Chem.* 2011, *21*, 12571.
[5] Zhang, Q.; Mochalin, V. N.; Neitzel, I.; Knoke, I. Y.; Han, J.; Klug, C. A.; Zhou, J. G.; Lelkes, P. I.; Gogotsi, Y. *Biomaterials* 2011, *32*, 87-94.
[6] Schrand, A. M.; Hens, S. A. C.; Shenderova, O. A. *Crit. Rev. Sol. State Mater Sci.* 2009, *34*, 18.
[7] Mochalin, V. N.; Shenderova, O.; Ho, D.; Gogotsi, Y. *Nature Nanotechnol.* 2012, *7*, 11-23.
[8] Danilenko, V. V. *Phys. Solid State* 2004, *46*, 595.
[9] Grainer, N. R.; Philips, D. S.; Johnson, J. D. *Nature* 1988, *333*, 440.
[10] Ho, D. *Nanodiamonds: Applications in Biology and Nanoscale Medicine*; Springer, 2009.
[11] Yu, S.-J.; Kang, M.-W.; Chang, H.-C.; Chen, K.-M.; Yu, Y.-C. *J. Am. Chem. Soc.* 2005, *127*, 17604.
[12] Fu, C.-C.; Lee, H.-Y.; Chen, K.; Lim, T.-S.; Wu, H.-Y.; Lin, P.-K.; Wei, P.-K.; Taso, P.-H.; Chang, H.-C.; Fann, W. *PNAS* 2007, *104*, 727.
[13] Mochalin, V. N.; Gogotsi, Y. *J. Am. Chem. Soc.* 2009, *131*, 4594.
[14] Chang, Y.-R.; Lee, H.-Y.; Chen, K.; Chang, C.-C.; Tsai, D.-S.; Fu, C.-C.; Lim, T.-S.; Tzeng, Y.-K.; Fang, C.-Y.; Han, C.-C.; Chang, H.-C.; Fann, W. *Nature Nanotechnol.* 2008, *3*, 284-288.
[15] Hui, Y. Y.; Cheng, C.-L.; Chang, H.-C. *J. Phys. D: Appl. Phys.* 2010, *43*, 374021.
[16] Marcon, L.; Riquet, F.; Vicogne, D.; Szunerits, S.; Bodart, J.-F.; Boukherroub, R. *J. Mater. Chem.* 2010, *20*, 8064-8069.
[17] Schrand, A. M.; Huang, H.; Carlson, C.; Schlager, J. J.; Osawa, E.; Hussain, S. M.; Dai, L. *J. Phys. Chem. B* 2007, *111*, 2.

[18] Balasubramanian, G.; Chan, I. Y.; Kolesov, R.; Al-Hmoud, M.; Tisler, J.; Shin, C.; Kim, C.; Wojcik, A.; Hemmer, P. R.; Krueger, A.; Hanke, T.; Leitenstorfer, A.; Bratschitsch, R.; Jelezko, F.; Wrachtrup, J. *Nature* 2008, *455*, 648.

[19] Ozawa, E. *Diamond Relat. Mater.* 2007, *16*, 2018-2022.

[20] Liu, Y.; Gu, Z.; Margrave, J. n. L.; Khabashesku, V. N. *Chem. Mater.* 2004, *16*, 3924-3930.

[21] Krueger, A.; Liang, Y.; Jarre, G.; Stegk, J. *J. Mater. Chem.* 2008, *16*, 2322.

[22] Krueger, A.; Lang, D. *Adv. Funct. Mater.* 2012, *22*, 890-906.

[23] Williams, O. A.; Hees, J.; Dieker, C.; Jager, W.; Kirste, L.; Nebel, C. E. *ACS Nano* 2010, *4*, 4824.

[24] Shimkunas, R. A.; Robinson, E.; Lam, R.; Lu, S.; Xu, X.; Zhang, X.-Q.; Huang, H.; Osawa, E.; Ho, D. *Biomaterials* 2009, *30*, 5720-5728.

[25] Huang, L.-C. L.; Chang, H.-C. *Langmuir* 2004, *20*, 5879-5884.

[26] Chao, J.-I.; Perevedentseva, E.; Chung, P.-H.; Liu, K.-K.; Cheng, C.-Y.; Chang, C.-C.; Cheng, C.-L. *Biophys. J.* 2007, *93*, 2199-2208.

[27] Nguyen, T.-T. B.; Chang, H.-C.; Wu, V. W.-K. *Diamond Relat. Mater.* 2007, *16*, 872-876.

[28] Perevedentseva, E.; Cai, P.-J.; Chiu, Y.-C.; Cheng, C.-L. *Langmuir* 2011, *27*, 1085-1091.

[29] Chung, P.-H.; Perevedentseva, E.; Tu, J.-S.; Chang, C. C.; Cheng, C.-L. *Diamond Relat. Mater.* 2006, *15*, 622-625.

[30] Bondar', V. S.; Pozdnyakova, I. O.; Puzyr', A. P. *Phys. Solid State* 2004, *46*, 758-760.

[31] Huang, H.; Pierstorff, E.; Osawa, E.; Ho, D. *Nano Lett.* 2007, *7*, 3305-3314.

[32] Lam, R.; Chen, M.; Pierstorff, E.; Huang, H.; Osawa, E.; Ho, D. *ACS Nano* 2008, *2*, 2095-2102.

[33] Li, J.; Zhu, Y.; Li, W.; Zhang, X.; Peng, Y.; Huang, Q. *Biomaterials* 2010, *31*, 8410-8418.

[34] http://www.nabond.com/06/03/2013.

[35] http://www.microdiamant.com/06/03/2013.

[36] http://www.nanodiamond.co.il/Products.htm/06/03/2013.

[37] Frenklach, M.; Howard, W.; Huang, D.; Yuan, J.; Spear, K. E.; Koba, R. *Appl. Phys. Lett.* 1991, *59*, 546.

[38] Daulton, T. L.; Kirk, M. A.; Lewis, R. S.; Rehn, I. E. E. *Nucl. Instrum. Meth. B* 2001, *175*, 12.

[39] Welz, S.; Gogotsi, Y.; McNallan, M. J. *J. Appl. Phys.* 2003, *93*, 4207.

[40] Gogotsi, Y. G.; Nickel, K. G.; Bahloul-Hourlier, D.; Merle-Mejean, T.; Khomenko, G. E.; Skjerlie, K. P. *J. Mater. Chem.* 1996, *6*, 595-604.

[41] Boudou, J. P.; Curmi, P. A.; Jelezko, F.; Wrachtrup, J.; Aubert, P.; Sennour, M.; Balasubramanian, G.; Reuter, R.; Thorel, A.; Gaffet, E. *Nanotechnology* 2009, *20*, 235602.

[42] Chang, Y.-R.; Lee, H.-Y.; Chen, K.; Chang, C.-C.; Tsai, D.-S.; Fu, C.-C.; Lim, T.-S.; Tzeng, Y.-K.; Fang, C.-Y.; Han, C.-C.; Chang, H.-C.; Fann, W. *Nature Nanotechnol.* 2008, *3*, 284.

[43] Krüger, A.; Katoako, F.; Ozawa, M.; Fujino, T.; Suzuki, Y.; Aleksesnkii, A. E.; Vul , A. Y.; Osawa, E. *Carbon* 2005, *43*, 1722.

[44] Shenderova, O. A.; Zhirnov, V. V.; Brenner, D. W. *Crit. Rev. Solid State Mater. Sci.* 2007, *27*, 38.
[45] Aleksenski, A. E.; Osipov, V. Y.; Dideikin, A. T.; Vul', A. Y.; Adriaenssens, G. J.; Afanas'ev, V. V. *Techn. Phys. Lett.* 2000, *26*, 819-821.
[46] Krüger, A.; Ozawa, M.; Jarre, G.; Liand, Y.; Stegk, J.; Lu, L. *Phys. Stat. Sol. (a)* 2007, *204*, 2881.
[47] Osswald, S.; Yushin, G.; Mochalin, V.; Kucheyev, S. O.; Gogotsi, Y. *J. Am. Chem. Soc.* 2006, *128*, 11635.
[48] Ushizawa, K.; Sato, Y.; Mitsumori, T.; Machinami, T.; Ueda, T.; Ando, T. *Chem. Phys. Lett.* 2002, *351*, 105.
[49] Ozawa, M.; Inaguma, M.; Takahashi, M.; Kataoka, F.; Krüger, A.; Ōsawa, E. *Adv. Mater.* 2007, *19*, 1201-1206.
[50] Osawa, E. *Diamond Relat. Mater.* 2007, *16*, 2018.
[51] Ozawa, E. Single-nano buckydiamond particles: Synthesis strategies, characterization methodologies and emerging applications; *Springer*: New York, 2010.
[52] Xu, X.; Yu, Z.; Zhu, Y.; Wang, B. *Diamond Relat. Mater.* 2005, *14*, 206-212.
[53] Xu, X.; Zhu, Y.; Wang, B.; Yu, Z.; Xie, S. *J. Mater. Sci. Technol.* 2005, *21*, 109.
[54] Treussarta, F.; Jacques, V.; Wu, E.; Gacoin, T.; Grangier, P.; Roch, J.-F. *Physica B* 2006, *376-377*, 926-92.
[55] Neugart, F.; Zappe, A.; Jelezko, F.; Tietz, C.; Boudou, J. P.; Krueger, A.; Wrachtrup, J. *Nano Lett.* 2007, *7*, 3588-3591.
[56] Gruber, A.; Dräbenstedt, A.; Tietz, C.; Fleury, L.; Wrachtrup, J.; von Borczyskowski, C. *Science* 1997, *276*, 2012-2014.
[57] Wee, T.-L.; Mau, Y.-W.; Fang, C.-Y.; Hsu, H.-L.; Han, C.-C.; Chang, H.-C. *Diamond Relat. Mater.* 2009, *18*, 567-573.
[58] Faklaris, O.; Garrot, D.; Joshi, V.; Druon, F.; Boudou, J.-P.; Sauvage, T.; Georges, P.; Curmi, P. A.; Treussart, F. *Small* 2008, *4*, 2236-2239.
[59] Pentecost, A.; Gour, S.; Mochalin, V.; Knoke, I.; Gogotsi, Y. *ACS Appl. Mater. Interfaces* 2010, *2*, 3289.
[60] Kulakova, I. I. *Phys. Solid State* 2004, *46*, 636.
[61] Jiang, T. L.; Xu, K.; Ji, S. F. *J. Chem. Soc., Faraday Trans.* 1996, *92*, 3401.
[62] Shenderova, O. A.; Koscheev, A.; Zaripov, N.; Petrov, I.; Skryabin, Y.; Detkov, P.; Turner, S.; Van Tendeloo, G. *J. Phys. Chem. C* 2011, *115*, 9827.
[63] Krueger, A.; Stegk, J.; Liang, Y.; Lu, L.; Jarre, G. *Langmuir* 2008, *24*, 4200.
[64] Arnault, J. C.; Petit, T.; Girard, H. A.; Chavanne, A.; Gesset, C.; Sennour, M.; Chaigneau, M. *Phys. Chem. Chem. Phys.* 2011, *13*, 11481.
[65] Michaelson, S.; Stacey, A.; Akhvlediani, R.; Prawer, S.; Hoffman, A. *Surf. Sci.* 2010, *604*, 1326.
[66] Girard, H. A.; Petit, T.; Perruchas, S.; Gacoin, T.; Gesset, C.; Arnault, J. C.; Bergonzo, P. *Phys. Chem. Chem. Phys.* 2011, *13*, 11517.
[67] Tu, J.-S.; Perevedentseva, E.; Chung, P.-H.; Cheng, C.-L. *J. Chem. Phys.* 2006, *125*, 174713.
[68] Nguyen, T.-T.-B.; Chang, H.-C.; Wu, V. W.-K. *Diamond Relat. Mater.* 2007, *16*, 872.
[69] Kong, X. L.; Huand, L. C. L.; Liau, S.-C. V.; Han, C.-C.; Chang, H. C. *Anal. Chem.* 2005, *77*, 4273.
[70] Chang, C.-K.; Wu, C.-C.; Wang, Y.-S.; Chang, H. C. *Anal. Chem.* 2008, *80*, 3791.

[71] Mitev, D.; Dimitrova, R.; Spassova, M.; Minchev, C.; Stavrev, S. *Diamond Relat. Mater.* 2007, *16*, 776-780.
[72] Cheng, J.; He, J.; Li, C.; Yang, Y. *Chem. Mater.* 2008, *20*, 4224-4230.
[73] Martin, R.; Alvaro, M.; Herance, J. R.; Garcia, H. *ACS Nano* 2010, *4*, 65-74.
[74] Martin, R.; Heydorn, P. C.; Alvaro, M.; Garcia, H. *Chem. Mater.* 2009, *21*, 4505.
[75] Krueger, A.; Liang, Y.; Jarre, G.; Stegk, J. *J. Mater. Chem.* 2006, *16*, 2322.
[76] Zheng, W.-W.; Hsieh, Y.-H.; Chiu, Y.-C.; Cai, S.-J.; Cheng, C.-L.; Chen, C. *J. Mater. Chem.* 2009, *19*, 8432-8441.
[77] Ciftan Hens, S.; Cunningham, G.; Tyler, T.; Moseenkov, S.; Kuznetsov, V.; Shenderova, O. A. *Diamond Relat. Mater.* 2008, *17*, 1858.
[78] Spitsyn, B. V.; Denisov, S. A.; Skorik, N. A.; Chopurova, A. G.; Parkaeva, S. A.; Belyakova, L. D.; Larionov, O. G. *Diamond Relat. Mater.* 2010, *19*, 123.
[79] Obraztsova, I. I.; Eremenko, N. K. *Russ. J. Appl. Chem.* 2008, *81*, 603-608.
[80] Ando, T.; Ishii, M.; Kamo, M.; Sato, Y. *J. Chem. Soc. Faraday Trans.* 1993, *89*, 1783-1789.
[81] Ida, S.; Tsubota, T.; Hirabayashi, O.; Nagata, M.; Matsumoto, Y.; Fujishima, A. *Diamond Relat. Mater.* 2003, *12*, 601-605.
[82] Ando, T.; Yamamoto, K.; Suehara, S.; Kamo, M.; Sato, Y.; Shimosaki, S.; Nishitani-Gamo, M. *J. Chin. Chem. Soc.* 1995, *42*, 285-292.
[83] Takimoto, T.; Chano, T.; Shimizu, S.; Okabe, H.; Ito, M.; Morita, M.; Kimura, T.; Inushibi, T.; Komatsu, N. *Chem. Mater.* 2010, *22*, 3462.
[84] Yeap, W. S.; Chen, S.; Loh, K. P. *Langmuir* 2009, *25*, 185.
[85] Girard, H. A.; Arnault, J. C.; Perruchas, S.; Saasa, S.; Gacoin, T.; Boilot, J.-P. *Diamond Relat. Mater.* 2010, *19*, 1117.
[86] Ray, M. A.; Tyler, T.; Hook, B.; Martin, A.; Cunningham, G.; Shenderova, O. A.; Davidson, J. L.; Howell, M.; Kang, W. P.; McGuire, G. *Diamond Relat. Mater.* 2007, *16*, 2087.
[87] Sotawa, K.-I.; Amamoto, T.; Sobana, A.; Kusakabe, K.; Imato, T. *Diamond Rel. Mater.* 2004, *13*, 145.
[88] Lisichkin, G. V.; Korol'kov, V. V.; Tarasevich, B. N.; Kulakova, I. I.; Karpukhin, A. V. *Russ. Chem. Bull.* 2006, *55*, 2212.
[89] Saito, T.; Ikeda, Y.; Egawa, S.-I.; Kusakabe, K.; Morooka, S.; Maeda, H.; Taniguchi, Y.; Fujiwara, Y. *J. Chem. Soc., Faraday Trans.* 1998, *94*, 929-932.
[90] Zhang, G.-J.; Song, K.-S.; Nakamura, Y.; Ueno, T.; Funatsu, T.; Ohdomari, I.; Kawarada, H. *Langmuir* 2006, *22*, 3728-3734.
[91] Szunerits, S.; Jama, C.; Coffinier, Y.; Marcus, B.; Delabouglise, D.; Boukherroub, R. *Electrochem. Commun.* 2006, *8*, 1185-1190.
[92] Coffinier, Y.; Szunerits, S.; Marcus, B.; Desmet, R.; Melnyk, O.; Gengembre, L.; Payen, E.; Delabouglise, D.; Boukherroub, R. *Diamond Relat. Mater.* 2007, *16*, 892.
[93] Koch, H.; Kulisch, W.; Popov, C.; Merz, R.; Merz, B.; Reithmaier, J. P. *Diamond Relat. Mater.* 2011, *20*, 254-258.
[94] Sotowa, K.-I.; Amamoto, T.; Sobana, A.; Kusakabe, K.; Imato, T. *Diamond Relat. Mater.* 2004, *13*, 145.
[95] Miller, J. B.; Brown, D. W. *Langmuir* 1996, *12*, 5809-5817.
[96] Hens, S. C.; Cunningham, G.; Tyler, T.; Moseenkov, S.; Kuznetsov, V.; Shenderova, O. *Diamond Relat. Mater.* 2008, *17*, 1858-1866.

[97] Chen, J.; Deng, S. Z.; Chen, J.; Yu, Z. X.; Xu, N. S. *Appl. Phys. Lett.* 1999, *74*, 3651-3653.
[98] Osipov, V. Y.; Enoki, T.; Takai, K.; Takahara, K.; Endo, M.; Hayashi, T.; Hishiyama, Y.; Kaburagi, Y.; Vul', A. Y. *Carbon* 2006, *44*, 1225-1234.
[99] Qiao, Z.; Li, J.; Zhao, N.; Shi, C.; Nash, P. *Chem. Phys. Lett.* 2006, *429*, 479-482.
[100] Zeppilli, S.; Arnault, J. C.; Gesset, C.; Bergonzo, P.; Polini, R. *Diamond Relat. Mater.* 2010, *19*, 846-853.
[101] Zang, J.; Wang, Y.; Bian, L.; Zhang, J.; Meng, F.; Zhao, Y.; Ren, S.; Qu, X. *Electrochim. Acta* 2012, *72*, 68-73.
[102] Zou, Q.; Li, Y. G.; Lv, B.; Wang, M. Z.; Zou, L. H.; Zhao, Y. C. *Inorg. Mater.* 2010, *46*, 127-131.
[103] Mykhaylyk, O. O.; Solonin, Y. M.; Batchelder, D. N.; Brydson, R. *J. Appl. Phys.* 2005, *97*, 074302.
[104] Liang, Y.; Meinhardt, T.; Jarre, G.; Ozawa, M.; Vrdoljak, P.; Schöll, A.; Reinert, F.; Krueger, A. *J. Colloid Interface Sci.* 2011, *354*, 23-30.
[105] Xu, X.; Yu, Z.; Zhu, Y.; Wang, B. *J. Solid State Chem.* 2005, *178*, 688-693.
[106] Chang, I. P.; Hwang, K. C.; Ho, J.-a. A.; Lin, C.-C.; Hwu, R. J.-R.; Horng, J.-C. *Langmuir* 2010, *26*, 3685-3689.
[107] Kuznetsov, O.; Sun, Y.; Thaner, R.; Bratt, A.; Shenoy, V.; Wong, M. S.; Jones, J.; Billups, W. E. *Langmuir* 2012, *28*, 5243−5248.
[108] Petit, T.; Arnault, J.-C.; Girard, H. A.; Sennour, M.; Bergonzo, P. *Phys. Rev. B* 2011, *84*, 233407.
[109] Qiao, Z.; Li, J.; Zhao, N.; Shi, C.; Nash, P. *Scripta Mater.* 2006, *54*, 225-229.
[110] Xie, F. Y.; Xie, W. G.; Gong, L.; Zhang, W. H.; Chen, S. H.; Zhang, Q. Z.; Chen, J. *Surf. Interface Anal.* 2010, *42*, 1514-1518.
[111] Hiraki, J.; Mori, H.; Taguchi, E.; Yasuda, H.; Kinoshita, H.; Ohmae, N. *Appl. Phys. Lett.* 2005, *86*, 223101.
[112] Qin, L.-C.; Iijima, S. *Chem. Phys. Lett.* 1996, *262*, 252-258.
[113] Roddatis, V. V.; Kuznetsov, V. L.; Butenko, Y. V.; Su, D. S.; Schlögl, R. *Phys. Chem. Chem. Phys.* 2002, *4*, 1964-1967.
[114] Weng, M.-F.; Chiang, S.-Y.; Wang, N.-S.; Niu, H. *Diamond Relat. Mater.* 2009, *18*, 587-591.
[115] Wei, L.; Zhang, W.; Lu, H.; Yang, P. *Talanta* 2010, *80*, 1298-1304.
[116] Zhan, Q.; Mochain, V. N.; Neitzel, I.; Knoke, I. Y.; Han, J.; Klug, C. A.; Zhou, J. G.; Lelkes, P. I.; Gogotsi, Y. *Biomaterials* 2011, *32*, 87.
[117] Park, C.-L.; Jee, A. Y.; Lee, M.; Lee, S.-G. *Chem. Commun.* 2009, 5576-5578.
[118] Barras, A.; Szunerits, S.; Marcon, L.; Monfilliette-Dupont, N.; Boukherroub, R. *Langmuir* 2010, *26*, 13168-72.
[119] Kruger, A.; Boedeker, T. *Diamond Relat. Mater.* 2008, *17*, 1367.
[120] Das, M. R.; Wang, M.; Szunerits, S.; Gengembre, L.; Boukherroub, R. *Chem. Commun.* 2009, 2753-2755.
[121] Liang, Y.; Ozawa, M.; Krueger, A. *ACS Nano* 2009, *3*, 2288-96.
[122] Rojas, S.; Gispert, J. D.; Martin, R.; Abad, S.; Menchon, C.; Pareto, D.; Victor, V. M.; Alvaro, M.; Garcia, H.; Herance, J. R. *ACS Nano* 2011, *5*, 5552-5559.
[123] Zhang, X.-Q.; Chen, M.; Lam, R.; Xu, X.; Osawa, E.; Ho, D. *ACS Nano* 2009, *3*, 2609-2616.

[124] Vial, S.; Mansuy, C.; Sagan, S.; Irinopoulou, T.; Burlina, F.; Boudou, J.-P.; Chassaing, G.; Lavielle, S. *ChemBioChem* 2008, *9*, 2133.

[125] Liu, K.-K.; Zhang, W.-W.; Wang, C.-C.; Chiu, Y.-C.; Chang, C.-L.; Lo, Y.-S.; Chen, C.; Chao, J.-I. *Nanotechnology* 2010, *21*, 315106.

[126] Barras, A.; Lyskawa, J.; Szunerits, S.; Woisel, P.; Boukherroub, R. *Langmuir* 2011, *27*, 12451.

[127] Yang, W.; Butler, J. E.; Russell, J. N.; Hamers, R. J. *Langmuir* 2004, *20*, 6778-6787.

[128] Dahoumane, S. A.; Nguyen, M. N.; Thorel, A.; Boudou, J. P.; Chehimi, M. M.; Mangeney, C. *Langmuir* 2009, *25*, 9633-8.

[129] Mangeney, C.; Qin, Z.; Dahoumane, S. A.; Adenier, A.; Herbst, F.; Boudou, J.-P.; Pinson, J.; Chehimi, M. M. *Diamond Relat. Mater.* 2008, *17*, 1881-1887.

[130] Pinson, J.; Podvorica, F. *Chem. Soc. Rev.* 2005, *34*, 429.

[131] Nakamura, T.; Ohana, T.; Hagiwarab, Y.; Tsubotab, T. *Phys. Chem. Chem. Phys.* 2009, *11*, 730-734.

[132] Jarre, G.; Liang, Y.; Betz, P.; Lang, D.; Krueger, A. *Chem. Commun.* 2011, *47*, 544-546.

[133] Chang, I. P.; Hwang, K. C.; Ho, J.-a. A.; Lin, C.-C.; Hwu, R. J. R.; Horng, J.-C. *Langmuir* 2009, *26*, 3685-3689.

[134] Moses, J. E.; Moorhouse, A. D. *Chem. Soc. Rev.* 2007, *36*, 1249-1262.

[135] Marcon, L.; Kherrouche, Z.; Lyskawa, J.; Fournier, D.; Tulasne, D.; Woisel, P.; Boukherroub, R. *Chem. Commun.* 2011, *47*, 5178.

[136] Barras, A.; Martin, F. A.; Bande, O.; Baumann, J.-S.; Ghigo, J.-M.; Boukherroub, R.; Beloin, C.; Siriwardena, A.; Szunerits, S. *Nanoscale* 2013, *5*, 2307–2316.

[137] Lin, C. C.; Yeh, Y. C.; Yang, C. Y.; Chen, C. L.; Chen, G. F.; Chen, C. C.; Wu, Y. C. *J. Am. Chem. Soc.* 2002, *124*, 3508.

[138] Durka, M.; Buffet, K.; Lehl, J.; Holler, M.; Nierengarten, J.-F.; Taganna, J.; Bouckaert, J.; Vincent, S. P. *Chem. Commun.* 2011, *47*, 1321-1323.

[139] Meinhardt, T.; Lang, D.; Dill, H.; Krueger, A. *Adv. Funct. Mater.* 2011, *21*, 494.

[140] Goldberg, K.; Krueger, A.; Meinhardt, T.; Kroutil, W.; Mautner, B.; Liese, A. *Tetrahedron Asymm.* 2008, *19*, 1171-1173.

[141] Purtov, K. V.; Petunin, A. I.; Burov, A. E.; Puzyr, A. P.; Bondar, V. S. *Nanoscale Res. Lett.* 2010, *5*, 631-636.

[142] Yeap, W. S.; Tan, Y. Y.; Loh, K. P. *Anal. Chem.* 2008, *80*, 4659-4665.

[143] Manakamana, M.; Cherredy, K.; Barras, A.; Marcon, L.; Boukherroub, R.; Szunerits, S. *unpublished results* 2012.

[144] Chen, M.; Zhang, X.-Q.; Man, H. B.; Lam, R.; Chow, E. K.; Ho, D. *J. Phys. Chem. Lett.* 2010, *1*, 3167.

[145] Alhaddad, A.; Adam, M.-P.; Botsoa, J.; Dantelle, G.; Perruchas, S.; Gacoin, T.; Mansuy, C.; Lavielle, S.; Malvy, C.; Treussart, F.; Bertrand, J. R. *Small* 2011, *7*, 3087.

[146] Shimkunas, R. A.; Robinson, E.; Lam, R.; Lu, S.; Xu, X.; Zhang, X.-Q.; Huang, H.; Osawa, E.; Ho, D. *Biomaterials* 2009, *30*, 5720.

[147] Smith, A. H.; Robinson, E. M.; Zhang, X. Q.; Chow, E. K.; Lin, Y.; Osawa, E.; Xie, J.; Ho, D. *Nanoscale* 2011, *3*, 2844-2848.

[148] Liu, Y. L.; Sun, K. W. *Nanoscale Res. Lett.* 2010, *5*, 1045-1050.

[149] Huang, L.-C., L.; Chang, H.-C. *Langmuir* 2004, *20*, 5879.
[150] Petráková, V.; Taylor, A.; Kratochvílová, I.; Fendrych, F.; Vacík, J.; Kucka, J.; Štursa, J.; Cígler, P.; Ledvina, M.; Fišerová, A.; Kneppo, P.; Nesládek, M. *Adv. Funct. Mater.* 2012, *22*, 812-819.

In: Applied Surface Chemistry of Nanomaterials
Editors: Mohamed M. Chehimi and Jean Pinson
ISBN: 978-1-62808-351-4

Chapter 2

DIAMOND NANOSTRUCTURES: TECHNOLOGIES, PROPERTIES AND ELECTROCHEMICAL APPLICATIONS

Nianjun Yang,* Waldemar Smirnov and Christoph E. Nebel
Fraunhofer-Institute for Applied Solid State Physics (IAF), Freiburg, Germany

1. INTRODUCTION

Diamond, which is one of the promising wide-band gap semiconductors, has been considered as an excellent material for many applications due to its unusual chemical and physical properties [1-6]. Much attention has thus been paid to its growth, characterization, applications, and devices realization [2-6]. Depending on the type of doping, diamond can be n- or p-type conductive, resulting in a wide variation of conductivity from insulating, semiconducting, to metallic.

Heavily boron-doped diamond films have been recognized as the best electrode materials for electrochemistry and biochemical applications. These diamond films have shown unique features such as i) high chemical stability in harsh environments and/or at high current densities as working electrodes; ii) weak or no surface bio-fouling; iii) biocompatibility; iv) low and stable capacitive currents in aqueous and organic solutions; and v) wide electrochemical potential window.

For example, the capacitive current of a hydrogen terminated diamond electrode (boron concentration: 2×10^{20} cm^{-3}) in 0.1 M pH 7 phosphate buffer is 10 times lower than that of a gold electrode and about 100 times lower than that of a glassy carbon electrode.

A typical electrochemical potential window for diamond electrodes is 3.25 V in neutral and basic solutions [5]. In organic solutions or ionic liquids, the electrochemical potential window of diamond is even wider. The diamond surface can be terminated with hydrogen, hydroxyl, and oxygen [1-3], which allows tuning the electronic properties of the solid/electrolyte interface with respect to energy alignments of interacting levels.

---

* Corresponding author: Nianjun Yang. E-mail: nianjun.yang@iaf.fraunhofer.de.

These terminations can be achieved by plasma treatment at high temperature under different atmospheres [1-3] or by electrochemical reduction/oxidation at high current densities in acid solutions [7]. Moreover, it is ultra-hard (50-150 GaP) [8]. Diamond is thus a promising candidate for applications in different nanostructures.

Nanostructures such as nanotextures and nanowires are known to have characteristics of low weight with sometimes extraordinary mechanical, electrical, thermal and multifunctional properties [9, 10]. By creating diamond based nanostructures, it is possible to utilize the properties of diamond for applications which require strong mechanical stability in combination with chemical stability. These nanostructures are attractive especially for giant surface-to-volume ratio enhancements [9, 10]. Fabrication, characterization and applications of diamond based nanostructures such as textures, wires, and pores are therefore attractive new research areas.

In this chapter, we summarize techniques applied for the fabrication of diamond based nanostructures, mainly nanotextures and nanowires. The characterizations of these nanostructures using atomic force microscope, scanning electron microscope, electrochemical techniques is then introduced.

In the last part of this chapter, the electrochemical and biochemical sensing application of these nanostructures are shown. As examples, direct electron transfer of redox protein cytochrome c and highly sensitive detection of DNA hybridization using diamond nanotextures as sensing electrodes are explained in detail.

Keywords: Carbon materials, Diamond films, Nanotechnology, Diamond nanostructures, Diamond nanotextures, Diamond nanowires, Diamond nanoparticles, Top-down technology, Etching mask, Bottom-up technology, AFM, STM, Bio-functionalization, Diazonium salts, Electrochemical sensing, DNA biosensors, Redox indicators, Sign-off sensors, Redox enzyme, Direct electron transfer, Cytochrome c, Electron transfer rate

2. TECHNOLOGIES

Generally, nanostructures such as nanotextures and nanowires can be fabricated [11-13] either by a bottom-up approach (self-assembly of small sized structures to form larger structures) or by a top-down approach (reduction of large systems down to small size).

The realization of diamond nanostructures started in 1997 by Shiomi [14], who demonstrated for the first time the formation of porous diamond films by reactive ion etching (RIE) using O_2. Later in 2000, nanostructured diamond honeycomb films have been prepared [15] by etching through a porous anodic alumina mask. These activities have been summarized by Shenderova et al. [16]. Growth induced formation of nano-scale tubular structures have been reported for the first time in 2003, applying a microwave plasma of hydrogen under a bias potential [17]. In 2008, Zou et al. reported about the fabrication of nano-pillar arrays using self-aligned Au nano-dots as etching mask in a bias-assisted reactive ion etching, applying a hydrogen/argon plasma [18].

In our group vertically aligned diamond nanostructures (nanotextures and nanowires) were fabricated by reactive ion etching (RIE) using O_2 and by using diamond nanoparticles [19-21] or self-formed nickel nanoparticles as etching mask [22, 23].

2.1. Top-Down Approach

In most cases, diamond nanostructures are produced by top-down approaches [14-32] where diamond is etched with reactive ions in a plasma using etching (hard) masks. Various masking materials have been applied, including Al, SiO_2, Au, Ni, Mo, polymers, oxides, nitrides, and diamond nanoparticles [14-29]. These masks were made either with help of lithographic techniques [24, 25], or by unintentionally sputtered material from the substrate holder onto the sample during the etching process [25], or by deliberately depositing thin metal layers [26-29]. By using diamond nanoparticles [19-21] and self-formed nickel nanoparticles [22, 23] as etching mask, vertically aligned diamond nanotextures [19-21] and diamond nanowires [22, 23] were generated. Thermal removal/etching of the non-diamond phase from nano-crystalline diamond films was applied to form diamond nanopores [32].

Figure 1 shows one example we applied to fabricate vertically aligned diamond nanotextures [19-21]. The substrate is a metallically boron-doped, single crystalline diamond film. It has atomically smooth surface and its root-mean-square (RMS) roughness is typically about 0.8 Å. This substrate was grown homoepitaxially on the Ib diamond substrate, using a micro-wave assisted chemical vapor deposition (CVD) technique [33, 34]. The Ib substrate is nitrogen-doped diamond and always generated with high-pressure-high-temperature (HPHT) technique. An etching mask from mono-dispersed diamond nanoparticles was applied [19-21]. Diamond nanoparticles have well-defined size and quality. They are typical 8-10 nm in diameter [35] and can be dispersed in water or ethanol by ultra-sonification (200 W, 20 kHz, 12 h) to form a pseudo-stable suspension [36, 37]. It is to note that the stability of the suspension is strongly affected by the pretreatment of diamond nanoparticles and the concentration of the suspension [38-40]. Then the diamond plate is immersed into the suspension (0.05% in ethanol, wt%) and sonicated (100 W, 10 minutes) to seed diamond nanoparticles on the diamond surface. The seeding density of diamond nanoparticles depends on the surface termination of nanoparticles, suspension quality (e.g. concentration, solvent), and time of sonication [38-40]. After the formation of diamond nanoparticles based etching mask, reactive ion etching (RIE) in an O_2 (97 %) / CF_4 (3 %) gas mixture is applied. The typical etching time is varied between 2 to 60 s [19].

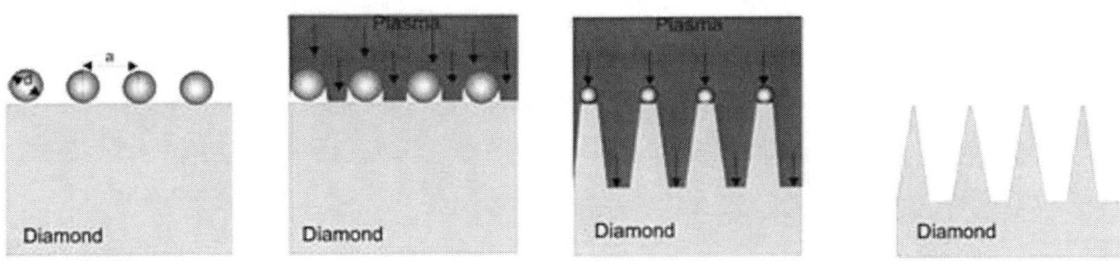

a b c d

[a]Nano-particle Seeding.
[b]ICP/RIE (t=0).
[c]ICP/RIE (t>0).
[d]Final Surface Structure.

Figure 1. Schematic illumination of fabricating vertically aligned diamond nanotextures using a top-down technology and using diamond nanoparticles as the etching mask.

The formation of diamond nanotextures using diamond nanoparticles as hard mask were characterized with tapping-mode atomic force microscope (AFM) with respect to the variation of surface morphology of atomically smooth diamond before and after RIE etching [19]. AFM experiments were conducted using ultra-sharp silicon cantilevers with a typical resonant frequency of 75 kHz and a spring constant of 3.5 N m^{-1}. Before imaging, the samples were cleaned in a mixture of sulfuric acid and nitric acid (V/V=3:1) at 200 °C for 1 h.

Figure 2 shows four 3-D AFM tapping mode images of diamond surface after RIE etching for 5, 10, 20 and 60 s. It is obvious that vertically aligned diamond nanotextures arise where diamond nanoparticles have been deposited. Geometrical properties of these textures depend on size of nanoparticles and etching rate.

Each of these surfaces has been evaluated using Fourier-analysis to determine the geometrical properties. For Fourier-analysis line scans of AFM topographies were used. The y-axis shows the amplitude of features and the x-axis the periodicity. To average over a large number of line scans the graphs were normalized. These analysis results are shown in Figure 3. An etching for 5 s (Figure 2a, 3a) generates short wires of less than 2.0 nm in length. The average distance between wires is in the range of 5 nm, varying from 2 to 9 nm. An etching for 10 s results in wires with diameters of 1 to 5 nm (Figure 2b, 3b). They are distributed regularly on the surface with about 11 nm distance in between wires. An etching for longer times results in deterioration of these properties.

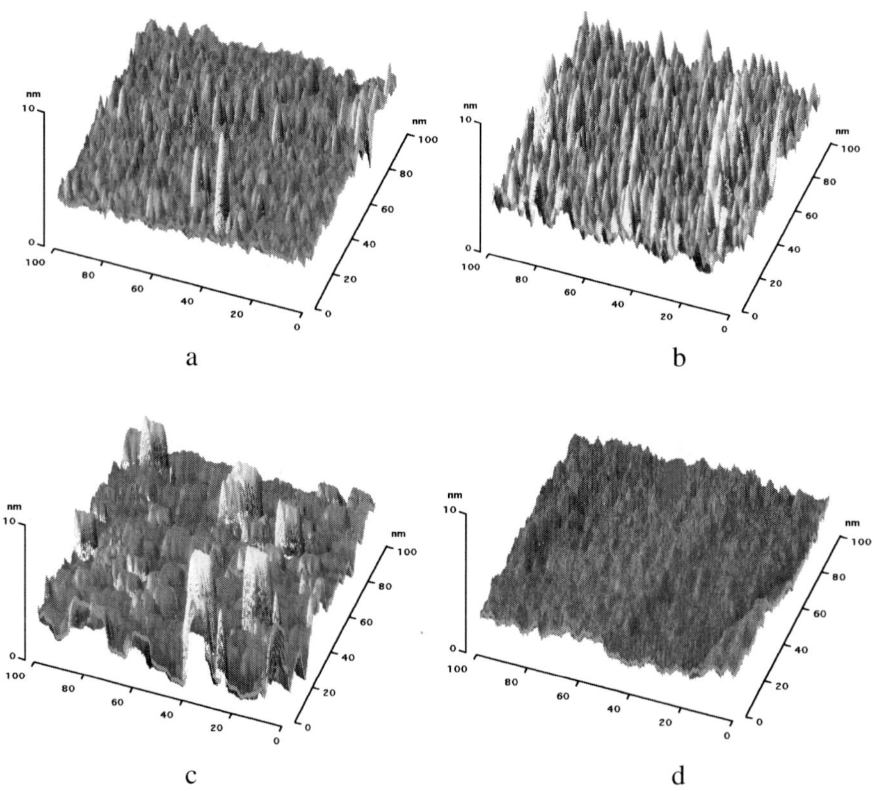

Figure 2. AFM tapping mode images of diamond surface after RIE etching with a gas mixture of 3% CF_4 and 97% O_2 for an etching time of (a) 5, and (b) 10, and (c) 20 and (d) 60 s.

For example, after 20 s etching, the wires become undefined, varying in shape, height and periodicity (Figure 2c, 3c). Finally, after 60 s etching the wire structure vanishes and a relative smooth surface is generated (Figure 2d, 3d). One optimized geometrical dimensions of 10 nm in length and an average separation of 11 nm were obtained with an etching time of 10 s (Figure 2b). Since these vertically aligned wires are short in length and smaller in diameter, they are therefore called diamond nanotextures.

The length of diamond nanowires can be increased to several micrometers if for example diamond nanoparticles are replaced by metal particles such as nickel, Au or Pt [22, 23]. A self-organized nickel nanoparticle array can be prepared by firstly evaporating a thin Ni layer (e.g. 1 nm in thickness) onto diamond. Then the nickel layer is annealed in vacuum at 700 °C for 5 min in an induction furnace. The heating and cooling ramp from room temperature to 700 °C takes 1 min and 30 min, respectively.

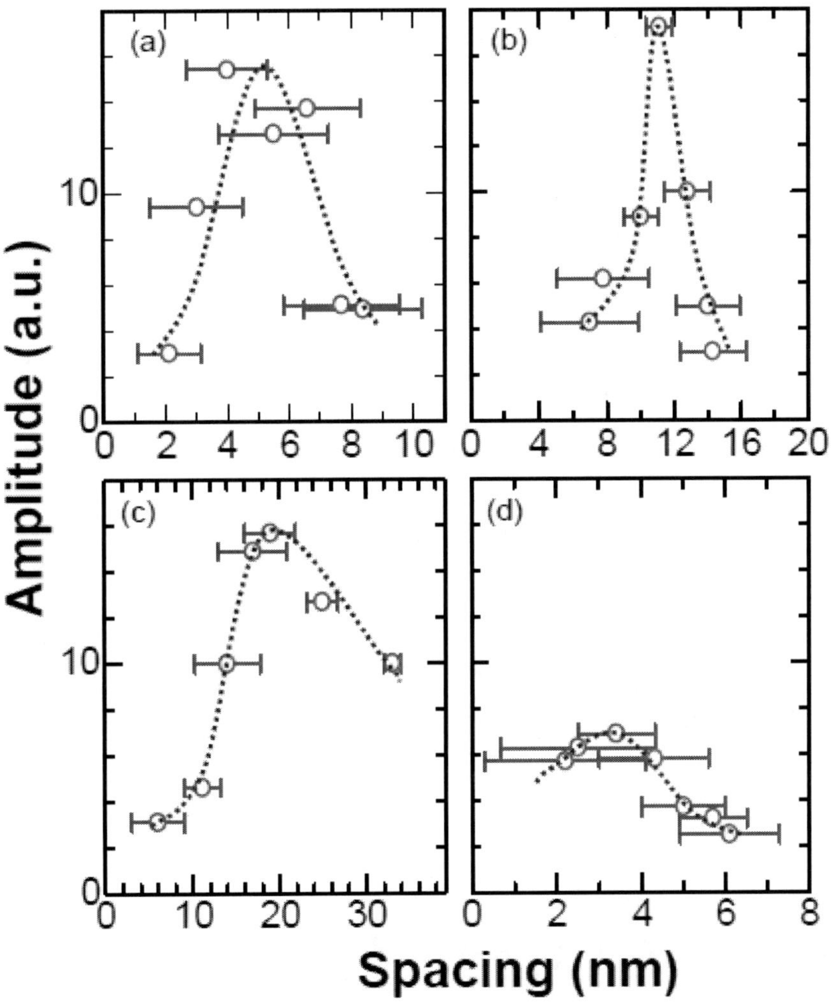

Figure 3. Surface analysis of diamond nanotextures fabricated with an etching time of (a) 5, and (b) 10, and (c) 20 and (d) 60 s in a gas mixture of 3% CF_4 and 97% O_2. The blue lines are for guiding the eyes.

Nickel nanoparticles of typically 10 nm to 30 nm in diameter are generated and distributed in a homogeneous pattern on the diamond surface.

After forming Ni nanoparticles based etching mask, a O_2 inductive coupled plasma (ICP) is applied to etch diamond. Etching parameters are required to be optimized. One example of these parameters is: a) oxygen gas flow of 20 sccm (standard cubic centimetres per minute), b) gas pressure of 5×10^{-2} mbar, c) etching time of 5 min, and d) bias power of 1000 W [19]. After the ICP process residual Ni nanoparticles are able to be removed by wet-chemical cleaning in aqua-regia.

Figure 4 shows typical SEM images of diamond nanowires fabricated using Ni nanoparticles as the etching mask [22]. The resulting diamond nanowires have typical diameters of 23(±7) nm (Figure 4a), which corresponds to the size of Ni nanoparticles. The density of nanowires is about 10^{11} cm^{-2}, which agrees with the distribution of Ni nanoparticles before etching. The length (height) of these nanowires is measured to be 1200(±200) nm (Figure 4b,c) and the aspect ratio is in the range of 50. Assuming all nanowires are cylindrically shaped, we calculate the surface enlargement which can be varied from 10 to 80, up to the dimensions of nanowires. Single, stand-alone, and conical nanowires are also noticed (Figure 4b, 4c). This may arise from plasma shielding during etching due to charging of the nanowires and reflected ions at nanowires [41]. It is worth to note that, the diameter (10 to 60 nm) and the length of nanowires (up to few μm) as well as the spacing (up to few hundreds nm) in between nanowires are able to be controlled by varying the size of Ni nanoparticles during the formation of mask and by changing the etching time during etching process [22].

2.2. Bottom-Up Approach

Diamond nanostructures, mainly diamond nanowires, can be also produced by the overgrowth of other nanostructures with diamond. In 2003, growth induced formation of nanoscale tubular structures with microwave plasma of hydrogen under a bias potential was reported for the first time [17].

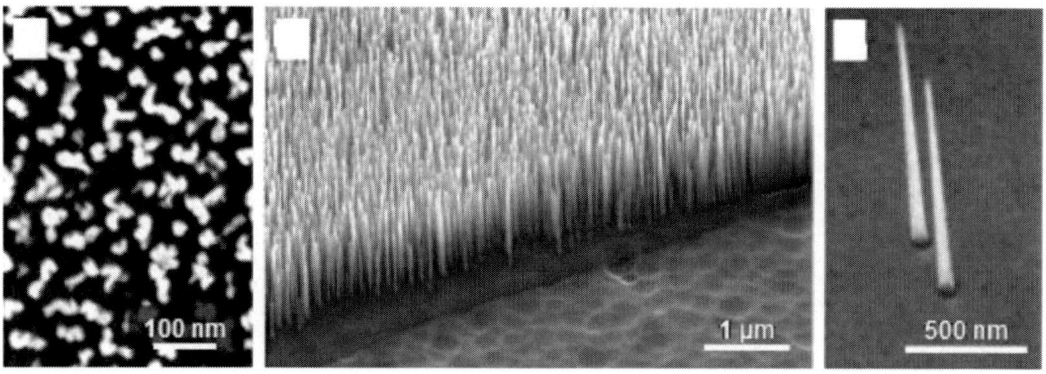

a b c

Figure 4. SEM images of diamond nanowires fabricated using Ni nanoparticles as the etching mask (a) in a top-view, (b) at a 45° tilt angle with respect to sample surface, and (c) in higher magnification for two individual nanowires.

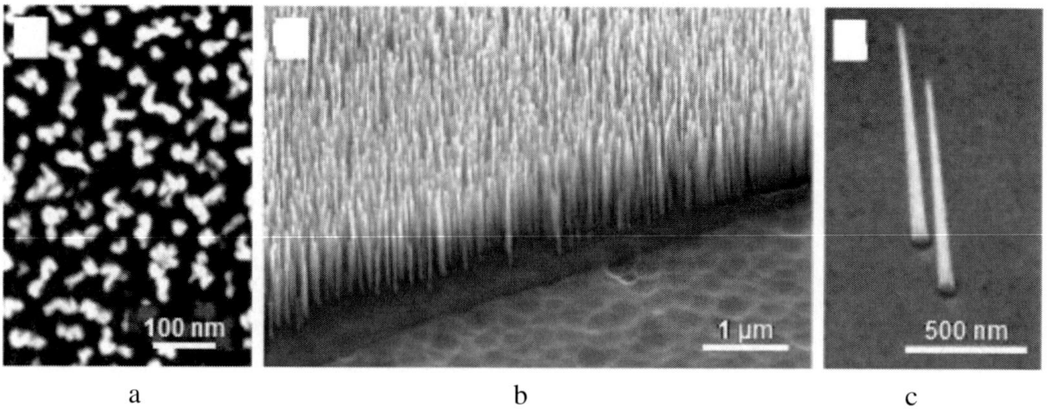

Later overgrowing diamond homoepitaxially on other materials like Si nanowires or carbon nanotubes has been applied for the fabrication of diamond nanostructures [42-45]. Recently it has been shown that diamond nanowires are able to be fabricated by use of anisotropic oxygen plasma etching and Ni hard mask [46].

3. PROPERTIES

3.1. Interfacial Electronic Properties

To apply diamond nanostructures such as nanotextures and nanowires for electrochemical and biochemical applications, their interfacial electronic properties need to be characterized. For this purpose, we have performed capacitance-voltage measurements on diamond nanotextures, fabricated using diamond nanoparticles as etching mask. A potentiostat was utilized with a conventional three-electrode configuration in an electrochemical cell at room temperature. Figure 5 shows the Mott-Schottky plots, re-drawn from the results of capacitance-voltage measurements recorded in 0.1 M pH 7.4 phosphate buffer at 1.0 kHz. We calculated the built-in potential and the variation of the area of diamond nanotextures by use of Mott-Schottky equation [47]:

$$\frac{1}{C^2} = \frac{2}{\varepsilon_0 \varepsilon_r e A^2 N_B}(V - V_{bi} - \frac{kT}{e}) \quad (1)$$

where ε_o is the dielectric constant, ε_r = 5.7 (diamond), A is the active area, N_B the boron acceptor density in diamond, V_{bi} the built-in potential, k the Boltzmann constant, T the temperature (300 K) and e the elementary charge. In these experiments, the acceptor concentration was fixed (N_B = const.) as parts of the same diamond wafer have been used. If N_B is constant, variations of slope arise from variations in active area. Smooth diamond shows a doping concentration of 7×10^{19} cm^{-3} and a built-in potential of 1.6 V, as the surface is hydrogen terminated. The depletion layer width at the interface is about 10 Å. After etching for 5, 10 and 20 s the Mott-Schottky graphs are changing significantly. The active area is increasing by RIE, showing a maximum of $2.1\times A_0$ after 10 s RIE, where A_0 is the smooth surface area. The built-in potential is changing as hydrogen is removed from the surface and surface defects give rise to surface Fermi-level pinning [1-3, 19]. It is interesting, that the highest barrier of 3.1 eV is generated by 10 s RIE, while shorter (5 s) or longer (20 s) etching times do not change the built-in potential.

3.2. Electrochemical Activity

The electrochemical activity of diamond nanowires fabricated using self-formed Ni nanoparticles as etching mask [22] was investigated with respect to their geometric/active area and voltammetric response of redox probes on nanowires based electrode [48]. The diamond nanowires investigated have the height of 1200 (±200) nm, width of 35 (±5) nm, and a density of ~10^{10} cm^{-2} [22].

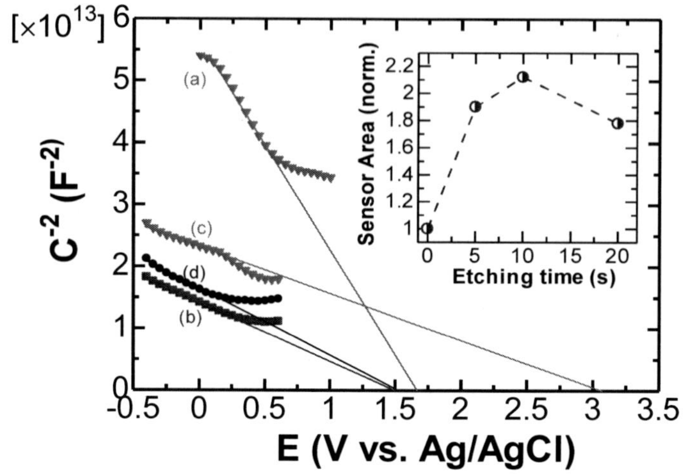

Figure 5. Mott-Schottky plots of a smooth diamond (a) and diamond nanotextures fabricated with an etching time of (b) 5 s, (c) 10 s, and (d) 20 s. They were re-drawn from the results of capacitance-voltage measurements in 0.1 M pH 7.4 phosphate buffer at a frequency of 1.0 kHz. The insert is the active area calculated as a function of the etching time.

To calculate geometric area of diamond nanowires based electrodes, cyclic voltammograms of these diamond nanowires based electrodes were recorded in 0.1 M KCl at different scan rates. Figure 6a shows voltammograms for two electrodes at a scan rate of 0.05 V s^{-1} as an example. If we assume that the normalized capacitance of diamond does not change before and after etching process and is not affected by the formation of the nanowires, the capacitive current will be proportional to the electrode geometric area [48]. The capacitive current of diamond nanowires based electrodes is more than 10 times larger than on the smooth diamond electrode. From this we can deduce, that the geometric area of diamond nanowires based electrodes is enhanced by a factor of 10. This is in good agreement with theoretical calculations based on the dimensions of these nanowires.

The electrochemical activity of diamond nanowires based electrodes was characterized using redox couples of $Fe(CN)_6^{3-/4-}$ as probes. The currents were normalized by the geometric areas of the electrode. As shown in cyclic voltammograms in Figure 6b, the peak difference between oxidation peak potential and the reduction peak potential is about 80 (±10) mV for both electrodes at a scan rate of 0.1 V s^{-1}, indicating a quasi-reversible electrode reaction on both electrodes. On both electrodes, the peak currents are proportional to the square-root of scan rates in the range of 0.01 to 0.5 V s^{-1}, indicating a diffusion-controlled electrode process [48]. For a diffusion-controlled electrode process, the peak current is known to be defined by scan rate, the diffusion coefficient of analytes, electrode active area, and other experimental conditions [48]. By plotting the peak current as a function of square-root of scan rates, the active electrode area can be calculated. The calculated electrode active area of the diamond nanowires based electrode is only 1.5 times larger than that of a smooth diamond electrode. This is much lower than the increase (10 times) of geometric area for these two electrodes. This is because the faradaic current of ferri-/ferro- cyanide is dominated by the diffusion lengths of the analytes, which are typically in the range of tens of micrometers.

Since the distance in between nanowires is much smaller than the value of diffusion lengths of analytes, diamond nanowires based electrode actually behaves as a macro-electrode.

3.3. Functionalization of Diamond Nanotextures

To realize electrochemical and bio-electrochemical applications with high selectivity and sensitivity and reproducibility, the substrate electrode always needs to be functionalized with different linker molecules [49-51]. Chemical synthesis, photochemical functionalization, and electrochemical grafting of linker molecules are three mostly applied approaches [52-7]. For example, after photochemical attachment of amine linkers to diamond [60-66], various electrochemical (e.g. field effect transistor [1-3, 60] and biochemical sensors [1-3, 60, 61] have been realized [60-66].

Electrochemical functionalization of diamond with nitrophenyl linker molecules can be applied by chemical or electrochemical reduction of diazonium salts [60-72], which has been investigated and used widely on other carbon related materials, such as glassy carbon [55-59] and silicon carbide [73, 74] etc. The attachment can be conducted in a potentiodynamic or potentiostatic mode. In the potentiodynamic mode, the potential applied to the electrode varies as a function of time and the normal technique is cyclic voltammetry [48]. By use of this mode, the growth of sub-monolayer, monolayers and multi-layer nitrophenyl is possible to be realized [75-79]. While in the potentiostatic mode, the potential for the functionalization of the electrode is fixed and chronoamperometry is the technique frequently applied [48]. In this way, nitrophenyl layers with different thickness and density can be grafted by selecting appropriate potentials for the reduction of diazonium salts, by variation of the attachment time, and by adjusting the surface termination of the electrode. For example, on partially oxidized diamond electrodes, nitrophenyl monolayer and sub-monolayers have been achieved [72].

The grafting of diamond nanotextures with diazonium salts has been performed with a conventional three-electrode configuration at room temperature [19, 75-79]. The reference electrode was Ag/Ag^+ (0.01 M Ag^+ in CH_3CN) electrode. The potentiostatic mode was selected with optimized attachment potential as detected by cyclic voltammetry [77]. The current-time curve for the deposition of nitrophenyl on diamond nanotextures (at -0.05 V *vs.* Ag/Ag^+ for 4 s) follows the Cottrell law with a time-dependence of $I \sim t^{-0.5}$. Here I is the current for the deposition of nitrophenyl and t is the deposition time. For comparison, attachment of nitrophenyl on smooth and oxidized diamond was conducted using the same experimental conditions. The current on diamond nanotextures is about two times smaller although the area is two times larger. The nitrophenyl density on the smooth diamond surface is 5×10^{14} cm^{-2} and the density of phenyl molecules on diamond nanotextures is about 2×10^{13} cm^{-2} [19, 78, 79].

The variation of redox currents of redox probes like $Fe(CN)_6^{3-/4-}$ on diamond nanotextures have been characterized before and after nitrophenyl attachment [19, 78, 79]. The voltammogram before nitrophenyl attachment shows a peak difference of 868 mV for the redox reaction of $Fe(CN)_6^{3-} + e^- \leftrightarrow Fe(CN)_6^{4-}$. This is due to the relative low boron doping concentration (7×10^{19} cm^{-3}) of the diamond electrode as well as to surface oxidation of diamond electrode during the fabrication process of nanotextures [19].

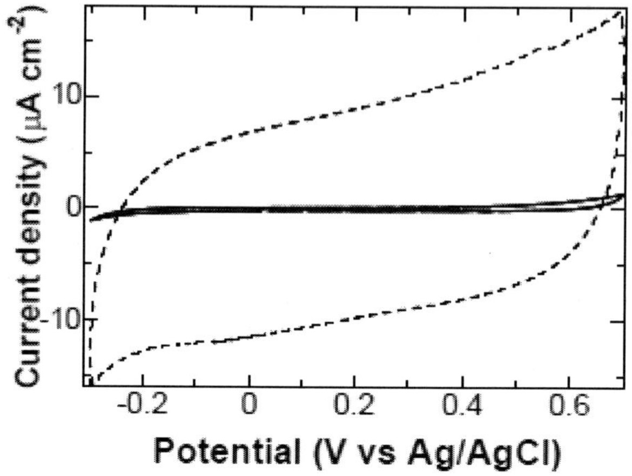

a

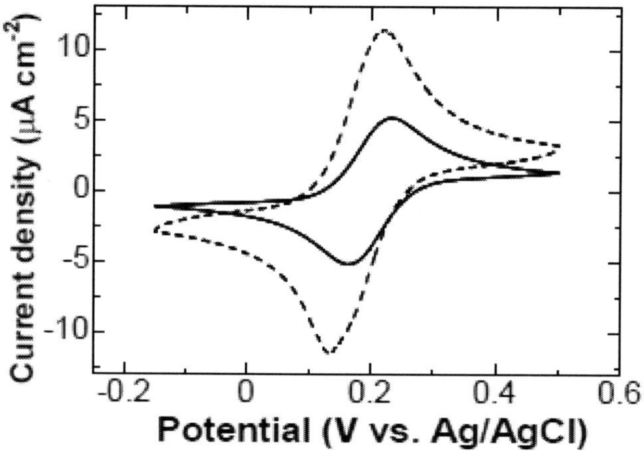

b

Figure 6. Cyclic voltammograms of (solid lines) smooth diamond electrode and (dashed lines) diamond nanowires based electrode in (a) 0.1 M KCl at a scan rate of 0.050 V s^{-1} and in (b) 1.0 mM $Fe(CN)_6^{3-/4-}$ in 0.1 M KCl at a scan rate of 0.1V s^{-1}.

A reduction of the difference of peak potentials down to 80 mV was achieved by increasing the boron doping concentration to 3×10^{20} cm^{-3} [3, 22]. Before nitrophenyl attachment the redox peak currents on diamond nanotextures are almost two times higher than those on a smooth diamond electrode, indicating a larger active electrode area of diamond nanotextures. After nitrophenyl attachment the peaks currents decreased by about 16% and the peak different (ΔEp) enlarged_to 1160 mV, indicating the formation of a thin and dispersed nitrophenyl layer on diamond nanotextures. These changes are a result of phenyl attachment as phenyl is an insulator giving rise to slow down electron transfer rates between transducer and redox molecules [56-59].

In order to identify locations of nitrophenyl attachment to diamond nanotextures, constant-current mode STM has been employed to image the diamond nanowires before and after grafting with nitrophenyl [16, 79].

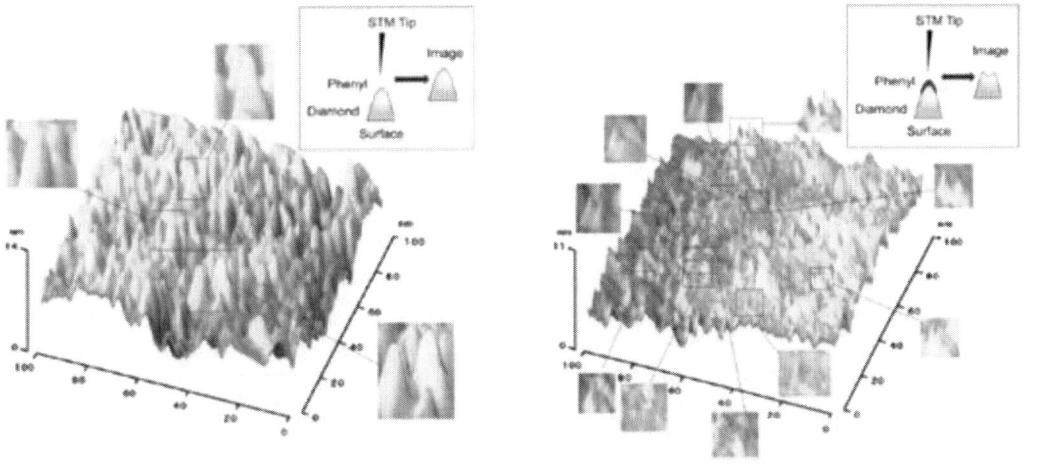

Figure 7. STM images of diamond nanotextures (a) before and (b) after the electrochemical grafting of a nitrophenyl film by a constant potential technique at -0.05 V *vs.* Ag/Ag^+ for 4 s.

As shown in Figure 7a, sharp tips of 3-10 nm in size and 8-10 nm in length are detected before nitrophenyl attachment. This result agrees with AFM data shown in Figure 2b. The STM image on phenyl-modified nanotextures (Figure 7b) shows a drastic change of tip features. A "double" or "triple" peak structure is detected, which is a result of constant tunnelling current mode applied. As nitrophenyl films are insulating, they gives rise to decreased STM tunnelling currents. To keep the tunnelling current constant, the STM tip will approach towards the surface which results in inverted cone shaped tips. Note that the molecular height of nitrophenyl is about 0.67 nm [70] while nanotextures are in the range of 10 nm. The short-time electrochemical modification of nanotextures gives rise to preferential phenyl bonding to the very tip of wires. In between, most of the surface remains uncoated. This is because on a nano-structured surface, the electric field concentrates at tips which stimulate current flowing preferentially through their tips. Therefore, this will lead to preferential bonding of nitrophenyl to the tips of wires, giving rise to preferential modifications of tip of nanotextures [19]. Note that formation of modified surfaces with low concentration is quite difficult with diazonium salts due to their spontaneous reaction with the substrates.

4. Applications

Diamond based electrodes have been used widely for different applications, such as electrochemical [1-6, 31, 44, 80-84] and biochemical [3-6, 75-79, 85-87] sensing applications, mechanical reinforcement [88], enhancements in thermal management [89], and devices design for microelectronics, for electron field emission, and for photonic applications

[90–93]. For example, on diamond nanostructures, electrochemical detection of dopamine in the presence of uric acid [31], non-enzymatic detection of glucose [44], and amperometric detection of catechol [80] have been realized.

For biochemical sensing applications, highly sensitive detection of DNA hybridization [75-79], direct electron transfer of cytochrome c [84, 85], and enhanced electron transfer of shewanella loihica PV-4 [83] have been reported.

4.1. Diamond Nanotextures for the Detection of DNA Hybridization

Diamond nanotextures have been utilized as a novel electrode for electrochemical detection of DNA hybridization [75-79]. As shown schematically in Figure 8, tips of diamond nanotextures are functionalized by electrochemical bonding of nitrophenyl linker molecules [19]. The nitrophenyl linkers are then electrochemically reduced to amino-phenyl groups, which is further utilized for the reaction with hetero bi-functional cross-linkers. The functionalized tips act as anchors for single-stranded marker DNA molecules which gives rise to nano-spacing between DNA molecules. The implemented vertically aligned diamond nanotextures give rise to geometrically controlled and diluted linker-molecules attachment without deterioration of the unique sensing properties of diamond (e.g. good chemical stability, low capacitive current, wide electrochemical potential window, ultra hardness), resulting in geometrically controlled bonding of DNA molecules to the nanostructured transducer to achieve a "like-in-solution" situation for DNA molecules [75-79]. Moreover, the enhancement of capacitance current is negligible due to slight surface area enhancements (only 2 times) [19].

DNA immobilization experiments were conducted in a way as shown previously [69]. Briefly, a constant potential of -0.05 V (*vs.* Ag/Ag^+) was applied on diamond nanotextures based electrode in 1.0 mM 4-nitrobenzene diazonium tetrafluoroborate solution for 4 s in a nitrogen-purged glove box. The electrolyte solution was 0.1 M tetrabutylammonium tetrafluoroborate (NBu_4BF_4) in acetonitrile.

After grafting, careful and copious cleaning of diamond electrodes with acetone and acetonitrile is required. Subsequently, 4-nitrophenyl groups ($–C_6H_5NO_2$) were electrochemically reduced from 0.5 V down to -1.5 V (*vs.* Ag/AgCl) to amino-phenyl groups ($–C_6H_5NH_2$) in 0.1 M KCl solution of $EtOH$-H_2O (V:V=1:9) mixture by cyclic voltammetry for 5 cycles at a scan rate of 0.2 V s^{-1} [19, 53-60]. The aminophenyl layer was then reacted with 14 nM solution of sulfosuccinimidyl 4- (N-maleimidomethyl)- cyclohexane-1-carboxylate (SSMCC) in 0.1 M pH 7 triethanolamine (TEA) buffer for 20 minutes at room temperature in a humid chamber, forming maleimide-modified diamond nanowires. Then dropping (2 - 4) µl thiol-modified probe DNA solutions (300 µM thiol DNA in 0.1 M pH 7 TEA buffer) directly onto the thiol-reactive maleimide terminated diamond surface and then keeping it for 1 h in a humid chamber at room temperature resulted in the immobilization of probe DNA. The 23-mer cancer marker cytokeratin 20 (CK20: 5'-HS-C_6H_{12}-T_9-CTG TTT TAT GTA GGG TTA GGT CA-3') was used as probe DNA. The immobilization time for probe DNA was controlled to be around 30 min and the molar density of probe DNA was around 10^{12} to 10^{13} cm^{-2} [6].

For DNA hybridization, 5 μl of saline-sodium phosphate-EDTA (SSPE) buffer (300 mM NaCl, 20 mM sodium phosphate, 2 mM EDTA, and 7 mM sodium dodecyl sulfate, pH 7.4) containing complementary DNA with different concentrations was deposited on the sensor surface for 1 h at room temperature in a humid cell.

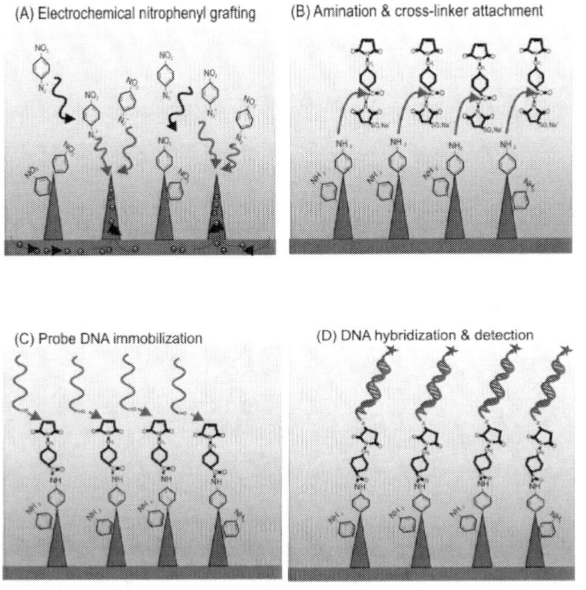

Figure 7. Schematic illustration of electrochemical detection of DNA hybridization on diamond nanotextures: (a) tip-functionalization with electrochemical grafting of 4-nitrobenzenediazonium tetrafluoroborate; (b) electrochemical conversion of nitrophenyl to aminophenyl and cross linker attachment; (c) probe immobilization, and (d) hybridization of complementary DNA and detection of hybridization event.

The sequences of the complementary and non-complementary DNA are, respectively, listed as 5'-Cy5-TGA CCT AAC CCT ACA TAA AAC AG-3' and 5'-Cy5-ACT TCC TTA CTA GTT TAT TCA CG-3', where Cy5 indicates the presence of a dye color center. After hybridization the samples were washed in de-ionized water for 1 h at 37 °C to remove non-intentional bonded DNA molecules.

We applied redox indicators like negatively charged redox probes for the detection of DNA hybridization [75, 78].

The sensing principle is explained in the following way. Thanks to diamond nanotextures and their tip-functionalization, the negatively charged redox molecules $Fe(CN)_6^{3-/4-}$ can still diffuse through the single stranded DNA layer (ss DNA) and reach diamond surface even after immobilization of probe DNA (ss DNA). This results in peak-shaped voltammograms. After hybridization with complementary DNA, the enlarged DNA molecule (double stranded DNA, ds DNA) repels negatively charged redox molecules due to electrostatic repulsive Coulomb force arising from the negatively charged sugar phosphate backbone of DNA. The spacing in between ds DNA molecules decreases and therefore prevents the diffusion of negatively charged redox molecules towards the electrode. This results in widening peak difference and weakening peak currents.

Figure 9 shows typical examples of (a) cyclic voltammetric (CV) response (scan rate: 100 mV s^{-1}), (b) differential pulse voltammetric (DPV) signals (scan rate: 100 mV s^{-1}, pulse height: 2.5 mV), and (c) electrochemical impedance spectra (Imp) in the regime 0.01 Hz to 10^6 Hz at -0.3 V (vs. Ag/AgCl) in pH 7.4 phosphate buffer as applied on probe DNA (ss-DNA) coated nanotextures (solid line) and after exposure to 10 nM of complementary DNA for 1 h (ds-DNA) (dashed line).

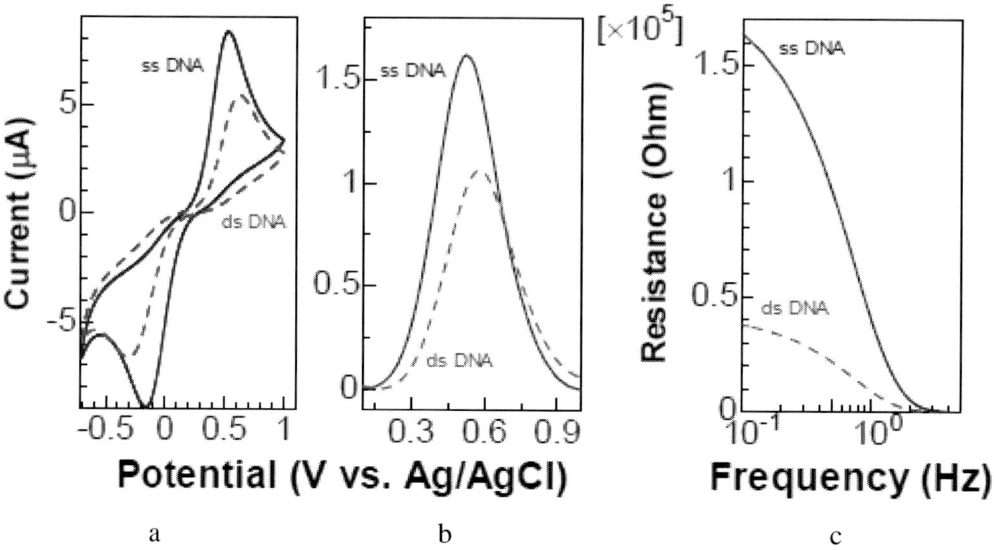

Figure 9. Typical examples of detection of DNA hybridization by cyclic voltammetry (a), differential pulse voltammetry (b), and electrochemical impedance microscopy (c). The concentration of target DNA was 10 nM.

Redox indicators of $Fe(CN)_6^{3-/4-}$ were utilized for these measurements. Hybridization leads to a decrease in resistance of DNA, which agrees with the results published in literature on diamond [75-70]. Large amplitude variations in peak current (a, b) and resistance (c) were detected before and after hybridization. The differences before and after DNA hybridization in peak currents from cyclic voltammograms and differential pulse voltammograms and from electrochemical impedance spectra were adopted for the detection of DNA sequences [75].

Sensitivity curves for DNA hybridization event on diamond nanotextures based electrode was checked by varying the concentration of complementary target DNA from 1 μM to 10 pM. The overall performance of the diamond nanotextures based DNA sensor was compared with those of published data available in literature using comparable DNA mere structures for planar Au [94-96], planar diamond [97, 98] and gold nanowires [99, 100]. Sensing by diamond nanotextures based electrode is about 100 to 1000 times better than using smooth surfaces from Au and diamond [75-79].

To identify the detection limit, differential pulse voltammograms of ss DNA modified diamond nanotextures before and after hybridization using 1.0 to 9.0 pM complementary DNA concentrations were recorded. A detection limit of about 2 pM on 0.03 cm^2 sensor area is detected using 100 μL solution but keeping the other parameters constant (hybridization temperature: 25 °C and hybridization time: 60 minutes). We expect that by miniaturization of sensor area a significantly lower detection limit can be achieved [20, 101, 102].

Stability measurements of probe DNA on diamond nanotextures were performed using 5.0 μL 1.0 μM complementary DNA solutions over extended cycles of hybridization/ denaturation treatments. Denaturation was performed in 8.3 M urea-solution for 30 minutes at

37 °C, followed by rinsing with de-ionized water. The cyclic voltammograms (scan rate: 0.1 V s^{-1}) and differential pulse voltammograms (scan rate: 0.1 V s^{-1}, amplitude: 2.5 mV) of 1.0 mM $Fe(CN)_6^{3-/4-}$ in pH 7.4 phosphate buffer on probe DNA modified diamond nanotextures before and after hybridization were recorded.

The anodic peak currents from above voltammograms were adopted as a parameter to describe the stability of probe DNA on diamond nanotextures. No degradation is detected over 30 cycles of DNA hybridization/denaturation [75-79], which is comparable to optical detected DNA stability on diamond [61]. The higher chemical stability of DNA on diamond than on other substrates like gold results from stronger covalent carbon-carbon bonds which has been demonstrated by AFM scratching experiments [6, 60-66]. Moreover, as diamond is chemically inert and ultra-hard, nanotextures from diamond will survive in harsh environments where other materials like gold and silicon will fail.

As for the discrimination of single-base mismatched complementary DNA, 5 µl of a 10 nM solution was used for hybridization under identical conditions [78, 79]. The sequence of single-base mismatched (shown in underline and italic) DNA is TGA CCT AAC C*A*T ACA TAA AAC AG. After exposure to single-base mismatched DNA, an amplitude decrease of 8% is detected with a small shift in peak potential. After hybridization with complementary DNA, the peak current decreases by about 40% and the peak potential shifts by 80 mV towards positive potential. It demonstrates single-base mismatch discrimination which is again promising for applications of diamond nanotextures in nucleic acid sensors.

Compared with self-assembled monolayers (SAMs) or dendritic modifications for DNA bonding, the special advantages of diamond nanotextures are the combination of modified surfaces (tips) and unmodified surfaces for sensing. Due to control of DNA spacing, the hybridization kinetics is optimized resulting in high sensitivity as well as in single base mismatch discrimination. The high sensitivity is promising for miniaturization of sensor area which is required for real device applications. Furthermore, this technique will be applied in ultramicro- and nano- electrode arrays to design oligonucleotide sensors [20, 101, 102] which suits demands in clinical environments where high through-put is required on a 24 hrs per day basis.

4.2. Direct Electrochemistry of Cytochrome C on Diamond Nanotextures

Cytochrome c (cyt c) is one of well-known and widely investigated redox systems from proteins. It has an approximately spherical shape of 4 nm in diameter. The theoretical formal redox potential of cyt c is 260 mV *vs.* Normal Hydrogen Electrode (NHE) in pH 7 buffer [101]. Since its redox activity is from the deeply buried Fe-based redox heme center and is affected greatly by its orientation on electrode, direct electrochemistry of cyt c has been thus widely investigated after covalent bonding to different electrode surface or some artificially designed interfaces [103, 104-114], for example, to self-assembled monolayers on flat or roughened (nanoparticles coated) gold electrodes [104-106].

Direct electrochemistry of cyt c has been also investigated [108-114] on planar (non-structured) diamond surface with or without chemical/physical modification of diamond surface. Nanotextured diamond electrode with geometrical properties close to protein dimensions is proved to be a better substrate electrode to realize direct electron transfer of cyt c [104-114]. No covalent bonding is required. This is because 3D entrapment for cyt c can be achieved on a nanotextured surface when the dimension of nanotextures is close to the geometries of protein. As shown in Figure 10, the "encavement" gives rise to the shortest tunneling distances and to possible electron transfer through different orientations and subsequently results in an enhancement of electron-transfer properties of cyt c.

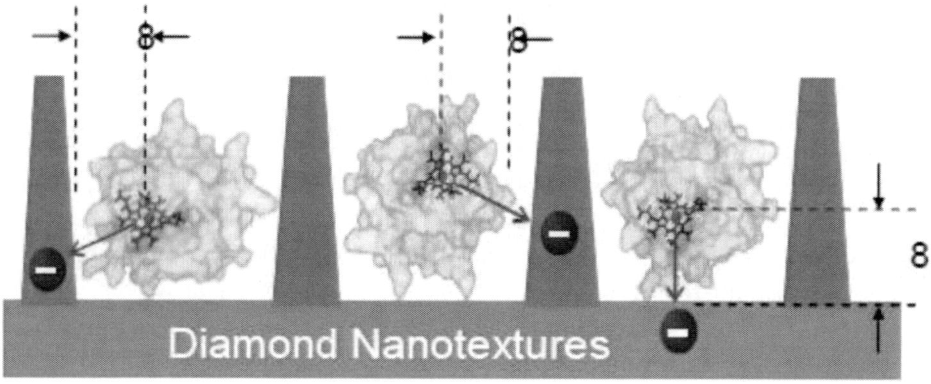

Figure 10. Schematic image of the electron transfer of trapped cyt c on a nanotextured diamond surface. 8 and 30 Å are the shortest and longest distances for tunneling from the redox heme center to the substrate. The red arrows are "guiding the eyes", showing the shortest distance for electron transfer between cyt c and diamond.

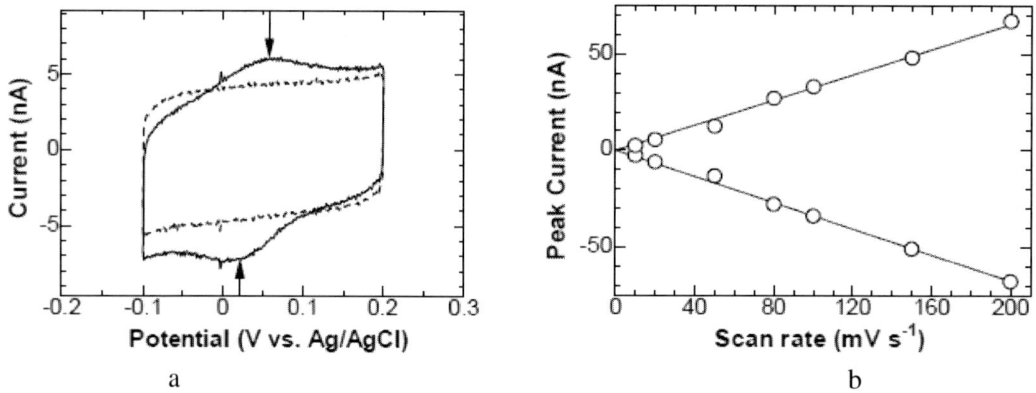

Figure 11. (a) Cyclic voltammograms of diamond textured surface in the presence of cyt c (solid line) and in the absence of (dashed line) 1.0 μM cyt c in 0.1 M pH 7 phosphate buffer measured at a scan rate of 0.02 V s^{-1}. The arrows are for eye-guiding to identify the peaks. (b) Peak currents of cyt c on diamond textured surface as a function of scan rate.

Figure 11(a) shows cyclic voltammograms on a nano-textured diamond surface in the presence of 1.0 μM cyt c (solid line) and in the blank buffer (dashed line). In the blank buffer, the voltammogram is featureless. After adding cyt c, a pair of well-defined redox peaks appears with equal magnitudes. The cathodic peak potential is around +20 mV and the anodic peak around 60 mV with respect to Ag/AgCl. The formal potential [48], $(E_{p,a}+E_{p,c})/2$, for the cyt c on textured surface is around 40 mV vs. Ag/AgCl electrode, 260 mV (vs. NHE), which is in agreement with data obtained on diamond [108-114]. The peak splitting is around 40 mV at a scan rate of 0.02 V s^{-1}, indicating a quasi-reversible electron transfer process. Moreover, the peak currents are proportional to scan rates (as shown in Figure 11b), indicating an adsorption-controlled process. This is different from the diffusion-controlled process reported on H-terminated as-grown rough diamond [108] and on polished diamond surface [109]. Since the OH-terminated diamond surface shows slight polarity and the thickness of the Helmholtz layer in our case is around 1 nm, the main force causing the adsorption of cyt c on OH-terminated surface is likely the electrostatic interaction between the negatively charged diamond surface dipoles with the positively charged lysine groups in the cyt c molecule at pH 7 phosphate buffer [85, 113, 114]. The trapping process as shown in Figure 10 might be another source to generate an adsorption-controlled electrode reaction.

For an adsorption-controlled process, the surface coverage of cyt c, Γ, was evaluated to be 4.2×10^{12} cm^{-2} using the equation of $I_p=(n^2F^2/4RT)vA\Gamma$, where I_p is the peak current, n (=1) is the electrons transferred, v is scan rate, A is surface area of the working electrode, and others are physical constants [48]. The surface coverage is slightly lower than that of a monolayer of cyt c (1×10^{13} cm^{-2}) on the flat surface [112], indicating a diluted cyt c monolayer on the textured surface. According to Laviron [116], the electron transfer rate constant (k_s) was estimated to be (1.43±0.05) s^{-1}. This value is higher than that of cyt c adsorbed on mesoporous niobium oxide film/ITO (0.28 s^{-1}) [117], NaY zeolite/GCE (0.78±0.04 s^{-1}) [118], Au nanostructures (1.13 s^{-1}) [119], colloid Au/carbon paste (1.21±0.08 s^{-1}) [120], L-cysteine/Au (1.25±0.10 s^{-1}) [121], and SiO_2/cyt c/SiO_2/diamond (1.39 s^{-1}) [111]. The enhancement of k_s results from the electrostatic and hydrophobic interaction of cyt c with diamond where electrostatic interaction controls coarse orientation of cyt c while hydrophobic interaction assists in the formation of the electron transfer complex, as confirmed by electrocatalytic reactions towards oxygen reduction as well as AFM tapping and scratching experiments in buffer [113, 114].

These interactions are favorable for the protein orientation and provide a narrow distance between the Fe redox centers to the diamond surface, and subsequently facilitates the electron transfer between cyt c and diamond electrode. Moreover, three-dimensional structure of the nano-textured diamond electrode provides a large contact area between cyt c ("3D entrapment") and diamond, resulting in an enhancement of the electron-transfer process. Please note that the sizes of pockets (spacing in between nanotextures) used here are in the range of 8-10 nm, which is slightly larger than the size of cyt c (3-4 nm). Higher k_s are expected when the spacing goes down to the size of cyt c.

Therefore, nano-textured diamond surfaces are promising to be applied as a novel platform to achieve direct electrochemistry of redox proteins or other big biomolecules and to construct novel biosensors.

SUMMARY AND OUTLOOK

Various diamond nanostructures such as nanotextures, nanowires, and nanopores can be fabricated by use of top-down or bottom-up approaches. The dimension of these nanostructures depends on the mask applied and the etching conditions. These nanostructures can be functionalized with different linkers for specified electrochemical and biochemical applications.

These nanostructures based electrochemical and biochemical sensors show better sensor performance and longer sensing life time than those on planar diamond based sensors. In summary, diamond nanostructures provide robust interfaces for a variety of applications like chemical and biochemical sensing.

In future, the effect of dimensions of these nanostructures (size, length, diameter, and spacing) and surface termination of these structures on their geometrical, electronic, optical, and electrochemical properties are required to be characterized in detail.

The application of three-dimensional diamond nanostructures with catalysts will be a new platform for industrial electrochemical applications, such as hydrogen evolution reaction, oxygen evolution reaction, carbon dioxide reduction, and fuel cell applications.

REFERENCES

[1] Nebel, C. E., Ristein, J. (eds.), *Thin-film Diamond I, Semiconductors and Semimetals Series*, Vol. 76, 2003, Elsevier, New York.

[2] Nebel, C. E., Ristein, J. (eds.), *Thin-film Diamond II, Semiconductors and Semimetals Series*, Vol. 77, 2004, Elsevier, New York.

[3] Fujishima, A., Einaga, Y., Rao, T. N., Tryk, D. A. (eds.), *Diamond Electrochemistry*, 2005, Elsevier, Tokyo

[4] Koizumi, S., Nebel, C. E., Nesladek, M. (eds.), *Physics and Applications of CVD Diamond*, 2008, Wiley-VCH, Weinheim.

[5] Brillas, E., Martinez-Huitle, C. A. (eds.), Synthetic Diamond Films, 2012, Wiley, Weinheim.

[6] Nebel, C. E., Rezek, B., Shin, D., Uetsuka, H., Yang, N. *J. Phys. D: Appl. Phys.* 2007, 40, 6443.

[7] Hoffmann, R., Kriele, A., Obloh, H., Hees, J., Smirnov, W., Yang, N., Nebel, C. E. *Appl. Phys. Lett.*, 2010, **97,** 052103.

[8] Blank, V., Popov, M., Lvova, N., Gogolinsky, K., Reshetov, V. *Diamond Relat. Mater.* 1998, 7, 427.

[9] Zheng, G., Patolsky, F., Cui, Y., Wang, W. U., Lieber, C. M. *Nature Biotechnol.* 2005, 23, 1294.

[10] Cui, Y., Wei, Q. W., Park, H., Lieber, C. M. *Science* 2001, 293, 1289.

[11] Li, S., He, P., Dong, J., Guo, Z., Dai, L. *J. Am. Chem. Soc.* 2005, 127, 14.

[12] Vaseashta, A., Dimova-Malinovska, D. *Sci. Technol. Adv. Mater.* 2005, 6, 312.

[13] Zhang, X. L., Wang, J. X., Wang, Z., Wang, S. C. *Sensors* 2005, 5, 580.

[14] Shiomi, H. *Jpn. J. Appl. Phys.* 1997, 36, 7745.

[15] Masuda, H., Watanabe, M., Yasui, K., Tryk, D., Rao, T., Fujishima, A. *Adv. Mater.* 2000, 12, 444.

[16] Shenderiva, O. A., Padgett, C. W., Hu, Z., Brenner, D. W. *J. Vac. Sci. Technol. B* 2005, 23, 2457.

[17] Kobashi, K., Tachibana, T., Yokota, Y., Kawakami, N., Hayashi, K., Yamamoto, K., Koga, Y., Fujiwara, S., Gotoh, Y., Nakahara, H., Tsuji, H., Ishikawa, J., Köck, F. A., Nemanich, R. J. *J. Mater. Res.* 2003, 18, 305.

[18] Zou, Y. S., Yang, Y., Zhang, W. J., Chong, Y. M., He, B., Bello, I., Lee, S. T. *Appl. Phys. Lett.* 2008, 92, 053105.

[19] Yang, N., Uetsuka, H., Osawa, E., Nebel, C. E. *Nano. Lett.* 2008, 11, 3572.

[20] Yang, N., Uetsuka, H., Williams, O. A., Osawa, E., Tokuda, N., Nebel, C. E. *phys. status solidi (a)* 2009, 206, 2048.

[21] Gao, F., Yang, N., Smirnov, W., Obloh, H., Nebel, C. E. *Electrochim. Acta* 2013, 90, 445.

[22] Smirnov, W., Kriele, A., Yang, N., Nebel, C. E. *Diamond Relat. Mater*. 2010, 19, 186.

[23] Gao, F., Yang, N., Obloh, H., Nebel, C. E. *Electrochem. Commun.* 2013, 30, 55.

[24] Hwang, D. S., Saito, T., Fujimori, N. *Diamond Relat. Mater*. 2004, 13, 2207.

[25] Ando, Y., Nishibayashi, Y., Sawabe, A. *Diamond Relat. Mater*. 2004, 13, 633.

[26] Li, C. Y., Hatta, A. *Diamond Relat. Mater*. 2006, 15, 357.

[27] Baik, E.-S., Baik, Y.-J., Lee, S. W., Jeon, D. *Thin Solid Films* 2000, 377–378, 295.

[28] Babchenko, O., Kromka, A., Hruska, K., Michalka, M., Potmesil, J., Vanecek, M. *Cent. Eur. J. Phys*. 2009, 7, 310.

[29] Janssen, W., Gheeraert, E. *Diamond Relat. Mater*, 2011, 20, 389.

[30] Lin, J. Y., Li, Z.-C., Chen, C.-Y., Chou, L.-J., Hwang, J.-C., Kou, C.-S. *Diamond Relat. Mater*, 2011, 20, 922.

[31] Wei, M., Terashima, C., Lv, M., Fujishima, A., Gu, Z.-Z. *Chem. Commun*. 2009, 45, 3624.

[32] Kriele, A., Williams, O. A., Wolfer, M., Hees, J., Smirnov, W., Nebel, C. E. *Chem. Phys. Lett.* 2011, 507, 253.

[33] Tokuda, N., Umezawa, H., Saito, T., Yamabe, K., Okushi, H., Yamasaki, S. *Diamond Relat. Mater*. 2007, 16, 767.

[34] Tokuda, N., Umezawa, H., Yamabe, K., Okushi, H., Yamasaki, S. *Jpn. J. Appl. Phys.* Part 1, 2007, 46, 1469.

[35] Kruger, A., Kataoka, F., Ozawa, M., Fujino, T., Suzuki, Y., Aleksenskii, A. E., Vul, A. Y., Osawa, E. *Carbon* 2005, 43, 1722.

[36] Williams, O. A., Douheret, O., Daenen, M., Haenen, K., Osawa, E., Takahashi, M. *Chem. Phys. Lett.* 2007, 445, 255.

[37] Williams, O. A., Daenen, M., Haen, D. J., Haenen, K., Maes, J., Moshchalkov, V. V., Nesladek, M., Gruen, D. M. *Diamond Relat. Mater*. 2006, 15, 654.

[38] Williams, O. A. *Diamond Relat. Mater.* 2011, 20, 621.

[39] Williams, O. A., Hees, J., Dieker, C., Jager, W., Kirste, L., Nebel, C. E. *ACS Nano* 2010, 4, 4824.

[40] Hees, J., Kriele, A., Williams, O. A. *Chem. Phys. Lett.* 2011, 509, 12.

[41] Rangelow, I. W. *Deep Etching Of Silicon*, Oficyna Wydawnicza Politechniki Wroclawskiej, Wroclaw, 1996.

[42] Yang, Q., Chen, W., Xiao, C., Hirose, A., Sammynaiken, R. *Diamond Relat. Mater.* 2005, 14, 1683.

[43] Chih, Y. K., Chueh, Y. L., Chen, C. H., Hwang, J., Chou, L. J., Kou, C. S. *Diamond Relat. Mater.* 2006, 15, 1246.

[44] Luo, D., Wu, L., Zhi, J. *ACS Nano* 2009, 8, 2121.

[45] Terranova, M. L., Orlanducci, S., Fiori, A., Tamburri, E., Sessa, V., Rossi, M., Barnard, A. S. *Chem. Mater.* 2005, 17, 3214.

[46] Janssen, W., Faby, S., Gheeraert, E. *Diamond Relat. Mater.* 2011, 20, 779.

[47] Finklea H. O. In: Finklea, H. O. (Ed.) *Semiconductor Electrodes*, Elsevier, Amsterdam, 1988, p. 27.

[48] Bard, A. J., Faulkner, L. R. *Electrochemical Methods: Fundamentals and Applications*, 2nd edition, 2001, Wiley, New York.
[49] Sassolas, A., Leca-Bouvier, B. D., Blum, L. J. *Chem. Rev.* **2008**, *108*, 109.
[50] Wang, J. *Chem. Rev.* 2008, 108, 814.
[51] Wilson, G. S., Hu, Y. *Chem. Rev.* 2000, 100, 2693.
[52] Mahouche-Chergui, S., Gam-Derouich, S., Mangeney, C., Chehimi, M. M. *Chem. Soc. Rev.* 2011, 40, 4143.
[53] Pinson, J., Podvorica, F. *Chem. Soc. Rev.* 2005, 34, 429.
[54] D. Bélanger, J. Pinson, *Chem. Soc. Rev*. 2011, 40, 3995.
[55] Allongue, P., Henry de Villeneuve, C., Cherouvrier, G., Cortes, R., Bernard, M. C. *J. Electroanal. Chem.* 2003, 550-551, 161.
[56] Delamar, M., Hitmi, R., Pinson, J., Savéant, J. M. *J. Am. Chem. Soc.* 1992, 114, 5883.
[57] Liu, Y.-C., McCreery, R. L. *J. Am. Chem. Soc.* 1995, 117, 11254.
[58] Allongue, P. A., Delamar, M., Desbat, B., Fagebaume, O., Hitmi, R., Pinson, J., Savéant, J. M. *J. Am. Chem. Soc.* 1997, 119, 201.
[59] Brooksby, P. A., Downard, A. J. *Langmuir* 2004, 20, 5038.
[60] Shin, D., Rezek, B., Tokuda, N., Takeuchi, D., Watanabe, H., Nakamura, T., Yamamoto, T., Nebel, C. E. *Phys. Status Solidi (a)* 2006, 203, 3245.
[61] Yang, W., Auciello, O., Butler, J. E., Cai, W., Carlisle, J. A., Gerbi, J. E., Gruen, D. M., Knickerbocker, T., Lasseter, T. L., Russell, Jr, J. N., Smith, L. M., Hamers, R. J. *Nature Mater.* 2002, 1, 253.
[62] Nebel, C. E., Shin, D., Takeuchi, D., Yamamoto, T., Watanabe, H., Nakamura, T. *Langmuir* 2006, 22, 5645.
[63] Yang, N., Uetsuka, H., Watanabe, H., Nakamura, T., Nebel, C. E. *Chem. Mater.* 2007, 19, 2852.
[64] Yang, N., Uetsuka, H., Watanabe, H., Nakamura, T., Nebel, C. E. *Diamond Relat. Mater.* 2008, 17, 892.
[65] Yang, N., Uetsuka, H., Watanabe, H., Nakamura, T., Nebel, C. E. *Diamond Relat. Mater.* 2008, 17, 1376.
[66] Nebel, C. E., Yang, N., Uetsuka, H. *Surf. Sci.* 2009, 603, L5.
[67] Kuo, T.-C., McCreery, R. L., Swain, G. M. *Electrochem. Solid St.* 1999, 2, 288.
[68] Wang, J., Firestone, M. A., Auciello, O., Carlisle, J. A. *Langmuir* 2004, 20, 11450.
[69] Shin, D., Tokuda, N., Rezek, B., Nebel, C. E. *Electrochem. Commun.* 2006, 8, 844.
[70] Uetsuka, H., Shin, D., Tokuda, N., Saeki, K., Nebel, C. E. *Langmuir* 2007, 23, 3466.
[71] Yang, N., Yu, J., Uetuska, H., Nebel, C. E. *Electrochem. Commun.* 2009, 11, 2237.
[72] Hoffmann, R., Obloh, H., Tokuda, N., Yang, N., Nebel, C. E. *Langmuir* 2012, 28, 47.
[73] Yang, N., Zhuang, H., Hoffmann, R., Smirnov, W., Hees, J., Jiang, X., Nebel, C. E. *Anal. Chem.* 2011, 83, 5827.
[74] Yang, N., Zhuang, H., Hoffmann, R., Smirnov, W., Hees, J., Jiang, X., Nebel, C. E. *Chem. Eur. J.* 2012, 21, 6514.
[75] Yang, N., Uetsuka, H., Osawa, E., Nebel, C. E. *Angew. Chem. Int. Edit.* 2008, 47, 5183.
[76] Nebel, C. E., Yang, N., Uetsuka, H., Osawa, E., Tokuda, N., Williams, O. *Japan New Diamond Forum* 2008, 24, 10.
[77] Yang, N., Uetsuka, H., Osawa, E., Nebel, C. E. *Diamond Relat. Mater.* 2009, 18, 592.
[78] Yang, N., Uetsuka, H., Nebel, C. E. *Adv. Funct. Mater.* 2009, 19, 887.

[79] Nebel, C. E., Yang, N., Uetsuka, H., Osawa, E., Tokuda, N., Williams, O. *Diamond Relat. Mater.* 2009, 18, 910.

[80] Lv, M., Wei, M., Rong, F., Terashima, C., Fujishima, A., Gu, Z.-Z. *Electroanalysis* 2010, 22, 199.

[81] Yang, Y., Oh, J.-W., Kim, Y.-R., Terashima, C., Fujishima, A., Kim, J. S., Kim, H. *Chem. Commun.* 2010, 46, 5793.

[82] Luo, D., Zhi, J. *Electrochem. Commun.* 2009, 11, 1093.

[83] Wu, W., Bai, L., Lin, X., Tang, Z., Gu, Z. *Electrochem. Commun.* 2011, 13, 872.

[84] Yang, N., Smirnov, W., Nebel, C. E. *Electrochem. Commun.* 2013, 27, 89.

[85] Yang, N., Smirnov, W., Kriele, A., Hoffmann, R., Nebel, C. E. *phys. status solidi (a)* 2010, 207, 2069.

[86] Yang, N., Hoffmann, R., Smirnov, W., Kriele, A., Nebel, C. E. *Electrochem. Commun.* 2010, 12, 1218.

[87] Harmer, R. J. *Nature* 2008, 545, 708.

[88] Uetsuka, H., Yamada, T., Shikata, S. *Diamond Relat. Mater.* 2008, 17, 728.

[89] Chen, M. Y., Chih, Y. K., Hwang, J., Kou, C. S., Chang, M. T., Chou, L. J. *Thin Solid Films* 2008, 516, 7595.

[90] Baik, E.-S., Baik, Y.-J., Jeon, D. *Diamond Relat. Mater.* 1999, 8, 2169.

[91] Kan, M.-C., Huang, J.-L., Sung, J., Chen, K.-H. *Thin Solid Films* 2004, 447–448, 187.

[92] Yamada, T., Hwang, D. S., Vinod, P. R., Makino, T., Fujimori, N. *Jpn. J. Appl. Phys.* 2005, 44, L385.

[93] Nebel, C. E. *Nature Photon* 2009, 3, 564.

[94] Numnuam, A., Chumbimuni-torres, K. Y., Xiang, Y., Bash, R., Thavarungkul, P., Kanatharana, P., Pretsch, E., Wang, J., Bakker, E. *J. Am. Chem. Soc.* 2008, 130, 410.

[95] Aoki, H., Tao, H. *Analyst* 2007, 132, 784.

[96] Fan, C., Plaxco, K. W., Heeger, A. J. *Proc. Nat. Acde. Sci. US* 2003, 100, 9134.

[97] Vermeeren, V., Bijnens, N., Wenmackers, S., Daenen, M., Haenen, K., Williams, O. A., Ameloot, M., vandeVen, M., Wagner, P., Michiels, L. *Langmuir* 2007, 23, 13193.

[98] Gu, H., Su, X-D., Loh, K. P. *J. Phys. Chem. B* 2005, 109, 13611.

[99] Lapierre-Devlin, M. A., Asher, C. L., Taft, B. J., Gasparac, R., Roberts, M. A., Kelley, S. O. Nano. Lett. 2005, 5, 1051.

[100] Gasparac, R., Taft, B. J., Lapierre-Devlin, M. A., Lazareck, A. D., Xu, J. M., Kelley, S. O. J. Am. Chem. Soc. 2004, 126, 12270.

[101] Smirnov, W., Yang, N., Hoffmann, R., Hees, J., Obloh, H., Müller-Sebert, W., Nebel, C. E. *Anal. Chem.* **2011**, *83*, 7438.

[102] Yang, N., Smirnov, W., Hees, J., Hoffmann, R., Kriele, A., Obloh, H., Müller-Sebert, W., Nebel, C. E. *phys. status solidi (a)* 2011, 208, 2087.

[103] Dolla, A., Blanchard, L., Guerlesquin, F., Bruschi, M. *Biochimie* 1994, 76, 471.

[104] Allen, H., Hill, O., Hunt, N. I., Bond, A. M. *J. Electroanal. Chem.* 1997, 436, 17.

[105] Taniguchi, I., Ishimoto, H., Miyagawa, K., Iwai, M., Nagai, H., Hanazono, H., Taira, K., Kubo, A., Nishikawa, A., Nishiyama, K., Dursun, Z., Hareau, G. P. J., Tazaki, M. *Electrochem. Commun.* 2003, 5, 857.

[106] Wang, L., Wang, E. *Electrochem. Commun.* 2004, 6, 49.

[107] Zhao, G.-C., Yin, Z.-Z., Zhang, L., Wei, X.-W. *Electrochem. Commun.* 2005, 7, 256.

[108] Haymond, S., Babcock, G. T., Swain, G. M. *J. Am. Chem. Soc.* 2002, 124, 10634.

[109] Marken, F., Paddon, C. A., Asogan, D. *Electrochem. Commun.* 2002, 4, 62.

[110] Baker, S. E., Colavita, P., Tse, K.-Y., Hamers, R. J. *Chem. Mater.* 2006, 18, 4415.
[111] Geng, R., Zhao, G., Liu, M., Li, M. *Biomaterials* 2008, 29, 2794.
[112] Zhou, Y., Zhi, J., Zou, Y., Zhang, W., Lee, S.-T. *Anal. Chem.* 2008, 80, 4141.
[113] Hoffmann, R., Kriele, A., Kopta, S., Smirnov, W., Yang, N., Nebel, C. E. *phys. status solidi (a)* 2010, 207, 2073.
[114] Hoffmann, R., Kriele, A., Obloh, H., Tokuda, N., Smirnov, W., Yang, N., Nebel, C. E. *Biomaterials* 2011, 32, 7325.
[115] Yang, N., Hoffmann, R., Smirnov, W., Nebel, C. E. *Diamond Relat. Mater.* 2011, 20, 269.
[116] Laviron, E. *J. Electronanal. Chem.* 1979, 101, 19.
[117] Xu, X., Tian, B., Kong, J., Zhang, S., Liu, B., Zhao, D. *Adv. Mater.* 2003, 15, 1932.
[118] Dai, Z., Liu, S., Ju, H. *Electrochima Acta* 2004, 49, 2139.
[119] Liu, H., Tian, Y., Deng, Z. *Langmuir* 2007, 23, 9487.
[120] Ju, H., Liu, S., Ge, B., Lisdat, F., Scheller, F. W. *Electroanalysis* 2002, 14, 141.
[121] Liu, Y. C., Cui, S. Q., Zhao, J., Yang, Z. S. *Eletrochemistry* 2006, 71, 121.

CSP2

In: Applied Surface Chemistry of Nanomaterials
Editors: Mohamed M. Chehimi and Jean Pinson
ISBN: 978-1-62808-351-4

Chapter 3

CHEMISTRY ON GRAPHENE

Cécilia Ménard-Moyon*
CNRS, Institut de Biologie Moléculaire et Cellulaire,
Laboratoire d'Immunopathologie et Chimie Thérapeutique, Strasbourg, France

I. INTRODUCTION

Thirty years ago the family of carbon allotropes was composed of graphite and diamond. Since, it has been extended by the discovery of fullerenes, [1] carbon nanotubes (CNTs), [2, 3] and recently graphene, which is the thinnest known material. Single-layer graphene was believed to be thermodynamically unstable under ambient conditions until Novoselov, Geim, and coworkers isolated single layers of graphene from crystalline graphite in 2004. [4] This discovery led to the distinction of the Nobel Prize in Physics in 2010 to both researchers.

Graphene is a single atomic plane of graphite. The two-dimensional layer is constituted of hexagonally arranged carbon atoms (Figure 1). Graphene is chemically unsaturated as it is composed of sp^2 carbon atoms. This new material belongs to the graphene-family nanomaterials comprising graphene oxide (GO), reduced graphene oxide (rGO), few-layer graphene constituted of 2 to 10 graphene layers, ultrathin graphite (more than 10 sheets but below 100 nm in thickness), as well as graphene nanosheets (GNS) and nano-GO (less than 100 nm lateral size dimension). [5]

So far, various methods have been developed for the synthesis of graphene and its derivatives, [6, 7, 8, 9, 10] including mechanical or liquid phase exfoliation of graphite, epitaxial growth, unzipping of CNTs, chemical vapor deposition (CVD), as well as exfoliation of GO and reduction. [11] These methods lead to graphene with different shape, size, and chemical composition. In 2004 Novoselov and coworkers obtained high quality single crystalline graphene with a lateral size up to millimeter-range by mechanical exfoliation of highly oriented pyrolytic graphite. [4] Graphene can also be prepared by liquid phase exfoliation of graphite. [12] Procedures using *N*-methylpyrrolidone [13] or aqueous

* Corresponding author: Dr. Cécilia Ménard-Moyon, CNRS, Institut de Biologie Moléculaire et Cellulaire, Laboratoire d'Immunopathologie et Chimie Thérapeutique, UPR 3572, 67000 Strasbourg, France, E-mail: c.menard@ibmc-cnrs.unistra.fr.

solution of sodium dodecylbenzene sulfonate (SDBS) [14] were found to be very efficient. Epitaxial growth of graphene on single crystal silicon carbide by annealing at high temperature has been reported. [15, 16] In this process graphene layers are formed by carbon segregation after evaporation of silicon atoms from the SiC surface. A strategy to obtain graphene nanoribbons of precise dimensions involves the longitudinally unzipping of CNTs. [17] Unzipping of multi-walled CNTs has been achieved through oxidation with potassium permanganate in high yield. [18] CVD allows to grow large-scale graphene by dissolving carbon atoms into a metal catalyst substrate, followed by precipitation after cooling. [19, 20] Graphene can also be obtained by reduction of GO, [21] but it still contains a significant amount of oxygen and probably a non-negligible number of defects, although thermal annealing of rGO allows restoration of the sp^2 carbon network of graphene.

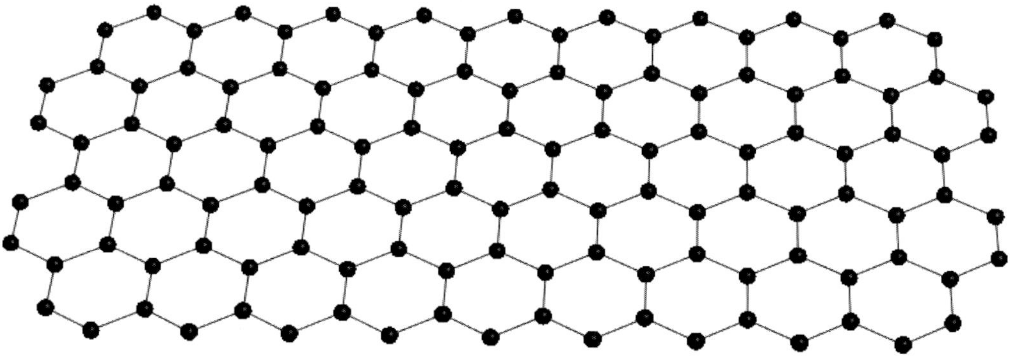

Figure 1. Structure of graphene.

The discovery of graphene has opened new opportunities in many research fields including physics, chemistry, and materials science. [22] Graphene has emerged as an extraordinary material and has attracted an extensive interest due to its exceptional electrical, optical, mechanical, and thermal properties. [23, 24, 25] Indeed, this new material exhibits high electrical and thermal conductivity, as well as mechanical flexibility. These properties make graphene a promising candidate for a wide range of applications in nanoelectronics, [26, 27] polymer nanocomposites, [28] and photovoltaics. [29]

However, applications of graphene are challenged by issues of production and processing. Indeed, the synthesis of single-layer graphene is difficult at ambient temperature. It is also rather difficult to avoid double or few-layered byproducts during the synthesis. In addition, due to their high specific surface, graphene sheets tend to form irreversible agglomerates through π-stacking and van der Waals interactions. The sheets can even restack irreversibly back to graphite, which really hinders its production, storage, and processability in terms of scalability, dispersibility, and stability. Therefore, chemical functionalization of graphene sheets is crucial to fully exploit their properties and prevent agglomeration. [30, 31, 32, 33]

Functionalization of graphene allows decreasing its hydrophobicity to improve dispersibility and processability in organic and aqueous solutions, which facilitate its processing by solvent-assisted techniques. Functionalization can modulate the mechanical and electronic properties by opening the energy gap in the electron spectrum. It improves processing of graphene into new materials and permits the combination of their outstanding properties with those of other materials or molecules, like electroactive or photoactive moieties. Functionalization is also a way to produce graphene from graphite, for instance by chemical reduction of graphite oxide.

There are some similarities between the reactivity of fullerenes, CNTs, and graphene. [34] Although many functionalization methods of fullerenes and CNTs have been applied to graphene, one major difference is the flat surface of graphene due to its single atomic layer. Indeed, functionalization can occur on both sides of the plane, while fullerenes and CNTs exhibit exo and endo faces. But, a perfect structure of a graphene sheet is considered to have a much lower chemical reactivity than fullerenes and CNTs. [35] The higher reactivity of fullerenes and CNTs can be explained by the curvature and the strain energy per carbon, which increases with decreasing diameter. Nevertheless, graphene can be functionalized under various conditions involving highly reactive species. It is postulated that reactions take place preferentially at defect sites.

One major issue with graphene, as it is the case with CNTs, is the variability in terms of structure (*i.e.* number of layers, lateral dimension) as well as chemical purity. In addition, the current synthesis methods produce graphene that contains unavoidable defects created unintentionally. [36, 37] These defects are edges, oxygen-containing species, and structural imperfections like vacancies, which complicate functionalization.

Pristine graphene is a semimetal or zero bandgap semiconductor, which hinders applications in nanoelectronics. To develop graphene-based nanodevices, it is essential to open the bandgap and precisely control the charge carrier type and density. Tailoring of the electronic properties is possible by either covalent or non-covalent (*e.g.* by stacking of aromatic molecules through π-π interactions) functionalization of graphene.

This chapter highlights the different methods developed to functionalize graphene. They can be classified into two categories: covalent and non-covalent functionalization. Section II focuses on covalent functionalization, in particular on binding of organic functionalities through derivatization of oxygen-containing species of GO (section II-1). The covalent attachment of hydrogen atoms, leading to graphane, and of halogen atoms is reported in sections II.2 and II-3, respectively. Covalent linking of highly reactive species based on direct addition of radicals, cycloaddition and electrophilic reactions is described in sections II-4 to II-6. [38] Section III reports non-covalent functionalization of graphene that does not disrupt the extended π-conjugation of graphene, unlike covalent functionalization. In particular, adsorption of aromatic compounds on the graphene surface by π-stacking and interactions with surfactants and polymers are described in details.

II. Covalent Functionalization of Graphene

Graphene can be functionalized at two different locations, namely at the edges and on the basal plane. During covalent addition, inner carbon atoms, strictly constrained in the graphene basal plane by surrounding carbon atoms and protected by the π-conjugation network, have to adopt a tetrahedral sp^3 geometry. This rehybridation causes strain in the basal plane. Thus, covalent addition on the basal plane generally requires large energy barriers and highly reactive species. Besides, the π-electron structure at the edges is different compared to the basal plane. The dangling bonds at edge sites, usually terminated by hydrogen atoms, are very reactive towards covalent addition reactions. Indeed, unlike inner carbon atoms, the carbons at the edges can adopt a tetrahedral sp^3 geometry more freely without causing additional strain.

Graphene has two edge geometries, namely zigzag and armchair. The zigzag configuration resembles a triplet carbene, while the armchair configuration resembles a singlet benzyne. [39] Therefore, the zigzag edges are more reactive than the armchair edges, because frontier molecular orbitals are mainly localized at zigzag edges with a state closer to the Fermi level. But, in reality functionalization of graphene cannot be completely controlled either on the basal plane or on the edges (zigzag or armchair) as different types of reactivity may be involved simultaneously.

In contrast to non-covalent functionalization, covalent addition of molecules to graphene is associated with rehybridization of one or more sp^2 carbon atoms of the π-network into sp^3 configuration, leading to a partial or total loss of π-π conjugation and modifications of the intrinsic physical and chemical properties of graphene. The electrical conductivity of graphene can be tuned by functionalization as a band gap can be opened by chemical covalent derivatization due to disruption of the crystallographic lattice, [40, 41, 42] which is particularly useful for the development of transistor devices. [43] It has been demonstrated that the addition of functional groups on graphene is reversible by thermal treatment, thereby restoring the intrinsic electronic properties of graphene. In addition, covalent functionalization allows stabilizing permanently graphene sheets in solution.

Different methods for covalent functionalization of graphene have been reported. Functionalization of GO is described in section II-1, in particular the condensation of isocyanate, the amidation and esterification of carboxylic acid groups, the derivatization of hydroxyl groups (e.g. by silanization, alkylation), as well as the opening of epoxides with amine derivatives and alkylation of GO in basic conditions. Hydrogenation of graphene leading to graphane and halogenation reactions are reported in sections II-2 and II-3, respectively. Addition of radicals using diazonium salts or arising from thermal decomposition of benzoyl peroxide or azo compounds is described in section II-4. Functionalization of graphene with other highly reactive species (e.g. dipoles, nitrenes, carbenes) by cycloaddition or electrophilic reactions is detailed in sections II-5 and II-6.

II.1. Derivatization of Graphene Oxide

One major limitation with graphene is its low solubility in organic solvents and water, which can hamper its manipulation for functionalization and limit full exploitation of its properties. To overcome this issue and increase the dispersibility of graphene, a pre-treatment based on oxidation can be performed, which induces exfoliation of graphite and leads to GO. The presence of polar oxygen-containing species in GO makes it more hydrophilic than pristine graphene. This is fundamental for further functionalization and processability. [44] For that reason, GO is often used as starting material for preparation of graphene derivatives thanks to the rich chemistry of epoxide rings, carboxyl, and hydroxyl groups. [45]

II.1.a.) Preparation of Graphene Oxide

GO is prepared mainly by two different processes based on i) the oxidation of graphite in various conditions (through Hummers, [46] modified Hummers, [47] Brodie, [48] Staudenmaier [49] methods or by electrochemical oxidation [50]) followed by exfoliation; [51, 52] ii) the direct oxidation of graphene with strong oxidants like concentrated acids, [53] which is commonly less used. The negatively charged oxygen-containing species on graphite oxide induce exfoliation of the graphene sheets to mono-layer graphene due to electrostatic repulsion. An alternative and original method for preparation of GO relies on lengthwise cutting and unraveling of the sidewalls of multi-walled CNTs by oxidation, leading to graphene oxide nanoribbons. [17,18]

GO has reactive oxygen functionalities, such as hydroxyl functions and epoxide groups on the basal plane, while carboxyl, carbonyl, phenol, lactone, and quinone groups are present at the edges (Figure 2). [54, 55, 56, 57] But, there is a lack of sufficiently sensitive analytical techniques for thorough and rigorous characterization of the GO structure. The stoichiometry of the different oxygenated species varies depending on the synthesis conditions, which generates heterogeneity of GO samples prepared by different methods. The chemical composition of GO is still strongly debated and different model structures of GO have been proposed in literature. [58, 59] There is still ambiguity about the presence [54] or absence [58] of carboxylic acids groups on the periphery of the basal plane of GO.

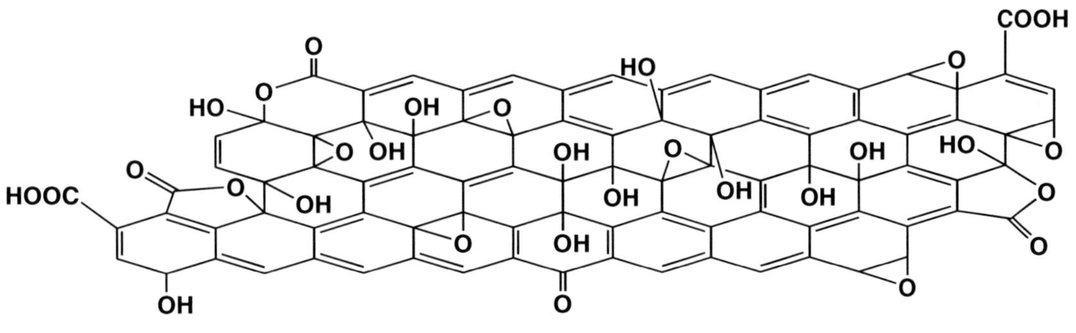

Figure 2. Structural model of GO (adapted from ref. [58]).

Carboxylic acid groups have been selectively stained with Pt-ammine complexes. Transmission electron microscopy reveals clusters at edges of graphene sheets, which indicates that some carboxyl functions exist mainly at the graphene edges. [60]

The main issue with GO is the disruption of the π-network and as a result GO is an insulator. In theory, a reductive treatment of GO should remove the functional groups and induce rehybridization of the sp^3 carbon atoms to sp^2, thus restoring aromaticity. But, it has to be noted that in reality after reduction of GO a significant number of oxygen-containing groups still remains on the graphene layer, which can alter the electronic properties of the reduced GO.

II.1.b.) Reaction of Hydroxyl and Carboxyl Groups with Isocyanate Derivatives

Stankovich *et al.* have demonstrated that the treatment of graphite oxide with organic isocyanates leads to functionalization of both the edge carboxyl and surface hydroxyl groups *via* formation of amides and carbamate esters, respectively (Scheme 1A). [61] The functionalized graphite oxide is completely exfoliated into individual functionalized GO sheets that form a stable dispersion in polar aprotic solvents. The level of functionalization is dependent on the reactivity of the isocyanate and can be controlled by the reaction time. The use of isocyanates containing functionalities like cyano, keto, and azidosulfonyl, opens different possibilities for further derivatization of the GO sheets. This method has been applied to change the surface properties of GO to allow homogeneous mixing with many organic polymers, thereby facilitating the preparation of graphene-polymer composites. [62, 63, 64]

In another study, graphite oxide hybrid porous materials have been constructed through the cross-linking of exfoliated GO sheets using the reaction of organic diisocyanates with the carboxyl and hydroxyl groups on both sides of the sheets. [65] Diisocyanate molecules serve as nanoscale spacers to cross-link the sheets, which could effectively limit their face-to-face stacking, leading to a lamellar nanoporous graphitic structure with potential application as hydrogen adsorbent material.

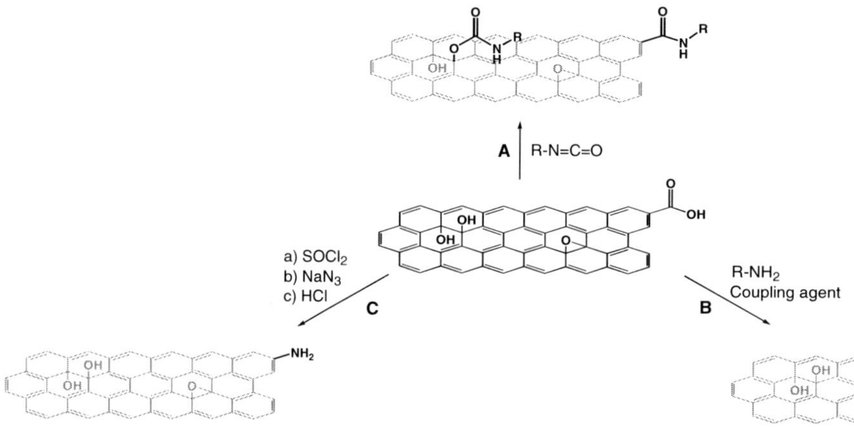

Scheme 1. Derivatization of GO by condensation with isocyanates (A), by amidation (B), and Curtius rearrangement (C). For the sake of clarity, only epoxide, hydroxyl, and carboxylic acid functions are represented on GO; only one hydroxyl group is derivatized.

II.1.c.) Selective Derivatization of Carboxylic Acid Functions

The carboxyl groups of GO can be derivatized by esterification or amidation (Scheme 1B). The coupling reaction requires activation of the carboxyl moieties using thionyl chloride or coupling agents such as *N*-hydroxysuccinimide (NHS), 1-ethyl-3-(3-dimethylaminopropyl)-carbodiimide (EDC), or 1-hydroxybenzotriazole hydrate (HOBt). Consecutive addition of amine or hydroxyl derivatives leads to the formation of amides [66] and esters, [67] respectively. This strategy has been first applied to increase dispersibility of GO by derivatization with octadecylamine. [68] It should be noted that multiple reactions may occur simultaneously as amine functions involved in the amidation process can also react with epoxides of GO *via* ring-opening (*cf.* section II-1.e). [69]

Amidation allows to prepare GO derivatives for various applications like polymer composites, [70, 71] optical limitation, [72, 73, 74] and optoelectronic devices. [75] This strategy has been extensively applied for functionalization of GO. For instance, GO has been derivatized with a photochromic molecule, azobenzene, by amidation. An optical modulation of the conductance of an azobenzene-GO film was observed upon ultraviolet irradiation, due to reversible trans-cis photoisomerization of the chromophore. [76] GO has also been combined by amidation with bioactive molecules such as antibodies [77, 78] or nucleic acids [79] for the development of biosensors. Derivatization with chitosan [80] or polyethyleneglycol (PEG) [81, 82] by formation of an amide linkage allows to impart hydrophilicity and prevent biofouling, thus increasing biocompatibility for drug delivery applications. Streptavidin has been attached to GO *via* amidation between biotin and GO, followed by complexation with streptavidin. [83] The streptavidin-GO complex has been used for affinity purification of biotinylated protein complexes.

Derivatization of carboxylic acid functions of GO has also been employed in polymer chemistry. GO has been functionalized by esterification with 2-bromoisobutyryl bromide, an atom-transfer radical-polymerization initiator (ATRP), which induced polymerization of poly(methyl methacrylate). [84] In another study, GO has been derivatized by esterification with 2,2,6,6-tetramethylpiperidine-1-oxyl (TEMPO), a radical scavenger species. [85] The GO sheets were simultaneously thermally reduced to graphene. The TEMPO-modified graphene were derivatized with a bromo-terminated polymer *via* atom transfer nitroxide radical coupling chemistry (ATNRC).

In addition to amidation and esterification, carboxyl acid groups of GO can be converted to amine functions by Curtius rearrangement (Scheme 1C). [86] The reaction proceeds *via* activation with thionyl chloride and subsequent reaction with sodium azide. The corresponding acyl azides are thermally decomposed to produce isocyanates. Acidic treatment of the isocyanate intermediates leads to formation of amines. The effect of the amino-derivatized GO on platelet reactivity has been studied and compared with GO to highlight the importance of functionalization and surface properties for biomedical applications.

II.1.d.) Selective Derivatization of Hydroxyl Groups

Functionalization of hydroxyl groups of GO has received little attention. Silanization has been applied to derivatize GO with *N*-(trimethoxysilylpropyl) ethylenediamine triacetic acid (EDTA) as chelating agent for use as adsorbent for Pb(II) removal [87] and as dopamine biosensor (Scheme 2A). [88] During the reaction, the trialkoxy groups of silane are hydrolyzed, generating Si-OH moieties, which react with hydroxyl functions of GO through Si-O-C bonds. The silanized GO is then reduced to obtain EDTA-graphene derivative.

Microwave irradiation has been used to perform silanization reactions under mild conditions and in short times. [89] In a comparative study, it has been shown that the grafting efficiency of silanization is much higher than that of amidation on GO. [90] This result could be explained by the higher quantity of hydroxyl groups compared to carboxylic acid functions on GO sheets.

Fisher esterification has been performed between the hydroxyl functions on graphite oxide and the carboxyl moieties on a fullerenoacetic acid derivative in the presence of a weak acid (*p*-toluenesulfonic acid) (Scheme 2B). [91] The attached fullerenes are able to weaken the van der Waals forces between the graphene layers, resulting in exfoliation of graphite oxide to monolayer GO, while preventing re-stacking of the graphene sheets.

Claisen rearrangement is another reaction that has been used for derivatization of hydroxyl functions on graphite oxide as it exploits the allylic character of the alcohol groups. The allylic hydroxyl functions on graphite oxide are converted to vinyl allyl ethers by treatment at 60 °C with *N,N*-dimethylacetamide dimethyl acetal (DMDA) as vinyl transfer reagent using the Eschenmoser-Claisen rearrangement variant. *In situ* sigmatropic-type transformation leads to carbon-bound *N,N*-dimethylamide groups (Scheme 2C). [92] The reaction conditions induce efficient exfoliation of the graphene sheets. It was found that DMDA serves as both vinyl transfer reagent and reducing agent for graphite oxide. The amide functions on the graphene surface can be saponified under strong basic conditions to form carboxylate moieties, resulting in an enhanced solubility of graphene in water.

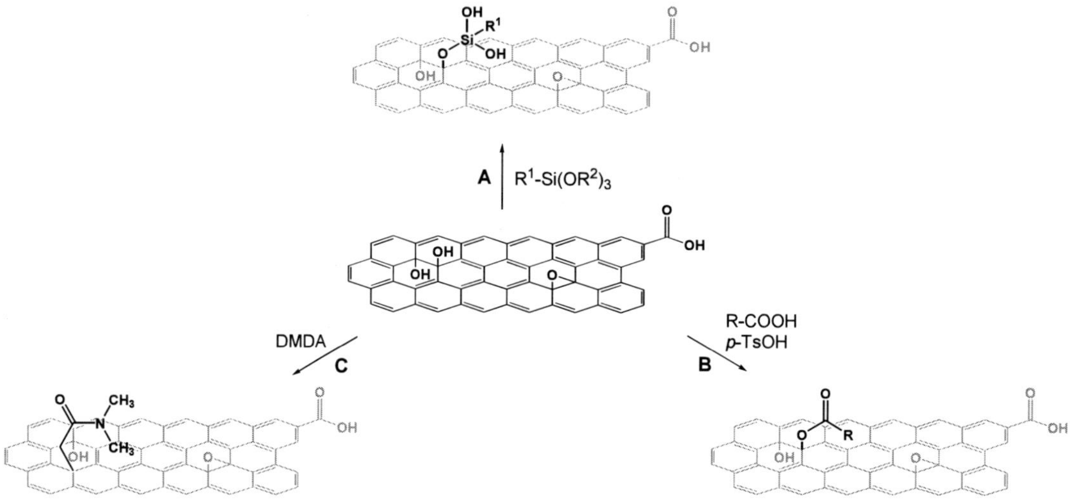

Scheme 2. Derivatization of GO by silanization (A), Fisher esterification (B), and Claisen rearrangement (C).

II.1.e.) Opening of Epoxide Rings

The epoxy groups on GO can be easily modified through ring-opening reactions under various conditions by nucleophilic attack at the α-carbon by amine derivatives bearing a lone pair of electrons (Scheme 3A). Compared to other methods this substitution reaction has the advantage to occur easily in mild conditions (at room temperature and in aqueous medium). This approach has been exploited for derivatization with 3-aminopropyltriethoxysilane for reinforcement of silica monoliths, [93] with an amine-terminated ionic liquid to improve dispersibility of GO, [94] and with cysteamine for voltammetric selective quantitative determination of Hg^{2+}. [95]

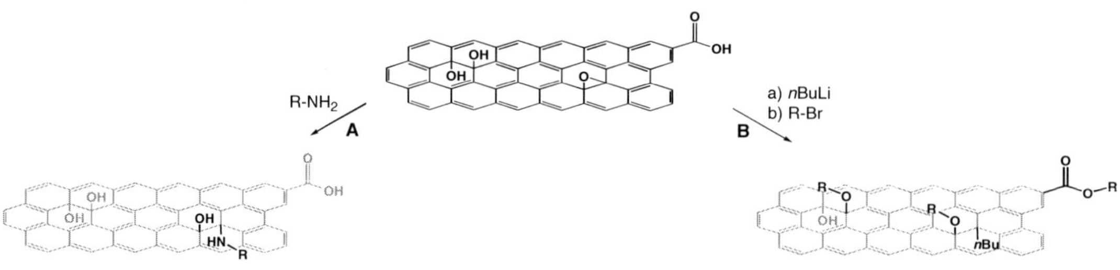

Scheme 3. Derivatization of GO by opening of epoxide rings (A) and alkylation in the presence of *n*BuLi (B).

GO can also react with sodium azide, generating amine functions after epoxide opening and successive reduction of the azides to amines. [96] In another study, rGO has been functionalized by ring opening of the residual epoxide functional groups with an amine derivative. [97] The advantage of this method relies on the use of residual epoxide rings on the graphene surface as reactive sites, which avoids causing further damage to the graphene layer.

II.1.f.) Alkylation of Graphene Oxide

Epoxide, carbonyl, carboxylic acid, hydroxyl functions, and double bonds on GO are reactive towards nucleophiles like *n*-butyllithium (*n*BuLi). [98] Deprotonation/carbometalation with *n*BuLi followed by electrophilic attack with 2-ethylhexyl bromide leads to alkylation, etherification, and esterification of GO (Scheme 3B). The covalent attachment of 2-ethylhexyl and *n*-butyl chains on GO results in improved dispersion properties. Raman spectroscopy confirmed the formation of sp^3 carbon atoms.

In another study, GO has been derivatized by nucleophilic substitution with a linear alkyl chain. [99] The treatment of GO with sodium hydroxide and dodecyl iodide in the presence of a phase transfer agent (tetra-*n*-octylammonium bromide) generates alkylated graphene with higher dispersibility, increased interfacial adhesion with polypropylene, and superior reinforcing effects compared to alkylated CNTs.

II.2. Hydrogenation of Graphene

Graphene can undergo hydrogenation, giving rise to fully hydrogenated graphene, the so-called graphane. Different methods have been reported for the synthesis of graphane, most rely on exposure of graphene to hydrogen plasma discharge [100, 101] or to hydrogen atoms generated *in situ* by electron beam-induced dissociation of hydrogen silsesquioxane. [102] Other methods have been investigated such as hydrogenation of graphene on transition metals [103] or electron irradiation of graphene having chemisorbed water. [104]

The electronic properties of graphene are drastically modified by hydrogenation. Indeed, graphene is a semi-metal, while graphane is an insulator. [105] The reaction of graphene with hydrogen is reversible, and graphane can be dehydrogenated by annealing at high temperatures to restore graphene. [100,102]

II.3. Halogenation

Graphene has been halogenated with fluorine and chlorine atoms. The stability of graphene halides decreases with increasing size of the halogen atom, in the order F > Cl > Br > I. [106] Fluorographene, namely graphene fluoride, is stable (even more stable than graphane), [107] whereas stoichiometric chlorographene is predicted unstable above 0 °C and graphene iodide decomposes spontaneously. [30] Fluorographene can be converted to graphene *via* halide exchange leading to intermediate graphene iodide.

II.3.a.) Fluorination

As hydrogenation, fluorination of graphene can reversibly modify the electrical conductivity of graphene. Fluorographene can be prepared by direct fluorination of graphene by treatment with XeF_2 [108, 109] and plasma (CF_4 [110] and SF_6 [111, 112]). [101] Alternative methods are based on exfoliation of graphite fluoride using sulfolane, [107] ionic liquids, [113] isopropanol, [114] or dimethylformamide. [115] Hydrothermal treatment of a mixture of GO and hydrofluoric acid also leads to fluorinated graphene with tunable degree of fluorination by varying the reaction conditions. [116]

The electrical properties of fluorographene are completely different compared to graphene and hydrogenated graphene as fluorographene is strongly insulating due to a wide gap opened in the electronic spectrum. Fluorination was found to be reversible after vacuum annealing. [112] Defluorination can also be induced by hydrazine treatment, thus restoring the electronic properties of graphene. [108]

An original approach to functionalize graphene with fluorine atoms on specific area of the graphene surface relies on laser irradiation of a fluoropolymer-covered graphene. [117] The photon-induced decomposition of the fluoropolymer generates fluorine radicals that react with graphene only in the laser-irradiated region.

The fluorine atoms on fluorinated graphene sheets can be substituted with amine derivatives. [110] Alternatively, reductive alkylation of fluorinated graphite with alkyl lithium reagents allows to obtain soluble alkylated graphene layers. [118, 119, 120]

II.3.b.) Chlorination

Grafting of chlorine atoms on CVD-grown graphene has been performed by gas-phase photochemical chlorination. [121] The level of functionalization is about 8 atom % chlorine. Electrical measurements of the partially chlorinated graphene show that the resistance increases over 4 orders of magnitude and a band gap appears. In another study, comparison between treatments of graphene prepared by mechanical exfoliation with hydrogen, fluorine, or chlorine plasma has been done. [122] In the case of hydrogen or fluorine plasma, a fast destruction of the graphene conjugated network is observed due to high degree of functionalization, imparting insulating properties. On the contrary, under chlorine plasma treatment, the reaction kinetics is slower and the chlorination can be controlled. As a result of moderate degree of functionalization, the graphene structure is not extensively damaged. The conductance of the chlorinated graphene is increased due to p-doping.

II.4. Radical Addition

As alternative to derivatization of the oxygenated groups of GO and halogenation reactions on graphene, a series of methods have been developed for direct functionalization of graphene, in particular the addition of radicals arising from diazonium salts, benzoyl peroxide, or azo compounds.

II.4.a.) Addition of Aryl Diazonium Salts

The diazotization of anilines leads to the formation of aryl diazonium salts, which are stable and can be isolated if the counterion is non-nucleophilic. In the presence of graphene, aryl diazonium salts are unstable; they are rapidly converted into aryl radicals with simultaneous release of nitrogen gas *via* an electron transfer from graphene to the diazonium salts. The aryl radicals add on the graphene surface by formation of a covalent bond (Scheme 4A).

Aryldiazonium salts have been widely used for functionalization of carbon nanotubes. [123] Tour and coworkers have applied this strategy on rGO dispersed in water with the aid of SDBS as surfactant. [52] Aryl diazonium salts para-substituted with different functional groups (Cl, NO_2, OCH_3, Br) were employed. Conditions have been optimized on graphene nanoribbons by *in situ* generating the aryl diazonium salts from the corresponding anilines, which prevents isolation of unstable diazonium salts. [124]

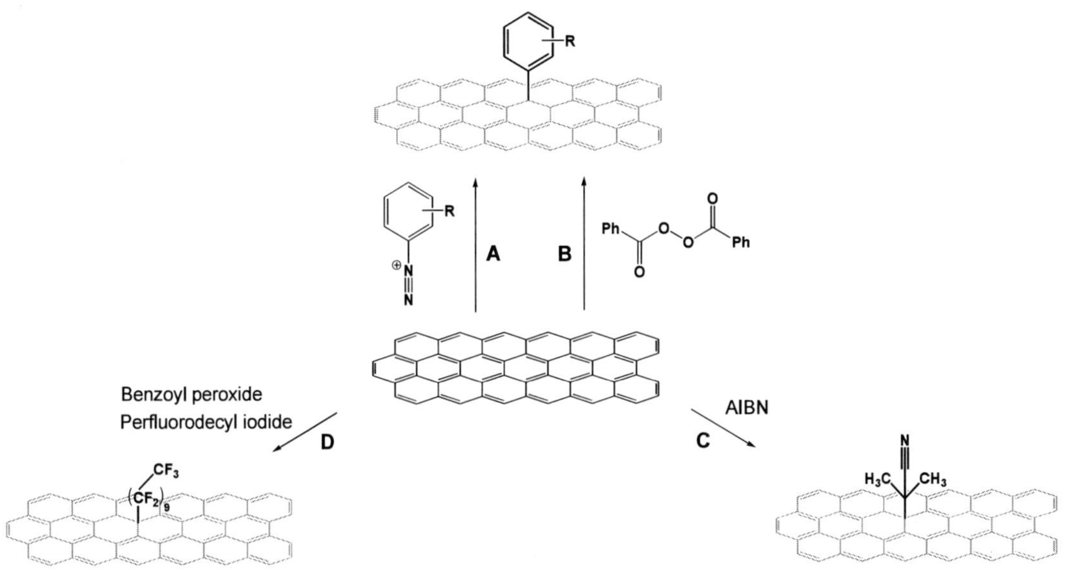

Scheme 4. Functionalization of graphene with radicals generated from aryldiazonium salts (A), benzoyl peroxide (B), AIBN (C), and a mixture of perfluorodecyl iodide and benzoyl peroxide (D).

Kinetic studies of diazonium addition on graphene nanoribbons have shown that the reaction at room temperature is fast (few minutes). [125] The group of Tour has also described a more environmentally friendly and potentially scalable one-step process based on mechanical assisted exfoliation and functionalization of graphene using aryl diazonium salts in ionic liquids. [126]

Electrochemistry is also an efficient approach for arylation of graphene. Indeed, electrochemical reduction of aryl diazonium salts generates aromatic radicals that attack the surface of CVD-grown graphene monolayer under mild conditions to form a uniform thin film within a few seconds. [127]

The nature of the functional groups in the aniline precursor can increase dispersibility and serve as anchor for further derivatization to allow different molecules to be coupled to graphene to tune their properties for various applications. For instance, polymer-functionalized rGO has been prepared by arylation to graft the initiator molecules, followed by polymerization of styrene. [128, 129, 130] In another study, functionalization of rGO with 4-mercapto-benzenediazonium has allowed to bound gold nanoparticles *via* anchoring to the thiol groups for fabrication of electronic devices. [131]

An original one-pot method has been developed by Hirsch and coworkers for exfoliation of graphite by reduction with a sodium/potassium alloy and subsequent functionalization of the negatively charged separated graphene sheets with aryl diazonium salts. [132, 133] Different levels of functionalization were obtained by variation of the diazonium salt concentration.

In a comparative study, graphene flakes have been derivatized with a conjugated polymer, *via* two different approaches: by arylation using diazonium salts or by amidation. [134] With regards to the level of functionalization, the amidation reaction led to 33% less functional groups on the graphene surface by comparison with the arylation process. This is most likely due to higher density of reactive sites for diazonium coupling in the graphene basal plane than for amidation with carboxylic acids at the edges.

Unlike oxidation of graphite to exfoliate the graphene sheets, the method with diazonium salts uses mild conditions as it is based on reduction by electrons that do not irreversibly damage the graphene structure. Indeed, thermal annealing allows to defunctionalize graphene and to regenerate completely intact hexagonal carbon network layers, contrary to GO, where a complete restoration is more complicated.

It has been shown that the chemical reactivity of graphene for the addition of diazonium salts is significantly influenced by the underlying substrate used for electronic purposes. [135, 136] In addition, the group of Strano has observed that the edges of single-layer and multilayer graphene are at least two times more reactive than the basal plane towards arylation using diazonium salts. [137] This result has been confirmed by other research groups, [138, 139] in particular by elemental mapping using energy filtered transmission electron microscopy, [140] showing that functional groups are mainly bound within 70 nm of the edges. The selectivity could arise from a higher exposition of the edges to reagents. Strano and coworkers have published a review on the reaction mechanism of graphene functionalization with diazonium salts. [141] It was reported that the chemical reactivity of graphene is influenced by different parameters such as the number of layers, the type of edges, the degree of strain and grain boundaries.

The electronic properties of graphene functionalized with aryl diazonium salts can be tuned [142, 143] either to a higher or lower level, depending on the functionalization degree. [125, 144, 145, 146] Furthermore, arylation of epitaxial graphene not only changes their electronic properties, but also imparts novel properties as it leads to disordered magnetism consisting of a mixture of ferromagnetic, superparamagnetic, and antiferromagnetic regions. [147]

II.4.b.) Functionalization with Radicals From Decomposition of Benzoyl Peroxide and Azo Compounds

A few studies have reported the use of radicals arising from thermal decomposition of benzoyl peroxide and azo compounds for the preparation of functionalized graphene. Due to weak oxygen-oxygen bond, benzoyl peroxide readily undergoes homolysis, forming free radicals $PhCOO^{\bullet}$, evolving to phenyl radicals by loss of carbon dioxide. Functionalization of graphite with phenyl radicals generated by thermal decomposition of benzoyl peroxide with subsequent sulfonation of the phenyl groups leads to exfoliated graphene (one to few layers) with functional groups only along the edges (Scheme 4B). [148] High level of exfoliation occurs during the sulfonation step. Alternatively, single- and few-layer graphene microscale films have been derivatized with phenyl radicals formed by a photochemical reaction between graphene and benzoyl peroxide. [120, 149] A mechanism based on an electron transfer from physisorbed benzoyl peroxide and photoexcited graphene was suggested, resulting in irreversible decomposition to a highly reactive phenyl radical.

Alternatively, azo compounds (R-N=N-R) can also be employed as radical precursors ($R^{\bullet}$) with concomitant release of nitrogen gas upon heating and/or by UV irradiation. For instance, azodiisobutyronitrile (AIBN) which thermally decomposes to isobutyronitrile radicals has been used for functionalization of graphene prepared by arc discharge (Scheme 4C). [150] The resulting cyano groups on the graphene surface can be converted to sodium carboxylate moieties by treatment with sodium hydroxide for immobilization of silver nanoparticles. The Ag-graphene hybrid material shows a good electrocatalytic activity for nitroaromatic molecules, which is promising for use as electrochemical sensor of nitroaromatic compounds.

Benzoyl peroxide can also serve as radical initiator to induce polymerization. Thermal decomposition of benzoyl peroxide has been used to form polystyrene-polyacrylamide block copolymer macroradicals that add covalently to the surface of rGO by *in situ* living free-radical polymerization. [151] The thermal decomposition of benzoyl peroxide has also been employed to initiate radical formation from perfluoroalkyl iodides that react with expanded graphite (Scheme 4D). [152] In this study, an alternative method was investigated based on UV photolysis of perfluorinated alkyl iodide. The addition of the generated radicals on microcrystalline graphite results in high degree of exfoliation, leading to functionalized single and few-layer graphene. Finally, benzoyl peroxide has also been used as radical initiator in the free radical functionalization of graphene with an alkene derivative prepared by reaction between poly(oxyalkylene)amines and maleic anhydride. [97]

II.4.c.) Sonochemical Preparation of Functionalized Graphene

A single-step sonochemical method has been developed for the preparation of polystyrene-functionalized graphene starting from graphite and styrene. [153] Irradiation of graphite flakes in styrene using ultrasound leads to mechanical exfoliation of graphite to single-layer and few-layer graphene, associated to concomitant functionalization with polystyrene by radical polymerization. The styrene acts as both solvent for graphite exfoliation and monomer.

Sonochemically initiated radical polymerization of polyvinyl alcohol (PVA) has been carried out by ultrasonication of graphene in an aqueous solution of PVA. [154] Ultrasonic irradiation is powerful enough to induce sonochemical degradation of the PVA solution, generating PVA macroradicals which are trapped by graphene and covalently grafted on its surface. The PVA-functionalized graphene is easy to disperse in a PVA matrix. Due to improved interface adhesion of PVA-graphene with PVA, the mechanical properties of films composed of functionalized graphene and PVA are significantly enhanced. This technique based on sonochemical preparation of polymer-graphene hybrid material is simple and could be extended to functionalization of graphene with other polymers.

II.5. Cycloaddition Reactions

II.5.a.) 1,3-Dipolar Cycloaddition

II.5.a$_1$.) 1,3-Dipolar Cycloaddition of Azomethine Ylides

1,3-Dipolar cycloaddition of azomethine ylides is a versatile reaction that has been extensively used for functionalization of a wide range of carbon nanostructures: fullerenes, [155] carbon nanotubes, [156, 157] carbon nano-onions, [158] and carbon nanohorns [159, 160] for various applications from nanomedicine [161] to photovoltaics. [162] Azomethine ylides are highly reactive species formed *in situ* by condensation of an α-amino acid with an aldehyde derivative, followed by thermal decarboxylation. Addition of the 1,3-dipole generates pyrrolidine rings on the graphene surface (Scheme 5A).

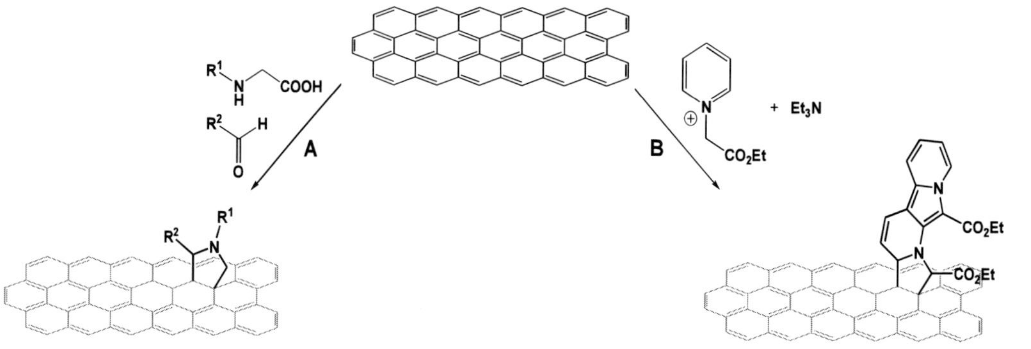

Scheme 5. Functionalization of graphene by 1,3-dipolar cycloaddition of azomethine ylides (A) and pyridinium ylides (B).

Georgakilas *et al.* first reported functionalization of monolayer and few-layer graphene with sarcosin (*i.e.* *N*-methylglycine) and 3,4-dihydroxybenzaldehyde, which facilitated dispersion of graphene in organic and aqueous solvents. [163]

The group of Prato [164] has also applied the 1,3-dipolar cycloaddition of azomethine ylides for functionalization of few-layer graphene with pyrrolidines bearing a triethylene glycol moiety terminated by ammonium groups. [165] The level of functionalization is assessed in this case by the colorimetric Kaiser test. [166] By binding gold nanorods to ammonium-derivatized graphene, it is possible to localize the functional groups and thus identify the reactive sites on the graphene surface. The gold nanorods, used as contrast markers, are distributed not only at the edges or in their proximity, but also over the graphene basal plane.

In another study, poly(amidoamine) (PAMAM) dendron moieties bearing ammonium groups have been introduced on few-layer graphene by two different reactions, *via* functionalization using 1,3-dipolar cycloaddition or by amidation on carboxylic acids generated during ultrasonication. [167] The ammonium functions can complex gold nanoparticles, which serve as contrast markers for the identification of the reactive sites. In the case of the 1,3-dipolar cycloaddition, the gold nanoparticles are distributed all over the graphene layers, whereas for the amidation the nanoparticles are deposited mostly at the edges, thus confirming previous observations that carboxylic groups are mainly present at the graphene edges.

A variety of substituted α-amino acids and aldehydes can be used as precursors for the 1,3-dipolar cycloaddition of azomethine ylides. For instance, 1,3-dipolar cycloaddition of exfoliated graphite in the presence of sarcosin and 4-formylbenzoic acid affords graphene bearing pyrrolidines with carboxyl functions used for covalent attachment of phthalocyanine by amidation. [168] In another study, single- and few-layer graphene flakes prepared by exfoliation of graphite and rGO have been directly derivatized with porphyrins [169] or polymers [170] by using sarcosin polymers or porphyrin derivatives bearing an aldehyde function, respectively. The porphyrin or phtalocyanine-graphene hybrids could find potential applications as solar cells.

Finally, graphene has been covalently functionalized with phenol groups by 1,3-dipolar cycloaddition without degrading its electronic properties. [171] The phenols were derivatized with 2-bromoisobutyryl bromide, an atom transfer radical polymerization initiator, which initiated polymerization of methyl methacrylate at room temperature.

II.5.a$_{2}$.) 1,3-Dipolar Cycloaddition of Pyridinium Ylides

1,3-Dipolar cycloaddition of pyridinium ylides onto rGO at 140 °C leads to covalent functionalization of rGO with indolizine moieties. [172] The pyridinium ylides, generated *in situ* by mixing triethylamine with pyridinium salts, add to graphene to form pyrrolidine rings. Subsequent addition of a second pyridinium ylide to the cycloadduct on the graphene surface results in the generation of indolizine rings (Scheme 5B). The functional groups have been localized with gold nanoparticles by reacting indolizine groups with *N,N*-dimethylethylenediamine and iodomethane to generate positive charges, followed by electrostatic interactions with negatively charged citric-stabilized gold nanoparticles. The indolizine functional groups were found to be distributed on the graphene surface. Fluorescence spectroscopy showed an efficient energy transfer from indolizine moieties to few-layer graphene, resulting in a strong quenching. The electrochemical properties of indolizine-functionalized graphene have been studied by cyclic voltammetry, demonstrating its efficient catalytic activity for acid ascorbic oxidation.

II.5.a$_{3}$.) "Click" Chemistry

The Cu(I)-catalyzed 1,3-dipolar Huisgen cycloaddition, the so-called "click" reaction, forming triazole rings between azides and alkynes, is a very versatile method widely used in organic chemistry. [173] This reaction has many advantages as it is rapid and efficient. It can be performed under mild conditions: in water and in the presence of a variety of functional groups. The resulting triazole ring is stable and inert to molecular oxygen. Graphene and GO can be derivatized with either alkyne or azide functions to allow "click" reaction with azide or alkyne derivatives, respectively (Scheme 6, A and B, respectively).

The "click" reaction has been successfully applied on graphene to link polymers [174] (*e.g.* polystyrene, [175] triblock copolymers, [176] polyfluorene, [134] polythiophene, [90] PEG [177]), deoxyribonucleic acid (DNA), [178] or photoactive molecules (*e.g.* diketopyrrolopyrrole, [179] porphyrin and phenanthroline [180]) for potential applications as composites, biosensors and in photovoltaics, respectively. The triazole moiety is a suitable linker as it retains conjugation and enables electronic transport between graphene and conjugated molecules or polymers.

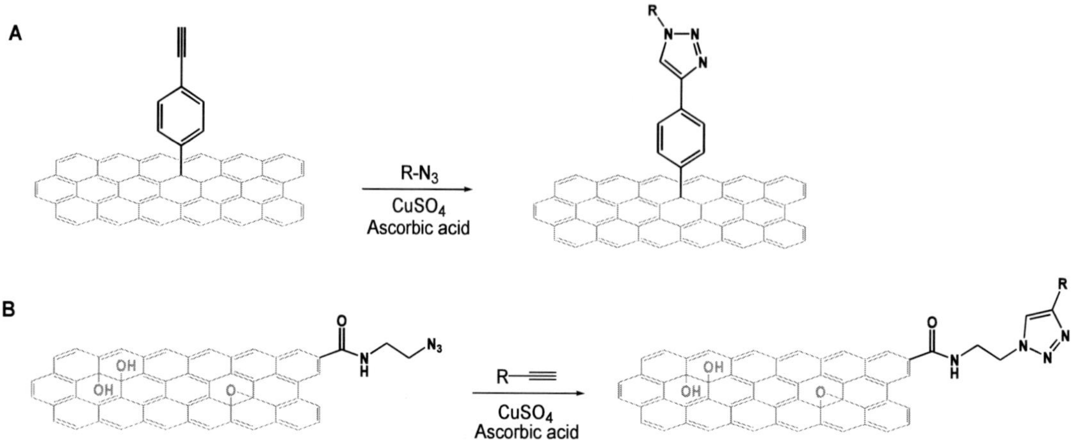

Scheme 6. Functionalization of alkyne-derivatized graphene (A) and azide-modified GO (B) by Cu(I)-catalyzed 1,3-dipolar Huisgen cycloaddition ("click" reaction).

Functionalization of GO with long alkyl chains [96] or polystyrene [175] by "click" reaction leads to exfoliation of the material to single layers. The "click" strategy has also been applied for anchoring CuPt nanorods on GO. [181] The surface of the CuPt nanorods was modified by ligand exchange with a molecule bearing an azide function for subsequent coupling to alkyne-derivatized GO by "click" chemistry. The CuPt nanorod-grafted GO was used as catalyst of a peroxidase-like *o*-phenylenediamine oxidation reaction in water.

II.5.b.) [2 + 1] Cycloaddition of Nitrenes

Nitrenes are highly reactive intermediates that can be generated *in situ* from azide derivatives upon photochemical or thermal activation associated with release of molecular nitrogen. They add to the π-electron system of graphene *via* [2 + 1] cycloaddition to form aziridine rings (Scheme 7A).

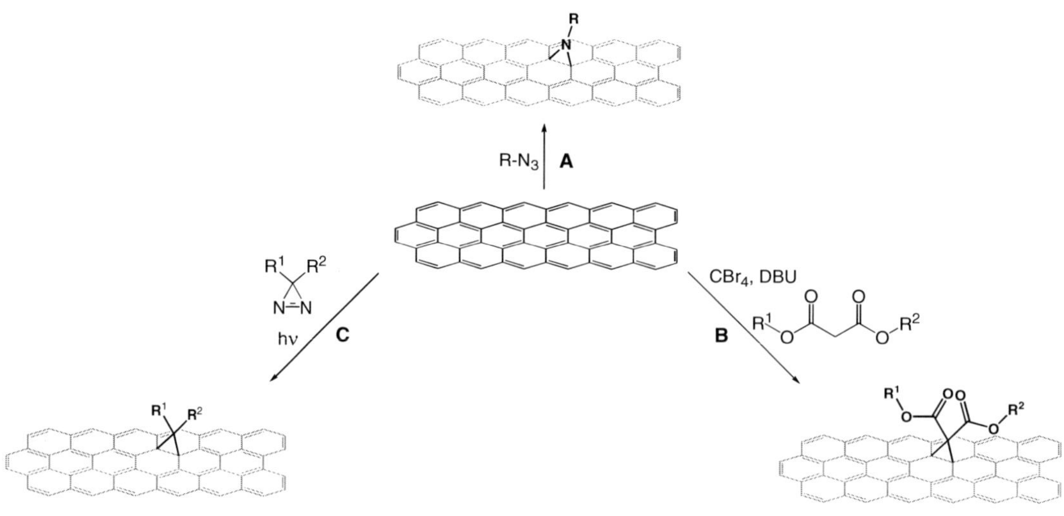

Scheme 7. Functionalization of graphene by [2 + 1] cycloaddition of nitrenes (A) and cyclopropanation using the Bingel reaction (B) or by diazirine irradiation (C).

Functionalized graphene nanosheets have been prepared by derivatization of GO using a variety of azide precursors, which allows to prepare individually dispersed graphene nanosheets with various functional groups such as hydroxyl, carboxylic acid, amine, long alkyl chain, and polymers like PEG and polystyrene. [182] Further derivatization of the functionalized graphene nanosheets is possible by ring-opening polymerization, amidation, ATRP, and growth of metal nanoparticles. The enhanced electrical conductivity of the functionalized graphene nanosheets by comparison with the starting GO is indicative of the reduction of GO in the functionalization process. This result indicates that the conjugated π-network of graphene has not been significantly altered by derivatization with nitrene, probably due to low level of functionalization, resulting in small conversion of sp^2 carbon atoms to sp^3. Few-layer graphene obtained from surfactant-aided exfoliation of graphite have been functionalized with various alkylazides (*e.g.* hexyl, dodecyl, undecanol, and undecanoic acid). [183] The carboxylic acid groups introduced on the graphene surface (2 - 3 layers) are used as markers for the reactive sites *via* binding of gold nanoparticles. Microscopy techniques show uniform distribution of the nanoparticles all over the graphene surface. The nitrene addition of azido-phenylalanine has been exploited to exfoliate micro-crystalline graphite to highly functionalized few-layer graphene ($n > 5$). [184] This strategy has the main advantage of avoiding the reduction step as it is the case with GO. Exfoliation of graphene can also be obtained by nitrene functionalization of rGO with tetraphenylethylene. [185] The π-electron rich property and high steric hindrance of tetraphenylethylene due to its twisted conformation favors disaggregation of the graphene layers. Functionalization by nitrene cycloaddition can tune the electrical properties of graphene. Indeed, by varying the amount of azidotrimethylsilane added in the reaction, the band gap of chemically modified epitaxial graphene could be controlled. [186] Cycloaddition of nitrenes can also be useful for electronic device fabrication. Indeed, a simple and efficient method to immobilize graphene on silicon substrates has been developed using perfluorophenylazide-functionalized wafers. By thermal treatment or UV irradiation, graphene sheets were covalently attached on the wafers *via* nitrene formation. [187, 188] The same strategy involving perfluorophenylazide derivatives has been applied in solution to functionalize few-layer graphene ($n = 4 - 5$). [189]

II.5.c.) Cyclopropanation

Cyclopropanation of few-layer graphene has been achieved by [2+1] cycloaddition using the Bingel reaction conditions under microwave irradiation. [190] The reaction takes place in the presence of a bromomalonate derivative formed *in situ* in the presence of a base and carbon tetrabromide (Scheme 7B). From a mechanistic point of view, a carbanion or enolate is generated by a base abstracting the acidic malonate proton, which reacts with the double bonds of graphene. The resulting carbanion displaces bromine atom leading to an intramolecular cyclopropane ring closure. Cyclopropanated graphene bearing the electroactive extended tetrathiafulvalene has been prepared by using a malonate derivative bearing the tetrathiafulvalene moiety. The redox potentials of the tetrathiafulvalene-graphene hybrid material have been determined by electrochemistry, suggesting the formation of a radical ion pair including one-electron reduction of graphene and one-electron oxidation of tetrathiafulvalene.

An alternative method of cyclopropanation of graphene relies on UV irradiation of diazirine rings and leads to formation of highly reactive carbenes with concomitant loss of nitrogen gas (Scheme 7C). This strategy has been applied for decoration of rGO with gold nanoparticles. [191]

II.5.d.) [4 + 2] Diels-Alder reaction

The group of Haddon has investigated the reactivity of different forms of graphene (*e.g.* exfoliated graphene and epitaxial graphene) towards the Diels-Alder reaction. [192, 193] They found that graphene can act as both diene and dienophile under appropriate conditions. The dual reactivity of graphene is attributed to the absence of an energy gap as the conduction and valence bands of graphene touch at the Fermi level (*i.e.* Dirac point). Graphene reacts at ambient temperature with tetracyanoethylene as dienophile, whereas the reaction with maleic anhydride acting as dienophile requires higher temperature (120 °C) (Scheme 8A).

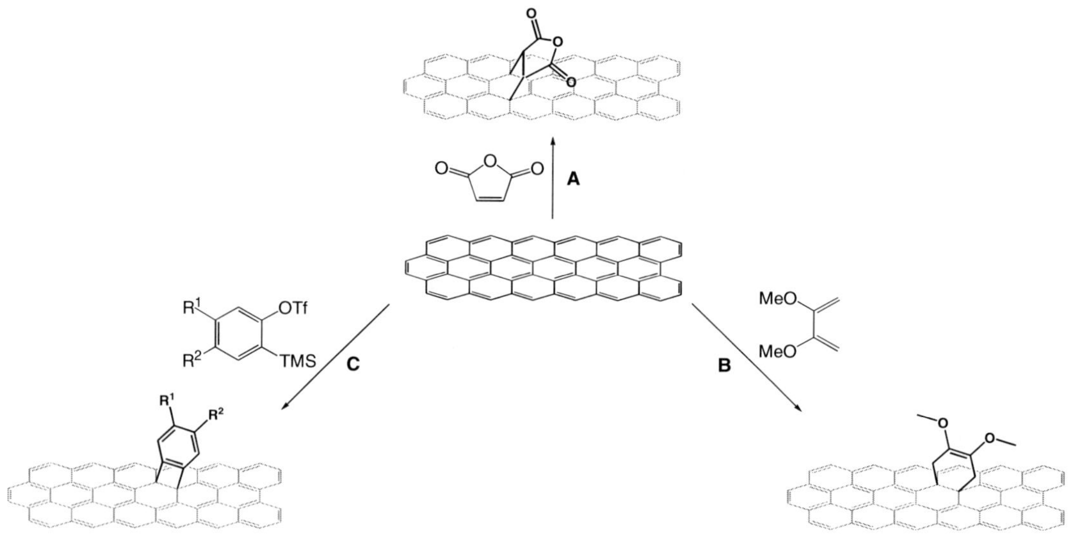

Scheme 8. Functionalization of graphene by Diels-Alder reaction with maleic anhydride (A) and with 2,3-dimethoxy-1,3-butadiene (B), and by [2 + 2] cycloaddition of arynes (C).

The formation of graphene adducts with electron-rich dienes 2,3-dimethoxy-1,3-butadiene (Scheme 8B) and 9-methylanthracene occurs at 130 °C and 50 °C, respectively. [192] The reversibility of the Diels-Alder reactions allows restoring the electronic properties of graphene, which is particularly interesting for reversible engineering of the electrical conductivity of graphene for electronic and optical applications.

In another study, rGO has been functionalized with cyclopentadienyl-capped PEG monomethyl ether at room temperature. [194] The resulting functionalized rGO showed improved dispersibility in various solvents. This strategy could be extended to the grafting of various polymers on graphene as the cyclopentadiene functionality could be readily introduced on other polymers, thus widening the applications of graphene for the preparation of composite materials.

II.5.e.) [2 + 2] Cycloaddition of Arynes

[2 + 2] Cycloaddition of arynes on graphene sheets prepared by arc discharge method has been investigated in only one study. [195] The reaction performed under mild reaction conditions results in the formation of a four-membered ring on the graphene surface. 2-Triflatophenyl silane has been used as benzyne precursor, which is obtained by removal of the two *ortho* substituents in benzene derivatives and is highly reactive towards the double bonds of graphene (Scheme 8C).

II.6. Electrophilic Substitution

II.6.a.) Friedel-Crafts Acylation

Friedel-Crafts acylation, which is an electrophilic aromatic substitution reaction, has been applied to obtain functionalized graphene. The treatment of 4-aminobenzoic acid in a mixture of polyphosphoric acid (PPA) and phosphorus pentoxide generates *in situ* carbonium ions (–C^{+}=O). These highly reactive species are able to add to defect sites located mainly at the edges of graphite, leading to high dispersibility in polar solvents and exfoliation of graphite to graphene sheets (n < 5) (Scheme 9A). [196] The reaction with 4-aminobenzoic acid has also been employed to rGO [197] and to GO where reduction occurred slowly during the functionalization process. [198]

Friedel-Crafts acylation is also possible with acid anhydrides, such as maleic anhydride, in the presence of aluminum chloride as Lewis acid. [199] In this study, exfoliated graphene sheets bearing short alkyl chains terminated by carboxylic acid functions have been prepared from graphite by Friedel-Crafts acylation with maleic anhydride. They were further used as biosensor for hydrogen peroxide and glucose detection.

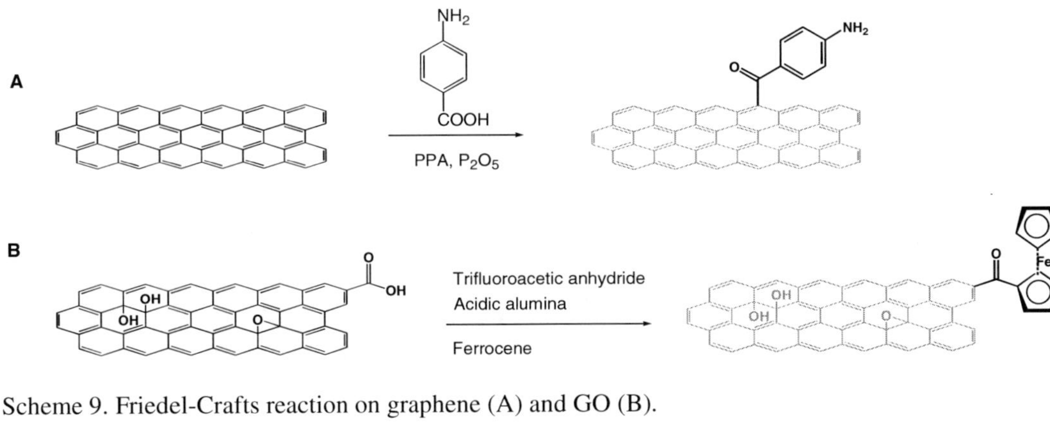

Scheme 9. Friedel-Crafts reaction on graphene (A) and GO (B).

Alternatively, carboxylic acid functions present on GO have been used as acylating agent for monoacylation of ferrocene by Friedel-Crafts reaction. [200] A green chemistry solvent free approach using solid phase acidic alumina, as Lewis acid, and trifluoroacetic anhydride at room temperature has been reported to covalently link ferrocene to GO and induce exfoliation to obtain single-layer GO sheets (Scheme 9B).

II.6.b.) Electrophilic Alkylation

Reduced GO can react sequentially with *n*-BuLi and bromotriethylamine by deprotonation/carbometalation and subsequent electrophilic attack to yield graphene covalently modified with triethylamine. [201] The Et_3N-graphene was used as solid basic catalyst for hydrolysis of ethyl acetate with a high catalytic activity.

Another strategy for electrophilic alkylation of graphene has been developed by Hirsch and coworkers. [202] Birch-type reduction by treatment of graphite with sodium/potassium alloy leads to reduction of graphite and exfoliation of the layers due to electrostatic interactions, yielding negatively charged single layers of graphene. The negative charges are trapped by hexyliodide acting as electrophile, which generates alkylated graphene by rehybridization of some sp^2 carbon atoms to sp^3 (Scheme 10). A similar strategy has been used by the group of Hirsch for functionalization of Na/K-reduced graphite intercalation compounds with diazonium salts. [132]

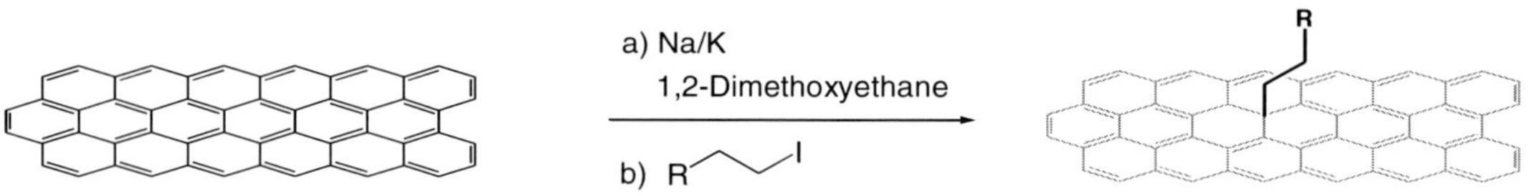

Scheme 10. Functionalization of graphene by electrophilic alkylation.

III. NON-COVALENT FUNCTIONALIZATION OF GRAPHENE

Non-covalent interactions mainly involve hydrophobic, van der Waals, and electrostatic forces. Non-covalent functionalization of graphene can be achieved by physical adsorption of aromatic molecules, surfactants or polymers. Graphene exhibits a high specific surface area due to its large two-dimensional plane and one-atom thickness, thus favoring physisorption of molecules. Non-covalent interactions do not convert carbon atoms from sp^2 to sp^3 hybridization, contrary to covalent functionalization. Therefore, the electronic π-network of graphene is not damaged. The electronic properties are not altered and functionalized graphene remains highly conductive. GO can also exhibit non-covalent binding *via* π-stacking on the areas of the π-network that are not oxidized or engaged in hydrogen bonding.

In section III-1 interactions of graphene with a variety of molecules by π-π interactions are detailed, in particular the adsorption of pyrene and perylene derivatives, and the impact on the modification of the electronic properties of graphene. Immobilization of aromatic drugs and interactions with DNA are also reported. In the different studies reported to illustrate π-π interactions of aromatic molecules with graphene, it should be noted that other interactions (*e.g.* hydrophobic, electrostatic) may contribute to the adsorption of these molecules on graphene. Finally, sections III-2 and III-3 focus on adsorption of surfactants on the graphene surface and coating with polymers, respectively.

III.1. π-Stacking Interactions

Acceptor and donor molecules are able to adsorb on graphene sheets by π-stacking, which opens opportunities to increase dispersibility and tune the electronic properties of graphene. A number of small π-orbital rich aromatic molecules such as pyrene, perylene, and pentacene have been used as exfoliating agent to obtain few-layer graphene or single sheets of graphene. The synergistic non-covalent π-stacking and charge-transfer interactions between the large planar surface of polycyclic aromatic hydrocarbons (PAHs) and the sp^2 hybridized sheets of graphene allow efficient separation of the graphene layers without disrupting the electronic conjugation of graphene. The aromatic compounds act as "molecular wedge" that intercalate between the graphene sheets. The presence of hydrophilic charged functional groups helps exfoliating graphene sheets in water by preventing both inter and intra π-stacking, leading to stable aqueous dispersion of graphene.

A series of aromatic compounds have been used to prepare stable aqueous solutions of single- and/or few-layer graphene, such as sodium dodecyl benzene sulfonate, [203, 204] a binol salt (containing two naphthalene moieties), [205] an anionic coronene derivative (*i.e.* the simplest six-fold symmetrical aromatic molecule after benzene), [206] pyridinium-based ionic liquids, [207] the electron acceptor 7,7,8,8-tetracyanoquinodimethane (TCNQ), [208] porphyrin derivatives, [209] or a Zn phtalocyanine-based poly(*p*-phenylene vinylene) oligomer. [210]

III.1.a) Adsorption of Pyrene Derivatives

Among PAHs, pyrene derivatives have been reported by various groups to stabilize graphene dispersions. For example, 1-pyrenecarboxylic acid (PCA) can exfoliate efficiently single-, few-, and multilayered graphene flakes on a large-scale from graphite powder. [211] The hydrophilic carboxylic group facilitates the formation of stable aqueous dispersions of graphene. This process is more direct and less damaging for graphene compared to the traditional approach based on oxidation and reduction of GO.

Water dispersible single- and bi-layered graphene sheets have been obtained by exfoliation of graphite flakes in the presence of pyrene-labeled single-stranded DNA (ss-DNA) prepared by reaction of NHS ester of pyrenebutanoic acid with amino-modified DNA. [212] The immobilized DNA could hybridize complementary DNA labeled with gold nanoparticles to produce gold nanoparticle-graphene nanocomposites.

Pyrene derivatives possessing a polar carbohydrate headgroup have been shown to self-assemble with GO and form gels in aqueous and organic solvents. [213] The driving forces for the gelation process are based on π-stacking and hydrogen bonding interactions.

Pyrene methylamine and sodium pyrenetetrasulfonate have also been used to exfoliate graphite powder to single-layer graphene sheets. [214] This method presents additional advantages by comparison with exfoliation of graphene with surfactants or polymers. Indeed, the pyrene derivatives act as healing agent during treatment by thermal annealing to repair some of the possible defects in the graphene sheets. In addition, they can also electrically "glue" adjacent graphene sheets together, thus improving the electric contacts between adjacent graphene sheets.

A more advanced structure based on four units of pyrene has been synthesized to efficiently exfoliate graphite into single- and double-layer graphene sheets in aqueous solution. [215] The aromatic molecule bears a hydrophilic oligoether dendron, facilitating dispersion in water, and may adopt a planar conformation on graphene sheets to maximize π-π interactions with the basal plane of graphene, while the tethered dendron increases the interlayer distance (Figure 3).

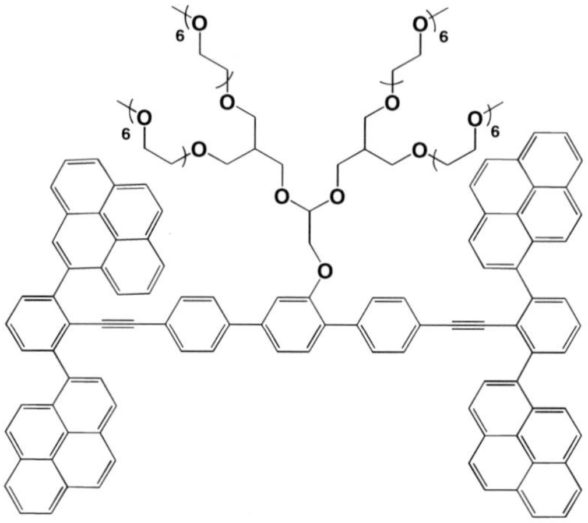

Figure 3. Chemical structure of an amphiphile composed of four pyrene units and a hydrophilic oligoether dendron. [215]

The surface diffusion of a tripod motif bearing three pyrene moieties and a redox-active Co(II) bis-terpyridyl complex adsorbed on a single-layer of graphene has been studied by scanning electrochemical microscopy. [216] The thermodynamic and kinetic binding parameters have been investigated electrochemically. [217] Desorption of the tripod from the graphene surface was 1000 times more slowly than model compounds bearing only one pyrene.

Green and coworkers have studied different parameters influencing the dispersion of expanded graphite with pyrene derivatives, in particular concentration, pH, the type of functional groups and their distance from the basal plane. [218]

The fluorescence properties of positively charged pyrene have been exploited to develop an optical graphene-based sensor for DNA. [219] Initially, the fluorescence of pyrene is quenched because of proximity with GO. By electrostatic complexation with double-stranded DNA (ds-DNA), the DNA-pyrene conjugate is released from the GO surface, turning on the quenched fluorescence of pyrene. A similar approach has been applied to provide an ideal platform for detection of cancer cell surface markers by adsorption of pyrene derivatized with a targeting ligand of cancer cells onto the surface of GO. [220] Upon competitive binding to the receptor, the pyrene derivative is desorbed from the graphene surface, resulting in the recovery of its fluorescence. The strategy of quenching and recovery of the fluorescence emission of dyes coating graphene has also been applied to visualize graphene sheets. [221]

Pyrenebutanoic acid derivatized with *N*-hydroxysuccinimide has been used to non-covalently anchor fluorescently labeled proteins on few-layer epitaxial graphene. The strategy relies on adsorption of pyrene on the graphene surface *via* π-π interactions and covalent specific binding of amines present in the protein structure to the reactive succinimide ester groups. Different proteins have been immobilized using this method, such as glucose oxidase and laminine for applications as glucose sensors and for tissue engineering, respectively. [222]

III.1.b.) Adsorption of Perylene Derivatives

Perylene derivatives are also excellent candidates for dispersion of graphene materials. Graphene decorated with a donor and an acceptor electron molecule has been prepared by adsorption of a mixture of an amphiphilic perylene and a pyrene derivative, yielding monolayer and double-layer rGO sheets. [223]

Hirsch and coworkers have demonstrated the potential of perylene bisimide derivatives for exfoliation and solubilization of graphene, driven by π-π and hydrophobic interactions. [133, 224] The ultrasonic treatment of graphite in an aqueous solution of dendronized perylene bisimide compounds leads to exfoliated sheets of few-layer and single-layer graphene (Figure 4). [225, 226]

The self-assembly of perylene-3,4,9,10-tetracarboxylic dianhydride (PTCDA) on epitaxial graphene grown on a silicon carbide surface has been studied by high vacuum molecular-resolution scanning tunneling microscopy (STM) at room temperature. [227] The adsorption of perylene and coronene tetracarboxylic diimide derivatives has also been studied by STM. [228] PTCDA forms stable, well-ordered islands and monolayers with a herringbone arrangement stabilized by hydrogen bonding and quadrupolar interactions between adjacent PTCDA, and π-stacking between the molecules and the graphene surface. The herringbone pattern closely resembles the (102) plane of the PTCDA bulk crystal structure.

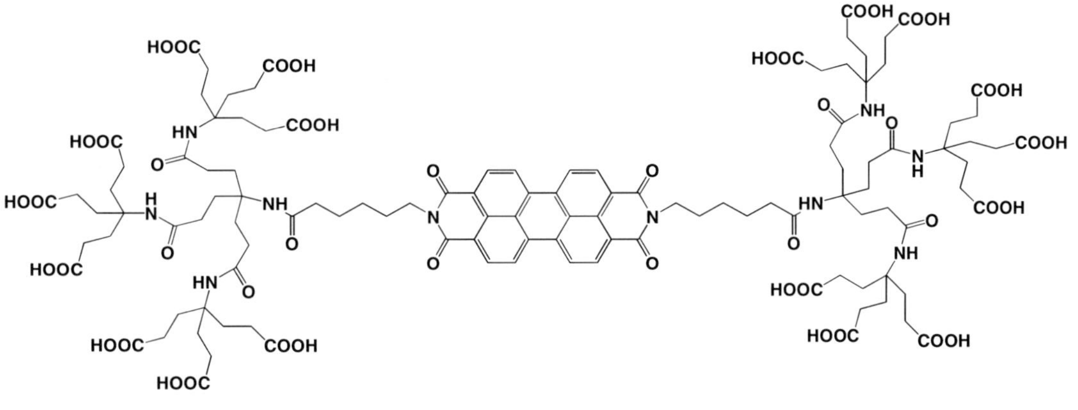

Figure 4. Chemical structure of a perylene-based bolaamphiphile (*i.e.* an amphiphilic molecule with hydrophilic head groups at both ends and a hydrophobic anchor group). [225]

III.1.c.) Modification of the Electronic Properties of Graphene

In addition to exfoliation, aromatic compounds have been used to improve the electrical conductivity of graphene thanks to charge transfer between the adsorbed aromatic rings and the graphene sheets. [229] The conductivity of a film made from an aqueous dispersion of graphene sheets and 1-pyrenebutyrate is 7 orders of magnitude higher than that of the GO precursor. [230]

In another study, electrical measurements have been performed on field-effect transistors (FET) of single-layer graphene with adsorbed aromatic molecules (*i.e.* 1,5-naphthalenediamine, 9,10-dimethylanthracene, 9,10-dibromoanthracene, and tetrasodium 1,3,6,8-pyrenetetrasulfonic acid). [231] The results indicate that the aromatic molecules with electron-donating groups induce n-doping, whereas electron-withdrawing groups exhibit p-doping on the FET. This observation has been confirmed in a separate study with different pyrene derivatives including 1-aminopyrene, 1-nitropyrene, 1-pyrenecarboxylic acid, and 1-pyrenesulfonic acid. [232]

A pyrene derivative bearing the photochromic spiropyran moiety has been adsorbed on graphene, causing n-doping on the resulting FET with interesting optoelectronic properties. [233] Indeed, the Dirac point of graphene can be controlled by light as spiropyran can be reversibly switched between two different conformations with a neutral form and a charge-separated form, generating a dipole field, upon UV and visible light irradiation, respectively.

III.1.d.) Adsorption of Drugs

The surface of GO can serve as platform for high density loading of aromatic and hydrophobic drugs like doxorubicin [82, 234]and camptothethin. [81] Aromatic drugs are insoluble in water, which hampers their use for an efficient treatment of a disease and generally leads to undesired effects. The resulting drug/GO conjugate exhibits better water solubility, while maintaining the therapeutic efficacy of the drug.

A photosensitizer molecule, chlorin e6 (Ce6), able to generate cytotoxic singlet oxygen under light excitation for photodynamic therapy (PDT), has been loaded on the surface of PEGylated GO. [235] The water solubility and cellular uptake of Ce6 is increased thanks to GO acting as nanocarrier. In addition, the PDT efficacy against cancer cells is enhanced due to photothermal effect of GO inducing mild local heating when exposed to a near-infrared laser at a low power density.

III.1.e.) Adsorption of DNA

Adsorption of ss-DNA onto the surface of rGO allows to prepare stable aqueous suspensions of individual graphene sheets. [236, 237] The driving force for non-covalent interactions between DNA and graphene relies on π-stacking involving both purine and pyrimidine bases of DNA. It was reported that ss-DNA adsorbed on graphene is protected from enzymatic cleavage by DNase I, which is interesting for biomedical applications. [238]

The interactions between ds-DNA and GO, implying π-stacking and hydrogen bonding between oxygen-containing groups of GO and DNA bases, have been thoroughly investigated. [239] Partial deformation of the double helix onto the GO surface also facilitates the adsorption of ds-DNA.

A simple 3D self-assembly method based on adsorption of ds-DNA on GO sheets has been developed to prepare DNA/GO composite hydrogels with high mechanical strength and dye loading capacity. [240] The heating process induces unwinding of the ds-DNA to ss-DNA which adsorbs strongly on the GO sheets.

III.2. Interactions with Surfactants

Non-covalent interactions of graphene with surfactants allow to exfoliate graphite to monolayer of few-layer graphene, generating stable aqueous dispersions of graphene whose basal planes are essentially free of structural defects. Alternative strategies rely on the use of surfactants as solubilizing agents during reduction of chemically exfoliated graphite oxide.

A few studies involving aromatic surfactants have already been reported in section III-1. Many surfactants have been investigated to disperse graphene in aqueous solution, such as cetyltrimethylammonium bromide (CTAB), [241] bile salt sodium cholate (SC) [242, 243] (Figure 5A) with the help of density gradient ultracentrifugation, [244] Pluronic triblock copolymers (Figure 5B), [245] and SDBS. [14, 203, 204]

Figure 5. Chemical structure of sodium cholate (A) and Pluronic triblock copolymer (B).

Coleman and coworkers have explored a series of ionic surfactants, such as sodium or lithium dodecyl sulfate, SDBS, CTAB, tetradecyltrimethylammonium bromide, SC, sodium deoxycholate, and sodium taurodeoxycholate, as well as non-ionic surfactants like IGEPAL CO-890 (*i.e.* polyoxyethylene (40) nonylphenyl ether), Triton X-100 (Figure 6A), Tween 20 (Figure 6B), and Tween 80. [246] In the case of ionic surfactants, the concentration of dispersed graphene is mainly controlled by the zeta potential of the graphene sheets coated with surfactant. Indeed, the concentration is proportional to the magnitude of the repulsive electrostatic potential barrier that stabilizes surfactant-decorated graphene sheets and prevents aggregation. For non-ionic surfactants, the concentration scales linearly with the magnitude of the steric repulsive potential barrier stabilizing the graphene sheets.

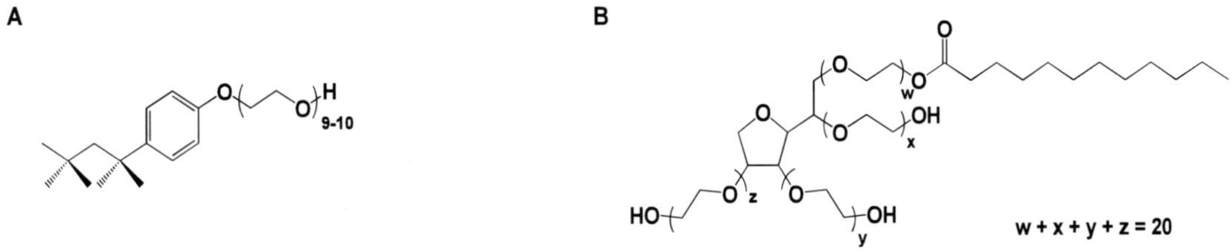

Figure 6. Chemical structure of Triton X-100 (A) and Tween 20 (B).

Another thorough study has been conducted with a wide range of non-ionic and ionic surfactants. [247] The non-ionic surfactants perform better than ionic surfactants; concentrations as high as ~ 1 $mg{\cdot}mL^{-1}$ of graphene in water can be reached.

Strano and coworkers have investigated the effects of anionic, cationic, and neutral surfactants (*i.e.* SC, SDS, CTAB, dodecyltrimethyl ammonium bromide, Triton X-100, and Tween 20) dispersing graphene in water on the transport properties of graphene transistors. [248] The authors related the change in transport characteristics to chemical properties of the surfactants.

III.3. Coating with Polymers

Besides adsorption of aromatic compounds *via* π-π interactions and coating with surfactants to improve dispersibility, graphene can be functionalized non-covalently with polymers, which allows combining the properties of both materials. [28, 249] Polymers can increase dispersibility of graphene in some solvents, improve mechanical strengthening in polymer/graphene composites, and/or add novel functionalities. In addition to reinforcement of the mechanical properties, graphene can impart electrical conductivity to the polymer matrix.

Among a variety of polymers, poly(*m*-phenylenevinylene-co-2,5-dioctoxy-*p*-phenylenevinylene) (PmPV) has been reported to readily exfoliate commercial expendable graphite to single-layer and few-layer graphene nanoribbons (Figure 7). [250]

Reduced GO sheets have been stably dispersed in water by functionalization with sulfonated polyaniline (SPANI), a conducting polymer, *via* π-π interactions. The electroactive composite film of SPANI-dispersed graphene exhibits both improved electrochemical stability and electrocatalytic activity. [251]

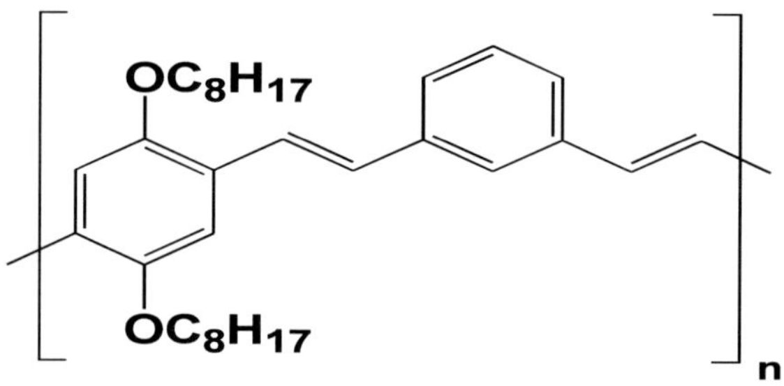

Figure 7. Chemical structure of PmPV.

The use of ionic liquid polymers (PIL) offers the possibility to reversibly transfer rGO sheets between aqueous and organic phases. [252] Indeed, upon anion exchange of the PIL, the surface properties of graphene can be switched from hydrophilic to hydrophobic. An alternative approach to transfer graphene from aqueous to organic phase relies on the carboxylate groups remaining after chemical reduction of GO. These functions can be exploited for complexation with protonated terminal amine moieties of polystyrene *via* electrostatic interactions, which facilitate phase transfer of the rGO sheets from aqueous to organic phase [253].

Nafion, a sulfonated tetrafluoroethylene based fluoropolymer-copolymer, which is constituted of a hydrophobic backbone and hydrophilic side chains, allows preparing stable dispersion of rGO for the preparation of conductive films for applications as electrochemical biosensors. [254] In this study, GO sheets were non-covalently functionalized by Nafion *via* hydrophobic interactions between the hydrophobic perfluoro backbone of Nafion and the hydrophobic part of the basal plane of GO, as well as hydrophilic interactions between the sulfonate groups of Nafion and the polar groups of GO (*i.e.* hydroxyl, epoxy, and carboxylic acid moieties).

A coil-rod-coil conjugated triblock copolymer, PEG/oligo(phenylene ethynylene), has been used to prepare amphiphilic graphene sheets having the capacity to be dissolved in both low polar organic solvents and highly polar water-miscible solvents. [255] The polymer is composed of a π-conjugated oligomer, interacting with the basal plane of rGO, as well as lipophilic side chains and two hydrophilic PEG coils, forming an amphiphilic outerlayer facilitating dispersion of rGO in both organic and water-miscible solvents.

GO has been simultaneously reduced and surface functionalized in one-step by poly(norepinephrine) (pNor), which is a catecholamine polymer. [256] The pNor coating affords multifunctionality as the functionalized rGO was found to be a versatile platform for surface-initiated polymerization or formation of metallic NPs.

Dai and coworkers have coated the surface of rGO with an amphiphilic PEGylated polymer to increase stability and biocompatibility of rGO in physiological media. [257] This biocompatible coating affords the possibility to further explore graphene properties for biomedical applications, [258] such as photothermal therapy.

CONCLUSION

Although the development of graphene chemistry has just begun to emerge, a series of functionalization strategies have been already reported. Such a rapid progress in the fundamental understanding of the chemistry of graphene has been possible due to knowledge accumulated from functionalization techniques elaborated for graphite, fullerenes, and carbon nanotubes. Functionalization of graphene allows increasing processability and imparting novel functionalities by tuning their properties or by combination with other molecules/materials.

Functionalization of graphene can be divided into two categories based on covalent addition and non-covalent interactions. Most methods have been inspired by fullerene and CNT chemistry. Derivatization of graphene by formation of covalent bonding usually involves highly reactive species (*e.g.* radicals, dipoles) and allows preparing stable conjugates. Nevertheless, depending on the level of functionalization, the electrical conductivity of the modified graphene can be drastically decreased, which reduces potential applications in nanoelectronics for instance. Formation of covalent bonds is associated with rehybridization of some sp^2 carbon atoms to sp^3, which disrupts the π-electronic conjugation of graphene. But, this limitation can be overcome by thermal treatment to remove covalently linked functional groups from the graphene surface, thus restoring the intrinsic properties of pristine graphene.

GO has been extensively used as graphene precursor due to its facile synthesis, increased dispersibility, and the possibility to further derivatize the oxygenated moieties. However, this approach is limited for some applications, like nanoelectronics, as some oxygen-containing species and defects still remain after reduction of GO, resulting in alteration of the electrical conductivity of graphene. Thermal annealing of rGO sheets can substantially restore the π-conjugated network of graphene, but complete restoration of the sp^2 carbon structure of graphene is still challenging. Efficient and non-destructive methods for the synthesis of graphene have to be developed and optimized.

Non-covalent functionalization of graphene *via* π-stacking, hydrophobic, and/or electrostatic interactions offers a softer alternative for exfoliation of graphene sheets and modification of their properties, as it does not induce the formation of defects. Aromatic compounds have a strong affinity for the basal plane of graphene due to π-π interactions.

Coating graphene with surfactants allows exfoliating graphite and preparing stable aqueous suspensions of graphene. Interactions of graphene with polymers broaden the scope of applications of graphene, in particular for the preparation of nanocomposite materials.

REFERENCES

[1] Kroto, H. W.; Heath, J. R.; O'Brien, S. C.; Curl, R. F.; Smalley, R. E. *Nature* 1985, *318*, 162.
[2] Iijima S. *Nature* 1991, *354*, 56.
[3] Monthioux, M.; Kuznetsov, V. L. *Carbon* 2006, *44*, 1621.
[4] Novoselov, K. S.; Geim, A. K.; Morozov, S. V.; Jiang, D.; Zhang, Y.; Dubonos, S. V.; Grigorieva, I. V.; Firsov, A. A. *Science* 2004, *306*, 666.
[5] Sanchez, V. C.; Jachak, A.; Hurt, R. H.; Kane, A. B. *Chem. Res. Toxicol.* 2012, *25*, 15.
[6] Rao, C. N.; Matte, H. S.; Subrahmanyam, K. S. *Acc. Chem. Res.* 2013, *46*, 149.
[7] Avouris, P.; Dimitrakopoulos, C. *Mater. Today* 2012, *15*, 86.
[8] Zhu, Y.; Murali, S.; Cai, W.; Li, X.; Suk, J. W.; Potts, J. R.; Ruoff, R. S. *Adv. Mater.* 2010, *22*, 3906.
[9] Wei, D.; Liu, Y. *Adv. Mater.* 2010, *22*, 3225.
[10] Allen, M. J.; Tung, V. C.; Kaner, R. B. *Chem Rev.* 2010, *110*, 132.
[11] Compton, O. C.; Nguyen, S. T. *Small* 2010, *6*, 711.
[12] Cui, X.; Zhang, C.; Hao, R.; Hou, Y. *Nanoscale* 2011, *3*, 2118.
[13] Hernandez, Y.; Nicolosi, V.; Lotya, M.; Blighe, F. M.; Sun, Z.; De, S.; McGovern, I. T.; Holland, B.; Byrne, M.; Gun'Ko, Y. K.; Boland, J. J.; Niraj, P.; Dueberg, G.; Krishnamurthy, S.; Goodhue, R.; Hutchison, J.; Scardaci, V.; Ferrari, A. C.; Coleman, J. N. *Nature Nanotech.* 2008, *3*, 563.
[14] Lotya, M.; Hernandez, Y.; King, P. J.; Smith, R. J.; Nicolosi, V., Karlsson, L. S.; Blighe, F. M.; De, S.; Wang, Z.; McGovern, I. T.; Duesberg, G. S.; Coleman, J. N. *J. Am. Chem. Soc.* 2009, *131*, 3611.
[15] Berger, C.; Song, Z.; Li, X.; Wu, X.; Brown, N.; Naud, C.; Mayou, D.; Li, T.; Hass, J.; Marchenkov, A. N.; Conrad, E. H.; First, P. N.; de Heer, W. A. *Science* 2006, *312*, 1191.
[16] Emtsev, K. V.; Bostwick, A.; Horn, K.; Jobst, J.; Kellogg, G. L.; Ley, L.; McChesney, J. L.; Ohta, T.; Reshanov, S. A.; Röhrl, J.; Rotenberg, E.; Schmid, A. K.; Waldmann, D.; Weber, H. B.; Seyller, T. *Nat. Mater.* 2009, *8*, 203.
[17] Jiao, L.; Zhang, L.; Wang, X.; Diankov, G.; Dai, H. *Nature* 2009, *458*, 877.
[18] Kosynkin, D. V.; Higginbotham, A. L.; Sinitskii, A.; Lomeda, J. R.; Dimiev, A.; Price, B. K.; Tour, J. M. *Nature* 2009, *458*, 872.
[19] Kim, K. S.; Zhao, Y.; Jang, H.; Lee, S. Y.; Kim, J. M.; Kim, K. S.; Ahn, J. H.; Kim, P.; Choi, J. Y.; Hong, B. H. *Nature* 2009 *457*, 706.
[20] Li, X.; Cai, W.; An, J.; Kim, S.; Nah, J.; Yang, D.; Piner, R.; Velamakanni, A.; Jung, I.; Tutuc, E.; Banerjee, S. K.; Colombo, L.; Ruoff, R. S. *Science* 2009, *324*, 1312.
[21] Park, S.; Ruoff, R. S. *Nat. Nanotechnol.* 2009, *4*, 217.
[22] Huang, X.; Yin, Z.; Wu, S.; Qi, X.; He, Q.; Zhang, Q.; Yan, Q.; Boey, F.; Zhang, H. *Small* 2011, *7*, 1876.

[23] Wei, W.; Qu, X. *Small* 2012, *8*, 2138.
[24] Rao, C. N.; Sood, A. K.; Subrahmanyam, K. S.; Govindaraj, A. *Angew. Chem. Int. Ed. Engl.* 2009, *48*, 7752.
[25] Singh, V.; Joung, D.; Zhai, L.; Das, S.; Khondaker, S. I.; Seal, S. *Prog. Mater. Sci.* 2011, *56*, 1178.
[26] Schwierz F. *Nat. Nanotechnol.* 2010, *5*, 487.
[27] Pumera, M. *Chem. Soc. Rev.* 2010, *39*, 4146.
[28] Huang, X.; Qi, X.; Boey, F.; Zhang, H. *Chem. Soc. Rev.* 2012, *41*, 666.
[29] Dai L. *Acc. Chem. Res.* 2013, *46*, 31.
[30] Georgakilas, V.; Otyepka, M.; Bourlinos, A. B.; Chandra, V.; Kim, N.; Kemp, K. C.; Hobza, P.; Zboril, R.; Kim, K. S. *Chem. Rev.* 2012, *112*, 6156.
[31] Yan, L.; Zheng, Y. B.; Zhao, F.; Li, S.; Gao, X.; Xu, B.; Weiss, P. S.; Zhao, Y. *Chem. Soc. Rev.* 2012, *41*, 97.
[32] Kuila, T.; Bose, S.; Mishra, A. K.; Khanra, P.; Kim, N. H.; Lee, J. H. *Prog. Mater. Sci.* 2012, *57*, 1061.
[33] Liu, J.; Tang, J.; Gooding, J. J. *J. Mater. Chem.* 2012, *22*, 12435.
[34] Loh, K. P.; Bao, Q.; Ang, P. K.; Yang, J. *J. Mater. Chem.* 2010, *20*, 2277.
[35] Jiang, D. E.; Sumpter, B. G.; Dai, S. *J. Chem. Phys.* 2007, *126*, 134701.
[36] Banhart, F.; Kotakoski, J.; Krasheninnikov, A. V. *ACS Nano* 2011, *5*, 26.
[37] Stadler, J.; Schmid, T.; Zenobi, R. *ACS Nano* 2011, *5*, 8442.
[38] Park, J.; Yan, M. *Acc. Chem. Res.* 2013, *46*, 181.
[39] Radovic, L. R.; Bockrath, B. *J. Am. Chem. Soc.* 2005, *127*, 5917.
[40] Bekyarova, E.; Sarkar, S.; Wang, F.; Itkis, M. E.; Kalinina, I.; Tian, X.; Haddon, R. C. *Acc. Chem. Res.* 2013, *46*, 65.
[41] Sreeprasad, T. S.; Berry, V. *Small* 2013, *9*, 341.
[42] Niyogi, S.; Bekyarova, E.; Hong, J.; Khizroev, S.; Berger, C.; de Heer, W.; Haddon, R. C. *J. Phys. Chem. Lett.* 2011, *2*, 2487.
[43] Sarkar, S.; Bekyarova, E.; Haddon, R. C. *Mater. Today* 2012, *15*, 276.
[44] Chen, D.; Feng, H.; Li, J. *Chem. Rev.* 2012, *112*, 6027.
[45] Dreyer, D. R.; Park, S.; Bielawski, C. W.; Ruoff, R. S. *Chem. Soc. Rev.* 2010, *39*, 228.
[46] Hummers W. S.; Offeman, R. E. *J. Am. Chem. Soc.* 1958, *80*, 1339.
[47] Cote, L. J.; Cruz-Silva, R.; Huang, J. *J. Am. Chem. Soc.* 2009, *131*, 11027.
[48] Brodie, B. C. *Ann. Chim. Phys.* 1860, *59*, 466.
[49] Staudenmaier, L. *Ber. Dtsch. Chem. Ges.* 1898, *31*, 1481.
[50] Hudson, M. J.; Hunter-Fujita, F. R.; Peckett, J. W.; Smith, P. M. *J. Mater. Chem.* 1997, *7*, 301.
[51] Luo, Z.; Lu, Y.; Somers, L. A.; Johnson, A. T. *J. Am. Chem. Soc.* 2009, *131*, 898.
[52] Lomeda, J. R.; Doyle, C. D.; Kosynkin, D. V.; Hwang, W. F.; Tour, J. M. *J. Am. Chem. Soc.* 2008, *130*, 16201.
[53] Subrahmanyam, K. S.; Vivekchand, S. R. C.; Govindaraj, A.; Rao, C. N. R. *J. Mater. Chem.* 2008, *18*, 1517.
[54] Lerf, A.; He, H.; Forster, M.; Klinowski, J. *J. Phys. Chem. B*, 1998, *102*, 4477.
[55] Eda, G.; Chhowalla, M. *Adv. Mater.* 2010, *22*, 2392.
[56] Li, X.; Zhang, G.; Bai, X.; Sun, X.; Wang, X.; Wang, E.; Dai, H. *Nat. Nanotechnol.* 2008, *3*, 538.
[57] Kim, F.; Cote, L. J.; Huang, J. *Adv. Mater.* 2010, *22*, 1954.

[58] Gao, W.; Alemany, L. B.; Ci, L.; Ajayan, P. M. *Nat. Chem.* 2009, *1*, 403.
[59] Szabó, T.; Berkesi, O.; Forgó, P.; Josepovits, K.; Sanakis, Y.; Petridis, D.; Dékány, I. *Chem. Mater.* 2006, *18*, 2740.
[60] Yuge, R.; Zhang, M.; Tomonari, M.; Yoshitake, T.; Iijima, S.; Yudasaka, M. *ACS Nano* 2008, *23*, 1865.
[61] Stankovich, S.; Piner, R. D.; Nguyen, S. T.; Ruoff, R. S. *Carbon*, 2006, *44*, 3342.
[62] Stankovich, S.; Dikin, D. A.; Dommett, G. H.; Kohlhaas, K. M.; Zimney, E. J.; Stach, E. A.; Piner, R. D.; Nguyen, S. T.; Ruoff, R. S. *Nature* 2006, *442*, 282.
[63] Liu, Q.; Liu, Z.; Zhang, X.; Yang, L.; Zhang, N.; Pan, G.; Yin, S.; Chen, Y.; Wei, J. *Adv. Funct. Mater.* 2009, *19*, 894.
[64] Wang, G.; Wang, B.; Park, J.; Yang, J.; Shen, X.; Yao, J. *Carbon* 2009, *47*, 68.
[65] Zhang, D. D.; Zu, S. Z.; Han, B. H. *Carbon*, 2009, *47*, 2993.
[66] Whitby, R. L. D.; Korobeinyk, A.; Glevatska, K. V. *Carbon*, 2011, *49*, 722.
[67] Salavagione, H. J.; Gómez, M. A.; Martínez, G. *Macromolecules* 2009, *42*, 6331.
[68] Niyogi, S.; Bekyarova, E.; Itkis, M. E.; McWilliams, J. L.; Hamon, M. A.; Haddon, R. C. *J. Am. Chem. Soc.* 2006, *128*, 7720.
[69] Mei, Q.; Zhang, K.; Guan, G.; Liu, B.; Wang, S.; Zhang, Z. *Chem. Commun.* 2010, *46*, 7319.
[70] Zhuang, X. D.; Chen, Y; Liu, G.; Li, P. P.; Zhu, C. X.; Kang, E. T.; Noeh, K. G.; Zhang, B.; Zhu, J. H.; Li, Y. X. *Adv. Mater.* 2010, *22*, 1731.
[71] Wu, C.; Huang, X.; Wang, G.; Wu, X.; Yang, K.; Li, S.; Jiang, P. *J. Mater. Chem.* 2012, *22*, 7010.
[72] Xu, Y.; Liu, Z.; Zhang, X.; Wang, Y.; Tian, J.; Huang, Y.; Ma, Y.; Zhang, X.; Chen, Y. *Adv. Mater.* 2009, *21*, 1275.
[73] Liu, Z. B.; Xu, Y. F.; Zhang, X. Y.; Zhang, X. L.; Chen, Y. S.; Tian, J. G. *J. Phys. Chem. B.* 2009, *113*, 9681.
[74] Liu, Y.; Zhou, J.; Zhang, X.; Liu, Z.; Wan, X.; Tian, J.; Wang, T.; Chen, Y. *Carbon* 2009, *47*, 3113.
[75] Yu, D.; Yang, Y.; Durstock, M.; Baek, J. B; Dai, L. *ACS Nano* 2010, *4*, 5633.
[76] Zhang, X.; Feng, Y.; Huang, D.; Li, Y.; Feng, W. *Carbon* 2010, *48*, 3236.
[77] Jung, J. H.; Cheon, D. S.; Liu, F.; Lee, K. B.; Seo, T. S. *Angew. Chem. Int. Ed. Engl.* 2010, *49*, 5708.
[78] Lee, J. S.; Joung, H. A.; Kim, M. G.; Park, C. B. *ACS Nano* 2012, *6*, 2978.
[79] Bonanni, A.; Ambrosi, A.; Pumera, M. *Chem. Eur. J.* 2012, *18*, 1668.
[80] Bao, H.; Pan, Y.; Ping, Y.; Sahoo, N. G.; Wu, T.; Li, L.; Li, J.; Gan, L. H. *Small* 2011, *7*, 1569.
[81] Liu, Z.; Robinson, J. T.; Sun, X.; Dai, H. *J. Am. Chem. Soc.* 2008, *130*, 10876.
[82] Sun, X.; Liu, Z.; Welsher, K.; Robinson, J. T.; Goodwin, A.; Zaric, S.; Dai, H. *Nano Res.* 2008, *1*, 203.
[83] Liu, Z.; Jiang, L.; Galli, F.; Nederlof, I.; Olsthoorn, R. C. L.; Lamers, G. E. M.; Oosterkamp, T. H.; Abrahams, J. P. *Adv. Funct. Mater.* 2010, *20*, 2857.
[84] Gonçalves, G.; Marques, P. A. A. P.; Barros-Timmons, A.; Bdkin, I.; Singh, M. K.; Emami, N.; Grácio, J. *J. Mater. Chem.* 2010, *20*, 9927.
[85] Deng, Y.; Li, Y.; Dai, J.; Lang, M.; Huang, X. *J. Polym. Sci., Part A: Polym. Chem.* 2011, *49*, 1582.

[86] Singh, S. K.; Singh, M. K.; Kulkarni, P. P.; Sonkar, V. K.; Grácio, J. J.; Dash, D. *ACS Nano* 2012, *6*, 2731.
[87] Madadrang, C. J.; Kim, H. Y.; Gao, G.; Wang, N.; Zhu, J.; Feng, H.; Gorring, M.; Kasner, M. L.; Hou, S. *ACS Appl. Mater. Interfaces* 2012, *4*, 1186.
[88] Hou, S.; Kasner, M. L.; Su, S.; Patel, K.; Cuellari, R. *J. Phys. Chem. C*, 2010, *114*, 14915.
[89] Melucci, M.; Treossi, E.; Ortolani, L.; Giambastiani, G.; Morandi, V.; Klar, P.; Casiraghi, C.; Samorì, P.; Palermo, V. *J. Mater. Chem.* 2010, *20*, 9052.
[90] Meng, D.; Sun, J.; Jiang, S.; Zeng, Y.; Li, Y.; Yan, S.; Geng, J.; Huang, Y. *J. Mater. Chem.* 2012, *22*, 21583.
[91] Zhang, Y.; Ren, L.; Wang, S.; Marathe, A.; Chaudhuri, J.; Li, G. *J. Mater. Chem.* 2011, *21*, 5386.
[92] Collins, W. R.; Lewandowski, W.; Schmois, E.; Walish, J.; Swager, T. M. *Angew. Chem. Int. Ed. Engl.* 2011, *50*, 8848.
[93] Yang, H.; Li, F.; Shan, C.; Han, D.; Zhang, Q.; Niu, L.; Ivaska, A. *J. Mater. Chem.* 2009, *19*, 4632.
[94] Yang, H.; Shan, C.; Li, F.; Han, D.; Zhang, Q.; Niu, L. *Chem. Commun.* 2009, *26*, 3880.
[95] Zhou, H.; Wang, X.; Yu, P.; Chen, X.; Mao, L. *Analyst* 2012, *137*, 305.
[96] Salvio, R.; Krabbenborg, S.; Naber, W. J.; Velders, A. H.; Reinhoudt, D. N.; van der Wiel, W. G. *Chem. Eur. J.* 2009, *15*, 8235.
[97] Hsiao, M. C.; Liao, S. H.; Yen, M. Y.; Liu, P. I.; Pu, N. W.; Wang, C. A.; Ma, C. C. *ACS Appl. Mater. Interfaces* 2010, *2*, 3092.
[98] Huang, Y.; Yan, W.; Xu, Y.; Huang, L.; Chen, Y. *Macromol. Chem. Phys.* 2012, *213*, 1101.
[99] Yun, Y. S.; Bae, Y. H.; Kim, D. H.; Lee, J. Y.; Chin, I. J.; Jin, H. J. *Carbon* 2011, *49*, 3553.
[100] Elias, D. C.; Nair, R. R.; Mohiuddin, T. M.; Morozov, S. V.; Blake, P.; Halsall, M. P.; Ferrari, A. C.; Boukhvalov, D. W.; Katsnelson, M. I.; Geim, A. K.; Novoselov, K. S. *Science* 2009, *323*, 610.
[101] Johns, J. E.; Hersam, M. C. *Acc. Chem. Res.* 2012, *46*, 77.
[102] Ryu, S.; Han, M. Y.; Maultzsch, J.; Heinz, T. F.; Kim, P.; Steigerwald, M. L.; Brus, L. E. *Nano Lett.* 2008, *8*, 4597.
[103] Ng, M. L.; Balog, R.; Hornekær, L.; Preobrajenski, A. B.; Vinogradov, N. A.; Mårtensson, N.; Schulte, K. *J. Phys. Chem. C* 2010, *114*, 18559.
[104] Jones, J. D.; Mahajan, K. K.; Williams, W. H.; Ecton, P. A.; Mo, Y.; Perez, J. M. *Carbon* 2010, *48*, 2335.
[105] Chandrachud, P.; Pujari, B. S.; Haldar, S.; Sanyal, B.; Kanhere, D. G. *J. Phys. Condens. Matter.* 2010, *22*, 465502.
[106] Karlický, F.; Zbořil, R.; Otyepka, M. *J. Chem. Phys.* 2012, *137*, 034709.
[107] Zbořil, R.; Karlický, F.; Bourlinos, A. B.; Steriotis, T. A.; Stubos, A. K.; Georgakilas, V.; Šafářová, K.; Jančík, D.; Trapalis, C.; Otyepka, M. *Small* 2010, *6*, 2885.
[108] Robinson, J. T.; Burgess, J. S.; Junkermeier, C. E.; Badescu, S. C.; Reinecke, T. L.; Perkins, F. K.; Zalalutdniov, M. K.; Baldwin, J. W.; Culbertson, J. C.; Sheehan, P. E.; Snow, E. S. *Nano Lett.* 2010, *10*, 3001.

[109] Nair, R. R.; Ren, W.; Jalil, R.; Riaz, I.; Kravets, V. G.; Britnell, L.; Blake, P.; Schedin, F.; Mayorov, A. S.; Yuan, S.; Katsnelson, M. I.; Cheng, H. M.; Strupinski, W.; Bulusheva, L. G.; Okotrub, A. V.; Grigorieva, I. V.; Grigorenko, A. N.; Novoselov, K. S.; Geim, A. K. *Small* 2010, *6*, 2877.

[110] Bon, S. B.; Valentini, L.; Verdejo, R.; Garcia Fierro, J. L.; Peponi, L.; Lopez-Manchado, M. A.; Kenny, J. M. *Chem. Mater.* 2009, *21*, 3433.

[111] Baraket, M.; Walton, S. G.; Lock, E. H.; Robinson, J. T.; Perkins, F. K. *Appl. Phys. Lett.* 2010, *96*, 231501.

[112] Yang, H.; Chen, M.; Zhou, H.; Qiu, C.; Hu, L.; Yu, F.; Chu, W.; Sun, S.; Sun, L. *J. Phys. Chem. C* 2011, *115*, 16844.

[113] Chang, H.; Cheng, J.; Liu, X.; Gao, J.; Li, M.; Li, J.; Tao, X.; Ding, F.; Zheng, Z. *Chem. Eur. J.* 2011, *17*, 8896.

[114] Cheng, S. H.; Zou, K.; Gutierrez, H. R.; Gupta, A.; Shen, N.; Eklund, P. C.; Sofo, J. O.; Zhu, J. *Phys. Rev. B* 2010, *81*, 205435.

[115] Bourlinos, A. B.; Safarova, K.; Siskova, K.; Zbořil, R. *Carbon* 2012, *50*, 1425.

[116] Wang, Z.; Wang, J.; Li, Z.; Gong, P.; Liu, X.; Zhang, L.; Ren, J.; Wang, H.; Yang, S. *Carbon* 2012, *50*, 5403.

[117] Lee, W.H.; Suk, J. W.; Chou, H.; Lee, J.; Hao, Y.; Wu, Y.; Piner, R.; Akinwande, D.; Kim, K. S.; Ruoff, R. S. *Nano Lett.* 2012, *12*, 2374.

[118] Worsley, K. A.; Ramesh, P.; Mandal, S. K.; Niyogi, S.; Itkis, M. E.; Haddon, R. C. *Chem. Phys. Lett.* 2007, *445*, 51.

[119] Chakraborty, S.; Guo, W.; Hauge, R. H.; Billups, W. E. *Chem. Mater.* 2008, *20*, 3134.

[120] Kelly, K. F.; Billups, W. E. *Acc. Chem. Res.* 2013, *46*, 4.

[121] Li, B.; Zhou, L.; Wu, D.; Peng, H.; Yan, K.; Zhou, Y.; Liu, Z. *ACS Nano* 2011, *5*, 5957.

[122] Wu, J.; Xie, L.; Li, Y.; Wang, H.; Ouyang, Y.; Guo, J.; Dai, H. *J. Am. Chem. Soc.* 2011, *133*, 19668.

[123] Doyle, C. D.; Rocha, J. D.; Weisman, R. B.; Tour, J. M. *J. Am. Chem. Soc.* 2008, *130*, 6795.

[124] Zhu, Y.; Higginbotham, A. L.; Tour, J. M. *Chem. Mater.* 2009, *21*, 5284.

[125] Sinitskii, A.; Dimiev, A.; Corley, D. A.; Fursina, A. A.; Kosynkin, D. V.; Tour, J. M. *ACS Nano* 2010, *27*, 1949.

[126] Jin, Z.; Lomeda, J. R.; Price, B. K.; Lu, W.; Zhu, Y.; Tour, J. M. *Chem. Mater.* 2009, *21*, 3045.

[127] Gan, L.; Zhang, D.; Guo, X. *Small* 2012, *8*, 1326.

[128] Fang, M.; Wang, K.; Lu, H.; Yang, Y.; Nutt, S. *J. Mater. Chem.* 2009, *19*, 7098.

[129] Fang, M.; Wang, K.; Lu, H.; Yang, Y.; Nutt, S. *J. Mater. Chem.* 2010, *20*, 1982.

[130] Wang, D.; Ye, G.; Wang, X.; Wang, X. *Adv. Mater.* 2011, *23*, 1122.

[131] Cui, P.; Seo, S.; Lee, J.; Wang, L.; Lee, E.; Min, M.; Lee, H. *ACS Nano* 2011, *5*, 6826.

[132] Englert, J. M.; Dotzer, C.; Yang, G.; Schmid, M.; Papp, C.; Gottfried, J. M.; Steinrück, H. P.; Spiecker, E.; Hauke, F.; Hirsch, A. *Nat. Chem.* 2011, *3*, 279.

[133] Hirsch, A.; Englert, J. M.; Hauke, F. *Acc. Chem. Res.* 2013, *46*, 87.

[134] Castelaín, M.; Martínez, G.; Merino, P.; Martín-Gago, J. Á.; Segura, J. L.; Ellis, G.; Salavagione, H. J. *Chem. Eur. J.* 2012, *18*, 4965.

[135] Farmer, D. B.; Golizadeh-Mojarad, R.; Perebeinos, V.; Lin, Y. M.; Tulevski, G. S.; Tsang, J. C.; Avouris, P. *Nano Lett.* 2009, *9*, 388.

[136] Wang, Q. H.; Jin, Z.; Kim, K. K.; Hilmer, A. J.; Paulus, G. L.; Shih, C. J.; Ham, M. H.; Sanchez-Yamagishi, J. D.; Watanabe, K.; Taniguchi, T.; Kong, J.; Jarillo-Herrero, P.; Strano, M. S. *Nat. Chem.* 2012, *4*, 724.

[137] Sharma, R.; Baik, J. H.; Perera, C. J.; Strano, M. S. *Nano Lett.* 2010, *10*, 398.

[138] Lim, H.; Lee, J. S.; Shin, H. J.; Shin, H. S.; Choi, H. C. *Langmuir* 2010, *26*, 12278.

[139] Koehler, F. M.; Jacobsen, A.; Ensslin, K.; Stampfer, C.; Stark, W. J. *Small* 2010, *6*, 1125.

[140] Sun, Z.; Kohama, S. I.; Zhang, Z.; Lomeda, J. R.; Tour, J. M. *Nano Res.* 2010, *3*, 117.

[141] Paulus, G. L.; Wang, Q. H.; Strano, M. S. *Acc. Chem. Res.* 2013, 160.

[142] Niyogi, S.; Bekyarova, E.; Itkis, M. E.; Zhang, H.; Shepperd, K.; Hicks, J.; Sprinkle, M.; Berger, C.; Lau, C. N.; deHeer, W. A.; Conrad, E. H.; Haddon, R. C. *Nano Lett.* 2010, *10*, 4061.

[143] Zhang, H.; Bekyarova, E.; Huang, J. W.; Zhao, Z.; Bao, W.; Wang, F.; Haddon, R. C.; Lau, C. N. *Nano Lett.* 2011, *11*, 4047.

[144] Bekyarova, E.; Itkis, M. E.; Ramesh, P.; Berger, C.; Sprinkle, M.; de Heer, W. A.; Haddon, R. C. *J. Am. Chem. Soc.* 2009, *131*, 1336.

[145] Huang, P.; Zhu, H.; Jing, L.; Zhao, Y.; Gao, X. *ACS Nano* 2011, *5*, 7945.

[146] Koehler, F. M.; Luechinger, N. A.; Ziegler, D.; Athanassiou, E. K.; Grass, R. N.; Rossi, A.; Hierold, C.; Stemmer, A.; Stark, W. J. *Angew. Chem. Int. Ed. Engl.* 2009, *48*, 224.

[147] Hong, J.; Niyogi, S.; Bekyarova, E.; Itkis, M. E.; Ramesh, P.; Amos, N.; Litvinov, D.; Berger, C.; de Heer, W. A.; Khizroev, S.; Haddon, R. C. *Small* 2011, *7*, 1175.

[148] Mukherjee, A.; Kang, J. H.; Kuznetsov, O.; Sun, Y.; Thaner, R.; Bratt, A. S.; Lomeda, J. R.; Kelly, K. F.; Billups, W. E. *Chem. Mater.* 2011, *23*, 9.

[149] Liu, H.; Ryu, S.; Chen, Z.; Steigerwald, M. L.; Nuckolls, C.; Brus, L. E. *J. Am. Chem. Soc.* 2009, *131*, 17099.

[150] Lu, X.; Qi, H.; Zhang, X.; Xue, Z.; Jin, J.; Zhou, X.; Liu, X. *Chem. Commun.* 2011, *47*, 12494.

[151] Shen, J.; Hu, Y.; Li, C.; Qin, C.; Ye, M. *Small* 2009, *5*, 82.

[152] Hamilton, C. E.; Lomeda, J. R.; Sun, Z.; Tour, J. M.; Barron, A. R. *Nano Res.* 2010, *3*, 138.

[153] Xu, H.; Suslick, K. S. *J. Am. Chem. Soc.* 2011, *133*, 9148.

[154] Shen, B.; Zhai, W.; Lu, D.; Wang, J.; Zheng, W. *RSC Adv.* 2012, *2*, 4713.

[155] Maggini, M.; Scorrano, G.; Prato, M. *J. Am. Chem. Soc.* 1993, *115*, 9798.

[156] Georgakilas, V.; Kordatos, K.; Prato, M.; Guldi, D. M.; Holzinger, M.; Hirsch, A. *J. Am. Chem. Soc.* 2002, *124*, 760.

[157] Karousis, N.; Tagmatarchis, N.; Tasis, D. *Chem. Rev.* 2010, *110*, 5366.

[158] Cioffi, C. T.; Palkar, A.; Melin, F.; Kumbhar, A.; Echegoyen, L.; Melle-Franco, M.; Zerbetto, F.; Rahman, G. M.; Ehli, C.; Sgobba, V.; Guldi, D. M.; Prato, M. *Chem. Eur. J.* 2009, *15*, 4419.

[159] Tagmatarchis, N.; Maigné, A.; Yudasaka, M.; Iijima, S. *Small* 2006, *2*, 490.

[160] Cioffi, C.; Campidelli, S.; Brunetti, F. G.; Meneghetti, M.; Prato, M. *Chem. Commun.* 2006, *28*, 2129.

[161] Bianco, A.; Kostarelos, K.; Partidos, C. D.; Prato, M. *Chem. Commun.* 2005, 571.

[162] Oelsner, C.; Herrero, M. A.; Ehli, C.; Prato, M.; Guldi, D. M. *J. Am. Chem. Soc.* 2011, *133*, 18696.

[163] Georgakilas, V.; Bourlinos, A. B.; Zboril, R.; Steriotis, T. A.; Dallas, P.; Stubos, A. K.; Trapalis, C. *Chem. Commun.* 2010, *46*, 1766.
[164] Quintana, M.; Vazquez, E.; Prato, M. *Acc. Chem. Res.* 2013, *46*, 138.
[165] Quintana, M.; Spyrou, K.; Grzelczak, M.; Browne, W. R.; Rudolf, P.; Prato, M. *ACS Nano* 2010, *4*, 3527.
[166] Kaiser, E.; Colescott, R. L.; Bossinger, C. D.; Cook, P. I. *Anal. Biochem.* 1970, *34*, 595.
[167] Quintana, M.; Montellano, A.; del Rio Castillo, A. E.; Van Tendeloo, G.; Bittencourt, C.; Prato, M. *Chem. Commun.* 2011, *47*, 9330.
[168] Ragoussi, M. E.; Malig, J.; Katsukis, G.; Butz, B.; Spiecker, E.; de la Torre, G.; Torres, T.; Guldi, D. M. *Angew. Chem. Int. Ed. Engl.* 2012, *51*, 6421.
[169] Zhang, X.; Hou, L.; Cnossen, A.; Coleman, A. C.; Ivashenko, O.; Rudolf, P.; van Wees, B. J.; Browne, W. R.; Feringa, B. L. *Chem. Eur. J.* 2011, *17*, 8957.
[170] Zhang, B.; Liu, G.; Chen, Y.; Zeng, L. J.; Zhu, C. X.; Neoh, K. G.; Wang, C.; Kang, E. T. *Chem. Eur. J.* 2011, *17*, 13646.
[171] Ou, B.; Zhou, Z.; Liu, Q.; Liao, B.; Yi, S.; Ou, Y.; Zhang, X.; Li, D. *Polym. Chem.* 2012, *3*, 2768.
[172] Wu, X.; Cao, H.; Li, B.; Yin, G. *Nanotechnology* 2011, *22*, 075202.
[173] Kolb, H. C.; Finn, M. G.; Sharpless, K. B. *Angew. Chem. Int. Ed. Engl.* 2001, *40*, 2004.
[174] Ye, Y. S.; Chen, Y. N.; Wang, J. S.; Rick, J.; Huang, Y. J.; Chang, F. C.; Hwang, B. J. *Chem. Mater.* 2012, *24*, 2987.
[175] Sun, S.; Cao, Y.; Feng, J.; Wu, P. *J. Mater. Chem.* 2010, *20*, 5605.
[176] Cao, Y.; Lai, Z.; Feng, J.; Wu, P. *J. Mater. Chem.* 2011, *21*, 9271.
[177] Jin, Z.; McNicholas, T. P.; Shih, C. J.; Wang, Q. H.; Paulus, G. L. C.; Hilmer, A. J.; Shimizu, S.; Strano, M. S. *Chem. Mater.* 2011, *23*, 3362.
[178] Wang, Z.; Ge, Z.; Zheng, X.; Chen, N.; Peng, C.; Fan, C.; Huang, Q. *Nanoscale* 2012, *4*, 394.
[179] Castelaín, M.; Salavagione, H.; Segura, J. L. *Org. Lett.* 2012, *14*, 2798.
[180] Wang, H. X.; Zhou, K. G.; Xie, Y. L.; Zeng, J.; Chai, N. N.; Li, J.; Zhang, H. L. *Chem. Commun.* 2011, *47*, 5747.
[181] Yang, H.; Kwon, Y.; Kwon, T.; Lee, H.; Kim, B. J. *Small* 2012, *8*, 3161.
[182] He, H.; Gao, C. *Chem. Mater.* 2010, *22*, 5054.
[183] Vadukumpully, S.; Gupta, J.; Zhang, Y.; Xu, G. Q.; Valiyaveettil, S. *Nanoscale* 2011, *3*, 303.
[184] Strom, T. A.; Dillon, E. P.; Hamilton, C. E.; Barron, A. R. *Chem. Commun.* 2010, *46*, 4097.
[185] Xu, X.; Lv, W.; Huang, J.; Li, J.; Tang, R.; Yan, J.; Yang, Q.; Qin, J.; Li, Z. *RSC Adv.* 2012, *2*, 7042.
[186] Choi, J.; Kim, K. J.; Kim, B.; Lee, H.; Kim, S. *J. Phys. Chem. C* 2009, *113*, 9433.
[187] Liu, L. H.; Yan, M. *Nano Lett.* 2009, *9*, 3375.
[188] Liu, L. H.; Yan, M. *J. Mater. Chem.* 2011, *21*, 3273.
[189] Liu, L. H.; Lerner, M. M.; Yan, M. *Nano Lett.* 2010, *10*, 3754.
[190] Economopoulos, S. P.; Rotas, G.; Miyata, Y.; Shinohara, H.; Tagmatarchis, N. *ACS Nano* 2010, *4*, 7499.
[191] Ismaili, H.; Geng, D.; Sun, A. X.; Kantzas, T. T.; Workentin, M. S. *Langmuir* 2011, *27*, 13261.

[192] Sarkar, S.; Bekyarova, E.; Niyogi, S.; Haddon, R. C. *J. Am. Chem. Soc.* 2011, *133*, 3324.
[193] Sarkar, S.; Bekyarova, E.; Haddon, R. C. *Acc. Chem. Res.* 2012, *45*, 673.
[194] Yuan, J.; Chen, G.; Weng, W.; Xu, Y. *J. Mater. Chem.* 2012, *22*, 7929.
[195] Zhong, X.; Jin, J.; Li, S.; Niu, Z.; Hu, W.; Li, R.; Ma, J. *Chem. Commun.* 2010, *46*, 7340.
[196] Choi, E. K.; Jeon, I. Y.; Bae, S. Y.; Lee, H. J.; Shin, H. S.; Dai, L.; Baek, J. B. *Chem. Commun.* 2010, *46*, 6320.
[197] Chua, C. K.; Pumera, M. *Chem. Asian J.* 2012, *7*, 1009.
[198] Liu, J.; Hou, L.; Peng, W.; Zhang, Q.; Zhang, X. *J. Mater. Sci.* 2012, *47*, 8052.
[199] Yang, H. W.; Hua, M. Y.; Chen, S. L.; Tsai, R. Y. *Biosens. Bioelectron.* 2013, *41*, 172.
[200] Avinash, M. B.; Subrahmanyam, K. S.; Sundarayya, Y.; Govindaraju, T. *Nanoscale* 2010, *2*, 1762.
[201] Yuan, C.; Chen, W.; Yan, L. *J. Mater. Chem.* 2012, *22*, 7456.
[202] Englert, J. M.; Knirsch, K. C.; Dotzer, C.; Butz, B.; Hauke, F.; Spiecker, E.; Hirsch, A. *Chem. Commun.* 2012, *48*, 5025.
[203] Chang, H.; Wang, G.; Yang, A.; Tao, X.; Liu, X.; Shen, Y.; Zheng, Z. *Adv. Funct. Mater.* 2010, *20*, 2893.
[204] Zeng, Q.; Cheng, J.; Tang, L.; Liu, X.; Liu, Y.; Li, J.; Jiang, J. *Adv. Funct. Mater.* 2010, *20*, 3366.
[205] Some, S.; Kim, Y.; Hwang, E.; Yoo, H.; Lee, H. *Chem. Commun.* 2012, *48*, 7732.
[206] Ghosh, A.; Rao, K. V.; George, S. J.; Rao, C. N. *Chem. Eur. J.* 2010, *16*, 2700.
[207] Cho, S. D.; Im, J. K.; Kim, H. K.; Kim, H. S.; Park, H. S. *Chem. Commun.* 2012, *48*, 6381.
[208] Hao, R.; Qian, W.; Zhang, L.; Hou, Y. *Chem. Commun.* 2008, *28*, 6576.
[209] Geng, J.; Jung, H. T. *J. Phys. Chem. C* 2010, *114*, 8227.
[210] Malig, J.; Jux, N.; Kiessling, D.; Cid, J. J.; Vázquez, P.; Torres, T.; Guldi, D. M. *Angew. Chem. Int. Ed. Engl.* 2011, *50*, 3561.
[211] An, X.; Simmons, T.; Shah, R.; Wolfe, C.; Lewis, K. M.; Washington, M.; Nayak, S. K.; Talapatra, S.; Kar, S. *Nano Lett.* 2010, *10*, 4295.
[212] Liu, F.; Choi, J. Y.; Seo, T. S. *Chem. Commun.* 2010, *46*, 2844.
[213] Cheng, Q. Y.; Zhou, D.; Gao, Y.; Chen, Q.; Zhang, Z.; Han, B. H. *Langmuir* 2012, *28*, 3005.
[214] Zhang, M.; Parajuli, R. R.; Mastrogiovanni, D.; Dai, B.; Lo, P.; Cheung, W.; Brukh, R.; Chiu, P. L.; Zhou, T.; Liu, Z.; Garfunkel, E.; He, H. *Small* 2010, *6*, 1100.
[215] Lee, D. W.; Kim, T.; Lee, M. *Chem. Commun.* 2011, *47*, 8259.
[216] Rodríguez-López, J.; Ritzert, N. L.; Mann, J. A.; Tan, C.; Dichtel, W. R.; Abruña, H. D. *J. Am. Chem. Soc.* 2012, *134*, 6224.
[217] Mann, J. A.; Rodríguez-López, J.; Abruña, H. D.; Dichtel, W. R. *J. Am. Chem. Soc.* 2011, *133*, 17614.
[218] Parviz, D.; Das, S.; Ahmed, H. S.; Irin, F.; Bhattacharia, S.; Green, M. J. *ACS Nano* 2012, *6*, 8857.
[219] Balapanuru, J.; Yang, J. X.; Xiao, S.; Bao, Q.; Jahan, M.; Polavarapu, L.; Wei, J.; Xu, Q. H.; Loh, K. P. *Angew. Chem. Int. Ed. Engl.* 2010, *49*, 6549.
[220] Wang, Z.; Huang, P.; Bhirde, A.; Jin, A.; Ma, Y.; Niu, G.; Neamati, N.; Chen, X. *Chem. Commun.* 2012, *48*, 9768.

[221] Kim, J.; Cote, L. J.; Kim, F.; Huang, J. *J. Am. Chem. Soc.* 2010, *132*, 260.
[222] Kodali, V. K.; Scrimgeour, J.; Kim, S.; Hankinson, J. H.; Carroll, K. M.; de Heer, W. A.; Berger, C.; Curtis, J. E. *Langmuir* 2011, *27*, 863.
[223] Su, Q.; Pang, S.; Alijani, V.; Li, C.; Feng, X.; Müllen, K. *Adv. Mater.* 2009, *21*, 3191.
[224] Backes, C.; Hauke, F.; Hirsch, A. *Adv. Mater.* 2011, *23*, 2588.
[225] Englert, J. M.; Röhrl, J.; Schmidt, C. D.; Graupner, R.; Hundhausen, M.; Hauke, F.; Hirsch, A. *Adv. Mater.* 2009, *21*, 4265.
[226] Kozhemyakina, N. V.; Englert, J. M.; Yang, G.; Spiecker, E.; Schmidt, C. D.; Hauke, F.; Hirsch, A. *Adv. Mater.* 2010, *22*, 5483.
[227] Wang, Q. H.; Hersam, M. C. *Nat. Chem.* 2009, *1*, 206.
[228] Pollard, A. J.; Perkins, E. W.; Smith, N. A.; Saywell, A.; Goretzki, G.; Phillips, A. G.; Argent, S. P.; Sachdev, H.; Müller, F.; Hüfner, S.; Gsell, S.; Fischer, M.; Schreck, M.; Osterwalder, J.; Greber, T.; Berner, S.; Champness, N. R.; Beton, P. H. *Angew. Chem. Int. Ed. Engl.* 2010, *49*, 1794.
[229] Zhang, Z.; Huang, H.; Yang, X.; Zang, L. *J. Phys. Chem. Lett.* 2011, *2*, 2897.
[230] Xu, Y.; Bai, H.; Lu, G.; Li, C.; Shi, G. *J. Am. Chem. Soc.* 2008, *130*, 5856.
[231] Dong, X.; Fu, D.; Fang, W.; Shi, Y.; Chen, P.; Li, L. J. *Small* 2009, *5*, 1422.
[232] Lee, J.; Hwang, E.; Lee, E.; Seo, S.; Lee, H. *Chem. Eur. J.* 2012, *18*, 5155.
[233] Jang, A. R.; Jeon, E. K.; Kang, D.; Kim, G.; Kim, B. S.; Kang, D. J.; Shin, H. S. *ACS Nano* 2012, *6*, 9207.
[234] Yang, X.; Zhang, X.; Liu, Z.; Ma, Y.; Huang, Y.; Chen, Y. *J. Phys. Chem. C* 2008, *112*, 17554.
[235] Tian, B.; Wang, C.; Zhang, S.; Feng, L.; Liu, Z. *ACS Nano* 2011, *5*, 7000.
[236] Patil, A. J.; Vickery, J. L.; Scott, T. B.; Mann, S. *Adv. Mater.* 2009, *21*, 3159.
[237] Premkumar, T.; Geckeler, K. E. *Prog. Polym. Sci.* 2012, *37*, 515.
[238] Tang, Z.; Wu, H.; Cort, J. R.; Buchko, G. W.; Zhang, Y.; Shao, Y.; Aksay, I. A.; Liu, J.; Lin, Y. *Small* 2010, *6*, 1205.
[239] Tang, L.; Chang, H.; Liu, Y.; Li, J. *Adv. Funct. Mater.* 2012, *22*, 3083.
[240] Xu, Y.; Wu, Q.; Sun, Y.; Bai, H.; Shi, G. *ACS Nano* 2010, *4*, 7358.
[241] Vadukumpully, S.; Paul, J.; Valiyaveettil, S. *Carbon* 2009, *47*, 3288.
[242] Lotya, M.; King, P. J.; Khan, U.; De, S.; Coleman, J. N. *ACS Nano* 2010, *4*, 3155.
[243] Lin, S.; Shih, C. J.; Strano, M. S.; Blankschtein, D. *J. Am. Chem. Soc.* 2011, *133*, 12810.
[244] Green, A. A.; Hersam, M. C. *Nano Lett.* 2009, *9*, 4031.
[245] Zu, S. Z.; Han, B. H. *J. Phys. Chem. C* 2009, *113*, 13651.
[246] Smith, R. J.; Lotya, M.; Coleman, J. N. *New J. Phys.* 2010, *12*, 125008.
[247] Guardia, L.; Fernández-Merino, M. J.; Paredes, J. I.; Solís-Fernández, P.; Villar-Rodil, S.; Martínez-Alonso, A.; Tascón, J. M. D. *Carbon* 2011, *49*, 1653.
[248] Shih, C. J.; Paulus, G. L.; Wang, Q. H.; Jin, Z.; Blankschtein, D.; Strano, M. S. *Langmuir* 2012, *28*, 8579.
[249] Kuilla, T.; Bhadra, S.; Yao, D.; Kim, N. H.; Bose, S.; Lee, J. H. *Prog. Polym. Sci.* 2010, *35*, 1350.
[250] Li, X.; Wang, X.; Zhang, L.; Lee, S.; Dai, H. *Science* 2008, *319*, 1229.
[251] Bai, H.; Xu, Y.; Zhao, L.; Li, C.; Shi, G. *Chem. Commun.* 2009, *7*, 1667.
[252] Kim, T.; Lee, H.; Kim, J.; Suh, K. S. *ACS Nano* 2010, *4*, 1612.

[253] Choi, E. Y.; Han, T. H.; Hong, J.; Kim, J. E.; Lee, S. H.; Kim, H. W.; Kim, S. O. *J. Mater. Chem.* 2010, *20*, 1907.

[254] Choi, B. G.; Park, H.; Park, T. J.; Yang, M. H.; Kim, J. S.; Jang, S. Y.; Heo, N. S.; Lee, S. Y.; Kong, J.; Hong, W. H. *ACS Nano* 2010, *4*, 2910.

[255] Qi, X.; Pu, K. Y.; Li, H.; Zhou, X.; Wu, S.; Fan, Q. L.; Liu, B.; Boey, F.; Huang, W.; Zhang, H. *Angew. Chem. Int. Ed. Engl.* 2010, *49*, 9426.

[256] Kang, S. M.; Park, S.; Kim, D.; Park, S. Y.; Ruoff, R. S.; Lee, H. *Adv. Funct. Mater.* 2011, *21*, 108.

[257] Robinson, J. T.; Tabakman, S. M.; Liang, Y.; Wang, H.; Casalongue, H. S.; Vinh, D.; Dai, H. *J. Am. Chem. Soc.* 2011, *133*, 6825.

[258] Liu, Z.; Robinson, J. T.; Tabakman, S. M.; Yang, K.; Dai, H. *Mater. Today* 2011, *14*, 316.

In: Applied Surface Chemistry of Nanomaterials
Editors: Mohamed M. Chehimi and Jean Pinson
ISBN: 978-1-62808-351-4

Chapter 4

CARBON NANOTUBES: SURFACE MODIFICATION AND APPLICATIONS

Mohamed M. Chehimi[1], Jean Pinson*[1,2] and Zakaria Salmi[1]
[1]Univ Paris Diderot, Sorbonne Paris Cité, ITODYS, CNRS UMR 7086, Paris, France
[2]Physico-Chimie des Electrolytes, des Colloïdes et Sciences Analytiques, ESPCI, CNRS UMR 7195, Paris Cedex 05, France

1. INTRODUCTION

Carbon nanotubes (CNT) present some extraordinary and very promising properties. [1] Their structure is that of a seamless tube made from a wrapped graphene sheet with very small diameters (typically in the nm range) and very high length (up to μm). Depending on the way the graphene sheet is rolled along two vectors (n, m) giving rise to armchair, zig-zag or chiral nanotubes (Figure 1) with metallic and semiconducting character; the band gap of semiconductiong CNT is defined by the values of n and m, it is of the order of 500 meV and varies inversely with diameter. [2] As the spectroscopic properties of CNT depend on the values of n and m, their values can be determined by Raman, UV-vis-nIR and photoluminescence. They can have a single wall (SWCNT), two walls (DWCNT) or multiple walls (MWCNT) (Figure 2). They present very low resistance; a perfect metallic CNT has a ballistic conductivity and its resistance is 6.5 Ω whatever its length; [3] thermal conductivity of SWNT and MWNT is 2400 and 1980 W/m K, [4] respectively, (by comparison the thermal conductivity of graphite along its planar direction is 1000 W/m K); tensile strength of MWCNT ranges from 11 to 63 GP, [5] by comparison that of high strength steel is around 4.1 GP.

However exploitation of these exceptional properties is quite difficult. CNT are synthesized by various methods such as arc discharge, laser ablation or chemical vapor deposition (CVD, that produces the largest quantities) techniques; it is now possible to produce separately SWCNT or MWCNT, but it is not yet possible to produce separately semiconducting or metallic CNT or even better a specific (n, m) CNT, that is a CNT with a

* E-mail: jean.pinson@espci.fr.

given diameter. In addition, CNT produced by CVD techniques always contain metallic nanoparticles that must be removed by acidic treatment and bundles of entangled CNT are always produced. [6] These bundles contain hundreds of close-packed CNT tightly bound by van der Waals attraction energy of 500 eV/µm of tube tube contact. [7] They must be debundled and dispersed to obtain separated CNT that are necessary for most applications. This is why surface modification methods are important as they provide, along with ultrasonication, the only means to obtain individualized CNT.

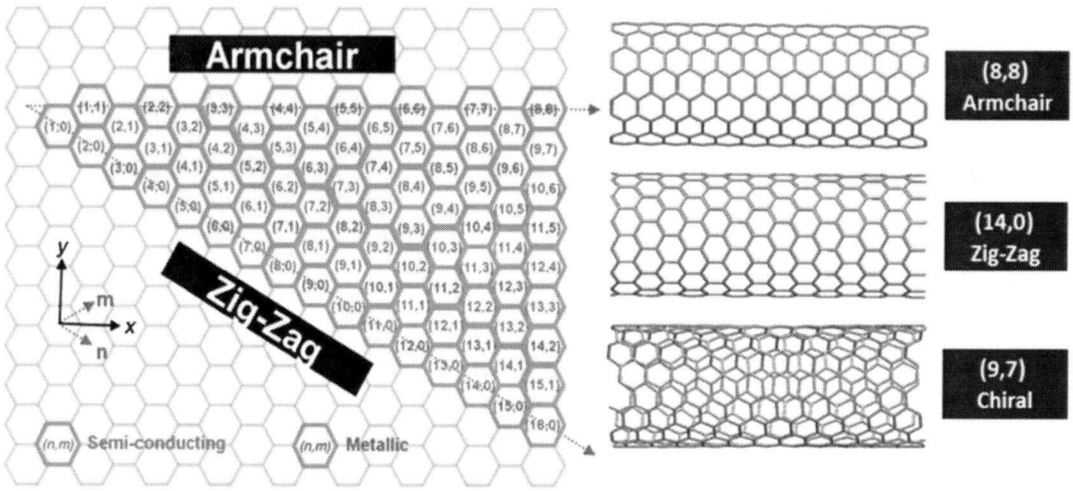

Figure 1. SWNT helicity map and examples of (n,m) chiral vectors that give rise to armchair, zig–zag and chiral nanotube structures with metallic or semi-conducting electronic character. Reproduced from ref. 2 by permission of the Royal Society of Chemistry.

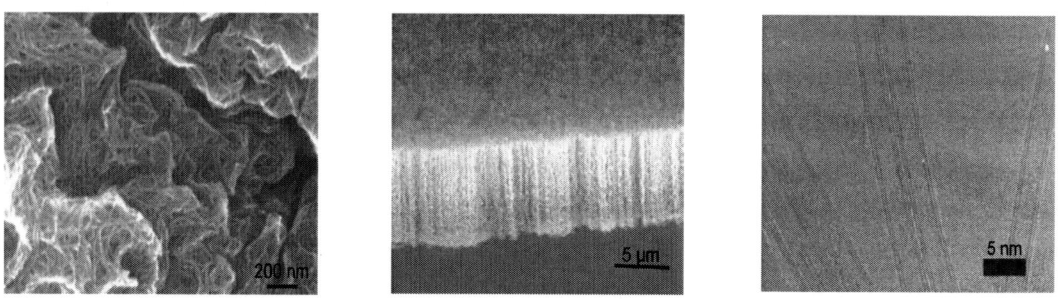

Figure 2. Images of a) as grown CVD CNT b) vertically grown CNT and c) DWCNT. [8]

SWCNT are pure surfaces with no matter inside therefore all their chemistry is a surface chemistry (except with the introduction of molecules such as fullerenes inside the tube). Surface modification is required in analytical methods for the attachment of a specific molecule, in biomedicine for the attachment of the bioactive material, in composites for binding CNT to the polymer, in nanoelectronics to attach the CNT in specific locations. Fortunately, as we will see below, CNT are very reactive species and a number of methods have been developed that provide functionalities on their surface. This chapter will describe the chemical reactions that are able to functionalize the CNT (covalent, non covalent modifications) as well as the applications of these methods in different fields.

CNT are the subject of a vast litterature (more than 91000 references are found under the heading "carbon nanotube") and several reviews have covered the topic of their surface modification. [9] Figure 3 gives an overview of the main methods that are used for surface modification

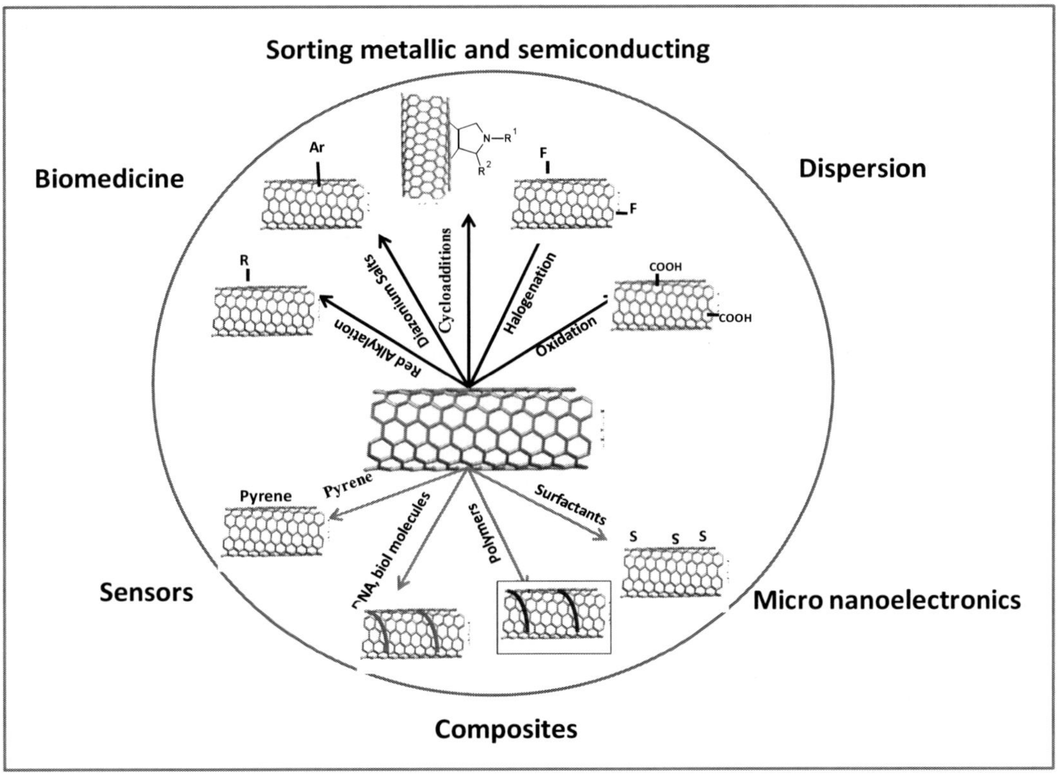

Figure 3. An overview of the modification of CNT and their applications

2. Surface Functionalization

Two different categories of methods have been used for the functionalization of CNT: the covalent functionalization and the non-covalent one where the modifying (macro)molecule or is only physically adsorbed on the nanotube.

2.1. Covalent Functionalization

In view of the many methods that have been used for the surface modification of CNT, these nano objects are as very reactive species, we will list below and discuss the most important of these methods. It should be stressed that covalent modification disrupts the sp^2 array and therefore decreases the conductivity of CNT, the structure of the nanotubes can be recovered by heating at high temperatures.

2.1.1. By Oxidation

Introduction of oxygen groups on the surface of CNT is probably the most widely used method of functionalization, it takes place at the end and on the walls of CNT; several methods are available both in solution and in the gas phase.

Oxidation in solution. This method is very popular, as it is easily implemented; it is also used to purify the as-synthetized CNT from metallic impurities. A number of oxidants are available [10, 11, 12]: HNO_3, HNO_3/H_2SO_4, $KMnO_4$, $(NH_4)_2S_2O_8$, H_2O_2 and dissolved O_3. Different groups are formed on the surface: COOH, C=O, OH; the relative percentage of these groups on the surface of MWCNT depends on the oxidant and the experimental conditions. [13, 14, 15] It was found that the percentage of the different oxygenated groups (carefully characterized by specific chemical reactions) depends on the oxidant, but not much of its concentration. COOH groups are mainly obtained with $KMnO_4$ (~ 3.8 atom %, by refluxing for 5 hours in 50 mM $KMnO_4$ in 0.5 M H_2SO_4) while C=O and OH are mainly obtained (~ 1 atom %) with $(NH_4)_2S_2O_8$. However, with SWCNT, it was found that carbonyls are the major functionalities after nitric acid or air/HCl treatment. In addition, SWCNT are ~5 times more functionalized by the second treatment (5 versus 1 atomic % oxygen functionalities). [16] Therefore, the nature of the functional groups seems to depend on the type of CNT. It should also be noted that these treatments lead to the shortening and thinning of the CNT [17, 18] and to the release of CO_2, indicating partial degradation of CNT [19] and the formation of "carboxylated carbons" [20].

Gas Phase Oxidation. Air oxidation of SWCNT (at 350°) is reported to be as efficient as solution oxidation but less damaging, [21] producing less debris and removing more carbonaceous impurities from the CNT surface. [22] The oxidation mechanism of DWCNT by atomic oxygen has been investigated by high resolution XPS, it starts by adsorption of oxygen, formation of C-O-C bonds, goes through the lattice deformation and formation of C=O bonds at latter stages. [23] Ozone can also be used for the functionalization of CNT. [24]

Plasma oxidation introduces oxygen group on the surface of MWCNT, to the point where they become superhydrophilic; [25] an oxidation mechanism has been proposed. [26]

Fenton reaction [27] (that furnishes hydroxyl radicals $OH^{\cdot}$ from H_2O_2) is quite efficient; it introduces 22 at % of oxygen and produces little damage to MWCNT.

Once the CNT have been oxidized the carboxylic groups can be easily transformed into esters by reaction with biodegradable polyol, [28] into acid chlorides, [29] or into amides [30] for the attachment of proteins [31] or lipids; [32] alcohol groups can be reacted with acylchlorides, [33] silanes [34], they also undergo biodegradation, for example with H_2O_2 in the presence of horseradish peroxidase. [35]

2.1.2. By Halogenation

Halogenation has been achieved mostly by fluorination, but also by chlorination, bromination and iodination. Fluorination can be achieved via different methods and reagents: heating in the presence of fluorine, [36] BrF_3 [37] or XeF_2 [38] gas, plasma treatments with CF_4 gas, [39] but also milder reagents such as HPF_6; [40] the stability of DWCNT fluorinated by different methods has been compared. [41] Chlorination has been achieved [42] with gaseous Cl_2; bromination with an aqueous bromine solution under microwave irradiation [43] but also by electrochemistry [44] and iodination [45] through a modified Hunsdiecker-type reaction. The F content [38] reaches $CF_{0.3\text{-}0.4}$. and the fluorination reaction can be reversed:

TG analysis and XPS examination of a fluorinated DWNT sample revealed that the fluorine removal begins at 100°C, and is nearly terminated at 300°C. [46] The mechanism of fluorination has been investigated, [38] it proceeds through attack of molecular fluorine and HF catalysis when using F_2 and through F' when using XeF_2.

Fluorinated CNT can be further modified, for example by reaction the amine group of (3-aminopropyl)triethoxysilane (APTES) to give *CNT*-NH-$(CH_2)_3$-Si$(OEt)_3$. [47]

2.1.3. By Cycloaddition

Azomethine ylides (obtained as non isolated intermediates by heating an amino acid and an aldehyde) can undergo a 1,3-cycloaddition reaction with all kinds of CNT; the reaction is shown in Figure 4 along with some possible substituents: oligoethyleneglycol or long alkyl chains on the aminoacid and phenyl or pyrenyl groups on the aldehyde' [9, 48, 49]

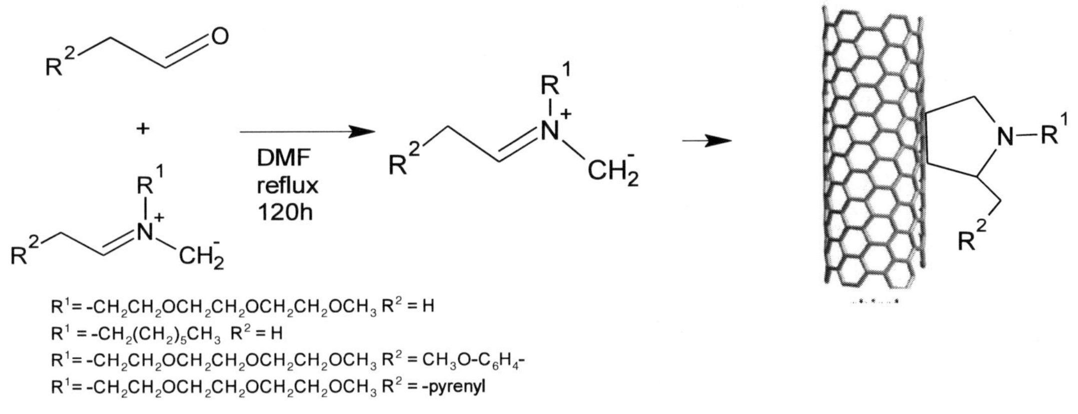

Figure 4. Cycloaddition of azomethine ylides.

This reaction has been used, for example, for the functionalization of MWCNT and SWCNT by 3,4-dihydroxyphenyl groups [50] (from 3,4-dihydroxybenzaldehyde), this permits the dispersion of these CNT in alcohol and water, and the formation of homogeneous composites with poly(ethylenevinylacetate) that give transparent films. Phtalocyanines have been attached to SWCNT by this method, [51] electron transfer from photoexcited phthalocyanines to the nanotube framework is observed in these ZnPc-SWNT ensembles. Dendrimers have been attached to CNT and further functionalized with porphyrins [52].

Nitrones, [53] nitrile oxides, [54] nitrile imines [55] and an oxazolone [56] can also be attached to CNT through 1,3-cycloaddition.

Diels Adler reaction of 2,3-dimethoxy-1,3-butadiene with SWCNT in the presence of $Cr(CO)_6$ has been achieved giving one functional group per 33 C atoms. [57] Under microwave irradiation electron-poor epoxides yield carbonyl ylides that react, without solvent, on SWCNT through a 1,3-cycloaddition reaction giving one functional group per 57 carbons. [58]

Carbenes and nitrenes can add to the conjugated wall of CNT along [2+1] cycloaddition reactions but also nucleophilic addition. [59] For example, diaryldiazomethane derivatives provide, under actinic irradiation, highly reactive intermediate carbenes that add to SWCNT (Figure 5). [60] Photolysis of 3-aryl-3-(trifluoromethyl)diazirine-modified gold nanoparticles resulted in the generation of reactive carbene intermediates that add to the CNT that becomes decorated by Au nanoparticles. [61] Under UV exposure, perfluoroarylazide (N_3-C_6F_4-C(=O)-R) provide nitrene intermediates that add to the surface of CNT. [62] A polymer with azide groups (poly(4-azidophenylmethacrylate-co-methylacrylate) has been attached to MWCNT and cotton fabrics. [63]

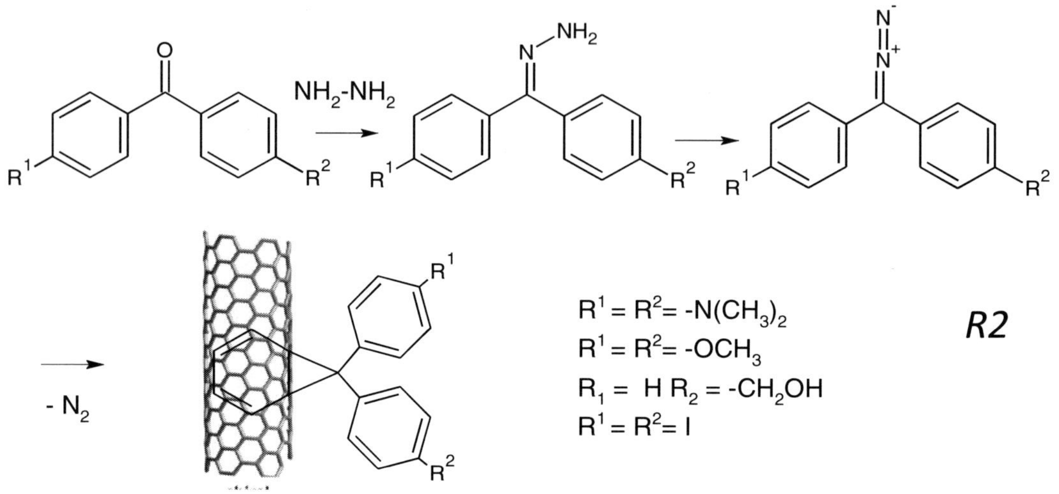

Figure 5. Addition of a carbene on a CNT (adapted from reference [**Error! Bookmark not defined.**]).

2.1.4. By Reaction with Diazonium Salts

Diazonium salts.ArN_2^+ have been widely used for the surface modification of a number of substrates including different forms of carbon, but also metals, semiconductors and polymers, this reaction can be performed by various methods including electrochemistry, solvolysis, and chemical reduction. This grafting reaction of diazonium salts has been assigned to the formation of radicals. [64] This same reaction has been widely applied to CNT. It should be emphasized that diazonium salts are stable only in aqueous acidic medium and in a few organic solvent among which acetonitrile (ACN). There are too many examples in the literature where diazonium salts are used in neutral or basic aqueous medium, in this case diazoates Ar-N=N-O^- are formed their relative concentration being pH dependent. [65] These diazoates dediazonize to aryl radicals as do diazonium salts but they differ through their interaction with CNT as they carry a negative charge (this is very important when theoretical simulations involving diazonium cations are compared with experiments involving diazoates). [66]

The modifications of CNT with diazonium salts have been performed by electrochemical reduction, [68] in this last case the CNT must be used as a buckypaper [69, 70, 71] obtained by filtration, as a paste in mineral oil or [72] as wet cast array, [73] in order to be able to establish an electrical contact. Up to one out of 20 carbons have been modified by this process. The chemical modification of individual single-walled carbon nanotubes has also been achieved by establishing a contact with a single SWCNT. [74] Instead of an electrode as a purveyor of electrons, it is possible to use a reducing reagent such as iron that permits the formation of variously substituted polyphenylene layers on MWCNT, [75] heating at 70°C, [76] microwave activation also permits to accelerate the surface reaction. [77]

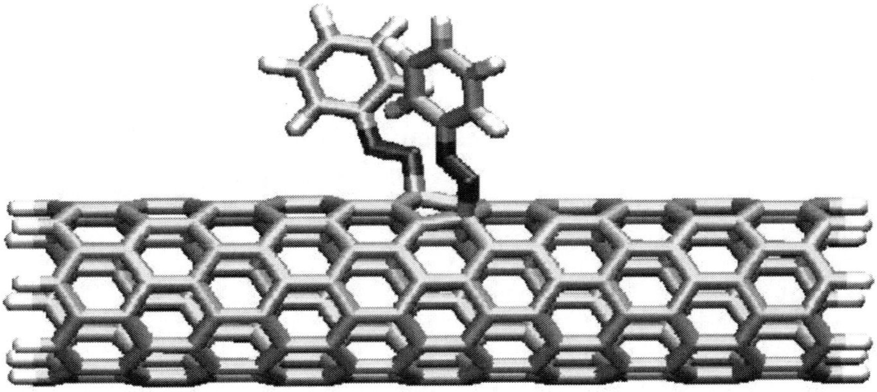

Figure 6. Modification of CNT by diazonium salts. Calculated image of a (8,0)SWCNT modified by two phenyl groups(optimized geometry). Reproduced from [67] by permission of John Wiley and Sons.

Spontaneous reaction has been achieved by mixing the diazonium salts with the suspension of CNT in ACN, [78] in aqueous solution, [79] in concentrated sulfuric acid and oleum (in the presence of radical precursors) [80, 81, 82] but also "on water" [83] and in absence of solvent [84] by vigorous stirring in the presence of isoamylnitrite and a substituted aniline. As stated in the introduction carbon nanotubes are produced as entangled bundles that are difficult to disperse and therefore to handle for further reactions; methods have been developed that aim at dispersing and at the same time modifying the surface of CNT. However, surfactants used for the dispersion of CNT interfere with the grafting of diazonium salts in a variety of ways: electrostatics, steric exclusion, and direct chemical modification of the reacting species, therefore they influence the course of the surface modification. [85, 86] Dispersion of CNT before reaction with diazonium salts can also be achieved by sonication for example in DMF, [87] N-methyl pyrrolidone [88] or dimethylacetamide [89] but this leads to the cleavage and shortening of the CNT. [90, 91]

Modelisation of the reaction has been performed by DFT methods: [67, 92] the binding energy of a single phenyl group adsorption on CNT has been calculated for example: metallic (4,4) SWCNT d = 0.542 nm and (10,10) SWCNT d = 1.356 nm E_{ads} = -1.74 and -0.81 eV respectively and semiconducting SWCNT (8,0) d = 0.626 nm E_{ads} = -1.48 eV. [92] By comparing different SWCNT it appears that bonding on metallic SWCNT is significantly stronger than that on semiconducting ones, and that the adsorption energy decreases with increased tube diameter. The addition of more aryl groups has been considered for a (8,0) tube, and the results indicate that the aryl groups prefer the pair arrangement with stronger binding energy (see Figure 6); the energy of the system is lowered by ~0.1 eV. [67]

A number of substituted aryl groups can be attached to CNT; in addition to simple substituents such as $-NO_2$, -COOH, -Cl, SH [93] more complicated groups such as -(C=O)-$NH[(CH_2)_2$-O-$]_2$-$(CH_2)_2$-$NH_3^+Cl^-$ (MWCT-NH_3^+) that permits to disperse the modified CNT [94],polyamines, [95] diphenyl amine, [96] pyridines, [97] bi- and terpyridines and their Ru complexes, [98] azobenzene, anthraquinone, [99] fluoroalcohols [89] and crown ethers.[91] Attachment of the monodiazonium salt of 1,4-benzenediamine [100] or ethylene dianiline [90], [101] $(NH_2$-C_6H_4-$CH_2)_2$ is particularly interesting as after diazotation and grafting of a first amino group they provide a free amino group that can be diazotized again for further attachments (see below).

Similarly to the surface modification of carbon and metals by diazonium salts, multilayers are obtained on CNT; it is however possible to obtain near monolayers by a formation-degradation approach that involves grafting a disulfide *CNT*-Ar-S-S-Ar-Ar-S-S-Ar and electrochemical cleavage of the -S-S- bond to give *CNT*-Ar-SH. [102]

It is possible to assess the ratio: [number of aryl groups attached to the surface/number of carbon atoms of the CNT] by XPS if this aryl group is substituted by a unique elemental marker (atom different from carbon). For example, with 4-fluorobenzenediazonium, one aryl group is obtained for 20-30 carbons, [69] but these values should be considered with care as the comparison of XPS ratios with the variation of the Raman D-band (that represents the amount of defects on the CNT sidewall) points to the formation of multilayers *CNT*-Ar-Ar-Ar previously observed on carbon and metals (Figure 7). [70] Moreover, if the CNT is wrapped by aryl layers, peaks from the nanotubes are attenuated. Simple models of quantification for homogeneous compounds are no longer valid and induce errors in quantifying the amount of top layer.

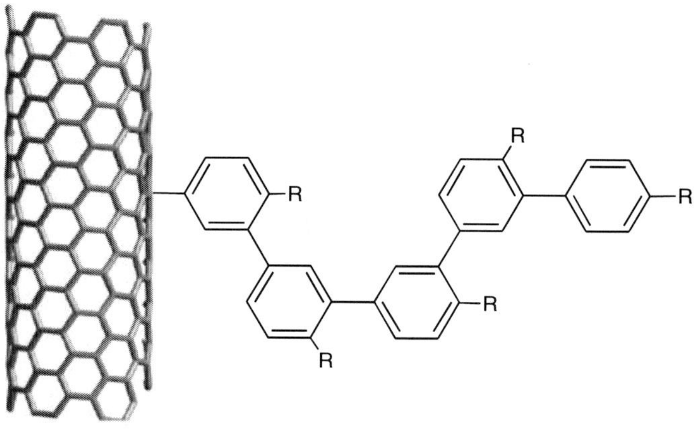

Figure 7. Aryl multilayers on a CNT.

Reversibility of the reaction. An interesting feature of the surface modification of CNT is the possibility to ungraft the aryl layer and to recover unmodified CNT, by heating at ~ 400°C; [69] or under UV light [103] the reaction can be followed by Raman spectroscopy and conductance measurements. [103]

Diazonium derivatives. By reaction with amines, diazonium groups ArN_2^+ provide triazenes [104] Ar-N=N-NR_2, these species are particularly interesting because they are stable in basic medium but convert to diazonium in acidic medium. In this way, it is possible to transform simultaneously the oxidized Si surface into SiH and attach various diazonium salts (in 2% HF). As stated above, in basic medium diazonium salts are transformed into diazoates Ar-N=N-O^-. For example, with much investigated 4-nitrobenzenediazonium at pH 9.4 equal concentrations of diazonium and diazoates are obtained [65] and with 4-chlorobenzenediazonium at pH=11 about half of the diazonium salts are transformed into diazoates. [66] These diazohydroxydes dediazonize spontaneously to the aryl radical $Ar^•$, they provide an easy method for surface modification. [105]

Multifunctionalization is possible, [106] by simultaneous or successive grafting of different diazonium salts to obtain *CNT*- (Ar_1, Ar_2, Ar_3), but this may lead to the formation of mixed multilayers a radical from a given diazonium attacking the grafted aryl group obtained from another diazonium salt *CNT*-Ar_1-Ar_2. It is also possible to prepare *CNT*-COOH by oxidation [106c] or to perform 1,3-dipolar cycloaddition of azomethine ylides and then add aryl groups from diazonium salts.[77, 107, 108]. As we will see below, this is important for biomedical applications of CNT were different grafted groups on the same tube can confer different properties to this tube.

Mechanism. When the surface modification of CNT is performed by electrochemistry or in the presence of a reducing agent, the reaction mechanism is similar to that observed on GC or metals (where aryl radicals have been evidenced by epr [109]); an electron transfer to the diazonium salt provides an aryl radical that reacts with the CNT as shown in Figure 8 (the evolution of Ar-CNT$^\bullet$ will be discussed below).

$$ArN_2^+ + 1\,e^- \longrightarrow Ar^\bullet + N_2 \quad R3$$

$$CNT + Ar^\bullet \longrightarrow Ar\text{-}CNT^\bullet \quad R4$$

Figure 8. Electrochemical grafting of CNT by diazonium salts.

The problem is more complicated when the reaction takes place spontaneously. Indeed, several mechanisms have been proposed; the most convincing one [110] is based on experiments that demonstrate the involvement of radicals by spin trapping, epr and partial kinetics reaction orders in water; there is therefore no doubt about the homolytic character of the reaction. By careful examination of the epr spectrum the signal could be assigned to *CNT*-Ar$^\bullet$. The mechanism has therefore been written as a chain reaction where Ar$^\bullet$ is the chain carrier:

$$ArN_2^+ + H_2O \longrightarrow Ar^\bullet + N_2 \qquad R5$$

$$Ar^\bullet + m/scCNT \longrightarrow m/scAr\text{-}CNT^\bullet \qquad R6$$

$$ArN_2^+ + m/scAr\text{-}CNT^\bullet \longrightarrow m/scAr\text{-}CNT^+ + Ar^\bullet + N_2 \qquad R7$$

Figure 9. Reaction mechanism of diazonium salts with SWCNT. m/scCNT stands for a mixture of metallic and semiconducting nanotubes.

Equation *R4* corresponds to a spontaneous homolytic dediazonation (although dediazonations in water are generally known to be mainly heterolytic [111]), this radical attacks a mixture of metallic and semiconducting CNT to give m/scAr-CNT (equation *R5*, Figure 9). The authors were also able to follow independently the reaction kinetics of m and scCNT and found that mCNT reacted 2 to 4 times faster than scCNT, producing mainly mAr-CNT. When all mCNT have been reacted, scCNT react more slowly, this is in agreement with the estimated reduction and oxidation potentials. The only drawback of this mechanism is that it ends (equation *R6*, Figure 9) with cationic species m/sc Ar-CNT$^+$ that cannot react through a nucleophilic reaction. [112] A possibility would be the attack of two aryl radicals on the same region of the CNT and pairing of the electrons, but this would shortcut the chain reaction. The mechanism becomes even more complicated in the presence of surfactants. [86] Selectivity between semiconducting and metallic SWCNT was observed to depend on the surfactant, the diazonium salt and its concentration, the reaction temperature, and pH. Additionally, the surfactant concentration was found to exert a significant influence on the selectivity rendering, for example, scCNT become more reactive than mCNT in the presence of sodium dodecylsulfonate. [86]

The selectivity of grafting between metallic and semiconducting CNT, as described above, has been investigated experimentally by several authors either on bulk mixtures [113] or on isolated nanotubes. [114, 115] However, in all these experiments the pH of the aqueous solution was not controlled at low pH and therefore the authors have observed the reaction of diazoates and diazonium salts.

Post functionalization: miscellaneous. As many substituents can be introduced para of the diazonium function, further functionalization of the CNT can be achieved; some recent examples are given below. Click chemistry has been widely used in conjunction with diazonium grafting to provide functionalized CNT. [9d] For example, SWCNT were modified by reaction with N_2^+-C_6H_4-C≡CH and clicked with an azido modified β-cyclodextrin, [116] or horseradish peroxidase. [117] The symmetrical reaction is possible: an azidophenyl modified SWCNT (*SWCNT*-C_6H_4-N_3) can react with a number ethynyl substituted compounds to provide dendrimer, ferrocene or β-D-glucopyranoside modified SWCNT. [118] By attaching –C_6H_4-COOH groups to CNT it is then possible to create amide bonds with the amino group of an antibody protein that can then recognize a biomarker for cancer. [119] In the same way, it is possible to attach a near–infrared fluorophore terminated by an amino group for photovoltaic applications. [120] The Suzuki reaction between SWCNT-C_6H_4-I and porphyrins has permitted their attachment and the quenching of their fluorescence. [121] Concerning post functionalization with grafted polymer for making nanocomposites, the reader is referred to Section 3.7

2.1.5. Other Radical Reactions

Other free radicals can also be used for the surface modification of CNT; for example radicals produced from the thermal decomposition of azo compounds [122] ($R_1R_2R_3$-C-N=N-C-$R_1R_2R_3$) react with MWCNT to give *MWCNT*- $R_1R_2R_3$.

2.1.6. Reductive Alkylation

SWCNT can be reduced by Li or Na in aprotic organic solvents, [123] liquid ammonia [124] (Figure 10) or ethylenediamine [125] to give soluble SWCNT$^-$ Li^+ or Na^+. Addition of n-butyl lithium, [126] potassium naphthenate, [127] lithium *n*-propylamide [128] is also possible to give *SWCNT-n*Bu, for example. The charged nanotubide salts: SWCNT$^-$ Na^+ can be easily dispersed in polar aprotic solvents [129] and water. [130] They can react with electrophiles such as alkyl halides (RX) to give *SWCNT*-R; [134] this reaction is analogous to the Birch reaction. The reaction occurs on sp^3 defect sites of SWCNT and propagates exclusively from these defects (the probability more than 1,300 times higher than otherwise random bonding to the 'π-electron sea' [131]). The same reaction has been carried out with aryl iodides (I-Ar-R; R = H, NH2, *t*-Bu, OMe, COOH), to give *CNT*-Ar-R. [132] Reaction of 6-bromohexanoic acid with reduced SWCNT provides *SWNT*-[$(CH_2)_5$COO]Na; the reaction is diameter dependent, smaller SWCNT being more reactive. This diameter-dependent alkylcarboxylation is attributed to diameter-dependent electron-transfer kinetics; it provides a -time consuming- method for the isolation of smaller diameter SWCNT. [130]

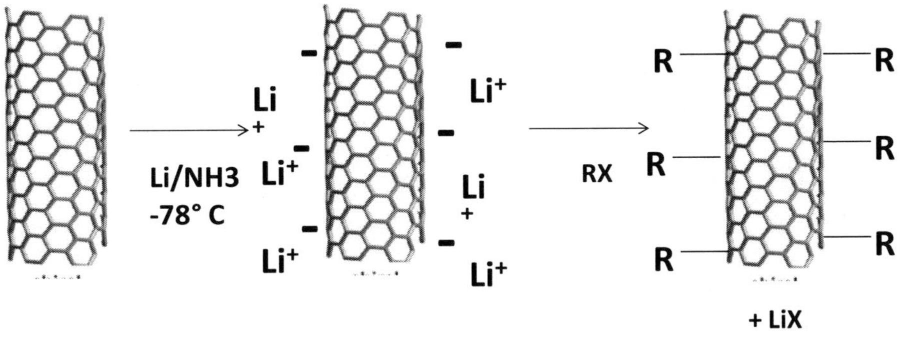

R8

Figure 10. Reductive alkylation. Adapted from reference [124b].

With n-butyl lithium, *SWCNT-n*Bu and *SWCNT*-(*n*Bu, R), can be obtained in the presence of an alkyl halide, RX. [126] By adding N-bromosuccinimide (Br-NSucc), dibenzoylperoxyde (Bz-O-O-Bz) or hydrazine to the reduced $SWCNT^- Li^+$ it is possible to obtain *SWCNT*-NSucc, and *SWCNT*-OBz; treatment of *SWCNT*-NSucc permits to obtain *SWCNT*-NH_2. [133] A modified procedure has permitted to increase the variety of substituents: in THF, the excess of Li is removed before adding the carbonyl-based electrophiles [acetophenone, 1,1,1-trifluoroacetone, methylbenzoate, methyl trifluoroacetate, 3,4-dichlorobenzoyl chloride] to give, for example, SWCNT-CH(OH)-C_6H_5 from acetophenone. [134]

Post functionalization of reduction modified CNT is possible for example, $SWCNT^- Li^+$ can be reacted with ω-bromocarboxylic acids to give *SWVNT*-$(CH_2)_n$-COOH that are further reacted with NH_2-PEG-OMe (PEG = poly(ethylene glycol)) to give SWCNT-$(CH_2)_n$-C(=O)-NH-PEG-OMe that can be dispersed in water. [135]

2.2. Noncovalent Modification

Surface modification of CNT is possible through weak non covalent interactions such as π-stacking, charge transfer and hydrophobic interactions. These non-covalent surface modifications permit to improve the dispersion of CNT without breaking and creating bonds on the CNT skeleton. [136]

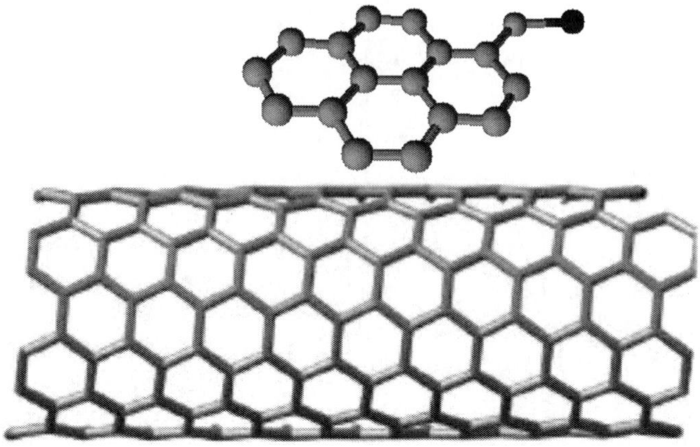

Figure 11. Non-covalent modification of a CNT with pyrenemethylamine.

2.2.1. Pyrene (Py) Derivatives

These compounds have been widely used for this purpose, they attach to the surface of CNT by π–π interaction (with a strength of 55 pN with graphite [137]) and can bear substituents (Figure 11) that permit the dispersion [138] of CNT or further chemical modifications; in addition, quenching of pyrene fluorescence and electron transfer between CNT and pyrene are of interest for further photovoltaic applications. In addition to pyrene, heterocycles such as Acridin Orange can also bind non-covalently to the surface of SWCNT. [139] SWCNT can be bonded to a water-soluble perylenediimide, a strong electron acceptor, by π– π and strong charge transfer interactions to give a stable electron donor-acceptor hybrid: SWCNT-perylenediimide. [140] $Py(CH_2)_nCOO^-$ is water soluble and provides *SWCNT*-$Py(CH_2)_nCOO^-$ that interacts with a porphyrins bearing 8 pyridinium groups: *SWCNT*-$Py(CH_2)_nCOO^-$ ZnP^{8+}, when this assembly is photoexcited with visible light, a rapid intrassembly charge separation causes the reduction of the electron-accepting SWNT and, simultaneously, the oxidation of the electron-donating ZnP^{8+}. [141] SWCNT have been modified with spiropyrans that undergo reversible transformations between the closed spiropyran and the opened merocyanine form, and are stable over many cycles of operation. [142]

Py-CH_2NH_2 has been attached to the sidewalls of SWCNT while the amino group can bind gold nanoparticles. [143]

2.2.2. DNA Wrapping

Bundled SWCNT associate with single-stranded DNA and the wrapped nanotubes can then be dispersed individually in water. Interaction occurs through π-stacking, resulting in helical wrapping to the surface (Figure 12). [144] The conformation of DNA (single strand versus hairpin DNA) influences its interaction with SWCNT. [145] It is possible to produce parallel SWNT arrays with sub-20 nm uniform spacing by using small structured DNA linkers that assemble solution dispersed CNT. [146]

2.2.3. Biological Molecules

Proteins can also interact non-covalently with SWCNT, [147] for example lysozyme a protein with an activity against *Staphylococcus aureus*, binds to SWCNT by direct molecular interaction between the indole moiety of tryptophan residues and the aromatic sidewall of SWNT. [148]

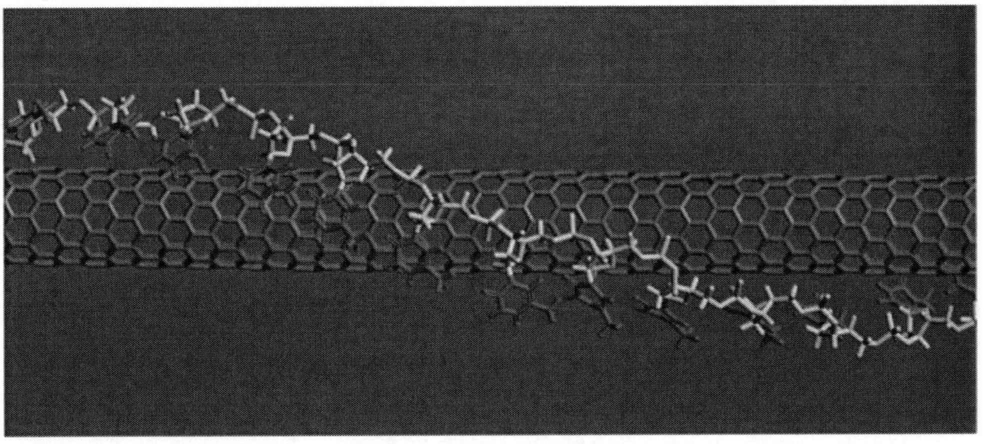

Figure 12. Binding model of a (10,0) carbon nanotube wrapped by a poly(T) sequence. Reproduced from reference [144] with permission of Nature Publishing Group.

2.2.4. Polymers Wrapping

Polymers wrapping on CNT and absorption of CNT in polymers has been reviewed, [149a] as well as the interaction of CNT with conjugated polymers. [149b] Charged polymers can at the same time interact with CNT and favor their dispersion in water through their charge: a polymer with tetraalkylammonium functional groups in the side chains of the macromolecule, creates a cloud of positive charge around SWCNT, which makes them hydrophilic and water dispersible. [150] Poly(pyridinium salt) [151] containing various aromatic diamine moieties and an ionic liquid-based polyether [152] [poly(1-glycidyl-3-methylimidazoliumchloride)] interact with SWCNT and permit their dispersion.

Poly[(*m*-phenylenevinylene)-*co*-(*p*-phenylenevinylene) substituted with alkoxy groups have been designed to wrap around SWCNT improving their dispersibility; in addition, the polymers can bear pseudorotaxanes in view of creating arrays of switches and actuators. [153] Even rigid linear poly(phenylenevinylene) that cannot wrap around the SWCNT can help in dispersing the CNT. [154] With block copolymers such as poly(3-hexylthiophene)-b-poly(acrylic acid), the poly(3-hexylthiophene) blocks bind to the surface of de-bundled carbon nanotubes through non-covalent π–π interactions, stabilizing the carbon nanotube dispersion, while the functional blocks located at the outer surface of the CNT confer specific properties to the CNT through the carboxylic groups. [155]

2.2.5. Surfactants

Comparative studies [156, 157] of the interaction between MWCNT and different surfactants in aqueous solution (sodium 4-dodecylbenzenesulfonate (NaDDBS), hexadecyl(trimethyl)amonium bromide (CTAB), sodium dodecane-1-sulfonate (SDS), Pluronic F68, Pluronic F127, and Triton X-100) have been performed; the modified CNT have been examined by XPS and thermogravimetry. It has been shown [156] that the tube diameter is an important parameter, that dispersion reaches a plateau upon increasing the concentration of the surfactant and that the best dispersion is obtained with CTAB at neutral pH. Non-ionic surfactant adsorption was negligibly sensitive to pH change from neutral. In contrast, both cationic and anionic surfactants were found to be poor dispersing aids for highly acidic solutions while, CTAB remained a good surfactant under strongly basic conditions [156] and Pluronic F88 permits a good dispersion of DWCNT in acidic aqueous solution. [190] The effect of SDS concentration has been investigated: as the SDS concentration in solution is increased, the ability of the SWCNT surface to absorb an increased surfactant loading becomes saturated as well as the ability for the SDS surface structures to reorganize in response to electrostatic screening. [158] Interaction between SWCNT and NaDBS is reported to rely on purely π–π stacking interactions, [159] but Coulomb interactions and Lewis acid–base complexation were found to play a key role in surfactant assisted MWCNT dispersibility. With gemini surfactants (surfactants that typically possess two hydrophobic tails and two polar head groups linked by a spacer), the dispersion is ensured by strong hydrophobic interactions between the two alkyl chains and the CNT backbone and the high charge capacity per single molecule of surfactant; this permits to use small amounts of surfactants. [160]

3. Applications

3.1. Dispersion of CNT

As indicated in the introduction, dispersion of CNT is a mandatory step for further applications of these species. It is by no way easy as one must struggle against a cohesion energy of 414 J nm^{-1} per (10,10) tube. [161] The terms solubilization and dispersion are both used in the literature, but as pointed in a recent review dispersion should be used for CNT in solvents, this review also discusses the different types of dispersions as well as the interactions that permit these dispersions. [162] Ultrasonication in different solvents can be used for the dispersion of CNT (for example in N-methylpyrrolidone [163] or DMF/N-methylpyrrolidone, [164] but also ionic liquids alone [165] or in aqueous solutions [166]). However, for long sonication times the tubes are damaged or cut. [167] Surface modification (covalent or non covalent) is therefore the main path to the dispersion, very often accompanied by an ultrasonication step. This topic has been recently thoroughly discussed in two reviews including suggestions for a better definition of dispersion and tables with dispersibility data. [162, 168] Dispersion of CNT in water is important for their further processing, it is possible with the help of surfactants, [169] proteins, [170] but also by covalent modification with PEG. [135] Most papers dealing with the dispersion of CNT only show images of homogeneous dispersion without giving dispersibility numbers. Table 2 gathers some quantitative dipersibility data that can be found in the literature. The values range from 0.2 to 12 mg mL^{-1} the best values being obtained either by reduction or by covalent modification with highly hydrophilic groups.

Surfactants. As indicated above, dispersion of CNT in water necessitates the use of surfactants, this topic has been reviewed. [171] The parameters involved in the process have been discussed. [161, 172] It was found that optimal dispersion may be accomplished using a tip sonicator operating at low power density (0.5 to 0.6 W mL^{-1} or below) for a sonication time sufficient to deliver an overall energy density of approximately 450 J mL^{-1}.

Centrifugation at ~ $40 \cdot 10^3$ g was found to provide the best metallic impurity removal, while ~ $120 \cdot 10^3$ g for 1–2 h affords the best compromise between concentration, aggregation state and average length of the nanotube ensemble. [172] A number of surfactants have been investigated for dispersion of carbon nanotubes, such as sodium dodecyl benzenesulfonate (SDBS), dodecyltrimethyl-ammonium bromide (DTAB), hexadecyltrimethylammonium bromide (CTAB), octylphenolethoxylate (Triton X-100), and sodium dodecyl sulfate (NaSDS)' [86, 156, 159b] cellulose xanthate. [173] Comparison of four surfactants: Triton X-100, Tween 20, Tween 80, and NaDDS indicated that Triton X-100 and NaDDS provide maximum and minimum dispersion, respectively. [159b] The mechanism of "unzipping" CNT bundles has been investigated by molecular mechanics: with the help of ultrasonication, a sufficient number of SDS molecules are found to penetrate into an initial gap between a single tube and other nanotubes in the bundle. Owing to further congregation of the surfactants at the gap site, the gap becomes enlarged until complete dispersion. [174]

Polymers can interact with CNT by non-covalent interactions as reported above and permit their dispersion in solvents -including water [175] and in polymers for the preparation of composites. The subject has been reviewed including polymer wrapping and adsorption, grafting from and grafting on. [176] A recent study [177] has shown that the dipersion of SWCNT in two different solvents (toluene, o-xylene) and polymers [poly(9,9-dioctylfluorene) and poly(9,9-dioctylfluorene-co-benzothiadiazole] depends not only on the type of solvent but also on the molecular weight of the polymer. The dispersibility increases with higher molecular weight polymers, but the selectivity between different (n, m)SWCNT decreases, the authors propose that viscosity is one of the factors influencing the apparent selectivity by changing the reaggregation rate of SWCNT. Dipersion in polymers can be achieved by polymerization of a monomer solution containing the CNT, for example, in a dispersion of SWCNT in aniline stabilized by the formation of donor acceptor complexes; electrochemical polymerization of aniline provides a SWCNT/polyaniline composite. [178] Dispersion in polymers can be improved by a dispersion technique that consists in the polymerization of a monomer in the presence of CNT in a bad solvent of the polymer. During its formation, the polymer precipitates and entraps all the CNT. [179]

DNA and other biological molecules have been used for the dispersion of CNT. [180] For example CNT were dispersed with 1,2-distearoyl-sn-glycero-3-phosphoethanolamine-N-[amino(polyethylene glycol)2000] for further binding of DNA. [181] The dispersion of CNT with the help of different proteins has been investigated; it was found that the main driving force is due to the hydrophobic interactions between the side-wall of the SWNT and the inner hydrophobic domain exposed to the solvent during the three-dimensional change of the protein induced by sonication. [182]

Covalent modification can be used for the dispersion of CNT, but at the difference of non-covalent functionalization defects are introduced in the array of sp^2 carbons and the reversibility of these reactions can only be achieved at rather high temperature. All the covalent surface reactions described above have been used for the dispersion of CNT. Carboxylation permits the dispersion of MWCNT in protic solvents [183] to produce heat transfer nanofluids; [184] highly carboxylated SWCNT can be dispersed in THF. [185] Photochemically induced dispersion of SWCNT can be achieved by anchoring 2,4,6-triarylpyrylium units after carboxylation. [186] Dispersion of CNT can also be achieved by reduction with metal–ammonia. [129]

Table 1. Some values for the dispersibility of CNT

Type of modification	Type of dispersion	Concentration of the dispersion mg mL^{-1}
Reduction [129]	$SWCNT^{n-}$	2-12 in NH_3
Oxidation [187]	*SWCNT*-COOH	0.25 in water
Oxidation [188]	*SWCNT*-COO-PEG	5.9 in water
Covalent modification [188]	*SWCNT*-CONH-$[C_6H_3(SO_3H)\text{-}NH]_n$	5.8 in water
Covalent modification [188]	*SWCNT*-COO-PEG	5.9 in water
Covalent modification [94]	*SWCNT*----NH_3^+	0.2 in water
Polymer [154]	*SWCNT/* poly(alkoxyphenylenevinylene)	~2 in $CHCl_3$
Polymer [189]	*SWCNT*/polyacrylic acid	0.4 at pH9
Polymer. [190]	*MWCNT*/alginic acid (polysaccharide)	3.2 in water
Polymer [191]	DWCNT /gum arabic	0.2 in water

3.2. Sorting Metallic and Semiconducting CNT

At the present time it is possible to produce (and to obtain commercially) SWCNT, DWCNT and MWCNT of variable length, but all the methods currently available produce a mixture (up to 40 species) of different (n,m)SWCNT and therefore of metallic and semiconducting SWCNT, for example, SWNTs prepared by the arc discharge or other methods generally contain around ~33% of metallic species. At the same time, there are many applications where metallic *or* semiconducting CNT are required for example in the micro/nanoelectronic applications. Therefore, expecting the discovery of new methods for the selective production metallic or semiconducting CNT, one has to sort metallic from semiconducting and then the different (n,m)SWCNT. There are different physical methods that permit to separate these species: gradient ultracentrifugation and dielectrophoresis; these methods have been recently reviewed. [192] Covalent and non-covalent surface functionalization have been investigated, a real separation between metallic and semiconducting species has been sometimes achieved (see below). The applications in electronics, optoelectronics and sensors of SWCNT sorted according to chirality, diameter or metallicity have been reviewed. [192]

Covalent grafting. Covalent modification of CNT with diazonium salts permits to distinguish metallic from semiconducting CNT, but most of these investigations are performed in pure water or in basic medium where diazonium salts and diazoates Ar-N=N-O^- are present, these last species are very unstable and rapidly decompose. For example, in water, at pH 10, the spontaneous reactivities of "diazonium salts" (of diazoates at this pH) were found to be greatest for semiconducting SWCNT having the largest band gaps. The magnitude of this band gap dependence varies according to the R-group of the salt, with R= OMe showing the strongest variation. For R = OH, acidification of the sample to pH 5.5 results in reversal of the structural trend, as smaller band gap SWCNT show slightly greater reactivities. [193] In another example, during the electrochemical grafting of SWCNT the changes in conductance of individually contacted SWCNT could be observed in water and at certain potentials the metallic SWCNT are modified much more strongly than their semiconducting counterparts. [194] Metallic SWCNT were selectively functionalized by reaction with the diazonium salt of 4-heptadecafluorooctylaniline and extracted in perfluorohexane leaving the semiconducting nanotubes in the aqueous layer. The metallic nanotubes obtained using this fluorous chemistry have a purity of at least 95%, and the fluorous groups on the surface can be removed by annealing at 600°C. Given its simplicity, this method can probably be scaled up. [195] SWCNT modified by reaction with 4-bromobenzenediazonium were separated by density gradient ultracentrifugation, to produce high yields of solution-processable scSWNT that were further used to fabricate Field Effect Transistors (FET) devices. [196] Concerning diazonium salt based selective separation of conducting and semiconducting carbon nanotubes the primary contribution comes from the interplay between the intrinsic electronic structure of the carbon nanotubes and that of the anion of the salt. [67] The reduction of SWCNT with Li also provides a method for the enrichment in metallic species. [130a]

Pyrene derivatives. A bis-pyrene molecule has been synthetized where two pyrene groups are linked by a specially designed tether. This molecule selectively disperses the scSWCNT in a THF−toluene mixture (80/20, v/v) and the metallic ones can be recovered by a simple centrifugation, the bis-pyrene molecule can be recycled. [197]

Surfactants. Gradient ultracentrifugation of dispersions of SWCNT in Pluronic F68 permits to recover semiconducting SWCNT with purity levels in excess of 99% while X-shaped Tetronic 1107 block copolymer displays an affinity for metallic SWNT, yielding metallic purity levels of 74%. [198]

Polymers Conjugated polymers have been used that selectively attach to (n,m)CNT. A water-soluble poly[(*m*-phenylenevinylene)-*alt*-(*p*-phenylenevinylene)] has been synthesized, with an unsymmetrical substitution pattern on the para-phenylene unit. The polymer chain has a tendency to fold in aqueous solution, thereby promoting helical conformation. The polymer is found to selectively disperse the single-walled nanotubes (SWNTs) of small diameters (d = 0.75−0.84 nm), this is assigned to the formation of a cavity that specifically hosts small diameter SWCNT. [199] Another polymer, poly-(N-decyl-2,7-carbazole) disperses in toluene almost exclusively semi-conducting SWNTs with differences of their chiral indexes (n-m) ≥ ≥ 2. [200] Various conjugated polymers consisting of alternating 9,9-dialkyl fluorene and 1,2,3-triazole units have been prepared by click chemistry, they are selective toward high chiral angle-SWCNT. [201] Selective dispersion of scSWNTs was achieved with regioregular poly(alkyllthiophene), [202] they could then be sorted by centrifugation. Copolymers composed of poly-9,9-dioctyl-2,7-fluorene and chiral binaphthol moieties with different composition ratios have been synthetized, they permit a simple one-pot separation of semiconducting (+)SWNT and (−)SWNT with a specific chirality. This recognition behavior was explained (by molecular mechanics simulations) by a cooperative effect of chiral and conformational interactions. [203] Semiconducting enriched SWCNT dispersion can be obtained with N-acetylated chitosan. [204]

Although this is not a strictly speaking a surface modification of CNT, it should be remarked that upon spin coating semiconducting SWCNT preferentially bind and align on amino terminated silanized Si surfaces while metallic SWCNT prefer a phenyl terminated surface. [205]

3.3. CNT for Micro and Nanoelectronics

CNT can be used in microelectronic applications [206]: by aligning a semiconducting SWCNT between source and drain on a Si back-gate, [207] it is possible to construct nano transistors; on the other hand metallic SWCNT can be used in the interconnection of transistors. These applications obviously rely on the separation of metallic and semiconducting SWCNT described above. For example, scSWCNT sorted by wrapping with a polyphenylenevinylene polymer provide performances similar to CVD grown CNT particularly as concerns the quality of the electrical contacts between CNT and metal electrodes. [206d] It is also possible to reduce the conductivity of metallic SWCNT in a mixture of both metallic and semiconducting SWNT by selective reaction of metallic SWCNT with 4-nitrobenzenediazonium salt [208] or radical initiators as dibenzoylperoxide; [209] their contribution to the transport can be completely suppressed so that the device now acts as a purely semiconducting FET.

Due to their high conductivity defectless metallic CNT could replace copper in the microelectronic interconnects, as a first step they must be aligned vertically or horizontally in the dielectric. This can be achieved by selective growth or by surface modification [210]. Ethylenedianiline (H_2N-C_6H_4-$(CH_2)_2$-C_6H_4-NH_2) is bonded first to the Si or Ti bottom of the trench (*Si, Ti*-C_6H_4-$(CH_2)_2$-C_6H_4-NH_2) and then to the DWCT (*Si, Ti*- C_6H_4-$(CH_2)_2$-C_6H_4-*DWCNT*) through diazonium chemistry, the reverse order of modification is also possible (H_2N-C_6H_4-$(CH_2)_2$-C_6H_4-*DWCNT* is attached to the surface); in this way the DWCNT are attached selectively to the bottom of the trench etched in SiO_2. [90] In another example, SWCNT were first chemically functionalized through diazonium chemistry with a hydroxamic acid end group that both renders the SWCNT water-soluble and discriminately binds the SWCNT to basic metal oxide surfaces (i.e., hafnium oxide (HfO_2)). The functionalized SWCNT are then assembled from an aqueous solution into narrow trenches etched into SiO_2films with HfO_2 at the bottom. [211] SWCNT dispersed with sodium cholate specifically align in hydrophilic HfO_2 modified trenches (*HfO_2*-H_2O_3P-C_6H_4COOH) surrounded by long alkyl chains hydrophobic regions. [212]

CNT also find applications in flexible electronics. [213] Transparent conducting films are useful for flexible transparent electronic devices, such as touch screens, e-papers, displays, transparent transistors, etc. SWCNT non-covalently modified by (poly(3,4-ethylenedioxythiophene)/poly(styrenesulfonate) (PEDOT/PSS) in dimethyl sulfoxide (DMSO) solvent have been used for preparing flexible transparent (90.5% at λ=550 nm) conducting films on polyethylene terephthalate with a with a low sheet resistance of 118 Ω/sq. [214]

3.5. Sensors

Many applications can be found concerning analytical applications of modified CNT. Their uses in electroanalysis, bioanalysis [215, 216] and in switches and FET [217] have been reviewed.

Gas sensors. [218] An artificial nose constructed from a modified SWCNT-FET platform was composed of polyethylene glycol (PEG)-coated regions to prevent non-specific absorption and chemically modified SWCNT regions. In these regions SWCNT was functionalized by non-covalent modification with 1,5-diaminonaphthalene, further attachment of the human olfactory receptor 2AG1 (hOR2AG1) was achieved through a glutaraldehyde linker. This artificial nose was tested by monitoring the source–drain current with different esters such as amylbutyrate; it displayed an ultrahigh sensitivity down to concentrations as low as 1 fM. [219] In a similar way, a highly sensitive and selective olfactory-nanovesicle-fused bioelectronics sensor based has been designed; it is based on a canine olfactory nanovesicle specific receptors–CNT hybrid structure. It was developed for the detection of hexanal, an indicator of the oxidation of food and can detect down to 1 fM of hexanal. [220]

Another interesting application of SWCNT based gas sensors is provided by the analysis of ethylene. The internal ethylene concentration in fruit can serve as an indicator for determining the time of harvest, while monitoring the atmospheric ethylene level in storage facilities and transportation is crucial for avoiding overripening of fruits. A sensor was constructed from an intimate mixture of SWNTs with a copper(I) complex based upon a fluorinated tris(pyrazolyl)borate ligand (known to form strong complexes with ethylene), which is able to interact covalently with the surface of carbon nanotubes. This sensor is able to detect sub-ppm of ethylene, a concentration that occurs during ripening of fruits. [221]

Other recent examples of CNT based sensors are gathered in Table 2.

Table 2. CNT-based sensors

Sensing system	Type of modification	Detected gas	Sensitivity	Referenc e
SWCNT-(octaethyl, tetraphenyl) (Fe, Ru)porphyrins	Non covalent	Acetone, methanol, ethanol, methylethylketone	sensitivity was limited to ~ 20% saturation	[222]
CNT-amylose, amylopectine	Non covalent	Methanol, water, toluene	-	[223]
MWCNT-NH$(CH_2)_2$-NH_2	Covalent MWCNT + $SOCl_2$ + amine	Formaldehyde	20 ppb	[224]
MWCNT paper// Pd nanoparticles	MWCNT-NH_2 crosslinked with benzoquinone	H_2	5%	[225]
MWCNT/polyurethane	Non covalent	Organics: $CHCl_3$, THF…	16 mbar	[226]
CNT/cholesteric liquid crystals	-	acetone	~ 10 ppm	[227]
SWCNT layer by layer	Non covalent SWCNT + sodium deoxycholate + poly(diallyldimethyl-ammonium chloride) solutions	Volatil Organic Compounds	-	[228]
CNT/polylactic acid	Non covalent	Volatil Organic Compounds	-	[229]

Biological molecules, Pollutants, Protons. Modified CNT can be deposited on electrodes, they increase the surface of the electrode and the modification improves the interaction with analytes. MWCNT modified by oxidation and immobilization of quercetin can be deposited on a carbon electrode; this modified electrode makes possible the detection of levodopa, uric acid and tyramine at concentration down to 0.38 μM and at potentials less positive than on bare GC eleclectrode. [230]

A microfluidic electrochemical sensor based on an indium tin oxide (ITO) microelectrode modified with oxidized CNT (*CNT*-COOH) was developed for single cells release monitoring. The sensitivity of this electrochemical sensor was improved by 2.5–3 orders of magnitude upon modification of the CNT, it was employed to monitor the dopamine release from single living rat pheochromocytoma (PC 12) cells. [231]

SWCNT were modified by reaction with the diazonium salt of a β-cyclodextrin derivative, after deposition on digitated electrodes the conductance of the device was measured. In the presence of Persistent Organic Pollutants (POPs) such as 2,4,5-trichlorobiphenyl (TCB) an inclusion complex is formed between SWCNT–□β-cyclodextrin and POP, and the conductance of the SWCNT–PCD hybrid networks decreases. [232]

Metallic SWCNT modified by reaction of the diazonium salt of N, N-diethylaniline can be used as resistive pH sensors. The attached organic moieties function as pH dependent carrier they create sp^3carbons acting as electron scattering sites, as a result of which the resistance of the modified tube varies as a function of pH. The degree of functionalization and hence the pH-sensitivity can be controlled by varying the parameters of electrochemical modification. [233]

Peptides, enzymes sensors. The modification of electrodes by deposition of CNT can increase the surface and the rate of electron transfer of the electrode and therefore the sensitivity of electrochemical sensors. These CNT can in their turn be modified by catalytic molecules for improving the selectivity. [234] For example sonicated SWCNT were adsorbed on the top of glassy carbon electrodes and covalently modified with aryl diazonium salts generated *in situ* from p-aminobenzoic acid and p-phenylenediamine, thus featuring negative or positive surface charges at acidic pH (3.5 and 4.5). These surface groups permit the adsorption of cellobiose dehydrogenase that catalyzes the deshydrogenation of lactose. The resulting electrode is stable and presents a high catalytic current density of 500 $\mu A\ cm^{-2}$. [235] SWCNT deposited on a carbon electrode were non-covalently modified with the electron transfer mediator Meldola blue and malic acid dehydrogenase; this modified electrode permits the detection of malic acid with a detection limit of 3μM. [236]

A FET was fabricated from SWCNT to produce a trinitrotoluene (TNT) sensor, the SWCNT were non covalently modified with a peptide, specially engineered by molecular mechanics calculations, for recognition of explosive molecules A very low pressure of TNT could be detected $P = 9.10^{-6}$ Torr. [237]

3.6. Biomedical Applications

From the point of view of biomedical or pharmacological applications, carbon nanotubes appear as very strange objects, the literature is split between the description of their toxicity and their potential biomedical/pharmaceutical applications.

Toxicity. The concern about CNT toxicity is associated with their structural resemblance to carcinogenic, long asbestos fibers. [238] Several reviews describe the toxicity of CNT and the origin of this toxicity. [239] After modification the toxicological properties and the biodegradability of CNT are modified as detailed in a recent review. [239b] For example,

MWCNT functionalized (by covalent azomethine ylide cycloaddition) with diethylenetriaminepentaacetic di-anhydride give a stable dispersion with high excretion rates. [240] The oxidative biodegradation depends on the substituents and oxidized CNT are more easily oxidized [35] than for example CNT-C_6H_4-SO_3H obtained from diazonium chemistry. [241] The toxological response to the deposition of MWCNT and modified MWCNT-COOH in mice lung has been analyzed. [242] Surface-modified multiwalled carbon nanotubes (MWNTs): ammonium-functionalized MWNT (*MWNT*---NH_3^+) and Pluronic F127 coated MWNT have been tested *in vitro*. Chemically functionalized MWCNT show significantly greater cellular uptake into lung epithelial cells compared to the non-covalently Pluronic F127-coated MWCNT. In spite of this, Pluronic F127 coated MWCNT exhibit enhanced cytotoxicity according to the modified lactate deshydrogenase assay. [243] Cellular cytotoxicity assays, revealed that the non covalent binding of blood proteins on the SWCNT surface can greatly alter their cellular interaction pathways and result in much reduced cytotoxicity for these protein-coated SWCNT. [244] In all these investigations, significant differences have been found between as-prepared and modified CNT.

Biomedicine. The prospects of modified carbon nanotubes in disease imaging [245] and therapy [246] have been recently discussed as theranostics, [247] as anticancer agents [248] as neurological agents [249] and for the delivery of therapeutic DNA agents [250]. The use of *CNT*-PEG for *in vitro* and *in vivo* therapeutic and imaging tools has been reviewed. [251]

MWCNT–NH_3^+ has been used for 3-D electron tomography techniques that offer optimum enhancement of contrast between carbon nanotubes and the plasma membrane to investigate the mechanisms involved in the cellular uptake of shortened of functionalised multi-walled carbon nanotubes. [252]

In view of imaging tumors sophisticated modifications of SWCNT was achieved. SWCNT constructs were covalently grafted by 1,3-cycloaddition of azomethide ylides with radiometal-ion chelates (1,4,7,10-tetraazacyclododecane-1,4,7,10-tetraacetic acid or desferrioxamine B) and the tumor neovascular-targeting antibody E4G10. The E4G10 antibody specifically targeted the monomeric vascular endothelial-cadherin epitope expressed in the tumor angiogenic vessels. [253]

For fluorescence targeted imaging of estrogen receptor overexpressed cancer cells, acid treated CNT were further modidied (π–π stacking) by poly(p-phenylene) decorated with poly(ethylene glycol). The obtained water self-dispersible, stable, and biocompatible *CNT*/PP-g-PEG conjugates were then bioconjugated to estrogen-specific antibody via −COOH functionalities present on the side-walls of CNT. [254]

In order to label stem cells SWCNT were non covalently modified with an amphiphilic polymer: amine-PEG-grafted poly(maleic anhydride-alt-1-octadecene); then SWNT-PEG with amine functional groups on PEG terminals is reacted with thiolated protamine via a bifunctional maleimide linker (protamine is an arginine-enriched protein with abundant positive charges known to enhance the cellular uptake of nanomaterials). [255]

The direct observation of drug release from carbon nanotube vehicles in living cells has been realized through a two-dye labeling approach. SWCNT are marked with fluorescein isothiocyanate (green fluorescence) attached to chitosan (non covalent modification), then a fluorescent anticancer drug doxorubicin (DOX, with red fluorescence) is attached to the SWCNT by means of π-stacking. In the cell DOX detaches from the SWCNT in an acidic environment due to the pH-dependent π–π stacking interaction between DOX and SWCNT. From observation of the two different kinds of fluorescence (green and red) that respectively represent the modified SWCNT and drug DOX, the process of drug release inside the living cell can be monitored under a confocal microscope (Figure 13). [256]

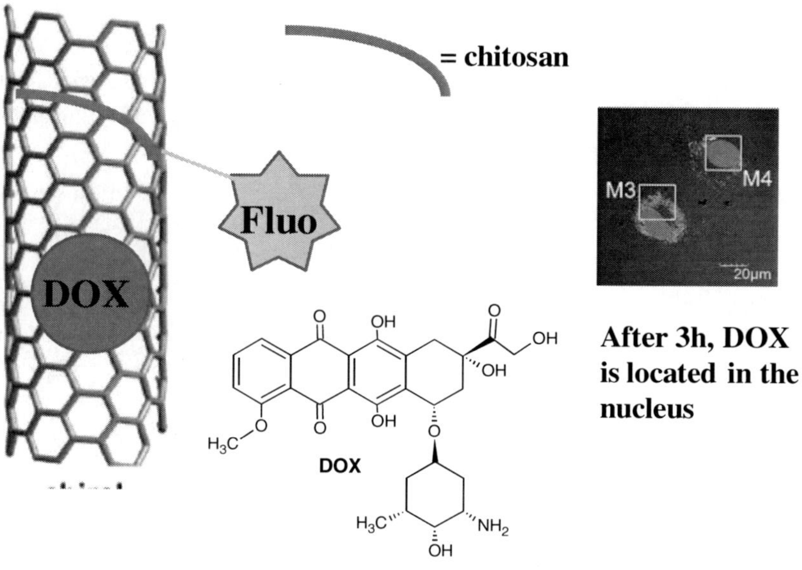

Figure 13. SWCNT modified with doxorubicin for release in the cell nucleus. Adapted from reference [256] with permission of John Wiley & Sons.

Among the many applications of CNT in biomedicine, drug [257] or RNA [258] delivery is particularly interesting. Doxorubicin, an anticancer agent, can be delivered in vitro to cancer cells thanks to a MWCNT construct. The MWCNT were functionalized with poly(acrylic acid) through free radical polymerization, decorated with magnetic nanoparticles (MN), modified with folic acid (FA), loaded with doxorubicin with high efficiency through $\pi-\pi$ interactions and hydrogen bonding (Figure 14); they show enhanced cytotoxicity toward cancer cells compared with free DOX. From transmission electron microscopy and confocal laser scanning microscopy, it was observed that the MWCNT-MN-FA-DOX - can be efficiently taken up by cancer cells with subsequent intracellular release of DOX, followed by transport of DOX into the nucleus with the nanocarrier left in the cytoplasm. [259]

SWNT can be used as antigen carriers to improve immune responses to peptides that are weak immunogens; for that purpose they were covalently $-NH_2$ modified (1,3-dipolar cycloaddition) for the attachment of COOH terminated peptides that were further reacted with specific antigens for delivery into dendritic cells. [260]

Polyethylene glycol (PEG) of different lengths and at various densities were reacted with poly(maleic anhydride-alt-1-octadecene)-PMHC18-; this amphiphilic polymer attaches covalently to SWCNT to give *SWCNT*- [PEG -PMHC18] some of which accumulated in the skin. After intravenous injection of these modified nanotubes in vivo photothermal ablation of tumors was performed by a low power NIR laser irradiation. [261]

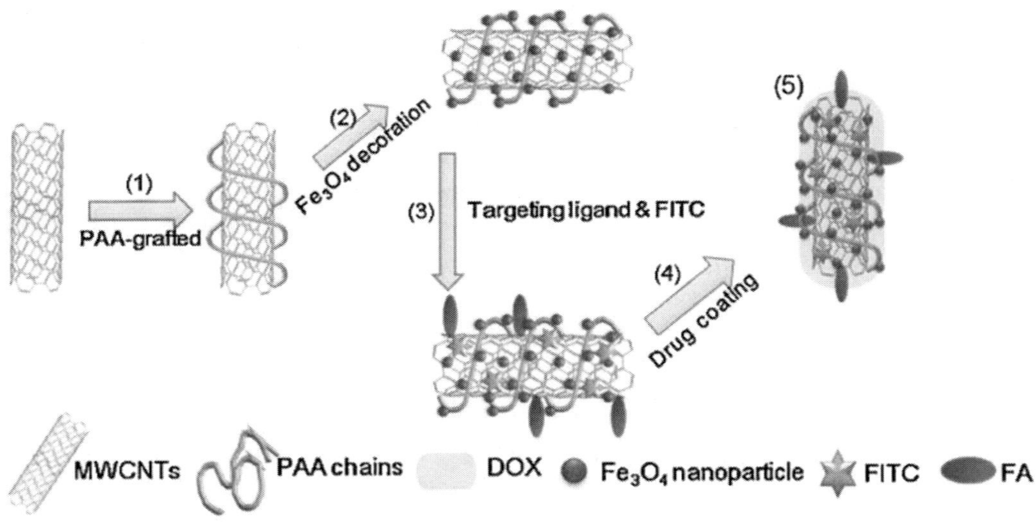

Figure 14. SWCNT for DOX delivery to cancer cells. Reproduced from ref. [259]; with permission of Elsevier.

Modified CNT can also be used for binding and transporting antibodies to cancer cells. For example, an anti MCU-1 was attached to D or MWCNT modified by anti-MUC1 that recognizes the mucin 1 (MUC1) receptor which is over-expressed in a series of human cancer cell. A first method uses tip –COOH groups of tip-oxidized CNT to attach the antibody by amidation through lysine groups; a second methods uses the 1,3-dipolar cycloaddition reaction that permits to attach triethylene glycol chain on the pyrrolidine ring with ammonium functions at the end the chain that can be further derivatized using the heterobifunctional crosslinker 3-maleimidopropionic NHS ester leading finally to the attachment of the antiboby through thioether groups. [262]

3.7. Carbon Nanotube-Polymer Nanocomposites through Covalent Grafting and Uses Thereof

In this section we shall review the making of CNT-polymer nanocomposites through covalent grafting of polymer chains. This is a topic of intense activity and has made the subject of several review papers. [263-266] Covalent grafting can be achieved through surface-confined polymerization, or using preformed polymers bearing chain ends or pendant reactive groups. The first approach is coined “grafting from” while the latter is termed “grafting onto”.

In this section we will describe the surface reactions conducted on CNT which permit further covalent attachment of polymers by the so-called “grafting from” and “grafting onto” strategies.

3.7.1. General Aspects

Several polymerization methods can be confined to carbon nanotubes pre-modified through a range of surface chemistries such as the Prato reaction, the reaction with diazonium salts or the oxidation followed by esterification to provide anchored ATRP (Atom Transfer Radical Polymerization) initiators. The surface reaction must permit to anchor either initiator or polymerizable groups. Note however that most of strategies aim at anchoring initiator groups in order to confine the polymerization process to the surface and limit the formation of free chains in solution. In the case of "grafting onto" much has been achieved in the recent years using clickable, nitrene-functionalized or diazotized polymers on the one hand, and polymeric radicals, on the other hand.

Selected patways are summarized in Figure 15. Reactive groups (R) permit to attach initiators (I) to the surface ; eventually the latter can be borne already by the coupling agents. Polymers (P) can be generated by surface-initiated polymerization techniques or attached to the surface if they are equipped with reactive moieties such as diazonium or nitrenes.

This section is not exhaustive and we shall restrict it to the most investigated approaches for making CNT-polymer nanocomposites through covalent bonding.

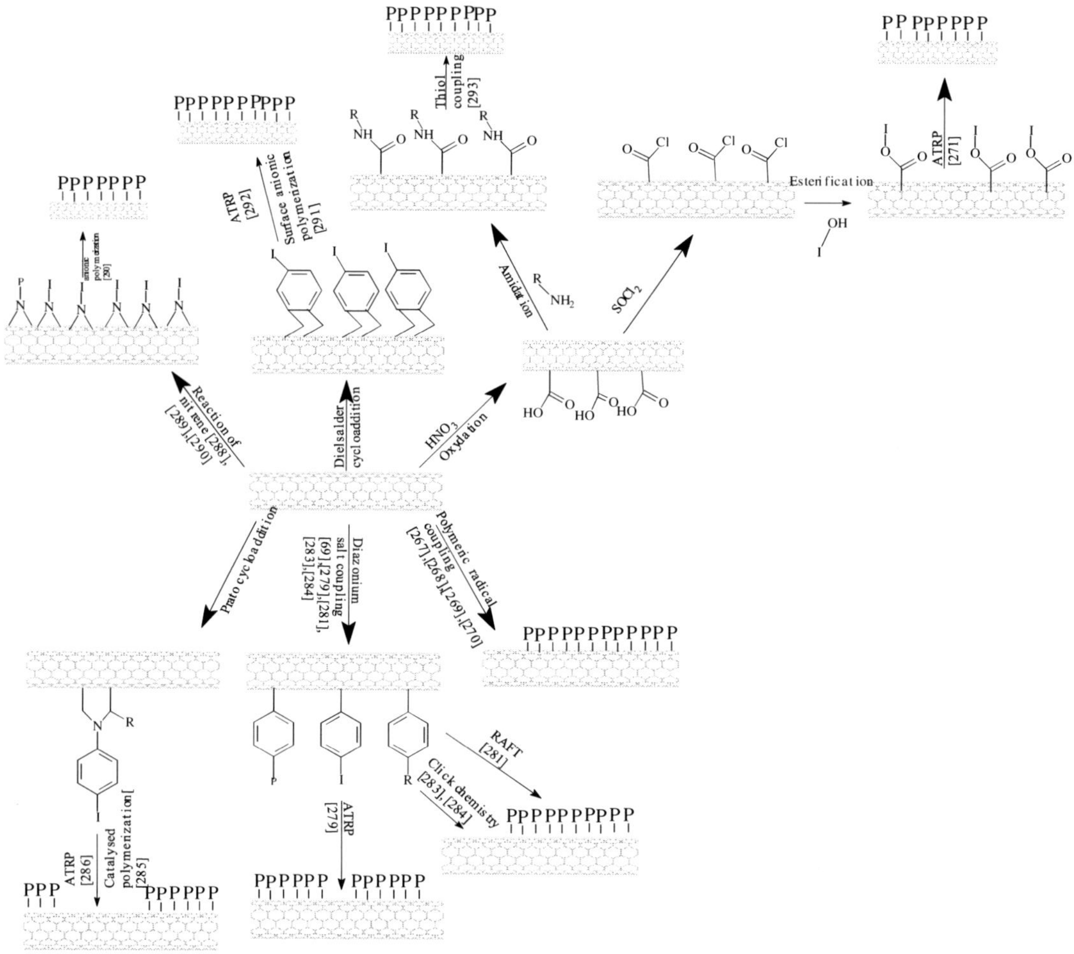

Figure 15. Selected pathways for the preparation of CNT-polymer nanocomposites. R: reactive group, I: initiator, P: polymer.

3.7.2. Reaction with Polymeric Radicals

The simplest method to make CNT-polymer nanocomposites is through a direct reaction between the polymer and the sidewalls of the nanotubes. This can be achieved through polymeric radicals generated from labile group-functionalized polymers or with the assistance of peroxides. For example, one can first prepare polymer chains by nitroxide-mediated radical polymerization (NMP) to yield polymer chains terminated by thermally labile 2,2,6,6-tetramethylpiperidinyl-1-oxy (TEMPO) end groups. Upon heating in the presence of the nanotubes these groups dissociate to give polymeric radicals reactive towards the double bonds available on the sidewalls. [267, 268] In this way, a range of NMP polymers, namely PS and poly(caprolactone-block-styrene), can be covalently grafted to MWCNT. [267] Alternatively, it is possible to functionalize a preformed polymer, e.g. polyethylene (PE), with TEMPO and react PE-TEMPO with MWCNT to obtain PE-functionalized MWCNT. [269] Another approach, consists in heating copolymers of α-methylstyrene (AMS) and a second monomer, glycidyl methacrylate (GMA) or styrene (S) in the presence of MWCNT in solution. [270] The decomposition of the AMS sequences provides macro-radicals which attack the double bonds on the MWCNT surfaces (Figure 16), hence the formation of MWCNT-grafted copolymer nanocomposites which were found to be soluble in DMF.

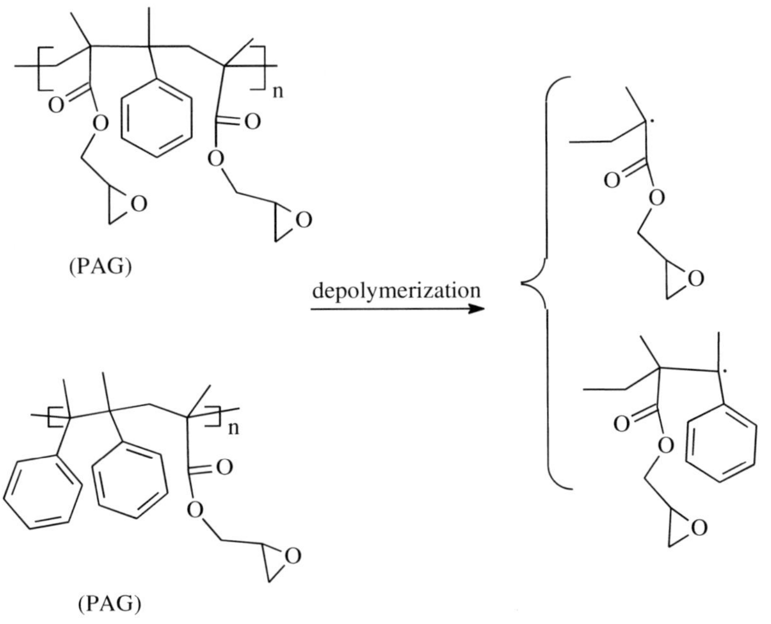

Figure 16. Depolymerization of poly(AMS-co-GMA) which results in macroradicals that are prone to react with double bonds from CNT. Adapted from [270].

3.7.3. Oxidation of CNT

Oxidation of carbon nanotubes in acidic media has been investigated by several authors as it permits to obtain short CNT with carboxylic acid groups. The latter can readily be esterified therefore providing surface grafted initiators of ATRP [271, 272], nitroxide-mediated radical polymerization (NMP) [273] and reversible addition–fragmentation chain transfer polymerization (RAFT). [274] In this regard, one of the most popular papers on CNT surface-initiated ATRP subject [271] described this classical but aggressive approach for activating single-walled carbon nanotubes (SWCNT) in strong acidic media (HNO_3) followed by attack of $SOCl_2$. The ATRP initiators were covalently attached to the SWCNT by esterification using 2-hydroxyethyl 2'-bromopropionate (Figure 17). Carbonyl chloride-functionalized MWCNT can also readily react with glycerol [272] and the OH-terminated nanotubes could then be reacted with 2-bromo-2-methylpropionyl bromide to provide carbon nanotube macroinitiators of ATRP.

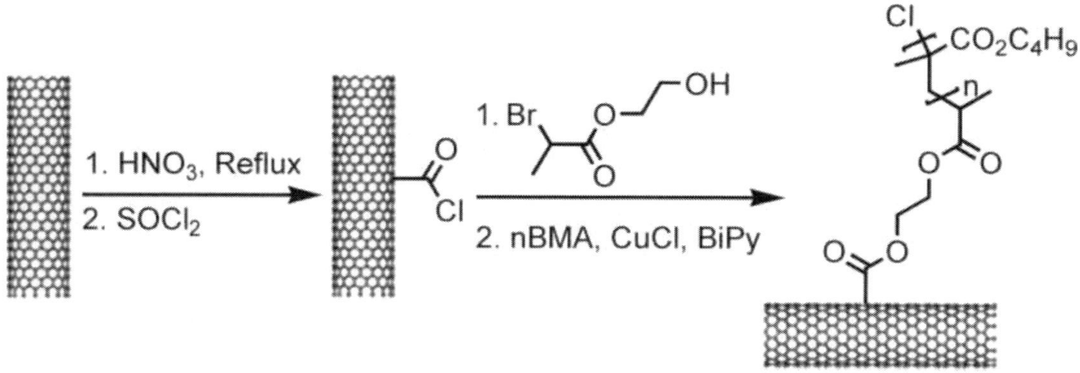

Figure 17. CNT-Polymer nanocomposite preparation by oxidation and further esterification of the Csp2 substrate. Reproduced from ref. [271] with permission of American Chemical Society.

Similar approaches were adapted in order to perform NMP and prepare dispersible MWCNT-PS hybrid. This primary nanocomposite was further used as macroinitiator for in-situ NMP of 4-vinyl pyridine in order to prepare MWCNT-(PS-block-P4VP). The diblock copolymer grafted MWCNT could be dispersed in chloroform but not in toluene where the P4VP block is insoluble. When hydrochloric acid was added to chloroform a phase separation occurred and the MWCNT-(PS-block-P4VP) nanocomposites locate at the chloroform-aqueous phase interface. In this interfacial region, P4VP blocks are soluble in either water or chloroform. In this way, grafted block copolymers with hydrophobic and hydrophilic blocks permit to tune the dispersibility of MWCNT. [273] Dispersibility of the nanotubes can also be temperature-controlled provided a thermoresponsive polymer, such as poly(*N*-isopropylacrylamide) (PNIPAM), is grafted by a surface-initiated polymerization technique (e.g. RAFT [274]).

Pristine MWCNT have naturally occurring carboxylic acid groups the surface concentration of which is sufficient for a polycondensation reaction with l-lactic acid, therefore providing biocompatible poly(L-lactic acid)-functionalized MWCNT. [275]

Oxidized CNT can be post-functionalized via amidation and esterification of the nanotube-bound carboxylic acids. [276] Interestingly, MWCNT-COOH can be post-functionalized in order to be amenable to Huisgen 1,3 cycloaddition *click* reaction to give clickable MWCNT-poly(azide) or MWCNT-poly(alkyne) hybrids. [277] The COOH groups generated by nitric acid treatment of the nanotubes were activated by propargyl alcohol therefore leading to clickable MWCNT-COO-C≡CH. Grafting polymer chains by click chemistry permitted to obtain MWCNT-polymer hybrids with a large concentration of alkyne or azide groups for a final "click" reaction between azide- or alkyne-functionalized PEGylated magnetic nanoparticles, respectively (Figure 18). The final MWCNT-polymer-Fe_3O_4 magnetic nanohybrids are robust and withstand ultrasonication; they exhibited superparamagnetic behavior at room temperature with no coercivity or remanence, and the saturation magnetization of nanohybrids was reduced compared with that of pure magnetic NPs.

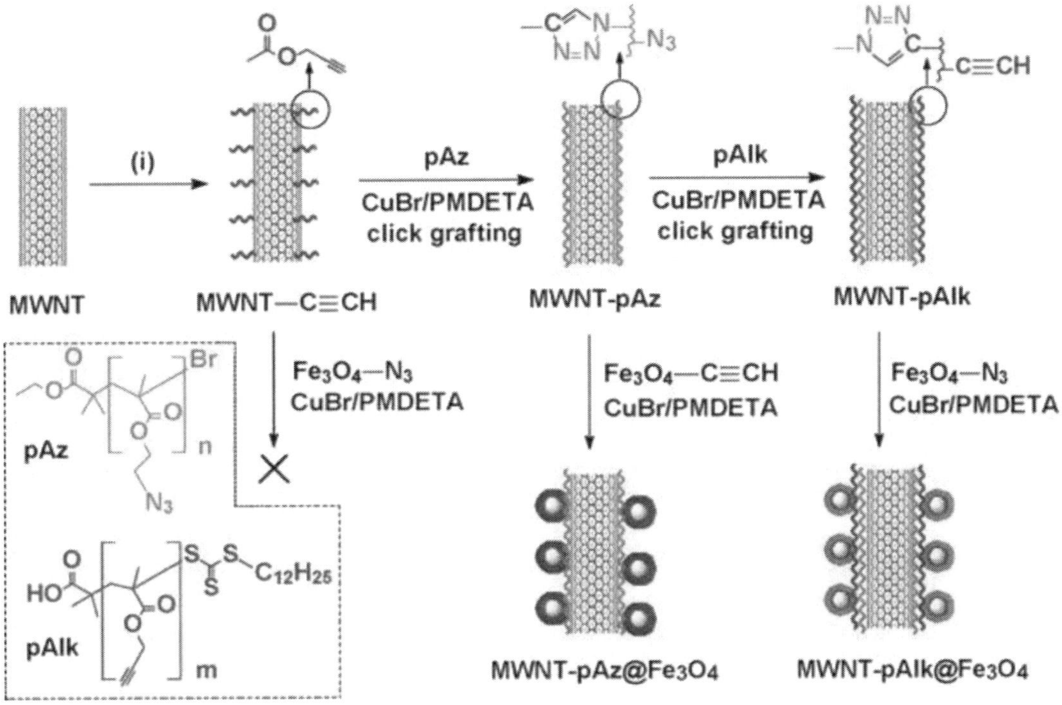

Figure 18. The synthesis procedures of 'clicked' magnetic nanohybrids. (i) (1) H_2SO_4/HNO_3, 90-133 °C, 100 min; (2) $SOCl_2$, 60°C, 24 h; (3) propargyl alcohol, triethylamine, $CHCl_3$, r.t., 24 h. Reproduced from ref. [277] by permission of the Royal Society of Chemistry.

3.7.4. Aryl Diazonium Salts

Due to the reactivity of sp2 carbon materials with aryl diazonium salts, making CNT-polymer nanocomposites using aryl diazonium coupling agents for the polymer has been much investigated. It arose as an original, elegant and versatile method. Indeed, aryl diazonium salts can be synthesized and isolated prior to use or can be generated in situ in the presence of the nanotubes (see section 2.1.4) to attach initiators of polymerization or functional groups reactive towards chain ends or pendant groups in prefabricated polymer chains. Diazotized polymers can also be prepared and reacted directly with CNT. [69]

Concerning the "grafting from" route for CNT-polymer nanocomposites, Tour et al. [278] have described the *grafting* of polystyrene *from* SWCNT via surface-initiated anionic polymerization of styrene. A bromo phenyl molecule was grafted to SWCNT through the in-situ generation and reduction of diazonium salt before switching the grafted bromophenyl to aryl lithium by reaction of alkyl-lithium (Figure 19). These species were found to be efficient anionic polymerization initiators at the surface of aryl-modified CNT.

CNT were modified by isolated or in-situ generated diazonium salts in order to obtain ATRP macroinitiators for the preparation of CNT-polymer hybrids using catalytic amount of copper salt (CuBr) and monomers under reflux in inert atmosphere. For example, chemical modification of SWCNT with 2-bromopropionate was followed by grafting poly(butyl acrylate), PBA, through ATRP [279]. In this approach, in-situ generated 4-hydroxyethylbenzenediazonium salt was reductively attached to the surface of the SWCNT to provide grafted 4-hydroxyethylphenyl groups which were esterified using 2-bromopropionyl bromide. This final reaction gives SWCNT-C_6H_4-$(CH_2)_2$-O-C(=O)$CHBrCH_3$ groups which act as macroinitiator of ATRP. The grafted chains could be directly visualized by AFM and the average surface grafting density (ρ_S = number of chains per nm² of tube surface) was found to be in the range 0.3-1.7 chain/nm² indicating possible close packed chains (($\rho_S \geq 1$ chain/nm²).

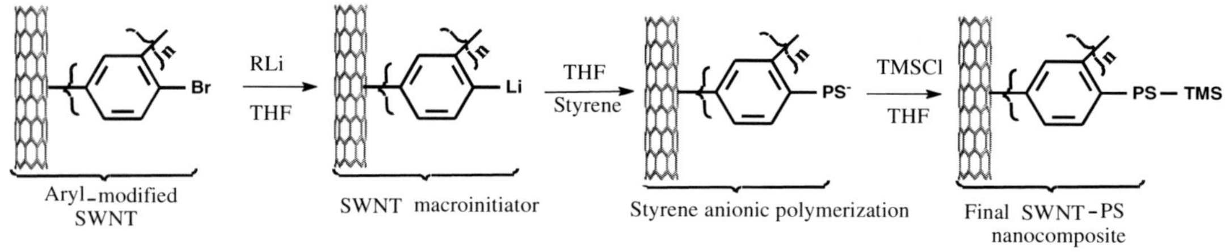

Figure 19. Schematic illustration of the anionic polymerization process at the surface of single-walled carbon nanotubes (SWCNT). TMS stands for the polymerization terminator trimethylsilyl. Adapted from ref. [278].

In another approach, Matrab et al. electrografted 4-(2-bromoethyl)phenyl groups to initiate the ATRP of MMA and styrene onto MWCNT forest. [280] Although the resulting grafted radical –C_6H_4-$CH_2CH_2^{\bullet}$ is unstable it nevertheless permitted to initiate the growth PS and PMMA chains as demonstrated by HRTEM visualization of the hairy CNT and the examination of the high resolution C1s and valence band regions from the polymer grafts. RAFT, was applied to graft polyacrylamide (PAM) to SWCNT [281], however not less than 6 steps were necessary to obtain CNT-PAM nanocomposite (Figure 20).

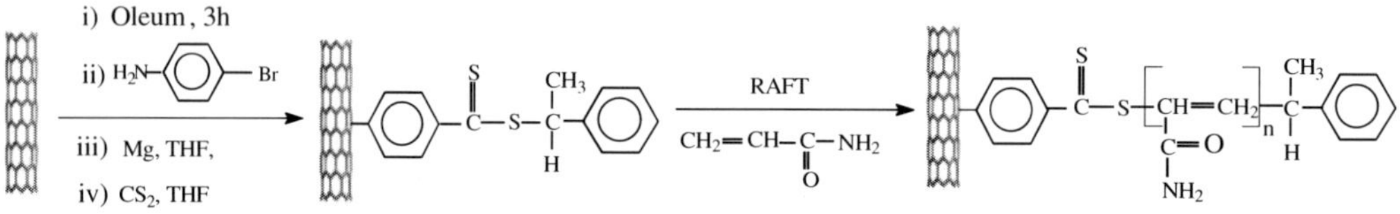

Figure 20. Process for the immobilization of a RAFT agent on SWCNT and further growth of PAM chains via RAFT polymerization (adapted from ref. [281]).

Firstly a bromo aniline molecule was diazotized *in situ* and reacted with carbon nanotubes then bromo functionalized carbon nanotube was treated with magnesium and with carbon disulfide (CS_2), and finally with (1-bromoethyl) benzene. The polymerization (under inert atmosphere) was conducted with the final RAFT agent (here the dithiocarbamate group grafted to SWCNT) ensuring control over the growth of the chains, azobisisobutyronitrile (AIBN) acting as a thermal radical initiator.

Diazonium salt chemistry permits also to modify CNT so that they are amenable to the *grafting onto* strategy using a reactive, preformed polymer. The direct diazotation and grafting on a carbon surface, of an aniline-terminated polymer was successfully achieved. [69]

The nitrophenyl-terminated PEG was obtained by reacting its alcohol derivative first with 4-toluenesulfonyl then with nitrophenol therefore providing nitrophenyl-functionalized PEG (PEG-C_6H_4-NO_2). The catalytic hydrogenation results in an aminophenyl-functionalized PEG (PEG-C_6H_4-NH_2) which was further diazotized (PEG-C_6H_4-N_2^+) in presence of CNT giving the grafted PEG-CNT hybrid. The whole process of diazotization of OH-terminated PEG and its reaction with a CNT is illustrated in Figure 21.

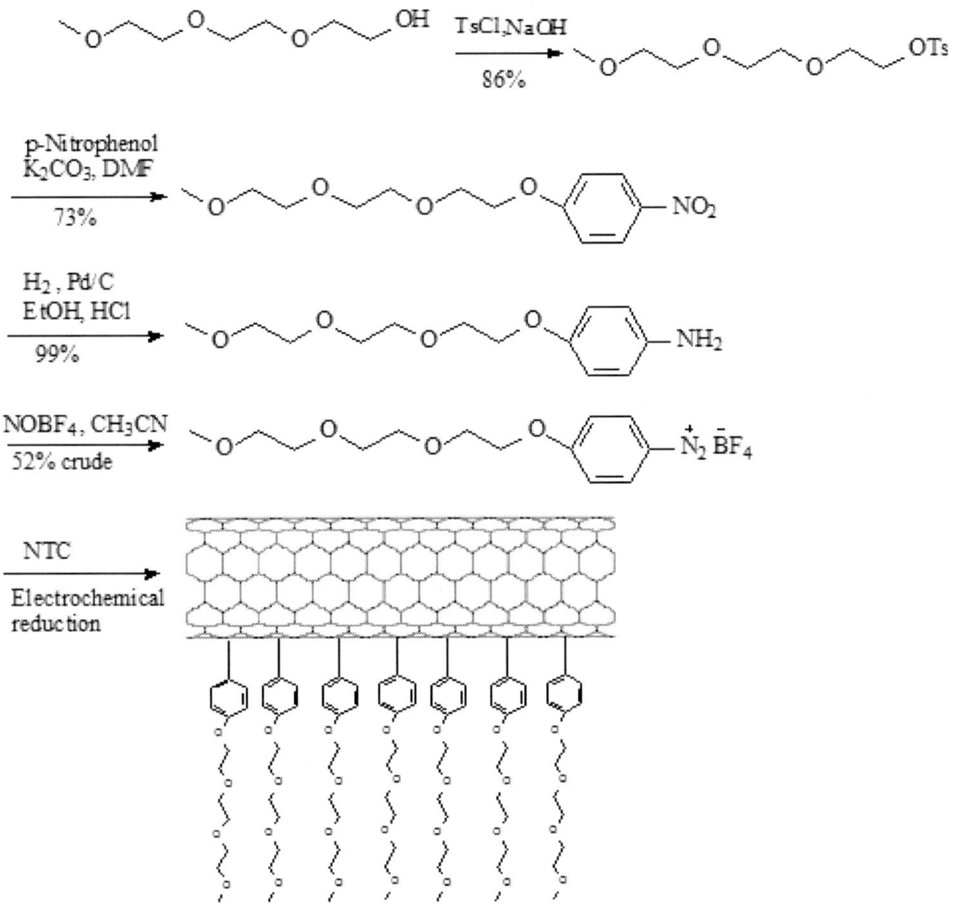

Figure 21. Method of preparation of diazotized PEG (ref. [69]) and its use to make *PEGylated* CNT.

Another widely used "grafting onto" technique is the Copper Azide Alkyne Cycloaddition (CuAAC). The combination of surface modification by diazonium salt and CuAAC permitted to successfully graft polystyrene chains onto CNT. [282] The carbon surface was modified by ethynyloxyphenyl groups from the aniline derivative precursor and reacted catalytically with azide-terminated polystyrene to obtain the nanocomposite CNT-PS with an interfacial triazole group between the polymer and the aryl layer topping the CNT. The CNT-PS are more organophilic than pristine CNT, the major application is dispersion in organic matrices.

Diazonium salts, carbon nanotubes and click chemistry can further be combined to create original catalytic heterostructures. In this way, azide-terminated poly(glycerol methacrylate) (noted POH-N_3) has been prepared by post-functionalization of poly(glycidyl methacrylate) prepared by ATRP. [283] The POH-N_3 was clicked to 4-alkynearyl functionalized MWCNT via CuAAC; the so-obtained hairy CNT were first oxidized then served for the in-situ formation and immobilization of Pd nanoparticle (Figure 22). The CNT@polymer@Pd heterostucture was successfully prepared with a substantial loading of 3 nm-sized Pd nanoparticles as judged from the intense Pd3d XPS peak doublet. The final heterostructure exhibited catalytic activity in the Suzuki coupling reaction between 4-bromobenzene and phenylboronic acid with 77% yield.

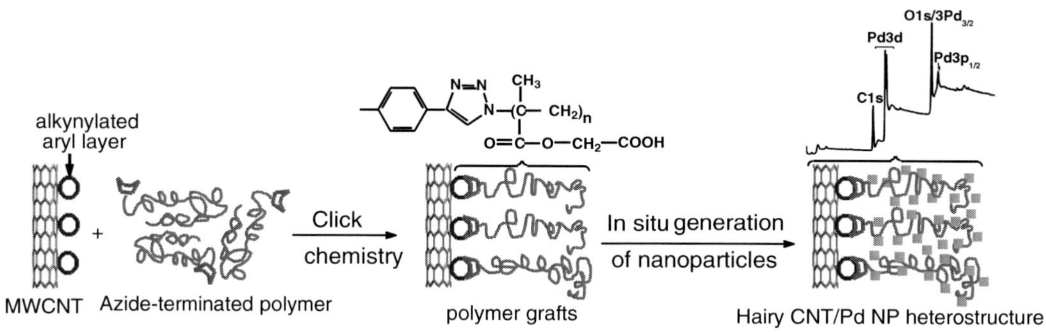

Figure 22. Synthetic route for the in situ grafting of alkynylated aryl groups and click reaction of PGMA on MWCNT. The glycerol groups from POH were further oxidized into COOH groups. XPS spectrum of the hairy MWCNT-immobilized Pd heterostructure is shown in insert. Reproduced and adapted from ref. [283]. Copyright © 2010 American Chemical Society.

In a follow up to this work, Salmi et al. [284] prepared azide-functionalized poly(4-vinyl pyridine), PVP-N_3, which was clicked to alkynylated MWCNT obtained by diazonium salt chemistry. The hairy MWCNT-PVP hybrid is pH sensitive and constitutes a platform for the immobilization of gold NPs.

3.7.5. Dipolar Cycloaddition for Grafting Polymers from

1,3 cycloaddition Prato reaction described above (Figure 4) has been exploited to prepare CNT-polymer hybrids either by chemically linking α-alkene groups [285] or by grafting ATRP initiators to on the sidewalls of SWCNT. [286] In the former approach, polyethylene (PE) chains were grown in-situ onto the sidewalls of SWCNT via catalyzed ethylene copolymerization with α-alkene functionalized SWCNT. [285] This is a less investigated approach as usually CNT-PE nanocomposites are made by mixing the matrix (PE) and the filler (CNT). In the second approach, Adronov and his co-workers [286] attached phenol groups via Prato reaction which further reacted with 2-bromoisobutyryl bromide to obtain SWCNT ATRP macroinitiators. PMMA and PtBA chains were grown by ATRP. SWCNT-PtBA hybrids, soluble in THF were further hydrolyzed to obtain SWCNT-PAA nanocomposites which turned to be soluble in water (Figure 23). In a mixture of water/dichloromethane, the SWCNT-PAA nanocomposites were selectively dispersed in the aqueous phase.

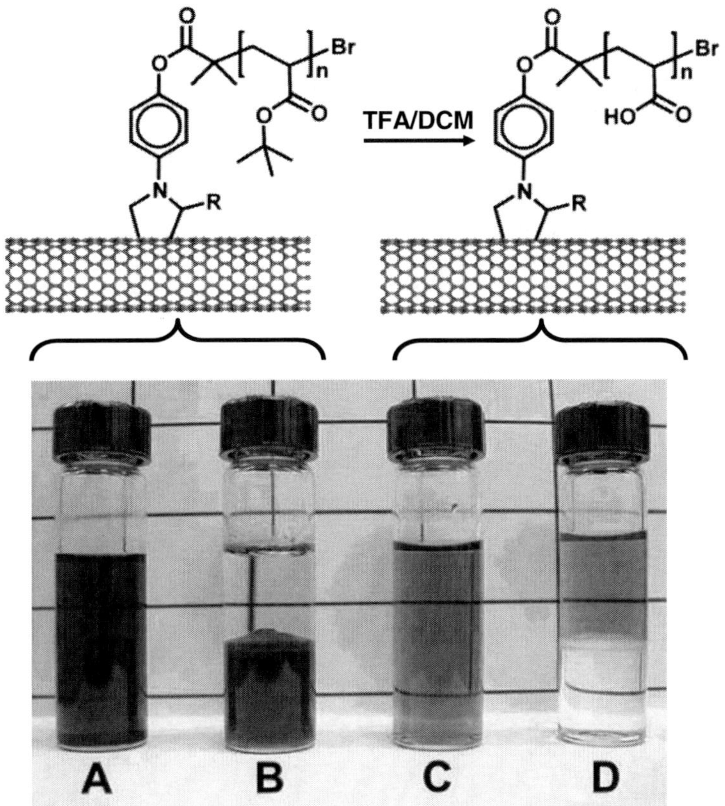

Figure 23. Upper panel: SWCNT-PtBA hybrids prepared by ATRP through the activation of the nanotubes by Prato reaction followed by esterification. Treatment in a mixture of trifluoroacetic acid (TFA) (1 mL) and anhydrous CH_2Cl_2 (DCM, 10 mL) leads to SWCNT-PAA. Lower panel: Solutions of SWCNT-polymer nanocomposites. (A) SWCNT-PtBA sample in THF; (B) SWCNT-PtBA sample in a mixture of H_2O/CH_2Cl_2; (C) SWCNT-PAA sample in H_2O; (D) SWCNT-PAA sample in a mixture of H_2O/CH_2Cl_2. Reproduced and adapted with permission of ACS from [286].

3.7.6. Reaction of Nitrenes with CNT

Nitrenes are highly reactive electrophilic species known to react with C-H, C=C and even OH and N-H. [287] They thus can be ideal reactive species for the functionalization of CNT sidewalls via either *grafting onto* or *grafting from* methods. Towards this end, one can prepare nitrene-functionalized polymers from the azido-precursors themselves prepared by ATRP or RAFT. Qin et al. [288] prepared PS by ATRP and the terminal C-Br bond reacted with NaN_3 to provide PS-N_3 chains which react with the CNT sidewalls by heating to 130°C under N_2 for 60 h in dichlorobenzene.

Poly(4-azidophenyl methacrylate-co-methyl methacrylate) permitted to disperse CNT by $\pi-\pi$ stacking. [63] The suspension was coated on cellulose fabric and illuminated with UV light therefore leading to the decomposition of azide groups into nitrenes which react with C=C bonds from the CNT and in the same time insert in the OH groups from the underlying cellulose. This elegant strategy permitted to fabricate highly stable superhydrophobic cotton; the stability being provided by the anchoring of the copolymer to the fabric and to the top CNT by stable covalent bonds.

Another application of nitrene-functionalized polymers concerns the RAFT preparation of poly(*N,N*-dimethylacrylamide)-b-poly(*N*-isopropylacrylamide-co-N-acryloxysuccinimide) (PDMA-b-P(NIPAM-co-NAS)), bearing an azide group at the PDMA end, and its covalent attachment to SWCNT. [289] The terminal azide groups are switched to nitrenes which readily react with sidewall C=C bonds (Figure 24). Interestingly, the SWCNT-copolymer was mixed with the free polymer and the whole system assembled in SWCNT-grafted micelles obtained by crosslinking the thermoresponsive core polymeric micelles.

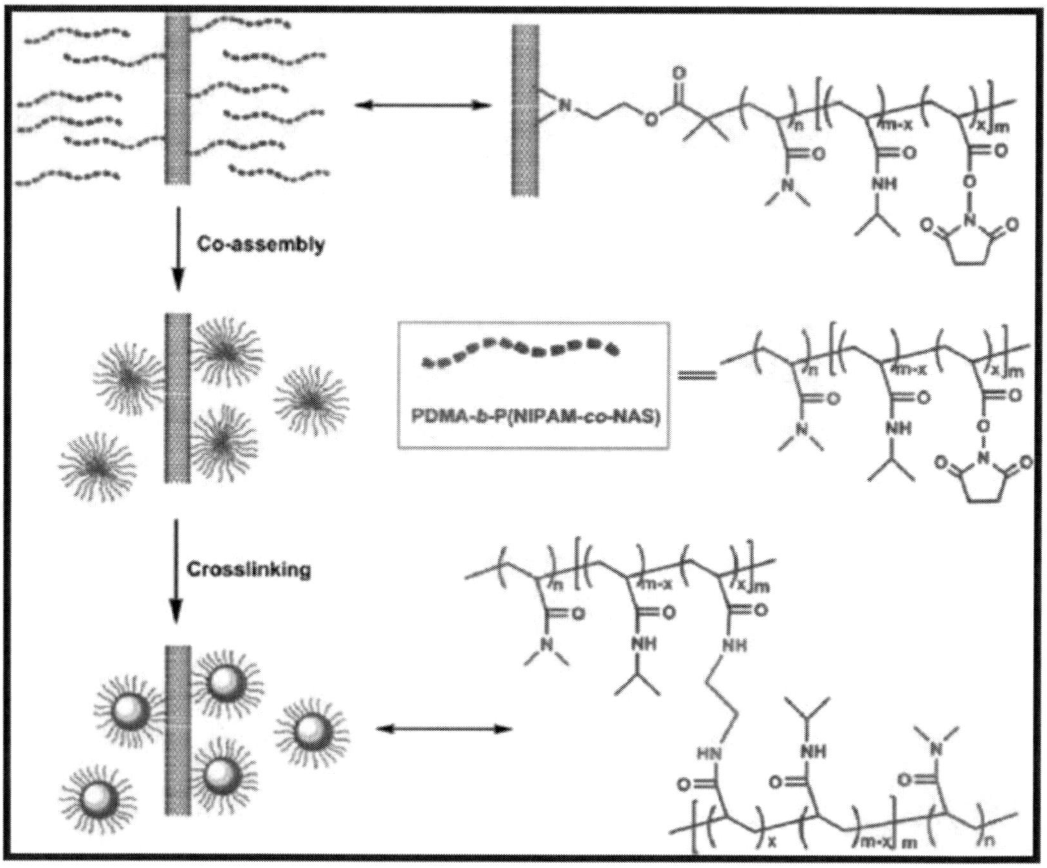

Figure 24. Grafting thermoresponsive copolymer to CNT through nitrene chemistry and subsequent formation of SWCNT-micelle by self-assembly of the SWCNT-copolymer hybrid with the free copolymer. Reproduced from ref. [289] with permissionof the American Chemical Society.

As far as surface-confined polymerization is concerned, biocompatible hyperbranched polyglycerol (HPG) were prepared in situ by anionic ring-opening polymerization conducted at the surface of MWCNT. [290] The macroinitiator MWCNT=N-CH_2CH_2OH was obtained by reaction of sonicated MWCNT with 2-azidoethanol at 160 °C for 12 h in *N*-methyl-2-pyrrolidone. Anionic polymerization of glycidol was initiated by MWCNT=N-CH_2CH_2OH resulting in the MWCNT-HPG hybrids. The latter served for anchoring Rhodamine 6B and obtaining fluorescent MWCNT potentially useful for drug delivery, cell imaging and bioprobing.

3.7.7. Diels-Alder Cyclobutene [2+4] Cycloaddition

This is another reaction for modifying CNT sidewalls that is off the beaten tracks though it has been claimed to ensure very high grafting density of polymer to MWCNT. Towards this end, poly(ε-caprolactone) (PεCL) was grafted to MWCNT which were first modified by 2-hydroxyethylbenzocyclobutene (BCB-EO) via a [2+4] cyclo-addition reaction. [291] Ring opening polymerization of ε-caprolactone was conducted in the presence of BCB-EO-modified MWNTs and the catalyst stannous octoate. Alternatively, the grafted-BCB-EO first reacted with cyclopentadienyltitanium trichloride ($CpTiCl_3$) and then the reactant was used to catalyze polymerization of ε-caprolactone (Figure 25). The grafting degrees of PCL prepared from BCB-EO modified MWNTs can reach at least 94%, much higher than those prepared using hydroxylated MWCNT prepared via nitrene chemistry or carboxyalation of MWCNT followed by esterification using glycol. The latter approaches either do not provide efficient macroinitiator or do not permit to evenly functionalize MWCNT sidewalls with initiator groups.

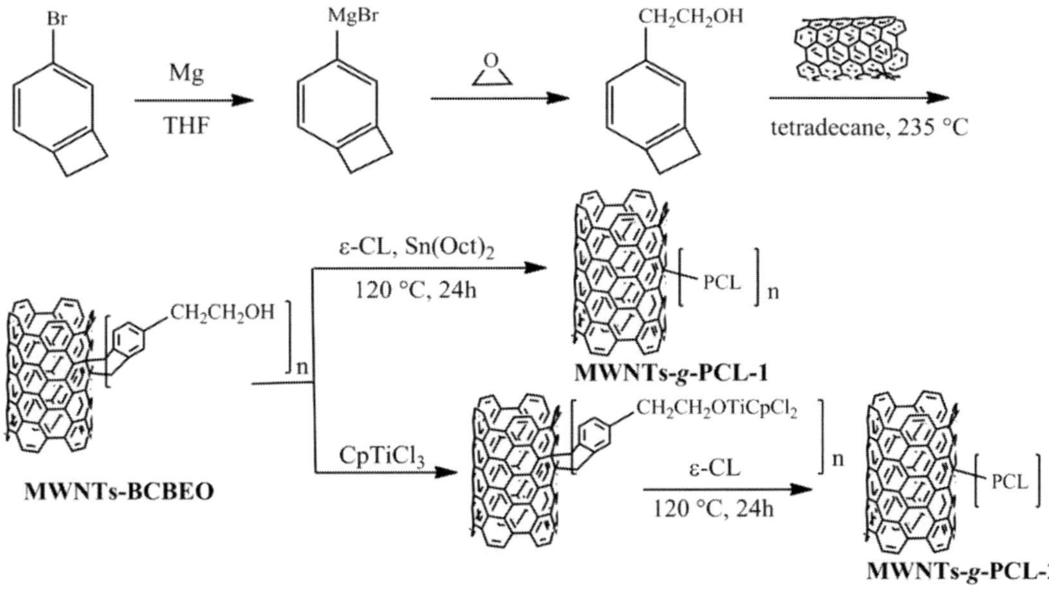

Figure 25. Routes for preparation of MWNTs-g-PCL hybrids starting from [2+4] cycloaddition. Reproduced from ref. [291] with permission of Elsevier.

Elsewhere, Sakellariou and co-workers [292] proposed a one-pot method for the grafting of binary polymer brushes at the surface of MWCNT. To do so, they used Diels–Alder cycloaddition [2+4] to graft ATRP and ring opening polymerization (ROP) initiators which served for the growth of PS or PMMA, and PεCL by surface-confined ATRP and ROP, respectively. SI-ROP was found to be faster than ATRP. The authors have also elegantly showed that sequential polymerizations of ε-CL and PS permitted to prepare MWNTs with multi-grafts of PS and PεCL forming Janus type structures when placed in a selective immiscible cyclohexane/acetonitrile solvent mixture.

Table 3. Selected methods of covalent modification of CNT by polymer grafts

Surface reaction	Coupling agent	Polymer	Grafting method	Polymer mass loading (wt/wt %) or grafting density (chains/nm²)	Application	Refs.
With polymer radicals	TEMPO	P(εCL-b-S)	Thermal dissociation of polymer chain end	6-12%	Dispersion in organic and acidic water, deposition on negatively charged surfaces	[267]
		P(AMS-co-GMA)	Thermal decomposition of a repeat unit sequence	29%	Dispersion in organic solvent	[270]
With $HNO_3/SOCl_2$	O=C-Cl tethered group	PS and PS-b-P4VP	Surface-initiated NMP	12-61%	Dispersion of CNT in aqueous and in organic solvents	[273]
		PNIPAM	Surface-initiated RAFT	35-87%	Thermoresponsive CNT	[274]
		Polyazide or polyalkyne	Click chemistry	60%	Superparamagnetic MWCNT@polymer @magnetic NP nanohybrid	[277]
With HNO_3	-COOH and post amidation in order to make reaction thiol-functionalized MWCNT	Thiol-terminated PHPMA prepared by RAFT	Thiol coupling	~35%	solubility of MWCNT in aqueous solutions	[293]
None	Native –COOH groups	PLLA	Surface-confined polycondensation	21-62%	Biocompatible MWCNT	[275]
With diazonium salts	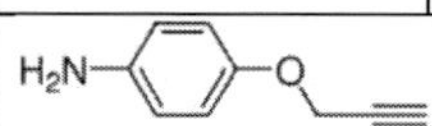	PS	Click chemistry	45% one polymer chain for every 200-700 carbon nanotube	Organosoluble CNT	[282]
	In-situ diazotization of p-aminophenyl propargyl ether followed by click coupling of azide-functionalized pyrrole	PPy	Surface oxidative polymerization of pyrrole	35-49%	NO_2 gas sensing	[294]
	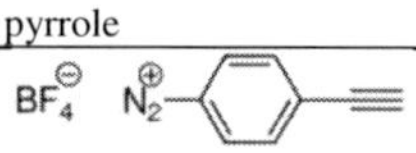	POH	Click chemistry	67%	Catalytic heterostructure	[283]
	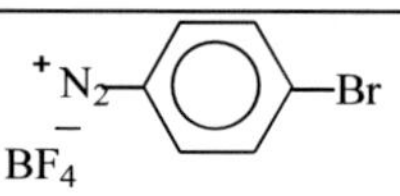	PS	Surface-initiated anionic polymerization		Nanocomposite	[278]
	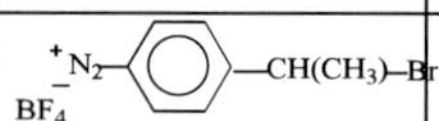	PAM	Surface-initiated RAFT	80%	Hydrosolubility of CNT	[281]
	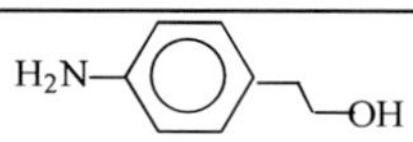	PBA	Surface-initiated ATRP	0.3 chains.nm^{-2}	NTC/Polymer nanocomposites	[279]
	Diazotized PEG	PEG	In-situ reduction of diazotized PEG on CNT	36 %	Dispersion of SWCNT in water. NB: these SWCNT-PEG were found hard to disperse by the authors.	[69]

Table 3. (Continued)

Surface reaction	Coupling agent	Polymer	Grafting method	Polymer mass loading (wt/wt %) or grafting density (chains/nm^2)	Application	Refs.
With ylides (Prato 1,3-cycloaddition)	Surface alkene group	PE	Catalyzed polymerization	2-8%	SWCNT-PE nanocomposite	[285]
	bromoisobutyryl	PMMA, PtBMA	Surface-initiated ATRP	7-21 chains per nanotube	Water soluble SWCNT	[286]
With nitrenes	Azide-functionalized polymer	PS	Thermal decomposition of azide-functionalized polymer into electrophilic nitrene-functionalized polymer	36-75%	Organic soluble SWCNT	[288]
		P(DMA-b-(NIPAM-co-NAS))	idem	26%	Crosslinked micelle-functionalized SWCNT	[289]
		P(azidophenylmethacrylate-co-MMA)	UV decomposition of azide-functionalized polymer into nitrene-functionalized polymer	17.5 %	Dispersion of CNT in cellulose	[63]
	N_3-CH_2CH_2OH (=N-CH_2CH_2OH)	HPG	Surface-initiated anionic polymerization	23-91%	Nanoplatform for Rhodamine 6B for biocompatible, fluorescent MWCNT-HPG useful in drug delivery and bioprobing	[290]
By Diels-Alder [2+4] cyclo addition	BCB-EO	PεCL	Surface-initiated anionic polymerization	94%	Dispersion in organic solvent	[291]
		PS and PεCL	Sequential or simultaneous surface-initiated ATRP of styrene and surface-initiated ROP of εCL	96%	Multi-polymer grafts for Janus MWCNT-polymer structures	[292]

HPG: hyperbranched polyglycerol, PAM : Polyacrylamide, P(AMS-co-GMA): poly(α-methylstyrene-co-glycidyl methacrylate), PBA: Poly(butyl acrylate), PεCL: poly(ε-caprolactone), P(εCL-b-S) : poly(ε-caprolactone-block-styrene), P(DMA-b-(NIPAM-co-NAS)): poly(*N,N*-dimethylacrylamide)-*b*-poly(*N*-isopropylacrylamide-*co*-*N*-acryloxysuccinimide), PE: polyethylene, PEG: poly(ethylene glycol), PHPMA: poly(N-(2-hydroxypropyl)methacrylamide), PLLA: poly(L-lactic acid), PMMA: Poly(methyl methacrylate), PNIPAM: Poly(*N*-isopropylacrylamide), PPy: polypyrrole, POH: polyglycerol, PS: Polystyrene, PtBMA : Poly (tert-butyl methacrylate),.

3.7.8. Overview of Methods for Making CNT/Polymer Nanocomposites through Covalent Grafting

As we have described above, several strategies can be applied to modify CNT by polymer grafts through covalent coupling of the polymer chains to SWCNT or to the outer tube of MWCNT. Table 3 reports selected examples of making CNT-polymer nanocomposites and their potential applications.

4. COMMERCIALLY AVAILABLE MODIFIED CARBON NANOTUBES

CNT Long, short S, D, MWCNT are now commercially available, but one can also find *S, D, MWCNT*-OH or –COOH; S, D, MWCNT [295], *S MWCNT*-NH_2; [295b,h]' *MWCNT*-SO_3H [295e] but also composites and ready for use dispersions. [295a]

CONCLUSION

Carbon nanotubes are the subject of very intense research, and one can appreciate from this review the many routes that are opening towards applications. Carbon nanotubes are now produced on an industrial scale, for example by Arkema (400 t of CNT per year), Bayer, and Nanocyl [296] most of this production is dedicated to composites some of them are already on the market. They are used for surface heating for automotive (seat or trim heating); panel heating in construction (floor, wall or object heating); antistatic coatings; battery electrodes. Applications in nano electronic will probably await the industrial production of separated m and scMWCNT. When one observes the superb results described above for biomedical applications, one can expect real progress in, for example, cancer curing. These academic and industrial achievements would not have been possible without understanding the surface and interface chemistry of carbon nanotubes.

ACKNOWLEDGMENTS

Zakaria Salmi wishes to thank the University Paris Diderot for the provision of a PhD scholarship.

REFERENCES

[1] (a) Reich, S.; Thomsen, C.; Maultzsch, J. *Carbon Nanotubes: Basic Concepts and Physical Properties*; VCH: Weinheim, 2004. (b) O'Connell, M. J. *Carbon Nanotubes, Properties and Applications,* Taylor and Francis, 2006. (c) Harris, P. J. F. *Carbon Nanotube Science Synthesis, Properties and Applications*, Cambridge University Press, 2009.(d) Jorio, A.; Dresselhaus, G.; Dresselhaus, M. S. (Eds.) *Carbon Nanotubes Advanced Topics in the Synthesis, Structure, Properties and Applications Topics in Applied Physics Series*, Springer, 2008, Vol. 111. (c) Marulanda, J. M. *Carbon*

Nanotubes, InTech, 2010.(e) Monthioux, M. *Meta-Carbon Nanotubes, Synthesis, Properties and Applications,* Wiley and sons, 2012.

[2] Hodge, S. A.; Bayazit, M. K.; Coleman, K. S.; Shaffer, M. S. P. *Chem. Soc. Rev*. 2012, *41*, 4409.

[3] Charlier, J.-C.; Blase, X.; Roche, S. *Rev. Mod. Phys.* 2007, *79,* 677.

[4] (a) Che, J; Cagin, T; Goddard, WA. *Nanotechnology* 2000, *11*, 65. (b) Berber, S.; Kwon, Y-K; Tomanek, D.; *Phys Rev Lett.* 2000,*84*,4613.

[5] Yu, M. F.; Lourie, O.; Dyer, M. J.; Moloni, K.; Kelly, T. F.; Ruoff, R. *Science* 2000, 287, 637.

[6] With exception of vertically aligned CNT (a) Hao, C; Ajit, R.; Jong-Beom, B.; Lin, Z.; Jia, Q.; Liming, D. *Mater Sci. Eng. Reports* 2010, *R70*, 63. (b) Choon-Ming, S.; Siang-Piao, C.; Rahman, M. A. *Carbon* 2011, *49*, 4613.

[7] Girifalco, L. A.; Hodak, M.; Lee R. S. *Phys. Rev. B* 2000, 62, 13104.

[8] Joyeux, X. Thesis, Université Joseph Fourier, Grenoble, France, 2009

[9] (a) Tasis, D.; Tagmatarchis, N.; Bianco, A.; Prato, M. *Chem. Rev.* 2006, *106*, 1105. (b) Peng, X.; Wong, S. S. *Adv. Mater.* 2009, 21, 625. (c) Karousis, N.; Tagmatarchis, N.; Tasis, D. *Chem. Rev.* 2010, *110*, 5366. (d) Clavé, G; Campidelli, S. *Chem. Sci.* 2011, 2, 1887. (e) Schnorr, J. M.; Swager, T. M. *Chem. Mater.*2011, *23*, 646.(f) Hodge, S. A.; Bayazit, M. K.; Coleman, K. S.; Shaffer, M. S. P. *Chem. Soc. Rev*. 2012, *41*, 4409.

[10] Mases, M; Noël, M.; Mercier, G.; Dossot, M.; Vigolo, B.; Mamane, V.; Fort, Y.; Soldatov, A. V.; McRae, E. *Phys Status Solidi B* 2011, *248*, 2552.

[11] Chen, J.; Chen, Q.; Ma, Q. *J. Colloid Interface Sci.* 2012, *370*, 32.

[12] Chiang, Y.-C.; Lin, W.-H.; Chang, Y.-C. *Appl. Surf. Sci.* 2011, *257*, 2401.

[13] (a) Wepasnick, K. A.; Smith, B. A.; Schrot; K. E.; Wilson, H. K.; Diegelmann, S. R.; Fairbrother, D. H. *Carbon* 2011, *49*, 24. (b) Wepasnick, K. A.; Smith, B. A.; Bitter, J. L.; Fairbrother, D. H. *Anal. Bioanal. Chem.* 2010, *396*, 1003.

[14] Zhang, J.; Zou, H.; Qing, Q.; Yang, Y.; Li, Q.; Liu, Z. *J. Phys. Chem. B* 2003, *107*, 3712.

[15] Datsyuk, V.; Kalyva, M.; Papagelis, K.; Parthenios, J.; Tasis, D.; Siokou, A.; Kallitsis, I.; Galiotis, C. *Carbon* 2008, *46,* 83.

[16] Dementev, N.; Feng, X.; Borguet E. *Langmuir* 2009, *25*, 7573.

[17] (a) Dumitrescu, I.; Wilson, N. R.; Macpherson, J. V. *J. Phys. Chem. C* 2007, *111*, 12944. (b) Avilés, F. Cauich-Rodríguez, J. V.; Moo-Tah, L.; May-Pat, A.; Vargas-Coronado, R. *Carbon* 2009, *47*, 2970.

[18] Forrest, G. A.; Alexander, A. J. *J. Phys. Chem. C* 2007, *111*, 10792.

[19] Hung, N. T.; Tuong, N. M.; Rakov, E. G. *Inorg. Mater.* 2010, *46*, 1077.

[20] Worsley, K. A.; Kalinina, I.; Bekyarova, E.; Haddon, R. C., *J. Am. Chem. Soc.*2009, *131,* 18153.

[21] Suri, A.; Coleman, K. S. *Carbon* 2011, *49*, 3031

[22] Ansón-Casaos, A.; Gonzàlez, M.; Gonzàlez-Domínguez, J. M.; Martínez, T. *Langmuir* 2011, *27*, 7192.

[23] Larciprete; R.; Gardonio, S.; Petaccia, L.; Lizzit, S. *Carbon* 2009, *47*, 2579.

[24] Simmons, J. M.; Nichols, B. M.; Baker, S. E.; Marcus, M. S.; Castellini, O. M.; Lee, C.-S.; Hamers, J.; Eriksson, M. A. *J. Phys. Chem. B* 2006, *110*, 7113.

[25] Lobo, A. O.; Ramos, S. C.; Antunes, E. F.; Marciano, F. R.; Trava-Airoldi, V. J.; Corat, E.J. *Mater. Lett.* 2012, 70, 89.

[26] Chen, C.; Liang, B.; Ogino, A.; Wang, X.; Nagatsu, M. *J. Phys. Chem. C* 2009, *113*, 7659.
[27] Bradley, R. H.; Cassity, K.; Andrews, R.; Meier, M.; Osbeck, S.; Andreu, A.; Johnston, C.; Crossley, A. *Appl. Surf. Sci.* 2012, *25*, 4835.
[28] Fernandez d'Arlas, B.; Goyanes, S.; Rubiolo, G. H.; Mondragon, I.; Corcuera, M. A.; Eceiza, A. *J. Nanosci. Nanotechnol.* 2009, *9*, 6064.
[29] Jung Sook, J.;. Byeong-Uk, N; Soon Huat, T.; Sung Chul, H. *Macromol. Symp.* 2007, 249/250, 276.
[30] Ramanathan, T.; Fisher, F. T.; Ruoff, R. S.; Brinson, L. C. *Chem. Mater.* 2005, 17, 1290.
[31] Jiang, K. Y.; Schadler, L. S.; Siegel, R. W.; Zhang, X. J.; Zhang, H. F. Terrones, M. *J. Mater. Chem.* 2004, *14*, 37.
[32] Dayani, Y.; Malmstadt, N. *Langmuir* 2012, *28*, 8174.
[33] Xiangdong, W.; Zhongjie, D.; Chen, Z.; We, Z.; Guozhi, X. *J. Polym. Mater.* 2010, *27*, 179.
[34] Ma, P. C.; Kim, J.-K.; Tang, B. Z. *Carbon* 2006, *44*, 3232.
[35] Russier, J.; Ménard-Moyon, C.; Venturelli, E.; Gravel, E.; Marcolongo, G.; Meneghetti, M; Doris, E.; Bianco, A. *Nanoscale* 2011, *3*, 893.
[36] (a) Mickelson, E. T.; Huffman, C. B.; Rinzler, A. G.; Smalley, R. E.; Hauge, R. H.; Margrave, J. L. *Chem. Phys. Lett.* 1998, *296*, 188.(b) Hayashi, T.; Shimamoto, D.; Kim, Y. A.; Murasami, H.; Okino, F.; Touhara, H.; Shimada, T.; Miyauchi, Y.; Maruyama, S.; Terrones, M.; Dresselhaus, M. S.; Endo, M. *ACS Nano* 2008,.*2*, 485.
[37] Lavskaya, Y. V.; Bulusheva, L. G.; Okotrub, A. V.; Yudanov, N. F.; Vyalikh, D.V.; Fonseca, A. *Carbon* 2009, *47*, 1629.
[38] Zhang, W.; Bonnet, P.; Dubois, M.; Ewels, C. P.; Guérin, K; Petit, E; Mevellec, J.-Y.; Vidal, L; Ivanov, D. A.; Hamwi A. *Chem. Mater.* 2012, *24*, 1744
[39] (a) Kalita, G.; Adhikari, S.; Aryal, H. R.; Ghimre, D. C.; Afre, R.; Soga, T.; Sharonc, M.; Umeno, M. *Physica E* 2008, *41*, 299.(b) Shoda, K.; Kohno, H.; Kobayashi, Y.; Takagi, D.; Takeda S. *J. Appl. Phys.* 2008, *104*, 113529.
[40] Maiti, J.; Kakati, N.; Lee, S. H.; Yoon, Y. S. *J. Fluorine. Chem.* 2012, *135*, 362.
[41] Bulusheva, L. G.; Fedoseeva, Y. V.; Okotrub, A. V.; Flahaut, E.; Asanov, I. P.; Koroteev, V. O.; Yaya, A.; Ewels, C. P.; Chuvilin, A. L.; Felten, X. A.; Van Lier, O. G.; Vyalikh D. V. *Chem. Mater.* 2010, *22*, 4197.
[42] Barthos, R.; Mehn, D.; Demortier, A.; Pierard, N.; Morciaux, Y.; Demortier, G.; Fonseca, A.; Nagy, J. B. *Carbon* 2005, *43*, 321.
[43] Colomer, J.-F.; Marega, R.; Traboulsi, H.; Meneghetti, M.;Van Tendeloo, G.; Bonifazi D. *Chem. Mater.*2009, *21*, 4747.
[44] Unger, E.; Graham, A.; Kreupl, F.; Liebau, M.; Hoenlein, W. *Curr. Appl. Phys.* 2002, *2*, 107.
[45] Coleman, K. S.; Chakraborty, A. K.; Bailey, S. R.; Sloan, J.; Alexander, M. *Chem. Mater.* 2007, *19*, 1076.
[46] (a) Bulusheva, L. G.; Gevko, P. N.; Okotrub, A. V.; Lavskaya, Y. V.; Yudanov, N. F.; Yudanova, L. I.; Abrosimov, O. G.; Pazhetnov, E. M.; Boronin, A. I.; Flahaut E. *Chem. Mater.* 2006,*18*,4967. (b) Kang, J. H.; Takhar, D; Kuznetsov, O. V.; Khabashesku, V. N.; Kelly, K. F. *Chem. Phys. Lett.* 2012, *534*, 43.

[47] Valentini, L.; Macan, J.; Armentano, I.; Mengoni, F.; Kenny, J. M. *Carbon* 2006, *44*, 2196.

[48] Georgakilas,V.; Kordatos, K.; Prato, M; Guldi, D. M.; Holzinger, M.; Hirsch, A. *J. Am. Chem. Soc.* 2002, *124,* 760.

[49] Tagmatarchis, N.; Prato, M. *J. Mater. Chem.* 2004, *14* , 437.

[50] Georgakilas, V; Bourlinos, A.; Gournis, D.; Tsoufis, T.; Trapalis, C.; Mateo-Alonso, A.; Prato, M. *J. Am. Chem. Soc.* 2008, *130,* 8733.

[51] Ballesteros, B.; de la Torre, G.; Ehli, C.; Aminur Rahman, G. M., Agulló-Rueda, F.; Guldi, D. M.; Torres, T. *J. Am. Chem. Soc.* 2007, *129*, 5061.

[52] Campidelli, S.; Sooambar, C.; Lozano Diz, E.; Ehli, C.; Guldi, D. M.; Prato, M. *J. Am. Chem. Soc.* 2006, *128*,12544.

[53] Giambastiani, G.; Cicchi, S.; Giannasi, A.; Luconi, L.; Rossin, A.; Mercuri, F.; Bianchini, C.; Brandi, A., Melucci, M.; Ghini, G.; Stagnaro, P.; Conzatti, L.; Passaglia, E.; Zoppi, M.; Montini, T.; Fornasiero P. *Chem. Mater.* 2011, *23*, 1923.

[54] (a) Alvaro, M.; Atienzar, P.; de la Cruz, P.; Delgado, J. L.; Troiani, V; Garcia, H.; Langa, F.; Palkar, A.; Echegoyen, L. *J. Am. Chem. Soc.* 2006, *128*, 6626. (b) Poplawska, M.; Zukowska, G. Z.; Cudzilo, S; Bystrzejewski, M. *Carbon* 2010, *48*, 1318.

[55] Alvaro, M.; de la Cruz, P.; Delgado, J. L.; Garcia L. H.; Langa, F. J. *J. Phys. Chem. B* 2004, *108*, 12691.

[56] Grassi, G.; Scala, A.; Piperno, A.; Iannazzo, D.; Lanza, M.; Milone, C.; Pistone, A.; Galvagno, S. *Chem. Commun.* 2012,*48*, 6836.

[57] Ménard-Moyon, C.; Dumas, F.; Doris, E.; Mioskowski, C. *J. Am. Chem. Soc.* 2006, *128*, 14764.

[58] Tagliapietra, S.; Cravotto, G.; Gaudino, E. C.; Visentin, S.; Mussi, V. *Synlett* 2012, *23*, 1459.

[59] Holzinger, M.; Vostrowsky, O.; Hirsch, A.; Hennrich, F.; Kappes, M.; Weiss, R; Frank, J. *Angew. Chem. Int. Ed.* 2001, *40*, 4002.

[60] Luksirikul, P.; Ballesteros, B; Tobias, G.; Moloneyd, M. G.; Green, M. L. H. *J. Mater. Chem.* 2011, *21*, 19080.

[61] Ismaili, H.; Lagugné-Labarthet, F.; Workentin, M. S. *Chem. Mater.* 2011, *23*, 1519.

[62] Pastine, S. J.; Okawa, D.; Kessler, B.; Rolandi, M.; Llorente, M.; Zettl, A.; Fréchet, J. M. J. *J. Am. Chem. Soc.* 2008, *130*, 4238.

[63] Li, G.; Wang, H.; Zheng, H.; Bai, R. *Langmuir* 2010, *26*, 7529.

[64] (a) *Aryl Diazonium Salts. New Coupling Agents in Polymer and Surface Science*, M. M. Chehimi, ed., Wiley-VCH, Weinheim, 2012. (b) Bélanger, D.; Pinson, J. *Chem. Soc.Rev.* 2011, *40*, 3995. (c) Gooding, J. J.; Ciampi, S. *Chem. Soc. Rev.* 2011, *40*, 2704. (d) Mahouche-Chergui, S.; Gam-Derouich, S.; Mangeney C.; Chehimi, M. M. *Chem. Soc. Rev.* 2011, *40*, 4143. (e) Aswal, D. K.; Koiry, S. P.; Jousselme, B.; Gupta, S. K.; Palacin, S.; Yakhmi, J. *Physica E* 2009, *41,* 325. (f) Barrière, F.; Downard, A. J. *J. Solid State Electrochem.* 2008, *12*, 1231. (g) McCreery, R. L. *Chem. Rev.* 2008, *108*, 2646. (h) Marquette, C. A.; Corgier, B. P.; Heyries, K. A.; Blum, L. J. *Frontiers in Bioscience* 2008, *13*, 3827. (i) Podvorica, F.; Pinson, J. *Chem. Soc. Rev.* 2005, *34,* 429.

[65] Bravo-Díaz, C. Diazohydroxides, *Diazoethers and Related Species* in *The Chemistry of Hydroxylamines, Oximes and Hydroxamic Acids*, vol. 2 (ed. Z. Rappoport), John Wiley & Sons, 2010 pp. 853 – 876.

[66] Usrey, M. L.; Lippmann, E. S.; Strano, M. S. *J. Am. Chem. Soc.* 2005, *127*, 16129.
[67] Sumpter, B. G.; Jiang D.-e.; Meunier, V. *Small* 2008, *4,* 2035.
[68] Balasubramanian, K.; Burghard, M. *J. Mater. Chem.* 2008, *18*, 3071.
[69] Bahr, J. L.; Yang, J.; Kosynkin, D. V.; Bronikowski, M. J.; Smalley, R. E.; Tour J. M. *J. Am. Chem. Soc.* 2001, *123*, 6536.
[70] Marcoux, P. R.; Hapiot, P.; Batail, P.; Pinson, J. *New. J. Chem.* 2004, *28*, 302.
[71] Steiner, S.; Busato, S.; Ermanni, P. *Carbon* 2012, *50*, 1713.
[72] Guo, D. J.; Li, H.L. *Carbon*, 2005, *43*, 1259.
[73] Ghanem, M. A.; Kocak, I.; Al-Mayouf, A.; AlHoshan, M.; Bartlett, P. N. *Electrochim. Act.* 2012, *68*, 74.
[74] (a) Kooi, S. E.; Schlecht, U.; Burghard, M.; Kern, K. *Angew. Chem. Int. Ed.* 2002, *41*, 1353.
[75] Gohier, A.; Nekelson, F.; Helezen, M.; Jegou, P.; Deniau, G.; Palacin, S.; Mayne-L'Hermite, M. *J. Mater. Chem.* 2011, *21*, 4615.
[76] Leinonen, J.; Lajunen, M. *J. Nanopart. Res.* 2012, *14*, 1064.
[77] Brunetti, F. G.; Antonia Herrero, M.; Muñoz, J. de M.; Díaz-Ortiz, J. A.; Alfonsi, J.; Meneghetti, M.; Prato, M.; Vàzquez, E. *J. Am. Chem. Soc.* 2008, *130,* 8094.
[78] Bahr J. L.; Tour, J. M. *Chem. Mater.* 2001, *13*, 3823.
[79] Dyke, C. A.; Tour, J. M. *Nanolett.* 2003, *3*, 1215.
[80] a) Hudson, J. L.; Casavant, M. J.; Tour J. M. *J. Am. Chem. Soc.* 2004, *126*, 11158. b) Stephenson, J. J.; Hudson, J. L.; Azad, S.; Tour J. M. *Chem. Mater.* 2006, *18*, 374.
[81] Xianhong,C.; Hanfeng, W; Wenbin, Z.; Feng, T.; Yang, X.; Chen, J. *Macromol. Chem. Phys.* 2008, *209*, 846.
[82] Khazaei, A.; Rad, M. N. S.; Borazjani, M. K.; Saednia, S.; Soudbard, D. *Synlett* 2011, *15*, 2145.
[83] Price B. K.; Tour, J. M. *J. Am. Chem. Soc.* 2006, *128*, 12899.
[84] Dyke, C. A.; Tour, J. M. *Chem. Eur. J.* 2004, *10*, 812.
[85] Hilmer, A. J.; McNicholas, T. P.; Lin, S.; Zhang, J.; Wang, Q. H.; Mendenhall, J. D.; Song, C.; Heller, D. A.; Barone, P. W.; Blankschtein, D.; Strano, M. S. *Langmuir* 2012, *28*,1309. (b) Lin, S.; Hilmer, A. J.; Mendenhall, J. D.; Strano, M. S.; Blankschtein, D. *J. Am. Chem. Soc.* 2012,*134*, 8194.
[86] Blanch, A. J.; Lenehan, C. E.; Quinton J. S. *J. Phys. Chem. C* 2012, *116,* 1709
[87] Peng, Z.; Holm, A.H.; Nielsen, L. S.; Pedersen, S. U.; Daasbjerg, K. *Chem. Mater.* 2008, *20*, 6068.
[88] Flavin, K.; Lawrence, K.; Bartelmess, J.; Tasior, M.; Navio, C.; Bittencourt, C.; O'Shea, D. F.; Guldi, D. M.; Giordani, S. *ACS Nano* 2011, *5*, 1198.
[89] Fifield, L. S.; Grate, J. W. *Carbon* 2010, *48*, 2085.
[90] Joyeux, X.; Mangiagalli, P.; Pinson, J. *Adv. Mater.* 2009, *21*, 4404.
[91] Lipińska, M. E.; Rebelo, S. L. H.; Pereira, M. F. R.; Gomes, J. A. N. F.; Freire, C.; Figueiredo, J. L. *Carbon* 2012, *50*, 3280.
[92] Zhao J.-x.; Ding Y.-h. *J. Phys. Chem. C* 2008, *112*, 13141.
[93] Mammeri, F.; , Ballarin, A.; Giraud, M.;, Vivet, L.; Herbst, F.; Connan, C; Chehimi, M. M.;, Brusatin, G.; Ammar, S. *MRS Proc.* 2011, *1359*, DOI: 10.1557/opl.2011.1017.
[94] Karousis, N.; Ali-Boucetta, H.; Kostarelos, K.; Tagmatarchis, N. *Mater. Sci. Eng. B* 2008, *152,* 8.

[95] Singh, P.; Samorí, C.; Toma, F. M.; Bussy, C.; Nunes, A.; Al-Jamal, K. T.; Ménard-Moyon, C.; Prato, M.; Kostarelos, K.; Bianco, A. *J. Mater. Chem.* 2011, *21,* 4850.
[96] Le Floch, F.; Thuaire, A.; Bidan, G.; Simonato J.-P. *Nanotechnology* 2009, 145705.
[97] Leinonen, H.; Lajunen, M. *Chem. Eng. J.* 2011, *172*, 557.
[98] Jousselme, B.; Bidan, G.; Billon, M.; Goyer, C.; Kervella, Y.; Guillerez, S.; Abou Hamad, E.; Goze-Bac, C.; Mevellec, J.-Y.; Lefrant, S. *J. Elelectroanal. Chem.* 2008, *621,* 277. (b) Tong, B.; Shi, J.; Zhao, W.; Shen, J.; Zhi, J.; Dong, Y. *J. Phys.Chem. C* 2010, *114*, 8040.
[99] Sadowska, K.; Roberts, K. P.; Wiser, R.; Biernat, J. F.; Jabłonowska, E.; Bilewicz, R. *Carbon* 2009, *47*, 1501.
[100] Ellison, M. D.; Gasda, P. *J. Phys. Chem.* C 2008, *112*, 738.
[101] Garrett, D. J.; Flavel, B. S.; Shapter, J. G.; Baronian, K. H. R.; Downard, A. J. *Langmuir* 2010, *26*, 1848.
[102] Peng, Z.; Holm, A. H.; Nielsen, L. T.; Pedersen, S. U.; Daasbjerg, K. *Chem. Mater.* 2008, *20*, 6068.
[103] Lim, J. K.; Yoo, B. K.; Yi, W.; Hong, S.; Paik, H.-j.; Chun, K; Kime, S. K.; Joo S.-W. J. *Mater. Chem.* 2006, *16*, 2374.
[104] (a) Hudson, J. L.; Jian, H.; Leonard, A. D.; Stephenson, J. J.; Tour, J. M. *Chem. Mater.* 2006, *18*, 2766. (b) Chen, B; Flatt, A. K.; Jian, H.; Hudson, J. L.; Tour, J. M. *Chem. Mater*. 2005, *17*, 4832.
[105] Podvorica, F. I.; Kanoufi, F.; Pinson, J.; Combellas, C. *Electrochim. Act.* 2009, *54,* 2164.
[106] Wang, F.; Deng, K.-J.; Zhou, L.; Zhao, J.-B.; Ke, X.-H.; Wen, L.-L. *J. Inorg. Organomet. Polym*. 2012, *22*, 1182. (b) Ménard-Moyon, C.; Fabbro, C.; Prato, M.; Bianco, A. *Chem. Eur. J.* 2011, *17*, 3222. (a) Stephenson, J. J.; Hudson, J. L.; Leonard, A. D.; Price, B. K.; Tour J. M. *Chem. Mater.* 2007, *19,* 3491.
[107] Rubio, N.; Herrero, M. A.; Meneghetti, M.; Diaz-Ortiz, A.; Schiavon, M.; Prato, M.; Vasquez, E. *J. Mater. Chem.* 2009, *19*, 4407.
[108] Mammeri, F.; Ballarin, A. Giraud, M.; Brusatin, G.; Ammar, S. *Colloids Surfaces A* in press, DOI: 10.1016/j.colsurfa.2013.03.026,
[109] Tessier, L.; Deniau, G., Charleux, B.; Palacin, S. *Chem. Mater*. 2009, *21*, 4261.
[110] Schmidt, G.; Gallon, C.; Esnouf, S.; Bourgoin, J.-P.; Chenevier, P. *Chem Eur. J.* 2009, *15*, 2101.
[111] Pazo-Llorente, R.; Bravo-Díaz, C.; Gonzalez-Romero, E. *Eur. J. Org. Chem.* 2004, 3221
[112] Dyke, C. A.; Stewart, M. P.; Maya, F.; Tour, J. M. *Synlett* 2004, 155.
[113] Nair, N.; Kim, W.-J.; Usrey, M. L.; Strano, M. S. *J. Am. Chem. Soc*. 2007, *129*, 3946.
[114] Abdula, D; Nguyen, K. T.; Shim, M. *J. Phys. Chem. C* 2007, *111,* 17755
[115] Balasubramanian, K.; Burghard, M.; Kern, K. *Phys Chem. Chem. Phys*. 2008, *10*, 2256.
[116] Guo, Z.; Liang, L.; Liang, J.-J.; Ma, Y.-F.; Yang, X.-Y.; Ren, D.-M.; Chen Y.-S.; Zheng, J.-Y. *J. Nanopart. Res*. 2007, *10*, 1077.
[117] Hayat, A.; Marty, J.-L.; Radi, A.-E. *Electroanalysis* 2012, *24*, 1446.
[118] (a) Tuci, G.; Vinattieri, C.; Luconi, L.; Ceppatelli, M.; Cicchi, S.; Brandi, A.; Filippi, J.; Melucci, M.; Giambastiani, G. *Chem. Eur. J.* 2012, *18*, 8454. (b) Coates, M.; Griveau, S.; Bedioui, F.; Nyokong, T. *Electroanalysis* 2012, *9*, 1833.

[119] Lerner, M. B.; D'Souza, J.; Pazina, T.; Dailey, J.; Goldsmith, B. R.; Robinson, M. K.; Johnson A. T. C. *ACS Nano* 2012, *6*, 5143.
[120] Flavin, K.; Lawrence, K.; Bartelmess, J.; Tasior, M.; Navio, C.; Bittencourt, C.; O'Shea, D. F.; Guldi, D. M.; Giordani, S. *ACS Nano* 2011, *5*, 1198.
[121] Cheng, F.; Adronov, A. *Chem. Mater.* 2006, *18*, 5389.
[122] Yang, Y.; Qiu, S.; Xie, X.; Wang, X.; Li, R. K. Y. *Applied Surf. Sci.* 2010, *256*, 3286.
[123] Penicaud, A.; Poulin, P.; Derre, A.; Anglaret, E.; Petit, P. *J. Am. Chem. Soc.* 2005, *127*, 8.
[124] (a) Liang, F.; Sadana, A. K.; Peera, A.; Chattopadhyay, J.; Gu, Z; Hauge, R. H.; Billups, W. E. *Nano Lett.* 2004, *4*, 1257. (b) Wunderlich, D.; Hauke, F.; Hirsch, A. *J. Mater. Chem.* 2008, *18*, 1493.
[125] Tang, X. L.; Jiao, Q. Z.; Cao, Y.; Zhang, P.; Liu, H. B.; Wu, H. Y.; Zhou, M. J.; Li, X. F.; Zhao, Y. *Chem. Lett.* 2009, *38*, 220·
[126] Maeda, Y.; Kato, T.; Hasegawa, T.; Kako, M.; Akasaka, T.; Lu, J.; Nagase, S. *Org. Lett.* 2010, *12*, 996.
[127] Voiry, D.; Roubeau, O.; Penicaud, A. *J. Mater. Chem.* 2010, *20*, 4385.
[128] Syrgiannis, Z.; Hauke, F.; Röhrl, J.; Hundhausen, M.; Graupner, R.; Elemes, Y.; Hirsch, A. *Eur. J. Org. Chem.* 2008, *15*, 2544.
[129] Fogden, S.; Howard, C. A.; Heenan, R. K.; Skipper, N. T.; Shaffer, M. S. P. *ACS Nano* 2012, *6*, 54.
[130] (a) Deng, S. L.; Brozen, A. H.; Zhang, Y.; Piao, Y.M.; *Chem. Comm.* 2011, *47*, 758. (b) Deng, S. L.; Piao, Y. M.; Brozena, A., H.; Wang, Y. H. *J. Mater. Chem.* 2011, *21*, 18568.
[131] Deng, S.; Zhang, Y.; H. Brozena, A.; Mayes, M. L.; Banerjee, P.; Chiou, W.-A.; Rubloff, G. W.; Schatz, G. C.; Wang, YH. *Nature Commun.* 2011, *2*, 382.
[132] (a) Chattopadhyay, J.; Sadana, A. K.; Liang, F.; Beach, J. M.; Xiao, Y.; Hauge, R H.; Billups W. E. *Org. Lett.* 2005, *7*, 4067.(b) Stephenson, J. J.; Sadana, A. K.; Higginbotham, A. L.; Tour J. M. *Chem. Mater.* 2006, *18*, 4658.
[133] Mukherjee, A.; Combs, R.; Chattopadhyay, J.; Abmayr, D. W.; Engel, P. S.; Billups, W. E. *Chem. Mater.* 2008, *20*, 7339.
[134] (a) Bayazit, M. K.; Suri, A.; Coleman, K. S. *Carbon* 2010, *48*, 3412. (b) Gebhardt, B.; Syrgiannis, Z.; Backes, C.; Graupner, R.; Hauke, F.; Hirsch, A. *J. Am. Chem. Soc.* 2011, *133*, 7985.
[135] Chattopadhyay, J.; de Jesus Cortez, F.; Chakraborty, S.; Slater, N. K. H.; Billups, W. E. *Chem. Mater.* 2006, *18*, 5864.
[136] Zhao, Y. L.; Stoddart, J. F. *Acc. Chem. Res.* 2009, *42*, 1161.
[137] Zhang, Y.; Liu, C.; Shi, W.; Wang, Z.; Dai, L; Zhang, X. *Langmuir* 2007, *23,* 7911.
[138] (a) Nakashima, N.; Tomonari, Y.; Murakami, H. *Chem. Lett.* 2002, *6*, 638 (b) Detriche, S.; Devillers, S.; Seffer, J.-F.; Nagy, J. B.; Mekhalif, Z.; Delhalle, J. *Carbon* 2011, *49,* 2935. (c) Yan, Y.; Yang, S.; Cui, J.; Jakisch, L.; Poetschke, P.; Voit, B. *Polym. Int.* 2011, *60*, 1425. (d) Wang, L.; Zhang, L.; Xue, X.; Geb, G.; Liang, X. *Nanoscale* 2012, *4*, 3983.
[139] Meng, Y.; Liu, F.; Hanc, J.; Sun, S.; Fan, J; Song, F.; Peng, X. *Mat. Sci. Eng. B* 2012, *177*, 887.
[140] Oelsner, C.; Schmidt, C.; Hauke, F.; Prato, M.; Hirsch, A.; Guldi, D. M. *J. Am. Chem. Soc.* 2011, *133*, 4580.

[141] Guldi, D. M.; Aminur Rahman, G. M.; Jux, N.; Balbinot, D.; Hartnagel, U.; Tagmatarchis, N.; Prato, M. *J. Am. Chem. Soc.* 2005, *127*, 9830.
[142] Setaro, A.; Bluemmel, P.; Maity, C.; Hecht, S; Reich S. *Adv. Funct. Mater.* 2012, *22*, 2425.
[143] Ou, Y.-Y.; Huang, M. H. *J. Phys. Chem. B* 2006, *110*, 2031.
[144] Zheng, M., Jagota, A.; Semke, E. D.; Diner, B. A.; Mclean, R. S.; Lustig, R. S.; Richardson, R. E.; Tassi, N. G. *Nature Mater.* 2003, *2*, 338.
[145] Hain, T. C.; Kröker, K.; Stich D. G.; Hertel, T. *Soft Matter* 2012, *8*, 2820.
[146] Han, S.-p.; Maune, H. T.; Barish, R. D.; Bockrath, M.; Goddard III, W. A. *Nano Lett.* 2012, *12*, 1129.
[147] Saifuddin, N.; Raziah, A. Z.; Junizah, A. R. *J. Chem.* 2013, 676815
[148] (a) Horn, D. W.; Tracy, K.; Easley, C. J.; Davis V. A. *J. Phys. Chem. C* 2012, *116*, 10341. (b) Calvaresi, M.; Hoefinger, S.; Zerbetto, F. *Chem. Eur. J.* 2012, *18*, 4308.
[149] (a) Liu, P. *Eur. J. Polym.* 2005, *41*, 2693. (b) Tuncel, D. *Nanoscale* 2011, *3*, 3545.
[150] Sinani, V. A.; Gheith, M. K.; Yaroslavov, A. A.; Rakhnyanskaya, A. A.; Sun, K.; Mamedov, A. A.; Wicksted, J. P.; Kotov, N. A. *J. Am. Chem. Soc.* 2005, *127*, 3463.
[151] Jo, T. S.; Han, H.; Bhowmik, P. K.; Ma, L. *Macromol. Chem. Phys.* 2012, *213*, 1378
[152] Gao, H.; Zhang, S.; Huang, D.; Zheng, L. *Colloid Polym. Sci.* 2012, *290*,757.
[153] Star, A.; Liu, Y.; Grant, K.; Ridvan, L.; Stoddart, J. F.; Steuerman, D. W.; Diehl, M. R.; Boukai, A.; Heath, J. R. *Macromolecules.* 2003, *36,* 553.
[154] (a) Chen, J.; Liu, H.; Weimer, W. A.; Halls, M. D.; Waldeck, D. H.; Walker, G. C. *J. Am. Chem. Soc.* 2002, *124*, 9034.(b) Chen, J.; Ramasubramaniam, R.; Xue, C.; Liu, H. *Adv. Funct. Mater.* 2006, *16*, 114.
[155] Zou, J.; Khondaker, S. I.; Huo, Q.; Zhai, L. *Adv. Funct. Mater.* 2009, *19*, 479.
[156] Clark, M. D.; Subramanian, S.; Krishnamoorti, R. J. *Colloid Interface Sci.* 2011, *354*, 144.
[157] Omastová, M.; Mičušík, M.; Fedorko, P.; , Chehimi, M. M.; Pionteck, *Mater. Sci. Forum* 2010, *636-637*, 676.
[158] Duque, J. G.; Densmore, C. G.; Doorn, S. K. *J. Am. Chem. Soc.* 2010, *132*, 16165.
[159] (a) Islam, M. F.; Rojas, E.; Bergey, D. M.; Johnson, A. T.; Yodh, A. G. *Nano Lett.* 2003, *3*, 269. (b) Rastogi, R.; Kaushal, R.; Tripathi, S. K.; Sharma, A. M.; Kaur, I.; Bharadwaj, L. M. *J. Colloid Interface Sci.* 2008, *328*, 421.
[160] Di Crescenzo, A.; Germani, R.; Del Canto, E.; Giordani, S ; Savelli, G.; Fontana, A. Eur J. Org. Chem. 2011, 29, 5641.
[161] Angelikopoulos, P.; Bock, H. *Phys. Chem. Chem. Phys.* 2012, *14*, 9546.
[162] Premkumar , T.; Mezzenga , R.; Geckeler, K. E. *Small* 2012, *8*, 1299.
[163] (a) Hasan, T.; Scardaci, V.; Tan, P. H.; Rozhin, A. G.; Milne, W. I.; Ferrari A. C. *J. Phys. Chem. C* 2007, *111*, 12594.(b) Barman, S. N.; LeMieux, M. C.; Baek, J.; Rivera, R.; Bao, Z. *ACS Appl. Mater. Interfaces.* 2010, *2*, 2672.
[164] Forney, M. W.; Poler, J. C. *J. Phys. Chem. C* 2011, *115*, 10531.
[165] (a) Gao, H.; Zhang, S.; Huang, D.; Zheng, L. *Colloid Polym. Sci.* 2012, *290*, 757. (b)López-Pastor, M.; Domínguez-Vidal, A.; Ayora-Cañada, M. J.; Simonet, B. M.; Lendl, B.; Valcárcel M. *Anal. Chem.* 2008, *80*, 2672.
[166] Jiao, J.; Zhang, H.; Yu, L.; Wang, X.; Wang, R. *J. Mol. Liquids* 2012, *171*, 6.

[167] (a) Cheng, Q; Debnath, S.; Gregan, E.; Byrne, H. J. *J. Phys. Chem.* C 2010, *114*, 8821.(b) Liu, T.; Luo, S.; Xiao, Z.; Zhang, C.; Wang, B. *J. Phys. Chem. C* 2008, *112*, 19193.
[168] Kim, S. W.; Kim, T.; Kim, Y. S.; Choi, H. S.; Lim, H. J.; Yang, S. J.; Park, C. R. *Carbon* 2012, *50*, 3.
[169] Bai, Y.; Park, I. S.; Lee, S. J.; Bae, T. S.; Watari, F.; Uo, M.; Lee, M. H. *Carbon* 2011, *49* , 3663.
[170] Crouzier, T.; Nimmagadda, A.; Nollert, M. U.; McFetridge, P. S. *Langmuir* 2008, *24*, 13173.
[171] Vaisman, L.; Wagner, H. D.; Marom, G. *Adv. Colloid Interface Sci.* 2006, *128–13*, 37.
[172] Blanch, A. J.; Lenehan, C. E.; Quinton, J. S. *Carbon* 2011, *49*, 5213
[173] Wei, B.; Guan, P.; Zhang, L.; Chen, G. *Carbon* 2010, *48*, 1380.
[174] Duan, W. H.; Wang, Q; Collins, F. *Chem. Sci.* 2011, *2*, 1407.
[175] Shieh, Y.-T.; Chen, J.-Y.; Twuc, Y.-K.; Chen, W.-J. *Polym. Int.* 2011, *61*, 554.
[176] Liu, P. *Eur. Polym. J.* 2005, *41*, 2693.
[177] Jakubka, F.; Schießl, S. P.; Martin, S.; Englert, J. M.; Hauke, F.; Hirsch, A.; Zaumseil, J. *ACS Macro Lett.* 2012, *1*, 815.
[178] Huang, J.-E.; Li, X.-H.; Xu, J.-C.; Li, H.-L. *Carbon* 2003, *41*, 2731.
[179] Thomassin, J.-M; Vuluga, D.; Alexandre, M.; Jérôme, C.; Molenberg, I.; Huynen, I.; Detrembleur, C. *Polymer* 2012, *53*, 169.
[180] Zheng, M.; Jagota, A.; Semke, E. D.; Diner, B. A.; McLean, R. S.; Lustig, S. R.; Richardson, R. E.; Tassi, N. G. *Nature Mater.* 2003, *2*, 338.
[181] Sanz, V.; Borowiak, E.; Lukanov, P.; Galibert, A. M.; Flahaut, E.; Coley, H. M.; Silva, S. R. P.; McFadden, J. *Carbon* 2011, *49*, 1775.
[182] Matsuura, K.; Saito, T.; Okazaki, T.; Ohshima, S.; Yumura, M.; Iijima, S. *Chem. Phys. Lett.* 2006, *429,* 497.
[183] Farbod, M.; Tadavani, S. K.; Kiasat, A. *Colloids Surf. A:* 2011, *384*, 685.
[184] Jyothirmayee Aravind, S S.; Baskar, P.; Baby, T. T.; Sabareesh, R K.; Das, S.; Ramaprabhu, S. *J. Phys. Chem. C* 2011, *115*, 16737.
[185] Gebhardt, B.; Hof, F.; Backes, C.; Müller, M.; Plocke, T.; Maultzsch, J.; Thomsen, C.; Hauke, F.; Hirsch, A. *J. Am. Chem. Soc.* 2011, *133*, 19459
[186] Alvaro, M.; Aprile, C.; Ferrer, B.; Garcia, H. *J. Am. Chem. Soc.* 2007*, 129,* 5647.
[187] Marsh, D. H.; Rance, G. A.; Zaka, M. H.; Whitby, R. J.; Khlobystov, A. N. *Phys. Chem. Chem. Phys*. 2007, *9*, 5490.
[188] Zhao, B.; Hu, H.; Yu, A.; Perea, D.; Haddon, R. C. *J. Am. Chem. Soc.* 2005, *127*, 8197.
[189] Saint-Aubin, K.; Poulin, P.; Saadaoui, H.; Maugey, M; Zakri, C. *Langmuir* 2009, *25*, 13206.
[190] Liu, Y.; Liang, P.; Zhang, H. Y.; Guo, D. S. *Small* 2006, *2*, 874.
[191] Datsyuk, V.; Landois, P.; Fitremann, J.; Peigney, A.; Galibert, A. M.; Soulaa, B.; Flahaut , E. *J. Mater. Chem*. 2009, *19*, 2729.
[192] Vijayaraghavan, A. *J. Mater. Chem.* 2012, *22*, 7083.
[193] Doyle C. D., R. Rocha, J.-D.; Weisman, R. B.; Tour J. M. *J. Am. Chem. Soc*. 2008, *130*, 6795.
[194] Balasubramanian, K.; Burghard, M.; Kern, K. *Phys. Chem. Chem. Phys*. 2008, *10*, 2256.
[195] Ghosh, S.; Rao, C. N. R. *Nano Res.* 2009, *2*, 183.

[196] Lee, C. W.; Han, X.; Chen, F.; Wei, J.; Chen, Y.; Chan-Park, M. B.; Li L.-J. *Adv. Mater.* 2010, *22*, 1278.
[197] Yang, F.; Anilkumar, P.; Anderson, A.; Cao, L.; Meziani, M. J.; Parenzan, A.; Sun, Y.-P. *J. Phys. Chem. C* 2012, *116*, 6800.
[198] Antaris, A. L.; Seo, J.-W. T.; Green, A. A.; Hersam, M. C. *ACS Nano* 2010, *4*, 4725.
[199] Chen, Y.; Xu, Y.; Perry, K.; Sokolov, A. P.; More, K.; Pang, Y. *ACS Macro Lett.* 2012, *1*, 701.
[200] Lemasson, F. A.; Strunk, T.; Gerstel, P.; Hennrich, F.; Lebedkin, S.; Barner-Kowollik, C.; Wenzel, W; Kappes, M. M.; Mayor, M. *J. Am. Chem. Soc.* 2011, *133*, 652.
[201] Gerstel, P.; Klumpp, S.; Hennrich, F.; Altintas, O.; Eaton, T. R.; Mayor, M.; Barner-Kowollik, C.; Kappes, M. M. *Polym. Chem.* 2012, *3*, 1966.
[202] (a) Park, S.; Lee, H. W.; Wang, H.; Selvarasah, S.; Dokmeci, M. R.; Park, Y. J.; Cha, S. N.; Kim, J. M.; Bao, Z. *ACS Nano* 2012, *6*, 2487.(b). Lee, H. W.; Yoon, Y.; Park, S.; Oh, J. H.; Hong, S.; Liyanage, L. S.; Wang, H.; Morishita, S.; Patil, N.; Park, Y. J.; Park, J. J.; Spakowitz, A.; Galli, G.; Gygi, F.; Wong, P. H.-S.; Tok, J. B.-H.; Kim, J. M.; Bao, Z. *Nature Comm* 2011, *2*, 541.
[203] Akazaki, K.; Toshimitsu, F.; Ozawa, H.; Fujigaya, T.; Nakashima, N. *J. Am. Chem Soc.* 2012, *134*, 12700.
[204] Najeeb, C. K.; Chang, J.; Lee, J.-H.; Lee, M.; Kim, J.-H. *J. Colloid Interface Sci.* 2011, *354*, 461.
[205] (a) LeMieux, M. C.; Roberts, M.; Barman, S.; Jin, Y. W.; Kim, J. M.; Bao, Z. *Science* 2008, *321*, 101.(b) Opatkiewicz, J. P.; LeMieux, M. C.; Vosgueritchian, M.; Barman, S. N.; Elkins, C. M.; Hedrick, J.; Bao, Z. *ACS Nano* 2012, *6*, 4845.
[206] (a) Sharma, P.; Ahuja, P. *Mater. Res. Bull.* 2008, *43*, 2517. (b) Cao, J.; Ionescu , A. M. *Solid-State Electronics* 2012, *74*, 121 (c) Mehlich, J.; Miyata, Y.; Shinohara, H.; Ravoo, B. J. *Small* 2012, *8*, 2258. (d) Cao, Q.; Han, S.-J.; Tulevski, G. S.; Franklin, A. D.; Haensch W. *ACS Nano* 2012, *6*, 6471 (e) Zhang, J.; Lin, A.; Patil, N.; Wei, H.; Wei, L.; Wong, H-S. P.; Mitra, S. *IEEE Trans. Comput.-Aided Design of Integrated Circuits Systems* 2012, *31*, 453.
[207] Balasubramanian, K.; Lee, E. J. H.; Weitz, R. T.; Burghard M.; Kern, K. *Phys. Stat. Sol. A* 2008, *205*, 633.
[208] Balasubramanian, K.; Sordan, R.; Burghard, M.; Kern, K. *Nano Lett.* 2004, *4*, 827.
[209] Zhao, J.; Gao, Y.; Lin, J.; Chen, Z.; Cui, Z. *J. Mater. Chem.* 2012, *22*, 2051.
[210] Xavier, J.; Ammar, S.; Dijon, J.; Pinson, J. *J. Mater. Chem.* 2010, *20,* 7197.
[211] Tulevski, S.G.; Hannon, J.; Afzali, A.; Chen, Z.; Avouris, P.; Kagan, C. R. *J. Am. Chem. Soc.* 2007, 129, 11964.
[212] Bardecker, J. A.; Afzali, A.; Tulevski, G. S.; Graham, T.; Hannon, J. B.; Jen, A. K.-Y. *Chem. Mater.* 2012, *24*, 2017.
[213] (a) Yang, S. B.; Kong, B.-S.; Jung, D.-H.; Baek, Y.-K.; Han, C.-S.; Ohe S.-K.; Jung, H.-T. *Nanoscale* 2011, *3*, 1361. (b) Gruner, G. *J. Mater. Chem.* 2006, *16*, 3533. (c) Brosseau, C. *Surf. Coatings Tech.* 2011, *206*, 753. (d) Tok, J. B.-H.; Bao Z. *Sci. China-Chem.* 2012, *55*, 718.
[214] Zhang, J.; Gao, L.; Sun, J.; Liu, Y.; Wang, Y.; Wang, J. *Diamond Rel. Mater.* 2012, *22, 82.*

[215] (a) Qureshi, A.; Kang, W. P.; Davidson, J. L.; Gurbuz, Y. *Diamond Relat. Mater.* 2009, *18*, 1401. (b) Wildgoose, G. G.; Banks, C. E.; Leventis, H. C.; Compton R. G. *Microchim Acta* 2006, *152*, 187.
[216] Cosnier, S.; Holzinger, M. *Chem. Soc. Rev.* 2011, *40*, 2146
[217] Liu, S.; Shen, Q.; Cao, Y.; Gan, L.; Wang, Z.; Steigerwald, M. L.; Guo, X. *Coord. Chem. Rev.* 2010, *254*, 1101.
[218] (a) Kauffman, D. R.; Star, A. *Angew. Chem , Int. Ed.* 2008, *47,* 6550. (b) Bondavalli, P.; Legagneux, P.; Pribat, D. *Sens. Actuators B* 2009, *140*, 304.
[219] Lee, S. H.; Jin, H. J.; Song, H. S.; Hong, S.; Park, T. H. *J. Biotechnol.* 2012, *157*, 467.
[220] Park, J.; Lim, J. H.; Jin, H. J.; Namgung, S.; Lee, S. H.; Park, T. H.; Hong, S. *Analyst* 2012, *137*, 3249.
[221] Esser, B.; Schnorr, J. M.; Swager, T. M. *Angew. Chem. Int. Ed.* 2012, *51*, 5752.
[222] Shirsat, M. D.; Sarkar, T.; Kakoullis Jr., J.; Myung, N. V.; Konnanath, B.; Spanias, A.; Mulchandani A. *J. Phys. Chem. C* 2012, *116*, 3845.
[223] Kumar, B.; Castro, M.; Feller J.-F. *Carbon* 2012, *50*, 3627.
[224] Xiea, H.; Shenga, C.; Chena, X.; Wanga, X.; Li; Z.; Zhou J. *Sens. Actuators B* 2012, *168*, 34
[225] Ventura, D. N.; Li, S.; Baker, C. A.; Breshike, C. J.; Spann, A. L.; Strouse, G. F.;. Kroto, H. W.; Acquah, S. F. A. *Carbon* 2012, *50,* 2666.
[226] Songmin, S; Wei, Z;. Tao, X.-m. *Sensors Actuators B* 2012, *166*, 330.
[227] Chang, C.-K.; Chiu, S.-W.; Kuo, H.-L.; Tang, K.-T. *Appl. Phys. Lett.* 2012, *100*, 043501.
[228] Kumara, B.; Park, Y. T.; Castroa, M.; Grunlan, J. C.; Feller J. F. *Talanta* 2012, *88*, 396.
[229] Kumar, B.; Castro, M.; Feller J. F. *Sens. Actuators B* 2012, *16,* 621.
[230] Raoof, J. B.; Ojani, R.; Amiri-Aref, M.; Baghayeri, M. *Sens. Actuators B* 2012, *166–167*, 508.
[231] Shi, B.-X.; Wang, Y.; Zhang, K.; Lam, T.-L.; Chan, H. L.-W. *Biosens. Bioelectron.* 2011, *26*, 2917.
[232] Kong, L.; Wang, J.; Meng, F.; Chen, X.; Jin, Z.; Li, M.; Liu, J.; Hun, X.-J., *J. Mater. Chem.* 2011, *21*, 11109.
[233] Maroto, A.; Balasubramanian, K.; Burghard, M.; Kern, K. *ChemPhysChem* 2007, *8*, 220.
[234] Zhang, M.; Smith, A.; Gorski, W. *Anal. Chem.* 2004, *76*, 5045.
[235] Tasca, F.; Harreither, W.; Ludwig, R.; Gooding, J. J.; Gorton, L. *Anal. Chem.* 2011, *83*, 3042.
[236] Arvinte, A.; Rotariu, L.; Bala, C.; Gurban A. M. *Bioelectrochemistry* 2009, *76*, 107.
[237] Kuang, Z.; Kim, S. N.; Crookes-Goodson, W. J.; Farmer, B. L.; Naik, R. R. *ACS Nano* 2010, *4*, 452.
[238] Hirotaka, N.; Shinya, T. *Cancer Sci.* 2012, *103*, 1378.
[239] (a) Sharifi, S.; Behzadi, S.; Laurent, S.; Laird Forrest, M.; Stroeve, P.; Mahmoudi, M. *Chem. Soc. Rev.* 2012, *41*, 2323. (b) Li, X.; Wang, L.; Fan, Y., Feng, Q.; Cui, F.-z. *J. Nanomater.* 2012, 548389. (c) Bianco, A.; Kostarelos, K.; Prato, M. *Chem. Comm.* 2011, *47*, 10182. (d) Kaiser, J. -P.; Roesslein, M.; Buerki-Thurnherr, T.; Wick, P. *Current Med. Chem.* 2011, *18*, 2115. (e) Cattaneo, A. G.; Gornati, R.; Sabbioni, E.; Chiriva-Internati, M.; Cobos, E.; Jenkins, M. R.; Bernardini, G. *J. Appl. Toxicol.* 2010, *30*, 730. (f) Shvedova, A. A.; Kisin, E. R.; Porter, D.; Schulte, P.; Kagan, V. E.; Fadeel,

B.; Castranova, V. *Pharmacol. Therap.* 2009, *121*, 192.(g) Handy, R. D.; Shaw, B. J. *Health Risk Soc.* 2007, *9*, 125. (h) Linkov, I.; Satterstrom, F. K.; Corey, L. M. *Nanomed.: Nanotech. Biol Med.* 2008, *4*, 167.

[240] Lacerda, L.; Soundararajan, A.; Singh, R.; Pastorin, G.; Al-Jamal, K.T.; Turton, J.; Frederik, P.; Herrero, M.A.; Li, S.; Bao, A.; Emfietzoglou, D.; Mather, S.; Phillips, W. T.; Prato, M.; Bianco, A.; Goinsand, B; Kostarelos, K. *Adv. Mater.* 2008, *20*, 225.

[241] Liu, X.; Hurtand, R. H.; Kane, A. B *Carbon* 2010, *48*,1961.

[242] (a) Ronzani, C.; Spiegelhalter, C.; Vonesch; J.-L.; Lebeau, L; Pons F. *Arch. Toxicol.* 2012, *86*, 137.(b) Wang, X.; Xia, T.; Addo Ntim, S.; Ji, Z.; Lin, S.; Meng, H.; Chung, C.-H.; George, S.; Zhang, H.; Wang, M.; Li, N.; Yang, Y.; Castranova, V.; Mitra, S.; Bonner, J. C.; Nel, A. E. *ACS Nano* 2011,*5*, 9772.

[243] Ali-Boucetta, H.; Al-Jamal, K. T.; Müller, K. H.; Li, S.; Porter, A. E.; Eddaoudi, A.; Prato, M.; Bianco, A.; Kostarelos, K. *Small* 2011, 7, 3230.

[244] Gea, C.; Dua, J; Zhaoa, L.; Wanga, L.; Liua, Y.; Lia, D.; Yanga, Y.; Zhoub, R.; Zhaoa, Y.; Chaid, Z.; Chen, C. *PNAS* 2011, *108*, 16968.

[245] (a) Pascu, S. I.; Arrowsmith, R. L.; Bayly, S. R.; Brayshaw, S.; Hu, Z. *Phil. Trans. R. Soc. A* 2010, *368*, 3683. (b) Liu, Z.; Yanga, K; Lee, S.-T. *J. Mater. Chem.* 2011, *21*, 586.

[246] (a) Kostarelos, K.; Bianco, A.; Prato, M. *Nature Nanotechnol.* 2009, *4*, 627. (b) Wick, P.; Clift, J. D.; Roesslein, M; Rothen-Rutishauser, B. *ChemSusChem* 2011, *4*, 905.

[247] (a) Liu, Z; Liang, X.-J. *Theranostics* 2012; *2*, 235. (b) Tan, A.; Yildirimer, L.; Rajadas, J.; De La Peña, H.; Pastorin, G.; Seifalian, A. *Nanomedicine* 2011, *6*, 1101.

[248] (a) Fabbro, C.; Ali-Boucetta, H.; Da Ros, T.; Kostarelos , K.; Bianco, A.; Prato, M. *Chem. Commun.* 2012, *48*, 3911. (b) Lay, C. L.; Liu, J.; Liu, Y. *Expert Rev. Med. Devices* 2011, *8*, 561. (c) Guven, A.; Rusakova, I. A.; Lewis, M. T. ; Wilson, L. J. *Biomaterials* 2012, *33*,1455. (d) Minati, L.; Antonini, V.; Dalla Serra, M.; Speranza, G. *Langmuir* 2012, *28*, 15900. (e) Chen, Mei-Ling; Chen, Xu-Wei; Wang, Jian-Hua *Langmuir* 2012, *28*,16469.

[249] Nunes, A.; Al-Jamal, K.; Nakajima, T; Hariz, M.; Kostarelos, K. *Arch. Toxicol.* 2012, *86*, 1009.

[250] Cheung, W.; Pontoriero, F.; Taratula, O.; Chen, A. M.; He, H. *Adv. Drug Delivery Rev.* 2010, *62*, 633

[251] Bottini, M.; Rosato, N.; Bottini, N. *Biomacromolecules* 2011, *12*, 3381.

[252] Al-Jamal, K. T.; Nerl, H.; Müller, K. H.; Ali-Boucetta, H.; Li, S.; Haynes, P. D.; Jinschek, J. R.; Prato, M.; Bianco, A.; Kostarelos, K.; Porter A. E. *Nanoscale* 2011, *3*, 2627.

[253] Ruggiero, A.; Villa, C. H.; Holland, J. P.; Sprinkle, S. R; May, C.; Lewis, J. S; Scheinberg, D. A.; McDevitt, M. R. *Int. J. Nanomedicine* 2010,5, 783.

[254] Yuksel, M.; Colak, D. G.; Akin, M.; Cianga, I.; Kukut, M.; Ilker Medine, E.; Can, M.; Sakarya, S.; Unak, P.; Timur, S.; Yagci, Y. *Biomacromolecules* 2012, *13*, 2680.

[255] Wang, C.; Ma, X.; Ye, S.; Cheng, L.; Yang, K; Guo, L.; Li, C.; Li , Y.; Liu, Z. *Adv. Funct. Mater.* 2012, *22*, 2363.

[256] (a) Li, R.; Wu, R.; Zhao, L.; Hu, Z.; Guo, S.; Pan, X.; Zou, H. *Carbon* 2011, *49*, 1797. (b) Kang, B.; Li, J.; Chang, S.; Dai, M.; Ren, C.; Dai, Y.; Chen, D. *Small* 2012, *8*, 777.

[257] Meng, L.; Zhang, X.; Lu, Q.; Fei, Z.; Dyson, P. J. *Biomaterials* 2012, *33*, 1689.

[258] Kirkpatrick, D. L.; Weiss, M.; Naumov, A.; Bartholomeusz, G.; Weisman, R. B.; Gliko, O. *Materials* 2012, *5*, 278.
[259] Lu, Y.-J.; Wei, K.-C.; Ma, C.-C. M.;Yang S.-Y.; Chen, J.-P. *Colloids Surf. B: Biointerfaces* 2012, *89*, 1.
[260] Villa, C. H.; Dao, T.; Ahearn, I.; Fehrenbacher, N.; Casey, E.; Rey, D. A.; Korontsvit, T.; Zakhaleva, V.; Batt, C. A.; Philips, M. R.; Scheinberg D. A. *ACS Nano* 2011, *5,* 5300.
[261] Liu, X.; Tao, H.; Yang, K.; Zhang, S.; Lee, S.-T.; Liu, Z. *Biomaterials* 2011, *32*, 144.
[262] Venturelli, E.; Fabbro, C.; Chaloin, O.; Ménard-Moyon, C.; Smulski, C. R.; Da Ros, T.; Kostarelos, K.; Prato, M.; Bianco, A. *Small* 2011, 7, 2179.
[263] Moniruzzaman, M.; Winey, K. I. *Macromolecules* 2006, *39*, 5194.
[264] Homenick, C. M.; Lawson, G.; Adronov, A. *Polym. Rev.* 2007, *47*, 265.
[265] Spitalsky, Z.; Tasis, D. ; Papagelis, K. ; Galiotis, C. Prog. Polym. Sci. 2010, *35*, 357.
[266] Sakellariou, G. ; Priftis, D. ; Baskaranw, D. *Chem. Soc. Rev.,* 2013, *42*, 677.
[267] Lou, X. D.; Detrembleur, C.; Pagnoulle, C.; Jerôme, R.; Bocharova, V.; Kiriy, A.; Stamm, M. *Adv. Mater.* 2004, *16*, 2123
[268] Liu, Y. Q.; Yao, Z. L.; Adronov, A. *Macromolecules* 2005, *38*, 1172.
[269] Akbar, S. ; Beyou, E. ; Chaumont, P. ; Mazzolini, J. ; Espinosa, E. ; D'agosto, F. ; Boisson, C. *J. Polym. Sci. A Polym. Chem*. 2011, *49*, 957.
[270] Jiang, S. ; Deng, J. ; Yang, W. *Macromol. Rapid Commun*. 2008, *29*, 1521.
[271] . Qin, K.; Qin, D.; Ford W.T; Resasco D.E.; and Herrera J.E. (2004) *J. Am. Chem. Soc.,* 2004, *126*, 170.
[272] Kong, H.; Gao, C.; Yan, D. *J. Am. Chem. Soc.* 2004, *126*, 412.
[273] Zhao, X.-D.; Fan, X.-H.; Chen, X.-F.; Chai, C.-P.; Zhou, Q.-F. *J. Polym. Sci. A Polym. Chem.*, 2006, *44*, 4656.
[274] Hong, C.-Y.; You, Y.-Z.; Pan, C.-Y. *Chem. Mater.* 2005, *17,* 2247.
[275] Song, W.; Zheng, Z.; Tang, W.; Wang, X. *Polymer* 2007, *48*, 3658
[276] Sun, Y.-P.; Fu, K.; Lin, Y.; Huang, W. *Acc. Chem. Res.*, 2002, *35*, 1096.
[277] He, H.; Zhang, Y.; Gao, C.; Wu, J. *Chem. Commun.* 2009, 1655.
[278] Tour, J. M., Hudson, J. L., Krishnamoorti, R., Yurelki, K., Mitchell, C. A. (2005) *Polymerization initiated at the sidewalls of carbon nanotubes*, WO/2005/030858.
[279] Wu, W., Tsarevsky, N.V., Hudson, J. L., Tour, J. M., Matyjaszewski, K., Kowalewski, T. *Small,* 2007, *3*, 1803.
[280] Matrab T., Chancolon, J.; Mayne-L'Hermite, M.; Rouzaud, J. N.; Deniau, G.; Boudou, J.-P., Chehimi, M. M.; Delamar, M. *Colloids Surf. A Physicochem. Asp.* 2006, *287,* 217.
[281] Wang, G.J.; Huang, S.Z.; Wang,Y.; Liu, L.; Qiu, J.; Li, Y. *Polymer,* 2007, *48*, 728.
[282] Li, H., Cheng, F., Duft, A. M., Adronov, A. *J. Am. Chem. Soc.* 2005, *127*, 14518.
[283] Chergui, S. M.; Ledebt, A.; Mammeri, F.; Herbst, F.; Carbonnier, B.; Ben Romdhane, H.; Delamar, M.; Chehimi, M. M. *Langmuir* 2010, *26*, 16115.
[284] Salmi, Z.; Epape, C.; Mahouche-Chergui, S.; Carbonnier, B.; Omastová, M.; Chehimi, M. M. *J. Colloid Sci. Biotechnol.* 2013, *2*, 53.
[285] Li, S.; Chen, H.; Bi, W.; Zhou, J.; Wang, H.; Li, J.; Cheng, W.; Li, M.; Li, L.; Tang, T. J. Polym. Sci. A Polym. Chem. 2007, *45*, 5459.
[286] Yao, Z.; Braidy, N.; Botton, G. ; Adronov, A. *J. Am. Chem. Soc.*, 2003, *125*, 16015.
[287] Li, G.; Zheng, H.; Wang, Y.; Wang, H.; Dong, Q.; Bai, R. *Polymer* 2010, *51*, 1940.

[288] Qin, S. H.; Qin, D. Q.; Ford, W. T.; Resasco, D. E.; Herrera, J. E. *Macromolecules*, 2004, *37*, 752.

[289] Li, Y.; Yang, D.; Adronov, A.; Gao, Y.; Luo, X.; Li, H. *Macromolecules*, 2012, *45*, 4698.

[290] Zhou, L.; Gao, C.; Xu, W. Macromol. Chem. Phys. 2009, *210*, 1011.

[291] Zhou, B.; Tong, Z.-Z.; Huang, J.; Xu, J.-T.; Fan, Z.-Q. *Mater. Chem. Phys*. 2013, *137*, 1053.

[292] Priftis, D.; Sakellariou, G.; Baskaran, D.; Mays, J. W.; Hadjichristidis, N. *Soft Matter*, 2009, *5*, 4272.

[293] You, Y.-Z.; Hong, Y.-Y.; Pan, C.-Y. *Macromol. Rapid Commun*. 2006, *27*, 2001.

[294] Rana, S.; Cho, J. W. *Nanotechnology* 2011, *22*, 275609 (7pp)

[295] (a) *http://www.nanocyl.com* (b) *http://www.cheaptubes.com*. (c) *http://www.nanotubes.com.cn* (d) *http://ssnano.com*. (e) *http://www.reinste.com*. (f) *http:*//www.*nano-lab.com*. (g) *http://www.nanotubes.com.cn*. (h) *http://www.nanoshel.com*

[296] (a)http://www.baytubes.com/.(b)*http://www.graphistrength.fr*.(c)*http://www.nanocyl.com*

In: Applied Surface Chemistry of Nanomaterials
Editors: Mohamed M. Chehimi and Jean Pinson
ISBN: 978-1-62808-351-4

Chapter 5

PROPERTY-ORIENTED SURFACE ENGINEERING OF CARBON NANOTUBES FOR POLYMER NANOCOMPOSITES

Jin-Kai Yuan, Sheng-Hong Yao, De-Long He and Jinbo Bai*
Laboratoire de Mécanique des Sols, Structures et Matériaux, Ecole Centrale Paris, CNRS UMR8579, Châtenay-Malabry Cedex, France

1. INTRODUCTION

Carbon nanotubes (CNTs) are *allotropes of carbon,* and can be envisioned as a rolled up graphene sheet into a seamless cylinder with fullerene caps for single-wall carbon nanotube (SWNT) or multiple sheets as in the case of multi-walled carbon nanotube (MWNT). [1] The unique structure endows CNTs exceptional thermal stability and remarkable mechanical, electronic, and structural properties. [2]

As an ideal quasi-1D structure with an atomically monolayered surface and extended curved π-bonding configuration, an individual SWNT exhibit semiconducting or metallic or semimetallic behavior, depending on the orientation of the graphene lattice with respect to the tube axis, which is called the chirality. Perfect MWNTs have rather similar electronic properties to SWNTs because of the weak coupling between the cylinders in MWNTs. They are typically metallic and can carry currents of up to 10^9 A cm^{-2}. [3] Phonons can also propagate easily along the carbon nanotube. Theory predicts their room temperature thermal conductivity can reach up to 6000 W $m^{-1}K^{-1}$. [4] While this has not been fully attained, the highest value around 3000 W $m^{-1}K^{-1}$ has been measured for individual MWNT, which is still greater than that of natural diamond and the basal plane of graphite (both 2000 W $m^{-1}K^{-1}$). [5] Besides the superlative electronic and thermal properties, CNTs are rather stiff and exceptionally strong when considering the cross-sectional area of the CNT walls only. The Young's modulus is measured ranging from 320 to 1470 GPa for SWNT [6] and from 270 to 950 GPa for MWNT. [7] The tensile test showed that during the loading process only the

* To whom correspondence should be addressed: Phone: +33-141131316. E-mail address:jinbo.bai@ecp.fr (J. Bai)

outmost layer breaks and the tensile strength can range from 11 to 60 GPa. [7] This strength is over 10-fold higher than any industrial fiber.

The unique combination of exotic electrical, thermal, and mechanical properties of CNTs along with their large shape anisotropy makes them the ideal reinforcing agent for manufacturing nanostuctured low-cost and high-performance polymer nanocomposites. [8-10] The first ever polymer nanocomposites with CNTs were reported in 1994 by Ajayan et al. [11] Since then, numerous research efforts have been devoted to fully harness the superb properties of individual CNT in polymer composites for applications ranging from electronics to automotive and aerospace sectors. [12] However, some long-standing problems still exist and sincerely hinder their applicability for practical devices. For instance, it is not so easy to realize the true potential of individual CNT in polymer composites due to the entangled or aggregated structure of pristine CNTs, especially in the case of SWNT-based composites. CNTs have been recognized as highly polaziable, smooth-sided materials with an attractive interaction potential of 0.5-2.0 eV per nanometer of tube-to-tube contact in vacuum. [13, 14] This extremely large cohesive force naturally induces CNTs to form twisted and bundled structures, and eventually harms their even dispersion in the host polymers. Furthermore, the smooth sidewalls of CNTs composed of a basal layer of carbon atoms usually lead to an intrinsically weak affinity to surrounding polymer matrices. In this case, embedded pristine CNTs only exhibit a fraction of their true potential, even worse weakens rather reinforces the final CNT composites. [15]

Surface engineering of CNTs have thus been extensively explored for promoting the dispersion of CNTs as well as improving the miscibility of CNTs with polymer matrices by establishing remarkable interfacial interaction through chemical bonding or physical/mechanical entanglements between modified CNTs and host polymers. Various surface engineering techniques have been developed in the last decades and could be basically divided into two categories: (i) chemical bonding of π-conjugated skeleton of CNTs with various reagents, thereby introducing functional moieties on the CNT surface, [15-18] (ii) wrapping or adsorption of mediating molecules through non-covalent interaction, such as van der Waals, CH-π, π-π, and hydrophobic interactions. [19-22] Although both methods can lead to better nanotube dispersion, and eventually to efficient thermodynamic wetting of nanotubes with polymer matrices, they actually infuluence the macroscopic properties of final composites in different manners. For example, a strong interfacial interaction may favor the load transfer and thus an improved elastic modulus and tensile strength in polymer composites, [23] while it may also force effective wrapping of polymer chains onto nanotube surfaces and decrease the tube-to-tube contact effectiveness. As a result, the composites show a limited improvement in the electrical conductivity [24] but excellent dielectric properties, e.g., high dielectric permittivity and low loss tangent. [25] Therefore, when optimizing CNT-based composites by surface engineering of nanotubes, the technique selection must be oriented by the specific physical property targeted to be enhanced. In this chapter, property-oriented surface engineering of CNTs for polymer composites is first reviewed by emphasizing the effect of interfacial interaction between CNTs and host polymers on a given property (e.g., mechanical, electrical and dielectric properties) of final composites. Then, an example of non-covalent modification of CNTs for dielectric polymer composites is presented in detail, showing how the surface engineering changed the interfacial interaction and thus favored the dielectric properties of the composites.

2. PROPERTY-ORIENTED SURFACE ENGINEERING OF CNTS

2.1. Mechanical Properties of CNT/Polymer Composites

To prepare polymer composites with superior strength and flaw-tolerance, CNTs have natural advantages over their micrometer-sized counterparts (e.g., carbon fiber) due to their unprecedented mechanical properties and high aspect ratio. However, a huge challenge still lies in enhancing the modulus and strength of composites because of the agglomeration tendency of CNTs and poor load transfer efficiency between the matrix and nanotubes. Although individual CNTs have Young's modulus of about 1 TPa, the polymer composites reinforced by randomly-dispersed CNTs have exhibited only a moderate enhancement in modulus and even more limited improvement in strength. [26] Their moduli and strengths are 2 orders of magnitude lower than those theoretically predicted by composite theory. [27] This is mainly attributed to the inefficient load-bearing status of CNTs in the polymer matrix due to two inherent problems of nanotubes. The first one is waviness. CNTs are normally curled and twisted because of their large flexibility, which largely harms their reinforcement effect in composites. [28, 29] The other is the limited interfacial shear strength between CNTs and polymer matrix. The atomically smooth walls of CNTs interact with the surrounding polymer matrix noncovalently, which results in ubiquitous interfacial slippages when local stress is increased to a certain value. [30] In this case, even if the carbon nanotubes are homogenously dispersed, they actually carry a load much lower than anticipated. The true potential of CNTs was not fully realized in this case.

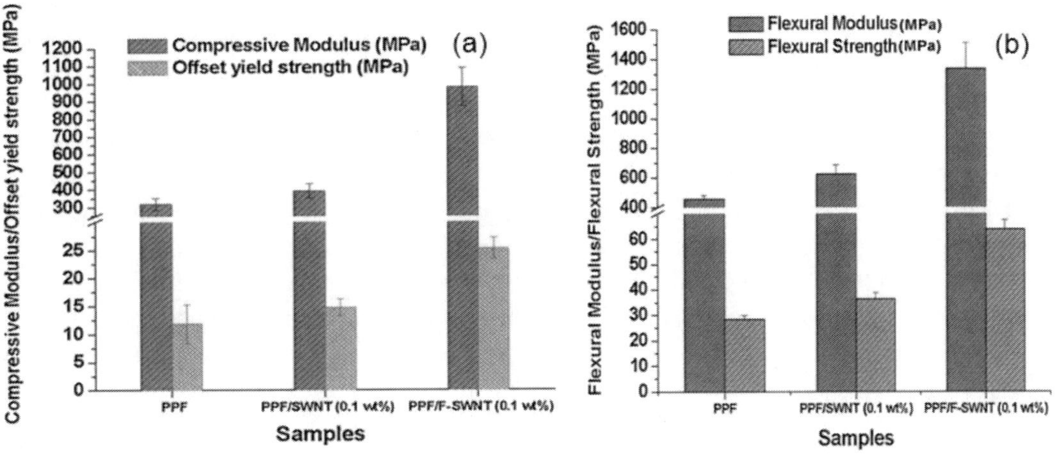

Figure 1. Comparisons of (a) compressive and (b) flexural mechanical properties of pristine and functionalized SWNT/PPF composites. Reprinted with permission from Ref. [18] Copyright 2012 Elsevier.

To improve the load-transfer efficiency, a great deal of efforts has been made either by designing load-transfer-favored structure of CNTs [26, 31, 32] or by grafting functional moieties or macromolecules on the CNTs surfaces. [18] Here we will only focus on the production of functionalized and polymer-grafted nanotubes, which have been largely exploited to improve mechanical properties of polymer composites. For instance, Shi et al.

[33] prepared functionalized SWNTs (F-SWNTs) from individual nanotubes by a diazonium-based method and then dispersed them in biodegradable polymer poly(propylene fumarate) (PPF). Fig. 1 shows the compressive and flexural mechanical properties of the composites with pristine and functionalized SWNT. It was found that the nanocomposites with 0.1 wt % F-SWNTs loading afforded a 3-fold increase in both compressive modulus and flexural modulus and a 2-fold increase in both compressive offset yield strength and flexural strength as compared to neat PPF matrix, while the use of pristine SWNTs gained less than 37% mechanical reinforcement at the same SWNT content. The extraordinary mechanical reinforcing effect of F-SWNT was attributed to the improved load transfer from the matrix to the nanotubes, which resulted from the strong interfacial interaction between F-SWNT and PPF matrix. Liu et al. [34] recently grafted n-hexa decylamine with acyl chloride functionalized MWNTs and then incorporated them into polyamide 6 (PA6) matrix. The resultant composites showed around 30% increase in Young's modulus, while it was only 24% for pristine MWNT-based nanocomposites.

It is obvious that the intermolecular interactions between CNTs and the polymer matrix are of the utmost importance for achieving high load-transfer capability and thus high mechanical strength of the composites. The organic functional groups covalently grafted on the CNT surfaces have been demonstrated much favorable to optimize the dispersion state of nanotubes and their miscibility with the polymer. Sahoo et al. [35] have selectively introduced three types of chemical functional groups, namely nitrophenyl ($C_6H_4NO_2$), benzoic acid (C_6H_4COOH), and aminophenyl ($C_6H_4NH_2$), on the sidewalls of MWNTs in order to systematically investigate the effects of these electron withdrawing and donating groups on the interfacial adhesion to a liquid crystalline polymer (LCP). The tensile test results (Fig. 2) showed that the chemically functionalized MWNTs had much more prominent reinforcing effect than raw MWNTs. The chemical moieties containing benzene rings attached to the naotube surface are testified to be able to favorably interact with the LCP chains via π-π stacking interactions. [35] Furthermore, as compared to the $-C_6H_4NO_2$ groups, $-C_6H_4COOH$ and $-C_6H_4NH_2$ functionalized MWNTs exhibited more significant influence on the mechanical behavior. The reason can be the formation of hydrogen bonds between the MWNTs and LCP chains, as schematically illustrated in Fig. 2A and B. Therefore, it can be concluded that the improvement in mechanical properties of the composites are very sensitive to the extent of interfacial interaction between nanotubes and the polymer matrix, namely the stronger the interfacial interaction, the more significant the mechanical reinforcing effect.

Besides the introducing of functional moieties, another area of intense research is the physically adsorbing or chemically bonding of macromolecules onto nanotube surface. Indeed, the addition of a whole polymer chain could be expected to have greater influence on the nanotube properties and their affinity to polymer matrices in comparison with the addition of low molecular weight functionalities. In recent years, many polymers including polythiophene [36] and poly(phenylene ethynylenes) [37] have been found to be capable of dispersing CNTs by forming π-π interfacial interactions. The mechanical properties of the resulting CNT/polymer composites could be effectively tailored by tuning the side-chain structure of a graft polymer dispersant. Yuan et al. [38] have synthesized three types of graft polymers, namely polyimide (PI) without side chain, polyimide-graft-glyceryl 4-nonylphenyl ether (PI-GNE), and polyimide-graft-bisphenol A diglyceryl acrylate (PI-BDA), which were then used to disperse SWNTs and to improve the affinity of nanotubes to cyanate ester (CE) polymer matrix. The interaction between SWNT and these polymer dispersants was well studied by molecular dynamics simulations. It was found that the adding of side chains could enhance the binding strength (π-π interaction) significantly so that the majority part of the PI-GNE and PI-BDA polymers would be well adsorbed onto the SWNT surface, not as the case of PI where only one or two rings make contacts with the SWNT surface (Fig. 3a-c). Because of the structural similarity of PI-BDA side chain to that of CE monomer, SWNTs/PI-BDA presented the more uniform dispersion state in CE matrix than their counterparts (Fig. 3d). Therefore, SWNTs/PI-BDA/CE composites have the largest improvement in Young's modulus (32%), tensile strength (47%) and toughness (71%) concurrently, as shown in Fig. 3e. Yuan's work successfully validates a principle that efficient load transfer and improved mechanical properties of the CNT composites can be achieved by judiciously selecting the side chains of the polymer dispersants.

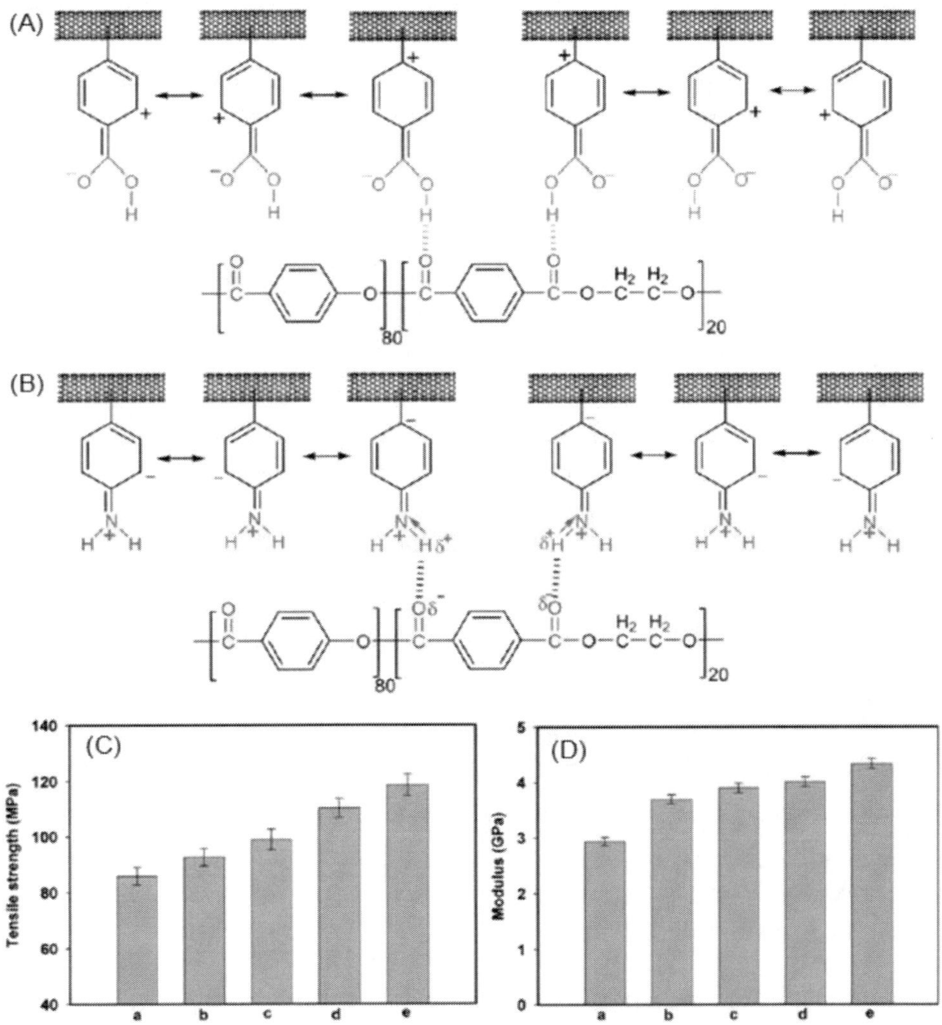

Figure 2. Formation mechanism of hydrogen bond (A) between a LCP chain and MWNT–C_6H_4COOH and (B) between a LCP chain and MWNT–$C_6H_4NH_2$. (C) Tensile strength and (D) modulus of MWNT/LCP composites: (a) neat LCP, (b) raw MWNT/LCP, (c) MWNT–$C_6H_4NO_2$/LCP, (d) MWNT–C_6H_4COOH/LCP, and (e) MWNT–$C_6H_4NH_2$/LCP composites. Reprinted with permission from Ref. [35], Copyright 2011 Royal Society.

As compared to the polymer-CNT physical adsorption, the formation of a polymer-CNT covalent bond induces more superior interfacial interactions. When properly designed, the resulting material can be a homogeneously blended composite in which the grafted polymer chains penetrate and become a part of the polymer matrix. In this way, the load transfer from matrix to the CNTs can be largely improved so that ultra-strong composites can be realized. Blake et al. [39] reported a generic route towards highly promising polymer reinforcement via the preparation of polymer-functionalized nanotubes using an organometallic approach. The process involves the binding of butyl groups to the nanotube surface as well as the subsequent coupling reaction between resultant *n*-butyllithium-functionalized MWNTs and chlorinated polypropylene (CPP) polymer (Fig. 4a). The polymer layer coated on the CNT surface can be clearly seen in transmission electron microscopy (TEM) image of the MWNT/CPP composites (Fig. 4b). Such covalent attachment of polymer molecules to nanotubes would not only facilitate processing to afford high composite homogeneity but also largely improve the interfacial stress transfer, especially when CPP was employed as polymer matrix. As compared to the neat polymer, Young's modulus and tensile strength of the composites were found to be increased by factors of 3 (from 0.22 to 0.68 GPa) and 3.8 (from 13 to 49 MPa), respectively (Fig. 4c).

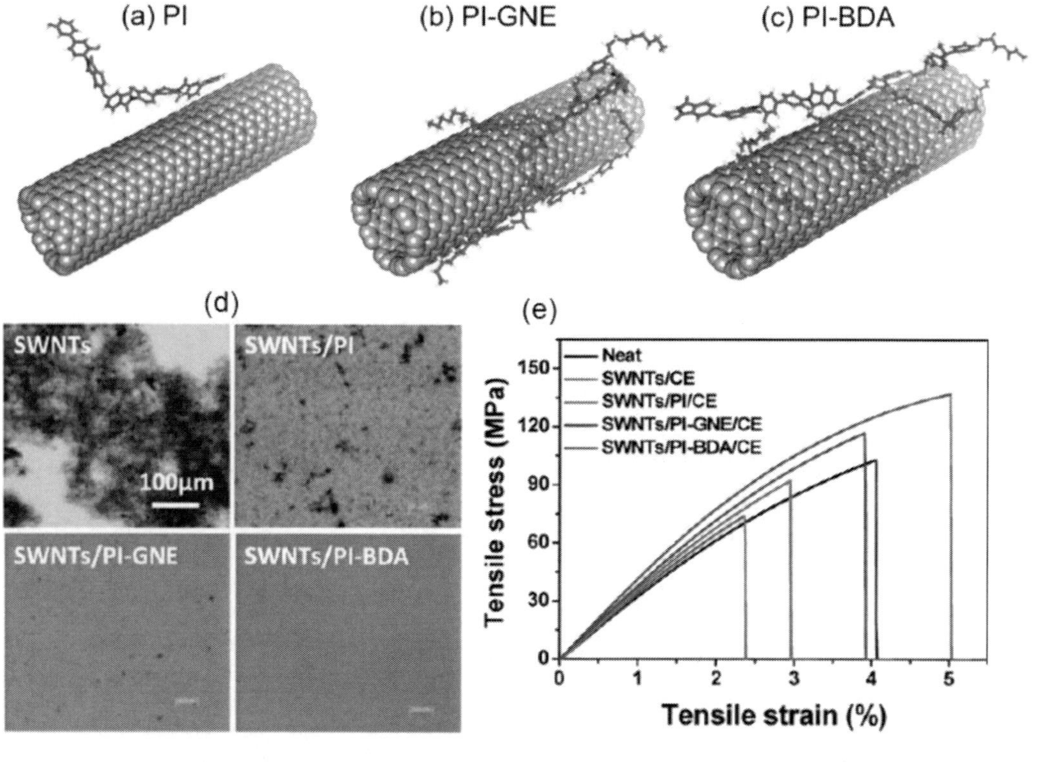

Figure 3. The binding patterns for different graft polymers on SWNT: (a) PI without side chains, (b) PI-GNE, and (c) PI-BDA. (d) Optical micrographs for SWNTs/CE, SWNTs/PI/CE, SWNTs/PI-GNE/CE, and SWNTs/PI-BDA/CE composites. Scale bars are 100 μm. (e) Stress-strain curves of neat CE, SWNTs/CE, SWNTs/PI/CE, SWNTs/PI-GNE/CE, and SWNTs/PI-BDA/CE composites. All the composites are loaded with 1 wt% SWNT. Reprinted with permission from Ref. [38], Copyright 2011 American Chemical Society.

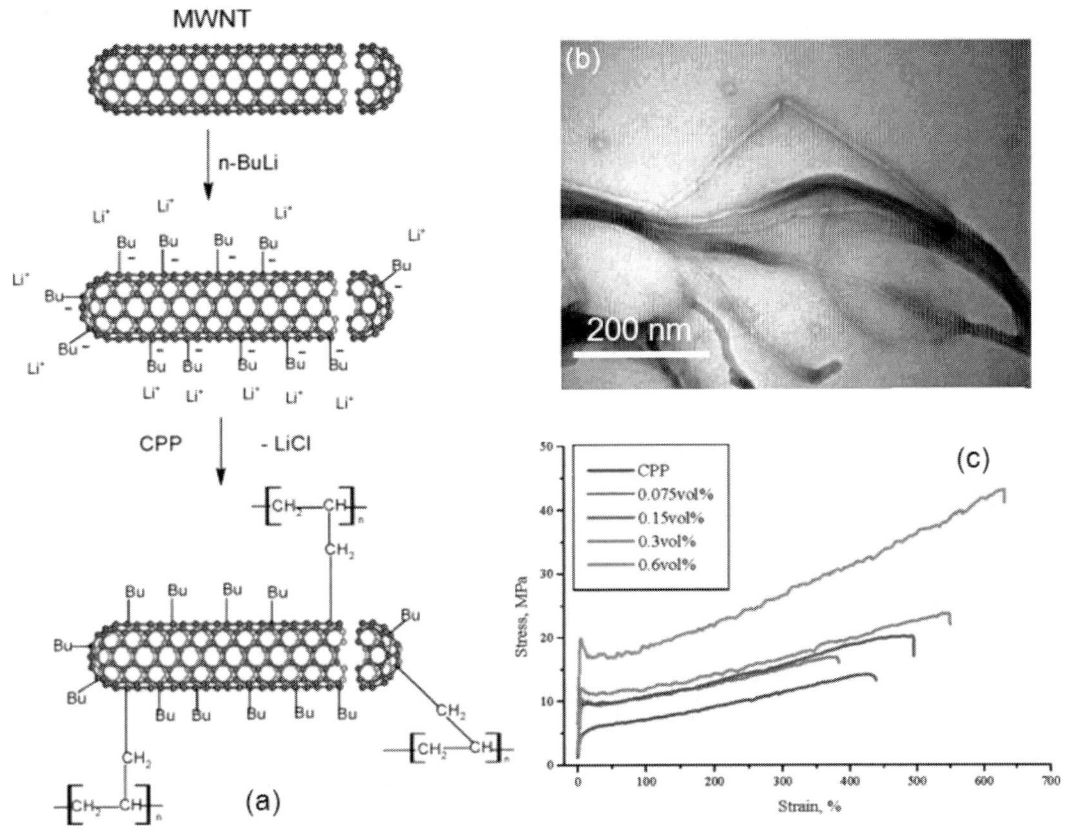

Figure4. (a) Schematic presentation for functionalization of MWNTs with CPP. (b) TEM images of MWNTs coated with the CPP. (c) Stress-strain curves for the MWNT/CPP composites studied. Reprinted with permission from Ref. [39], Copyright 2004 American Chemical Society.

Apart from grafting a given polymer chains onto the CNT surface covalently, there is another approach to establish polymer-CNT covalent bond, which involves growing polymers from CNT surfaces via *in-situ* polymerization of monomers initiated by chemical species immobilized on the CNT sidewalls. The high reactivity of monomers and rich choice of CNT functionalities allows for efficient, designable, and tailored grafting feasible, [40] which enables the nanoscale manipulation of the polymer-CNT interfacial interaction by systematically engineering the surface chemistry of nanotubes. Gao et al. [41] fabricated continuous composite fibers from an *in-situ* polymerization of caprolactam with SWNTs possessing a range of carboxylic acid (SWNT-COOH) and amide (SWNT-$CONH_2$) functionalities (Fig. 5a), showing the power of carbon nanotube chemistry in controlling the grafting polymerization reaction, the interfacial affinity, and the mechanical performance of the composite. The results showed in Fig. 5b demonstrated that SWNTs bearing a higher concentration of carboxylic acid groups allow for a stronger SWNT-nylon interfacial interaction, which consequently gives size to a larger Young's modulus and mechanical strength. They also compared the mechanical behavior of composites prepared using SWNT-COOH and SWNT-$CONH_2$, and found that the $CONH_2$ group changed the grafting polymerization chemistry and thereby the SWNT-nylon interfacial interaction, consequently increased the composite flexibility and toughness (Fig. 5c). This work indeed demonstrates an effective process to engineer the polymer-CNT interface by using CNT surface chemistry.

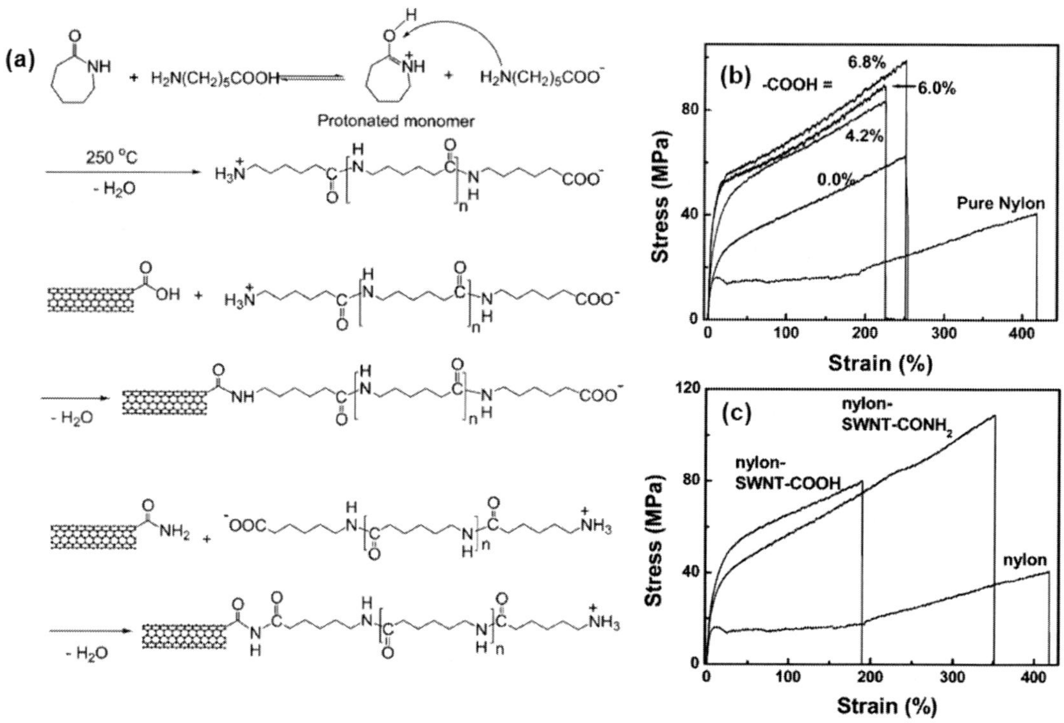

Figure 5. (a) Synthesis process of SWNT/nylon composite by starting from SWNT-COOH and SWNT-$CONH_2$. (b) Stress-strain curves of composite fibers loaded with 0.5 wt % SWNTs bearing 0.0%, 4.2%, 6.0%, and 6.8% carboxylic acid groups. (c) Comparison of stress-strain profiles of composite fibers with SWNT-COOH and SWNT-$CONH_2$ at a loading of 0.5 wt %. Reprinted with permission from Ref. [41], Copyright 2006 American Chemical Society.

In terms of the mechanical reinforcement, an ultimate interfacial adhesion of CNTs to the surrounding polymer matrix is usually required to fully realize the mechanical potential of nanotubes by enhancing the load transfer efficiency. Generally speaking, any surface treatment of CNTs to improve the CNT-polymer contact can be employed to meet this end. While the covalently grafting of functional groups or macromolecules based on the CNT surface chemistry has been demonstrated to be one of the most favorable methodologies because of strong interfacial binding strength as well as the large tunability of surface chemistry of CNTs.

2.2. Electrical Properties of CNT/Polymer Composites

The electrically conductive polymer composites made from an insulating matrix and highly conductive fillers has significant advantages in a variety of electronic applications, such as electrostatic discharge dissipation, electromagnetic interference (EMI) shielding and electrostatic painting. [42] CNTs have clearly demonstrated their capability as conductive filler to reinforce the electrical properties of polymer composites, in which, the formation of three-dimensional conductive CNT network results in a dramatic increase in the electrical conductivity. Numerous previous studies show that the percolation threshold and conductivity level of composites much relies on the aspect ratio of CNTs, alignment and dispersion of nanotubes, and the local intertube contact resistance. [40] Here we only focus on the issues much related to the surface engineering of CNTs, namely the disentaglement of CNT bundles in polymer matrices and the resistance of intertube junctions in the network.

The recent theoretical and experimental investigations indicate that the dispersion state of CNTs alters much on the conductivity of their composites. [10, 43] In principle, all the verified methods in the last section to improve CNT dispersion state can be used to prepare electrically conductive CNT composites. However, it should be emphasized that the non-covalent modification is much preferable, because it preserves the graphene structure of the nanotubes and thus their intrinsic electrical properties. Yuan et al. [44] demonstrated an effective way to achieve high-conductivity composites by using a novel poly(amic acid) (PAA) containing a rigid backbone with hydroxyl pendant groups, as both the nanotube dispersant and the matrix precursor (Fig. 6a). The resulting MWNT/PI composites afforded a percolation threshold as low as 0.48 wt %. While the nanotube content exceeded the percolation threshold and went up to 30 wt%, the electrical conductivity reached a value as high as 38.8 S cm^{-1} (Fig. 6b). It was speculated that the superlative electrical performance is mainly attributed to the good nanotube dispersion, even at the loading as high as 30 wt%. This was achieved through the use of a single polymer to perform the dual functions of nanotube dispersant and matrix precursor.

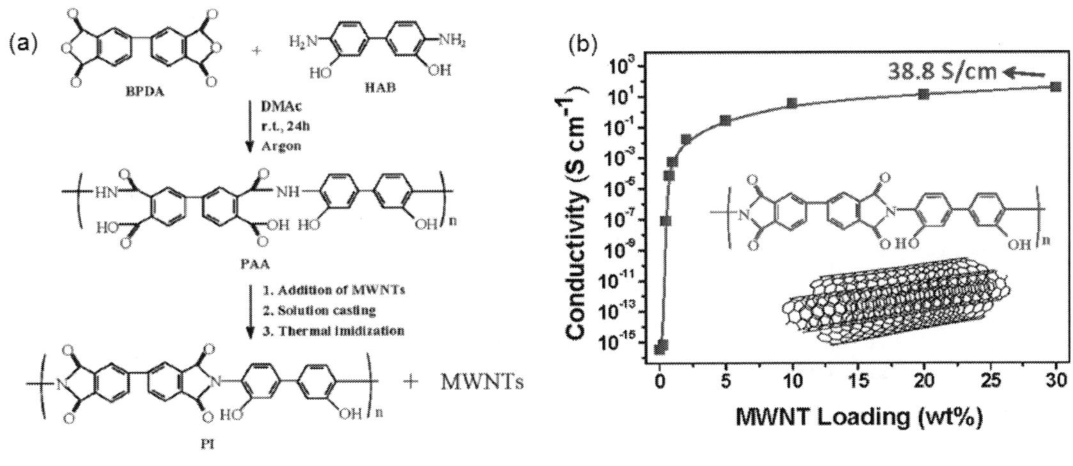

Figure 6. (a) Synthesis process of MWNT/PI composite and PI matrix's precursor PAA. PAA here serves as polymer dispersant to debundle and disperse nanotubes. (b) DC conductivity versus MWNT mass fraction for MWNT/PI composites. Reprinted with permission from Ref. [44], Copyright 2011 American Chemical Society.

As compared to the non-covalent modification of CNTs, the covalent functionalization allows for more significant and stable binding between CNTs and polymer matrix, yet it involves the sp^2 structure of CNTs in modification reactions so as to lower the electrical conductivity of the nanotubes, and consequently compromise the electrical performance of their composites. Therefore, defunctionalization of well-dispersed CNTs with surface moieties in composites may on one hand take full advantage of chemical modifications to optimize CNT dispersion as well as not be troubled with sacrificing inherent conductivity of individual nanotubes on the other. Chang et al. [45] reported a general approach to thermally induce defunctionalization of MWNTs in composites. They first produced functionalized MWNTs with small molecules N-(4-Hydroxyphenyl) maleimide (NHMI) and bis(3-furfuryl-3,4-dihydro-2H-1,3-benzoxazinyl)isopropane (BPA-FBz) through Diels-Alder (DA) reaction for the preparation of MWNT/poly(vinylidene fluoride) (PVDF) composites (Fig. 7a). And then thermally-induced defunctionalization was performed to enhance the composite conductivity by recovering the properties of CNT surfaces. It was found that the thermal treatment increase the conductivity of functionalized MWNT composites by at least 5 orders of magnitude (Fig. 7b-c) without any reduction of the mechanical properties. The reason could be the recovery of the conductivity of MWNTs through retro-DA reaction and the formation of a connected network of nanotubes in the composites.

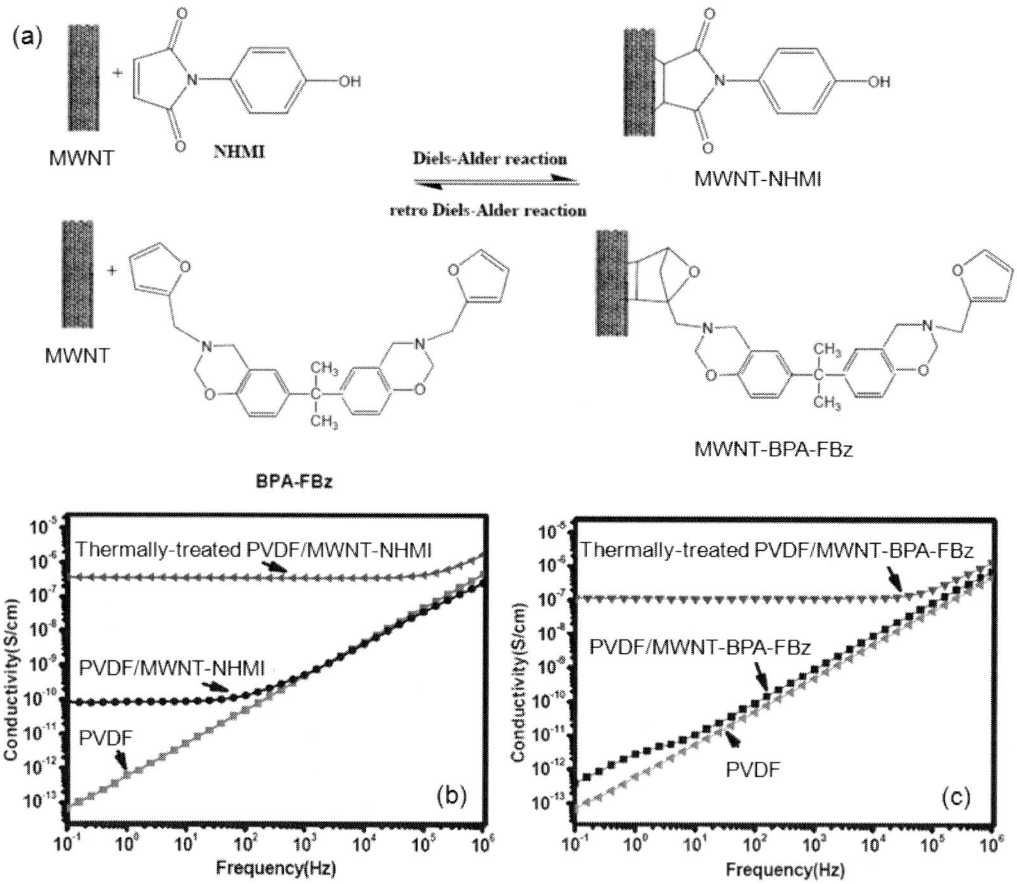

Figure 7. (a) The process of functionalization and defunctionalization of MWNTs through DA and retro DA reactions. (b) Electrical conductivities of PVDF film and MWNT-NHMI/PVDF composites films with 1.0 wt % of additives. (c) The comparison of conductivity of PVDF and their composites with MWNT-BPA-FBz fillers at loading of 1.0 wt %. For both types of composites, the thermal treatments have been done at 160 °C for 3 h. Reprinted with permission from Ref. [45], Copyright 2011 American Chemical Society.

Another more fundamental challenge associated with enhancing the electrical conductivity of CNT/polymer composites lies in reducing the intertube contact resistance in the CNT network, which is usually caused by separation of nanotubes by the insulating polymeric medium. It has been demonstrated that even with an optimized dispersion, the CNT/polymer composites often show a lower electronic performance when compared to CNT films. [46, 47] Previously, conductive polymers were employed to disperse CNTs and to minimize the contact resistance inside composites. Ma et al. [47] significantly improved the conductivity of SWNT networks by *in-situ* polymerization of a highly conductive self-doped conducting polymer (polyaniline boronic acid, PABA) around and along single stranded DNA (ss-DNA) dispersed and functionalized SWNTs (ss-DNA/SWNTs). Due to the electron richness of the ss-DNA/SWNTs (Fig. 8a), the pernigraniline state of PABA prepared during *in-situ* polymerization would be readily reduced to the stable and conductive emeraldine state. The resultant highly conductive polymer skin could not only decrease the contact resistance but also effectively assemble the SWNTs into a highly conductive network. In this way, the percolation threshold was largely reduced and the conductance of the composite above the percolation threshold was increased by 2 orders of magnitude when compared to the network of SWNTs alone (Fig. 8b).

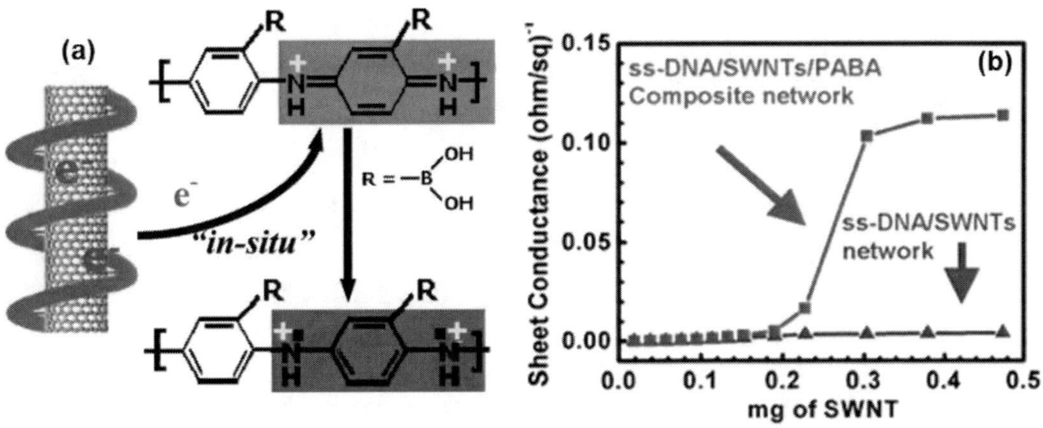

Figure 8. (a) Schematic of ss-DNA functionalized SWNT. (b) Comparison of conductance of *in-situ* polymerized composite and ss-DNA/SWNTs network, measured by a four probe approach. Reprinted with permission from Ref. [47], Copyright 2008 American Chemical Society.

Despite that enhanced conductivity was achieved by employing conductivity polymer to reduce the contact resistance, the ability to utilize polymers with other functionalities remains severely limited. To impart new and transformative functionalities to composites without sacrificing their conductivity or ease of manufacturing, Li et al. [48] have demonstrated a scalable method to prepare high-conductivity multifunctional SWNT/polymer composites. They first dispersed SWNTs in a high molecular weight polymeric derivative of cellulose (sodium carboxymethyl cellulose, CMC). The resulting SWNT/CMC films were not conductive, as the intertube resistance is very high due the presence of CMC between SWNTs (Fig. 9a). Acid treatment was then performed on the films to remove the CMC and dope the SWNTs. This leads to a denser network with direct electrical contacts between the SWNTs (Fig. 9b). Polymers with different functionalities can then be selected based on the intended application and penetrated into the highly conductive SWNT network to produce a freestanding SWNT/polymer multifunctional composite where the SWNT network is still highly conductive (Fig. 9c). This method enables the formation of SWNT network prior to the incorporation of a functional polymer and thus preserves the superb electrical contact between nanotubes by preventing the presence of polymer between nanotubes. Therefore the low intertube junction resistance affords a superb electrical performance of the final SWNT/polymer composites.

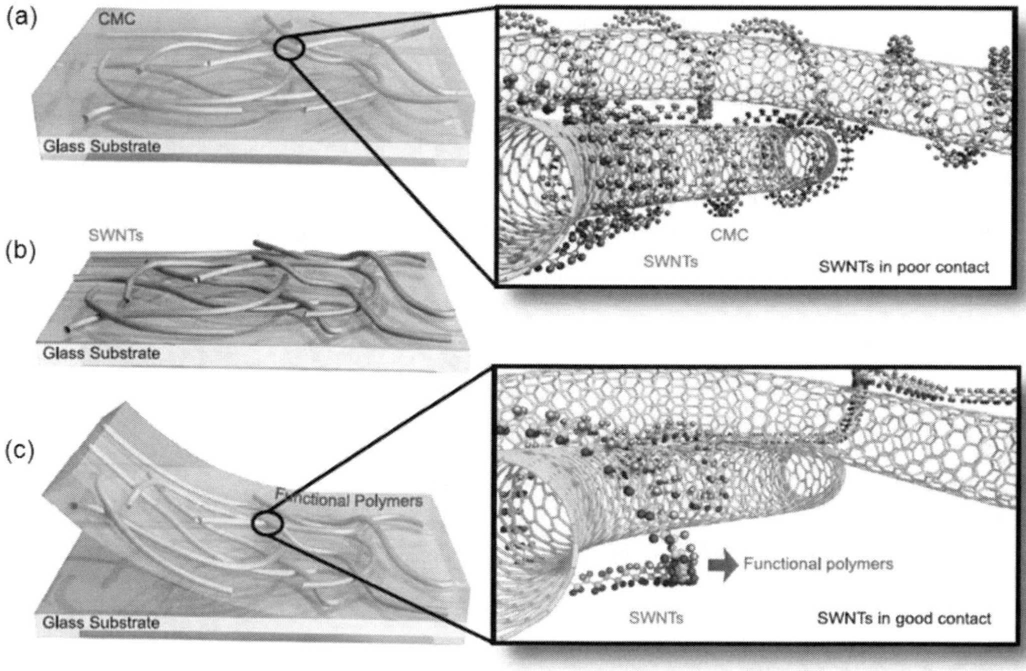

Figure 9. Schematics showing the process for the preparation of highly conductive SWNT/polymer composites. (a) SWNT/CMC films are produced using Mayer rod coating of SWNTs in CMC dispersion on glass substrates. (b) CMC in the Mayer rod casted films was removed by subsequent acid treatment, leading to a densely packed nanotube network. (c) Selective functional polymers were Mayer rod casted into the densely packed nanotube network to form freestanding multifunctional composite films. Reprinted with permission from Ref. [48], Copyright 2012 American Chemical Society.

By showing the same prerequisite with improving mechanical properties, enhancing the electrical performance also demands an even dispersion of CNTs in polymer matrix. To meet this end, non-covalent surface engineering generally has inherent advantages over covalent ones because of well preserving nearly all the properties of individual nanotube due to the absence of disruption of sp^2 hybridization structure. However, both methods have a fundamental weakness in terms of the high intertube resistance as a result of the presence of polymer between nanotubes. More attentions should be paid in the future to optimize the electrical behavior of CNT/polymer composites by tuning the CNT-CNT contact resistance based on the surface engineering of CNTs.

2.3. Dielectric Properties of CNT/Polymer Composites

High-dielectric-permittivity polymer composites may find applications in actuators for artificial muscles, [49] capacitors for energy storage, [24, 50] and active materials for EMI shielding. [51, 52] Electrically conductive CNTs, owing to their large dielectric permittivity and high aspect ratio, have been demonstrated attractive fillers for the preparation of high-permittivity composites. [9, 24, 50, 53, 54] The dielectric permittivity, ε_r, of the composites can be enhanced significantly following the power law: [53, 55]

$$\varepsilon_r / \varepsilon_m = \left| \frac{f_c - f_{CNT}}{f_c} \right|^{-q} \tag{1}$$

where ε_m is the dielectric permittivity of the matrix and q is an exponent of about 1. f_c and f_{CNT} are the percolation threshold and CNT loading, respectively. Additionally, the high thermal conductivity of CNTs would improve the heat dissipation and quickly release the undesired heat generated by loss tangent, thus preventing materials from dielectric breakdown. More importantly, these improvements are achieved at a low loading because of the high aspect ratio of nanotubes. [56]

The critical behavior of the dielectric permittivity near percolation can be explained by employing the microcapacitor model. Namely, each two neighboring nanotubes can be treated as a local capacitor with the CNTs as two electrodes and a very thin layer of dielectric in between, and a network of local capacitors expands between two virtual electrodes with increasing the filler content. Each microcapacitor contributes an abnormally large capacitance. Such large capacitance can then be correlated with a significant increase in the intensity of local electric field when CNTs are very close to each other near f_c. The significant increase in the intensity of the local electric field promotes the migration and accumulation of the charge carriers at the polymer-CNT interfaces. These charges are generated either by surface plasma resonance or by charge injection from the external electrodes, depending on the relaxation times of the two phases. This interfacial polarization, also known as the Maxwell-Wagner-Sillars (MWS) effect, is responsible for the significant increase in permittivity at low frequency. [57] In this regard, to increase the dielectric performance, it is important for the CNT/polymer composites to have huge internal surface areas and strong interfacial affinity. The former would provide numerous sites for the MWS effect while the latter might reinforce this effect. [53]

The large specific surface area of CNTs could naturally give a huge internal surface when they are individually dispersed in the polymer matrix. However, pristine CNTs usually exist in the form of bundles because of the strong interaction among nanotubes. As aforementioned, a great effort has been put to find effective protocols to disperse nanotubes into polymer matrices. Surfactants have been demonstrated to allow for an improvement in dispersion of the nanotubes, ultimately improving the overall quality of the composites. Unfortunately, residual free surfactant usually causes unexpectedly deteriorated dielectric properties at the applied fields, i.e., high leakage current and loss tangent. [9] An alternative approach is thus to strongly bind surface modifiers to CNTs via robust chemical bonds.

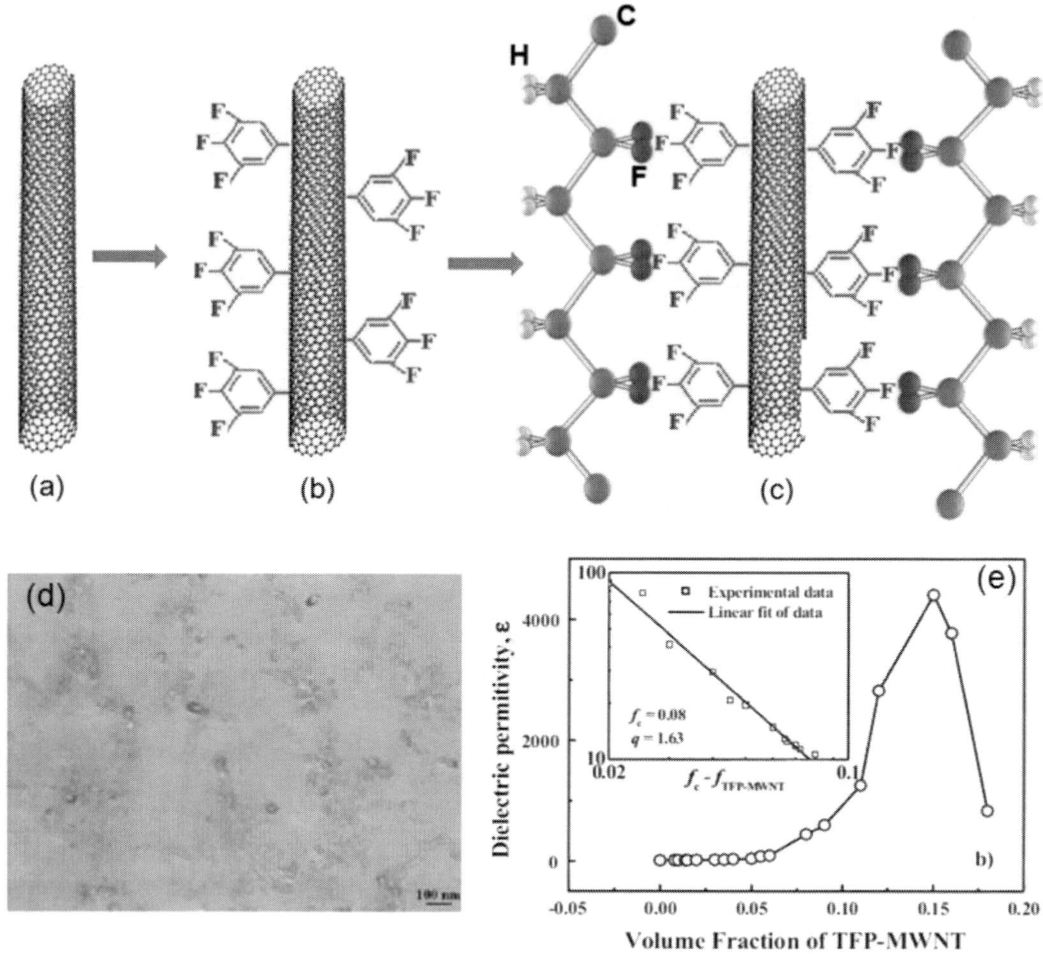

Figure 10. Schematic illustrations of (a) single pristine CNT, (b) TFP- functionalized CNT, (c) TFP-modified CNT in the PVDF matrix, showing a strong compatibility between nanotube and polymer. (d) TEM image illustrating the morphology of a section of the TFP-CNT/PVDF nanocomposite with a TFP-CNT volume fraction of 8 vol%. (e) Dependence of the dielectric permittivity of the TFP-CNT/PVDF nanocomposites on the TFP-CNT volume fraction, measured at room temperature and 10^3 Hz. The inset in (e) shows the best fit of the dielectric permittivity to the percolation theory. (d) and (e) are reprinted with permission from Ref. [53], Copyright 2007 John Wiley and Sons.

The implantation of functional complexes on the surfaces of CNTs has large impact on the interfacial morphologies and thus dielectric performances of the polymer composites. Li et al. [58] have compared pristine, carboxylic and ester functionalized MWNT/PVDF composites and found the dielectric permittivity in the carboxylic functionalized MWNT/PVDF composites was largely enhanced as compared with the others, reaching a value as high as 3600 with 8 vol% MWNT at 1 kHz. This could be attributed to the enhanced interfacial interaction between CNT and polymer after the surface modification. Dang et al. [53] have chemically modified MWNT fillers with 3,4,5-trifluorobro-mobenzene and then compounded them with PVDF to create dielectric composites (Fig. 10). The dielectric permittivity can reach 4000 when the volume fraction of trifluorophenyl (TFP)-functionalized MWNT is 0.15. The very high dielectric permittivity originated from the MWS effect at percolation, and was attributed to a thin insulating layer of PVDF that was part of the TFP-MWNT. The nomadic charge carriers were blocked at the internal interfaces between the TFP-MWNT and the PVDF, and the large π-orbital of the MWNT provided the nomadic electrons with large domains. Strongly electrophilic F groups on the TFP-MWNTs further reinforced the MWS effect.

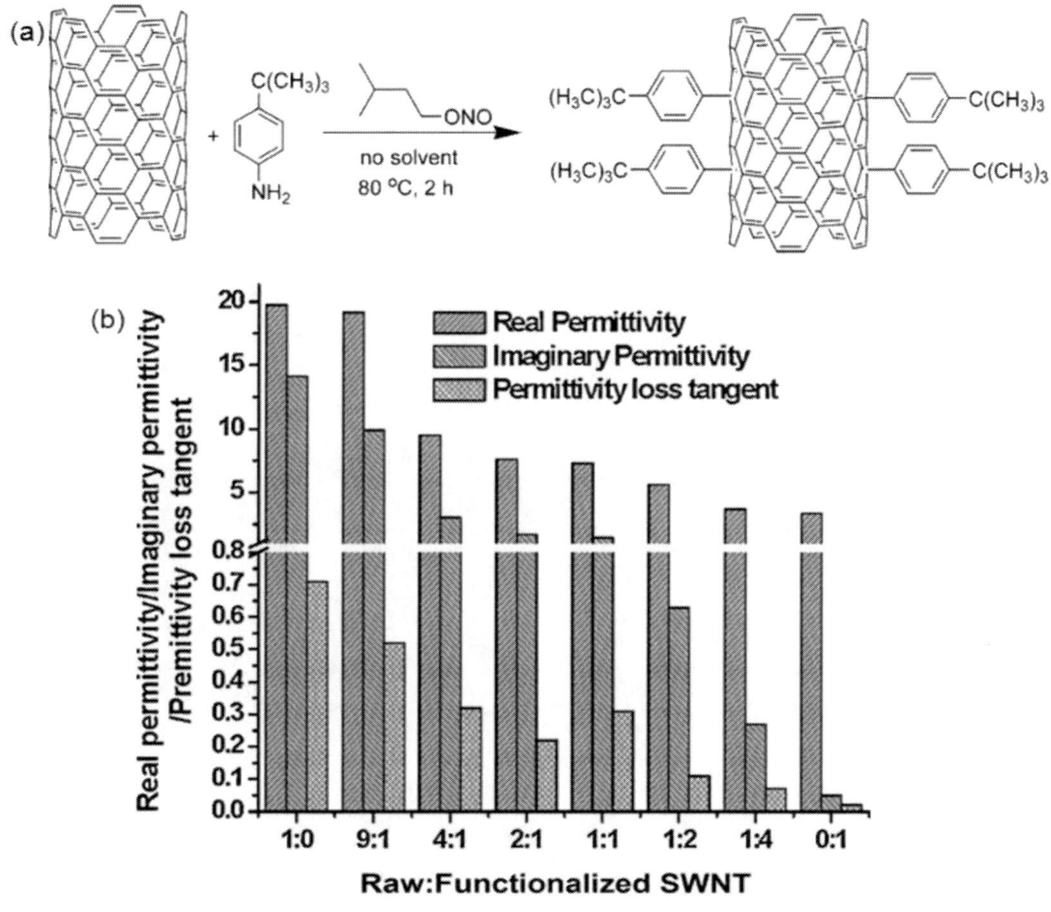

Figure 11. (a) The process of solvent-free SWNT functionalization. Reprinted with permission from Ref. [59], Copyright 2007 American Chemical Society. (b) Comparison of dielectric permittivity of composites with different raw to functionalized SWNT ratios. Reprinted with permission from Ref. [18] Copyright 2012 Elsevier.

Chemical modification of CNTs indeed favors the nanotube dispersion and interfacial affinity, consequently leads to a higher maximum dielectric permittivity near f_c when compared to the pristine CNT-filled composites. However, the functionalized CNT composites always show a relatively higher percolation threshold because of the fact that chemical functionalizing disrupts the conducting network in the sidewalls of CNTs and changes dramatically their electronic properties. Therefore, it was found that when far below f_c, the dielectric permittivity of composites which contain functionalized CNTs will be considerably lower than ones with only raw CNTs. [54] By using this feature, Higginbotham et al. [59] successfully generated a polymer composite with tailored permittivity by mixing tert-butyl aniline functionalized SWNT (Fig. 11a) and raw SWNT in silicone elastomer matrix. By simply varying the ratios of the two types of nanotubes, the dielectric permittivity of the resulting composite can be tuned to any desired value between 20 and 3, as shown in Fig. 11b.

Although high permittivity values could be realized in the uniform composites with f_{CNT} approaching the percolation threshold, the remarkable permittivity improvement is always accompanied by an unacceptable dielectric loss due to the tunneling or ohmic conduction between the neighboring CNTs. And also the dielectric properties near f_c are very sensitive to the variation in the microstructure of composites. To overcome these issues, the most straightforward approach is to introduce additional interlayers or insulating shells on the CNTs to prevent them from connecting with each other directly. The rationales are as follows: i the tunneling current between the neighboring CNTs can be then suppressed and hence lead to lower dielectric loss; ii the interlayers make control of the detailed and sensitive microcapacitor structures near f_c easier, which generally gives rise to a considerably expanded compositional window for percolative composites.

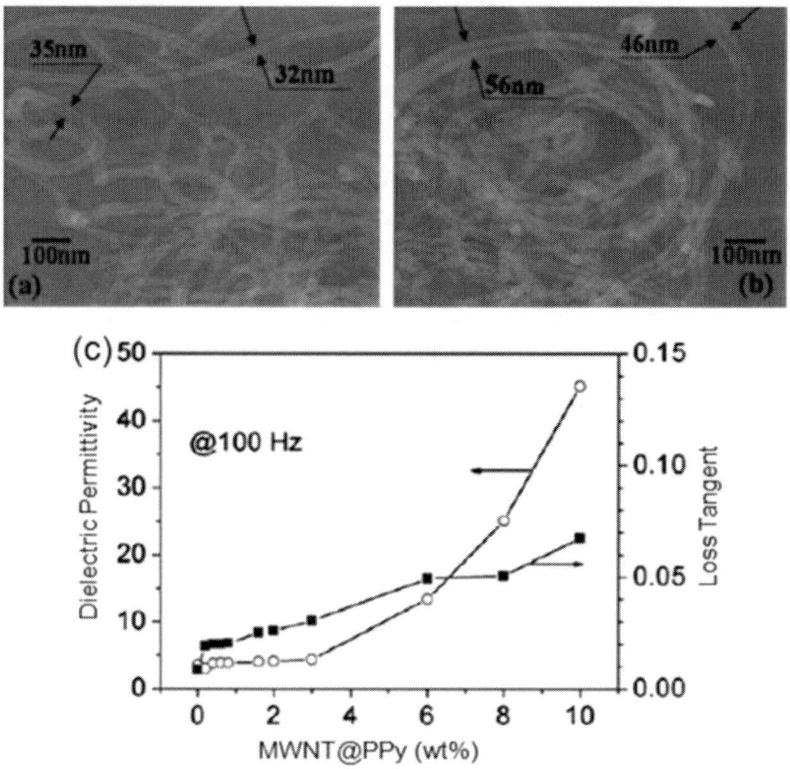

Figure 12. TEM images of (a) uncoated MWNTs and (b) PPy-coated MWNTs. (c) Dielectric permittivity and loss tangent of PPy-coated MWNT/polystyrene as a function of the filler concentration, measured at 100 Hz and room temperature. Reprinted with permission from Ref. [25] Copyright 2009 Elsevier.

A number of researchers have actively investigated the effect of shell on CNT surfaces on the dielectric behavior of CNT/polymer composites. Yang et al. [25] successfully coated the MWNTs with organic shells polypyrrole (PPy) by an inverse microemulsion polymerization. TEM images of MWNT gave the thickness of the PPy layer in the order of 10 nm (Fig. 12). Such PPy-coated MWNT/polystyrene composites presented a high dielectric permittivity (~44), low loss tangent (<0.07), and high energy density (4.95 J/cm^3). The significantly enhanced dielectric performance originated from the organic shell PPy, which not only enable good dispersion of MWNTs in the polymer matrix but also screen charge movement to shut off leakage current as a barrier.

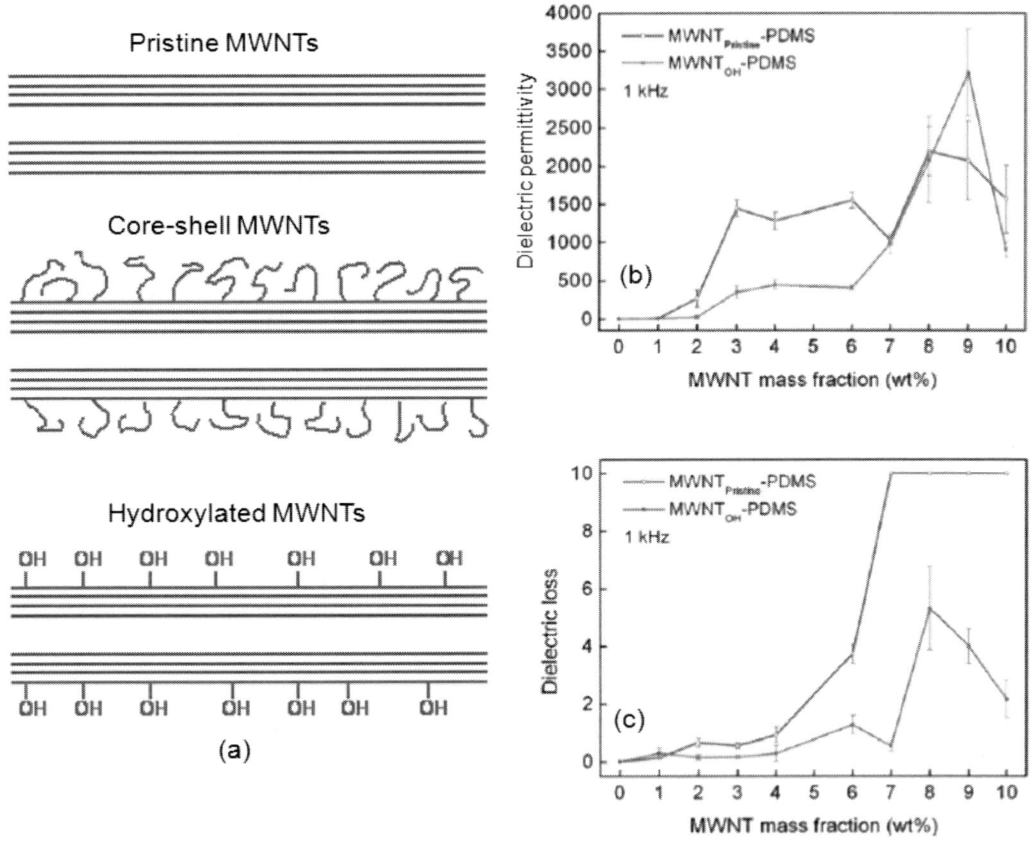

Figure 13. (a) Schematics of unmodified MWNTs, typical polymer-coated MWNTs core-shell structures, and the simple core-shell hydroxylated MWNT. Comparison of (b) dielectric permittivity and (c) dielectric loss of PDMS composites with hydroxylated MWNTs and pristine MWNTs. Reprinted with permission from Ref [54], Copyright 2009 American Chemical Society.

Kohlmeyer et al. [54] produced hydroxylated MWNTs as a much simpler core-shell filler to increase the dielectric permittivity and reduce the dielectric loss of polydimethylsiloxane (PDMS) composites. The outer graphene layer of hydroxylated MWNTs becomes nonconducting after covalent functionalization and serves as the interlayer shell, whereas the unfunctionalized inner graphene layers remain electrically conducting and act as the core (Fig. 13a). As compared to the pristine MWNT/PDMS composites, the core-shell MWNT/polymer composite has a higher MWNT loading tolerance while still exhibits a relatively low level of electrical conductivity. This allows for a much higher dielectric permittivity and a similar or lower dielectric loss (Fig. 13b).

Remarkable progress made before indicates the most important issues for improving dielectric properties of CNT/polymer composites are associated with the polymer-CNT interfaces and the interactions between the adjacent CNT particles. Quantum tunneling between adjacent CNTs is the main mechanism for electrical conductance, even when the content of CNTs is well above the percolation threshold. Increasing the contact resistivity associated with interfaces via CNT surface engineering should be an efficient way to achieve CNT/polymer composites with improved dielectric permittivity and suppressed dielectric loss. With maintaining the superlative properties of CNTs, non-covalent wrapping of polymer macromolecules onto individual CNT surface is much desirable to fully realize the true potential of CNTs in dielectric CNT/polymer composites.

We present here a high-dielectric-permittivity composite prepared by mixing the pristine MWNTs with PVDF matrix in the melt state. The high temperature and high shear forces are expected to cause chemical and/or physical interactions between CNTs and macromolecules. This non-covalent interaction may improve the overall CNT dispersion via effective polymer wrapping onto CNTs. [19-22] The separation of individual nanotubes enhances the intertube contact resistivity and thus forms numerous microcapacitors with neighboring CNTs as electrode and polymer in between as dielectric. Consequently, the true potential of pristine nanotubes in composites could be realized, leading to a high-dielectric-permittivity composite with low level of conductivity.

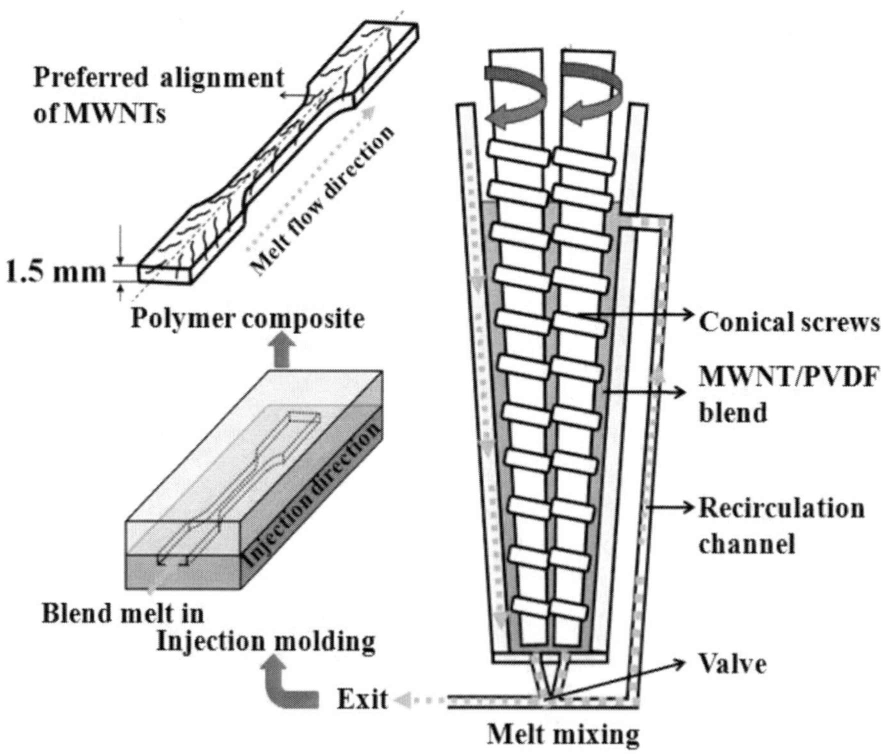

Figure 14. The MWNT/PVDF composites are prepared by melt mixing using a co-rotating, conical, twin-screw microcompounder, followed by injection molding.

The MWNT/PVDF composites were prepared using a co-rotating twin-screw microcompounder (DSM Xplore Microcompounder, 5 cm^3) with two stainless steel screws and a bypass allowing continuous recycling of the material at the head of the mixing chamber. Procedures were illustrated in Fig. 14. The mixing was done at 200 °C, for over 30 min and at a mixing speed of 60 rpm in Ar atmosphere. The final dumbbell-shaped sample with a thickness of 1.5 mm were prepared by injection molding of the mixed composites (Micro 5cc Injection Molder, DSM) using a press at 1.6 MPa for 1 min, while the temperature was kept at 60 °C. Then, the temperature was decreased to room temperature.

For the dielectric characterization, specimens with a dimension of about 10 mm×10 mm×1.5 mm were cut from the as-prepared dumbbell-shaped samples. Electrodes were painted on the sample using silver paste. The dielectric properties were measured in the frequency range of 1~10^6 Hz using a broadband dielectric spectrometer (Novocontrol BDS 20). Raman spectroscopy was performed using a microspectrometer (Horiba Jobin-Yvon LabRam) with a resolution of 1 cm^{-1}. The morphologies of the MWNT/PVDF composites and MWNTs were observed using a scanning electron microscope (SEM, LEO Gemini 1530) and a TEM (Jeol 1200EX) respectively.

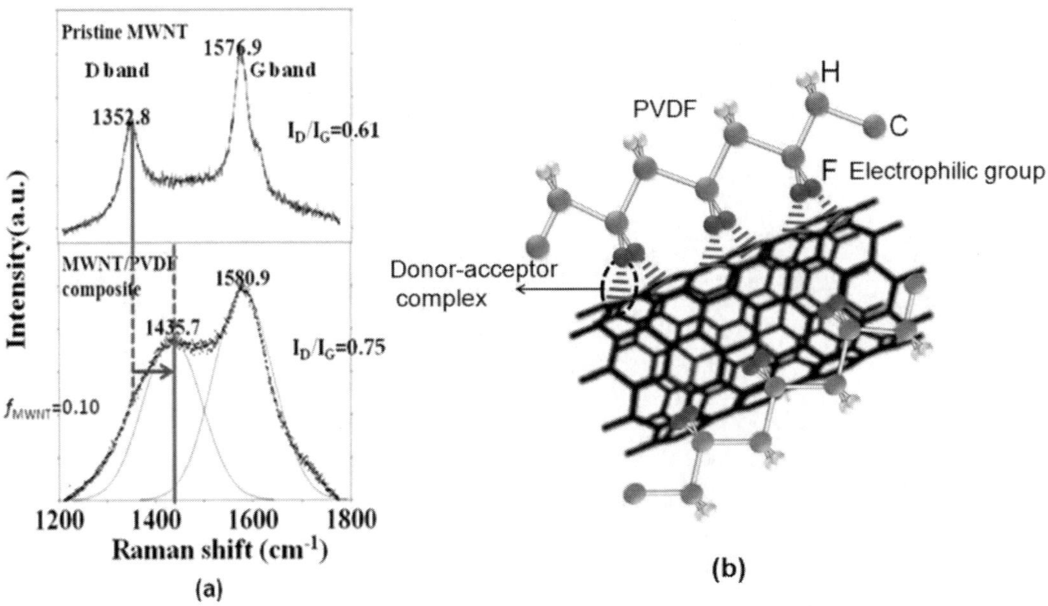

Figure 15. (a) Raman spectra of the raw MWNT and their PVDF composite with f_{MWNT}=10 vol%, (b) schematic image of donor-acceptor complexes between PVDF chains and nanotubes.

It has been proved that the CNTs are electron-rich molecules and would interact with any electron-deficient molecules to form donor-acceptor complexes, [60] much like the way of the π-π interaction formation between electron-rich molecules. [61, 62] The PVDF used in this study possesses electrophilic fluorine groups, which can sufficiently interact with nanotubes through noncovalently modifying the CNT surfaces. The interaction between nanotubes and PVDF chains can be verified by the Raman spectroscopy, as shown in Fig. 15a. The Raman spectrum of pristine nanotubes exhibited two typical peaks. The one centered at 1352.8 cm^{-1} is the disorder-induced D band caused by the multiple phonon scattering of defects or amorphous carbon. The other located at 1576.9 cm^{-1} is the G band and derives from the stretching of conjugated double bonds. [25, 63] However, The frequency of the D band and G band of the MWNTs in the composite are shifted up by 83 cm^{-1} and 4 cm^{-1} relative to that of the pristine MWNTs, respectively. The up-shift and width change of D and G band peaks can be a strong evidence for the remarkable interfacial interaction between PVDF chain and nanotubes. [64-66] This interaction is presumably as a consequence of the formation of donor-acceptor complexes. In the literature, it has been demonstrated that when CNTs are modified with bromine, an electron acceptor, the G band shifts up by 24 cm^{-1}. [67] The up-shift of the G band observed here for the MWNTs in PVDF indicates the interaction between the polymer and the MWNTs is possibly via the strongly electrophilic F groups of PVDF chain. [54] So it is reasonable to conclude that the donor-acceptor complexes are achieved between the delocalized 'π-electron' clouds of MWNTs and the fluorine of PVDF chains, as illustrated in Fig. 15b. Moreover, the I_D/I_G (intensity ratio of D-band to G-band) is usually employed to characterize defects and impurities in CNTs. [68] The value of I_D/I_G in our case changes from 0.61 to 0.75 after melt-blending MWNTs with PVDF matrix. This slight increase should be attributed to the shortage of nanotube and formation of additional defects duo to the high shear force during the melt-mixing.

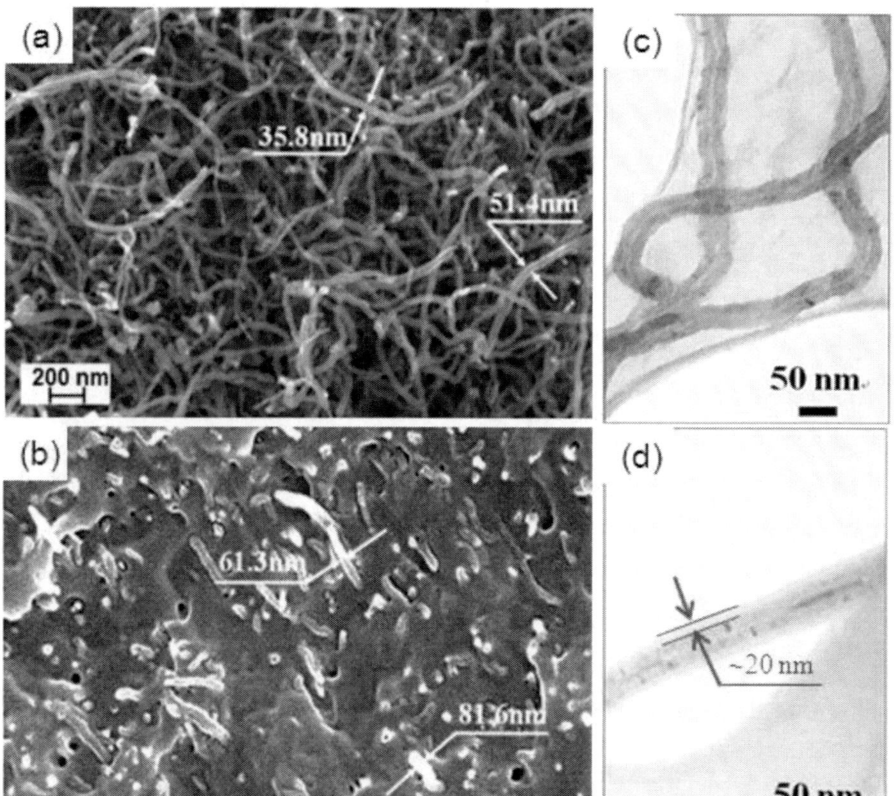

Figure 16. SEM morphologies of (a) pristine MWNTs and (b) fractured surface of the PVDF composite with CNT content at 10 vol%. TEM images of (c) pristine MWNT and (d) MWNT with a thin PVDF layer.

Figure 16a-b shows the SEM images of the pristine MWNTs and their PVDF composite. It is clearly seen that the diameter of the pristine MWNTs is in the range of 35.8-51.4 nm. However, after melt-mixing with PVDF, the morphology of the MWNTs in the composite obviously changes, and the diameter becomes larger, i.e., about 61.3-81.6nm. This suggests that the host polymer chains have effectively wrapped on the MWNTs' surfaces due to the non-covalent interaction between them. The thickness of the PVDF layer is ranging from 10 to 30 nm. In the literature, a similar change of the CNT diameter has also been observed by Ding et al. [20] in MWNT/polycarbonate system, which was taken as evidence to verify the polymer adhesion on the CNTs' surfaces. To further verify the existence of PVDF layer on the nanotube surfaces, TEM characterization of the MWNTs before and after melt-blended with PVDF has also been performed and the images are shown in the Fig. 16c-d. We can clearly see that a PVDF layer with an average thickness of 20 nm coats on MWNTs. Additionally, the SEM image of the MWNT/PVDF composite shows an individual CNT dispersion. This can be attributed to the significant wrapping of PVDF chains on the CNTs. As the non-covalent modification of CNTs can effectively disentangle MWNT bundles and sufficiently avoid the MWNT reaggregation, thus ensure an individual MWNT dispersion in the final composites.

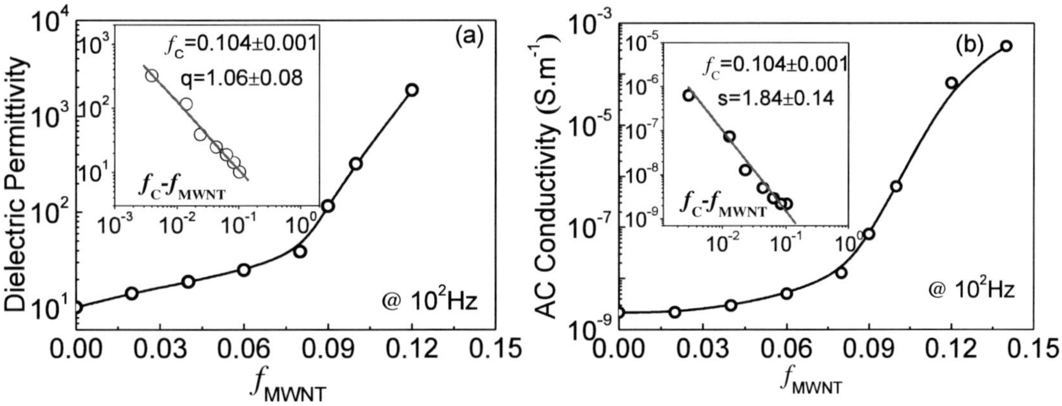

Figure 17. (a) The dielectric permittivity and (b) AC conductivity of the MWNT/PVDF composites as a function of the MWNT volume fraction, fMWNT, measured at room temperature and 100 Hz. The insets in (a) and (b) show the best fits of the dielectric permittivity to Eq. 1 and the AC conductivity to Eq. 2, respectively.

The individual CNT dispersion can form rich interfaces in the composites so as to fully take advantage of the CNT potential to realize microcapacitors. This is much desirable to improve the dielectric properties of the composites, especially when a strong interaction is achieved between the CNTs and polymer matrix. So we conducted dielectric characterizations on the as-prepared composites to study the effect of the surface engineering of CNTs on the dielectric properties of the composite. Fig. 17a shows the dependence of the dielectric permittivity values for the MWNT/PVDF composites on the MWNT volume fraction (f_{MWNT}) at room temperature. It can be clearly seen that the dielectric permittivity enhances significantly only when f_{MWNT} approaches the percolation threshold, f_c, which could be largely attributed to the MWS effect. [24] In our case, the relaxation time of PVDF is about several orders of magnitude larger than MWNT. Therefore, charge carriers would be blocked at internal PVDF-CNT interfaces due to the MWS effect, accordingly improving the dielectric permittivity significantly. Additionally, the non-covalent interaction at interfaces may intensify the MWS effect. In detail, the delocalized 'π-electron' clouds of MWNTs provide large domains for nomadic electrons, which are strongly attracted by the electrophilic F groups, giving rising to a significantly reinforced MWS effect. [24] Furthermore, the much thin PVDF layer on MWNTs could fully use the large specific surface area of nanotubes to provide a huge interfacial area for the reinforced MWS effect. In order to estimate f_c, the experimental permittivity data at 100 Hz are fitted to the Eq. 1. The best linear fit of the data to the log–log plot of the power law gives f_c=0.104±0.001, q=1.06±0.08 (see the inset in Fig. 17a). The exponent q is nearly the same with the universal one ($q \approx 1$). [53, 55] However, the f_c (10.4 vol%) is much larger than that usually reported in the literature for most CNT/polymer composites. [22, 53, 69, 70] We presume that the high f_c may be due to the following factors: i) A preferred alignment of MWNTs (as seen in Fig. 16b) is induced by the geometry of the extruder. This would reduce the number of MWNT-MWNT contact points. ii) the conduction effectiveness of the CNTs are decreased as a consequence of the presence of PVDF layer coated on nanotube surfaces. Therefore a higher MWNT content is needed to reach the percolation threshold. iii) the shortage of MWNTs caused by the high shear force results in a slight decrease in aspect ratio and a lower aspect ratio usually gives rise to a higher percolation threshold. [71]

Figure 17b presents the dependence of AC conductivity of the MWNT/PVDF composites on f_{MWNT}, measured at room temperature and 100 Hz. The conductivity, σ_c, increase significantly when f_{MWNT} approaches f_c. The variation shows a good agreement with the following power law. [53, 55]

$$\sigma_c \propto \sigma_{MWNT}(f_c - f_{MWNT})^{-s} \text{ for } f_{MWNT} < f_c \quad (2)$$

The best fit of the conductivity data at 10^2 Hz to the log–log plot of the power law gives f_c=0.104±0.001, s=1.84±0.14 (see the inset in Fig. 17b). The percolation threshold is the same with that obtained from the linear fit of permittivity data. The exponent s is larger than the universal one (s=0.8-1). [53, 55] This discordance has been also reported in previous composites [53, 72] and could be possibly due to the fact that the electrical connectivity in composites is not achieved by direct CNT-CNT contacts but by the tunneling. Balberg et al.[73] have proposed that a large inter-particle distance distribution can give rise to non-universal high s values. The absence of physical contacts between the nanotubes can be attributed to the existence of the much thin PVDF layer on the CNT surfaces, which prohibits the direct contacts between the nanotubes. This feature is much desirable for the dielectric application of the composites. As f_{MWNT} falls into the neighborhood of percolation threshold ($0.09 < f_{MWNT} < 0.12$), the maximal conductivity of composites in this region is still below 6.6×10^{-5} S.m^{-1}. Such semi-conductive composites with high permittivity can be supposed to serve as dielectric materials in capacitors. [72]

CONCLUSION

Creating an even dispersion and an optimized CNT-polymer contact are keys for fully realizing the true potential of CNTs in polymer composites. Tremendous efforts have been devoted to meet this end by tuning the surface characteristics of CNTs through introducing functional groups or macromolecules onto the CNT surfaces. The information regarding the relationship between the nature of modified surface and composite properties would be very helpful for engineering CNT surface to fit a desired end use. Therefore, we first reviewed the state of the art in the area of property-oriented surface engineering of CNTs. In terms of the mechanical reinforcement, the covalently grafting of functional groups or macromolecules is preferable because of the strong interfacial binding strength as well as the large tunability of CNT surface chemistry. However, the covalent modification of CNTs is sometimes detrimental to the electrical and dielectric properties as a result of the disruption of electronic structure of graphene sidewalls. Thus, homogenously distributing pristine CNTs into host polymer while controlling the intertube resistivity seem to be an efficient way to tailor electrical conductivity and/or dielectric permittivity of the composites. We presented an example of non-covalent modification of CNT surfaces by melt-blending pristine MWNTs within PVDF in molten state. Such surface engineering breaks through the common bottleneck of limited interfacial interactions displayed in other C–moieties based composites. The resultant MWNT/PVDF composite exhibited a high dielectric permittivity of ~1870 and a low conductivity of 6.6×10^{-5} $S.m^{-1}$. These composites could serve as dielectrics in the high-performance embedded capacitors.

ACKNOWLEDGMENT

Authors are grateful to Mrs. F. Garnier, Mr. P. Haghi-Ashtiani and Mrs. P. Gemeiner for the SEM and TEM investigations and Raman measurements, respectively.

REFERENCES

[1] Iijima, S. *Nature* 1991, *354*, 56-58.
[2] De Volder, M. F. L.; Tawfick, S. H.; Baughman, R. H.; Hart, A. J. *Science* 2013, *339*, 535-539.
[3] Wei, B. Q.; Vajtai, R.; Ajayan, P. M. *Appl. Phys. Lett.* 2001, *79*, 1172-1174.
[4] Berber, S.; Kwon, Y. K.; Tomanek, D. *Phys. Rev. Lett.* 2000, *84*, 4613-4616.
[5] Kim, P.; Shi, L.; Majumdar, A.; McEuen, P. L. *Phys. Rev. Lett.* 2001, *87*, 215502.
[6] Yu, M. F.; Files, B. S.; Arepalli, S.; Ruoff, R. S. *Phys. Rev. Lett.* 2000, *84*, 5552-5555.
[7] Yu, M. F.; Lourie, O.; Dyer, M. J.; Moloni, K.; Kelly, T. F.; Ruoff, R. S. *Science* 2000, *287*, 637-640.
[8] Breuer, O.; Sundararaj, U. *Polym. Compos.* 2004, *25*, 630-645.
[9] Dang, Z. M.; Yuan, J. K.; Zha, J. W.; Zhou, T.; Li, S. T.; Hu, G. H. *Prog. Mater. Sci.* 2012, *57*, 660-723.
[10] Bauhofer, W.; Kovacs, J. Z. *Compos. Sci. Technol.* 2009, *69*, 1486-1498.

[11] Ajayan, P. M.; Stephan, O.; Colliex, C.; Trauth, D. *Science* 1994, *265*, 1212-1214.
[12] Baughman, R. H.; Zakhidov, A. A.; de Heer, W. A. *Science* 2002, *297*, 787-792.
[13] Dyke, C. A.; Tour, J. M. *J. Phys. Chem. A* 2004, *108*, 11151-11159.
[14] Lu, W.; Chou, T. W. *J. Mech. Phys. Solids* 2011, *59*, 511-524.
[15] Song, Y. S.; Youn, J. R. *Carbon* 2005, *43*, 1378-1385.
[16] Kim, S. W.; Kim, T.; Kim, Y. S.; Choi, H. S.; Lim, H. J.; Yang, S. J.; Park, C. R. *Carbon* 2012, *50*, 3-33.
[17] Grossiord, N.; Loos, J.; Regev, O.; Koning, C. E. *Chem. Mater.* 2006, *18*, 1089-1099.
[18] Roy, N.; Sengupta, R.; Bhowmick, A. K. *Prog. Polym. Sci.* 2012, *37*, 781-819.
[19] Zhang, Z. N.; Zhang, J.; Chen, P.; Zhang, B. Q.; He, J. S.; Hu, G. H. *Carbon* 2006, *44*, 692-698.
[20] Ding, W.; Eitan, A.; Fisher, F. T.; Chen, X.; Dikin, D. A.; Andrews, R.; Brinson, L. C.; Schadler, L. S.; Ruoff, R. S. *Nano Lett.* 2003, *3*, 1593-1597.
[21] Potschke, P.; Fornes, T. D.; Paul, D. R. *Polymer* 2002, *43*, 3247-3255.
[22] Wong, M.; Paramsothy, M.; Xu, X. J.; Ren, Y.; Li, S.; Liao, K. *Polymer* 2003, *44*, 7757-7764.
[23] Coleman, J. N.; Khan, U.; Blau, W. J.; Gun'ko, Y. K. *Carbon* 2006, *44*, 1624-1652.
[24] Yuan, J. K.; Yao, S. H.; Dang, Z. M.; Sylvestre, A.; Genestoux, M.; Bai, J. *J. Phys. Chem. C* 2011, *115*, 5515-5521.
[25] Yang, C.; Lin, Y. H.; Nan, C. W. *Carbon* 2009, *47*, 1096-1101.
[26] Ma, W.; Liu, L.; Zhang, Z.; Yang, R.; Liu, G.; Zhang, T.; An, X.; Yi, X.; Ren, Y.; Niu, Z.; Li, J.; Dong, H.; Zhou, W.; Ajayan, P. M.; Xie, S. *Nano Lett.* 2009, *9*, 2855-2861.
[27] Coleman, J. N.; Khan, U.; Gun'ko, Y. K. *Adv. Mater.* 2006, *18*, 689-706.
[28] Bradshaw, R. D.; Fisher, F. T.; Brinson, L. C. *Compos. Sci. Technol.* 2003, *63*, 1705-1722.
[29] Fisher, F. T.; Bradshaw, R. D.; Brinson, L. C. *Compos. Sci. Technol.* 2003, *63*, 1689-1703.
[30] Ajayan, P. M.; Tour, J. M. *Nature* 2007, *447*, 1066-1068.
[31] Ci, L. J.; Bai, J. B. *Adv. Mater.* 2004, *16*, 2021-2024.
[32] Ci, L. J.; Zhao, Z. G.; Bai, J. *Carbon* 2005, *43*, 883-886.
[33] Shi, X. F.; Hudson, J. L.; Spicer, P. P.; Tour, J. M.; Krishnamoorti, R.; Mikos, A. G. *Biomacromolecules* 2006, *7*, 2237-2242.
[34] Park, J. J.; Park, D. M.; Youk, J. H.; Yu, W. R.; Lee, J. *Carbon* 2010, *48*, 2899-2905.
[35] Sahoo, N. G.; Cheng, H. K. F.; Bao, H.; Pan, Y.; Li, L.; Chan, S. H. *Soft Matter* 2011, *7*, 9505-9514.
[36] Kim, K. H.; Jo, W. H. *Macromolecules* 2007, *40*, 3708-3713.
[37] Chen, J.; Ramasubramaniam, R.; Xue, C.; Liu, H. *Adv. Funct. Mater.* 2006, *16*, 114-119.
[38] Yuan, W.; Li, W.; Mu, Y.; Chan-Park, M. B. *ACS Appl. Mater. Interfaces* 2011, *3*, 1702-1712.
[39] Blake, R.; Gun'ko, Y. K.; Coleman, J.; Cadek, M.; Fonseca, A.; Nagy, J. B.; Blau, W. J. *J. Am. Chem. Soc.* 2004, *126*, 10226-10227.
[40] Spitalsky, Z.; Tasis, D.; Papagelis, K.; Galiotis, C. *Prog. Polym. Sci.* 2010, *35*, 357-401.
[41] 1. Gao, J. B.; Zhao, B.; Itkis, M. E.; Bekyarova, E.; Hu, H.; Kranak, V.; Yu, A. P.; Haddon, R. C. *J. Am. Chem. Soc.* 2006, *128*, 7492-7496.

[42] Yang, S. Y.; Lozano, K.; Lomeli, A.; Foltz, H. D.; Jones, R. *Compos. Part a-Appl. S.* 2005, *36*, 691-697.
[43] Li, J.; Ma, P. C.; Chow, W. S.; To, C. K.; Tang, B. Z.; Kim, J. K. *Adv. Funct. Mater.* 2007, *17*, 3207-3215.
[44] Yuan, W.; Che, J.; Chan-Park, M. B. *Chem. Mater.* 2011, *23*, 4149-4157.
[45] Chang, C. M.; Liu, Y. L. *ACS Appl. Mater. Interfaces* 2011, *3*, 2204-2208.
[46] Bekyarova, E.; Itkis, M. E.; Cabrera, N.; Zhao, B.; Yu, A. P.; Gao, J. B.; Haddon, R. C. *J. Am. Chem. Soc.* 2005, *127*, 5990-5995.
[47] Ma, Y.; Cheung, W.; Wei, D.; Bogozi, A.; Chiu, P. L.; Wang, L.; Pontoriero, F.; Mendelsohn, R.; He, H. *ACS Nano* 2008, *2*, 1197-1204.
[48] Li, X.; Gittleson, F.; Carmo, M.; Sekol, R. C.; Taylor, A. D. *ACS Nano* 2012, *6*, 1347-1356.
[49] Zhang, Q. M.; Bharti, V.; Zhao, X. *Science* 1998, *280*, 2101-2104.
[50] Yuan, J. K.; Li, W. L.; Yao, S. H.; Lin, Y. Q.; Sylvestre, A.; Bai, J. *Appl. Phys. Lett.* 2011, *98*,032901.
[51] Li, N.; Huang, Y.; Du, F.; He, X. B.; Lin, X.; Gao, H. J.; Ma, Y. F.; Li, F. F.; Chen, Y. S.; Eklund, P. C. *Nano Lett.* 2006, *6*, 1141-1145.
[52] Yang, Y. L.; Gupta, M. C. *Nano Lett.* 2005, *5*, 2131-2134.
[53] Dang, Z. M.; Wang, L.; Yin, Y.; Zhang, Q.; Lei, Q. Q. *Adv. Mater.* 2007, *19*, 852-857.
[54] Kohlmeyer, R. R.; Javadi, A.; Pradhan, B.; Pilla, S.; Setyowati, K.; Chen, J.; Gong, S. *J. Phys. Chem. C* 2009, *113*, 17626-17629.
[55] Nan, C. W. *Prog. Mater. Sci.* 1993, *37*, 1-116.
[56] Yuan, J. K.; Yao, S. H.; Sylvestre, A.; Bai, J. *J. Phys. Chem. C* 2012, *116*, 2051-2058.
[57] Nan, C. W.; Shen, Y.; Ma, J. *Annu. Rev. Mater. Res.* 2010, *40*, 131-151.
[58] Li, Q.; Xue, Q. Z.; Hao, L. Z.; Gao, X. L.; Zheng, Q. B. *Compos. Sci. Technol.* 2008, *68*, 2290-2296.
[59] Higginbotham, A. L.; Stephenson, J. J.; Smith, R. J.; Killips, D. S.; Kempel, L. C.; Tour, J. M. *J. Phys. Chem. C* 2007, *111*, 17751-17754.
[60] Baskaran, D.; Mays, J. W.; Bratcher, M. S. *Chem. Mater.* 2005, *17*, 3389-3397.
[61] Zhang, J.; Lee, J. K.; Wu, Y.; Murray, R. W. *Nano Lett.* 2003, *3*, 403-407.
[62] Li, H. P.; Zhou, B.; Lin, Y.; Gu, L. R.; Wang, W.; Fernando, K. A. S.; Kumar, S.; Allard, L. F.; Sun, Y. P. *J. Am. Chem. Soc.* 2004, *126*, 1014-1015.
[63] Gogotsi, Y. Carbon Nanomaterials. Taylor and Francis: New York, 2006.
[64] Owens, F. J.; Jayakody, J. R. P.; Greenbaum, S. G. *Compos. Sci. Technol.* 2006, *66*, 1280-1284.
[65] Chen, G. X.; Li, Y. J.; Shimizu, H. *Carbon* 2007, *45*, 2334-2340.
[66] Zhao, Q.; Wagner, H. D. *Philosophical Transactions of the Royal Society of London Series a-Mathematical Physical and Engineering Sciences* 2004, *362*, 2407-2424.
[67] Rao, A. M.; Eklund, P. C.; Bandow, S.; Thess, A.; Smalley, R. E. *Nature* 1997, *388*, 257-259.
[68] Venkateswaran, U. D.; Rao, A. M.; Richter, E.; Menon, M.; Rinzler, A.; Smalley, R. E.; Eklund, P. C. *Phys. Rev. B* 1999, *59*, 10928-10934.
[69] Zhang, S. H.; Zhang, N. Y.; Huang, C.; Ren, K. L.; Zhang, Q. M. *Adv. Mater.* 2005, *17*, 1897-1901.
[70] Wang, L.; Dang, Z. M. *Appl. Phys. Lett.* 2005, *87*, 042903.

[71] Yao, S. H.; Dang, Z. M.; Jiang, M. J.; Xu, H. P.; Bai, J. B. *Appl. Phys. Lett.* 2007, *91*, 212901.

[72] Yuan, J. K.; Dang, Z. M.; Yao, S. H.; Zha, J. W.; Zhou, T.; Li, S. T.; Bai, J. *J. Mater. Chem.* 2010, *20*, 2441-2447.

[73] Balberg, I. *Carbon* 2002, *40*, 139-143.

METALS

In: Applied Surface Chemistry of Nanomaterials
Editors: Mohamed M. Chehimi and Jean Pinson
ISBN: 978-1-62808-351-4

Chapter 6

ATOMIC SIZED METALLIC POINT CONTACTS: FABRICATION METHODS AND CHARACTERISTICS OF QUANTIZED CONDUCTANCE

Jalal Ghilane and Jean Christophe Lacroix
Nano-Electro-Chemistry group, Univ Paris Diderot, Sorbonne Paris Cité,
Paris Cedex, France

1. INTRODUCTION

The growing activity in the area of nanometer scale science and technology can be traced to the quest for the miniaturization of electronic devices, driven by the need for faster and more powerful electronics. In addition to the importance of developing new generation electronics, materials on the nano-scale often exhibit interesting quantum phenomena. Although we are far from replacing silicon technology, it is important to unravel the quantum properties of conductors with nanometer dimensions and to discover new challenges which must be faced upon reducing the size of the material.

An important example is the electrical conductance of a metallic wire with a diameter of a few atoms. The electrical and mechanical properties of a piece of any metal are not different, whether its size is millimeters or kilometers. However, when its size approaches the atomic scale the familiar Ohm's Law, $G = \sigma S/L$, becomes invalid (G is the conductance, σ is the conductivity of the sample, from which we learn that the conductance of a conductor scales directly proportional to its transverse area S and inversely proportional to its length L). The reason is that the distance an electron travels between two scattering events is typically much larger than the atomic size. The electrons traverse an atomic-sized conductor ballistically, and the resistance becomes independent of its length. The character of the resistance changes conceptually and it is necessary to invoke the wave nature of the electrons in the conductor for a proper description.

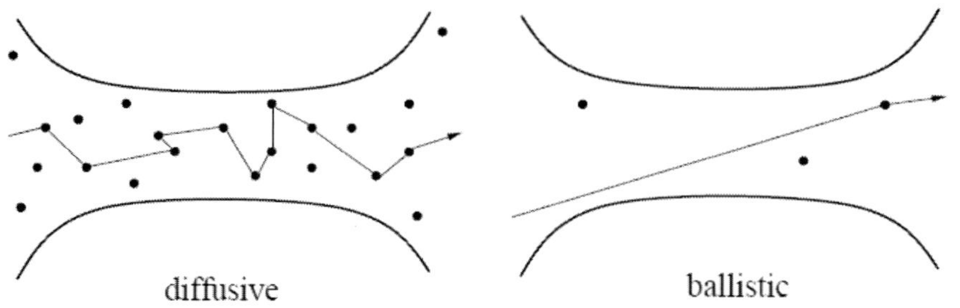

Figure 1. Schematic illustration of a diffusive (left) and ballistic (right) conductor.

Electron transport through single atoms, which are contacted to metallic leads, is thus different from the familiar diffusive transport in macroscopic metallic conductors (Figure 1). In diffusive transport, the scattering of electronic states from defects and lattice vibrations limits the mobility of electrons and, thus, determines the conductivity of a material or the conductance $G = I/V$ of a sample with given dimensions (I is the current, V the voltage drop at the sample). In microscopic structures, the characteristic dimension of a sample may become smaller than the elastic mean free path of the electrons and electron scattering occurs dominantly at the edges of the structure, which is typical of ballistic transport.

A metallic wire with a diameter of a few atoms is an extreme case, where the typical dimensions are comparable to the Fermi wavelength and a fully quantum mechanical description is required. Using a suitable set of states, so-called eigenchannels, and the conductance of a junction may be written in the form $G = G_0\sum_i T_i$ where T_i are transmission probabilities of eigenchannel "i" and $G_0 = 2e^2/h$ is the quantum of conductance [1, 2]. The transmission probabilities T_i may acquire values between 0 and 1, depending on how strongly a particular state is backscattered at the quantum contact. Much current research aims at measuring, understanding and controlling these transmission probabilities and their variation with studied metal, electrode material, electrode magnetization, contact chemical environment, etc. etc.

The experimental investigation of these phenomena requires tools for manipulation and characterization of structures at the atomic and molecular scale. In laboratories worldwide there is rapid progress in this area. A considerable amount of experimental and theoretical work on atomic contacts also named quantum point contacts (QPC) has been done. The observation of atomic point contact was followed mainly by transport measurement (variation of the conductance versus times and/or distance and the corresponding $I(V)$ curves). Experiments have been performed in various environments ranging from ambient conditions over liquid cells to ultra-high vacuum, and from ambient to cryogenic temperatures. Several reviews of various aspects of this field are available [3, 4]. This chapter uses many examples already reviewed and does not pretend to be exhaustive.

2. Methods for Preparing Atomic Point Contact

Several methods have been attempted to meet the challenge of connecting two metallic wires at the atomic scale level [3, 4]. In general, the strategies used could be defined in two distinct categories: mechanical and non-mechanical methods. In the following we will first describe the mechanical methods which are based on dynamic mode such as mechanically controlled break junction (MCBJ) and scanning tunneling microscopy (STM). Following this, the non-mechanical methods such as surface diffusion mediated deposition, electrochemical deposition and etching, and self-terminated electrochemical methods will be described.

2.1. Mechanical Methods

The most commonly used methods for constructing and studying atomic point contact are the scanning tunneling microscopy break junction (STM-BJ) and the mechanically controllable break junction (MCBJ) methods.

2.1.1. Scanning Tunneling Microscopy Break Junction (STM-BJ)

The scanning tunneling microscope (STM) [5] is a versatile tool that allows studying the topography and electronic properties of a metal or semiconductor surface with atomic resolution, and it is also ideal for studying atomic-sized contacts. In its normal topographic mode a sharp needle (the tip) is scanned over the sample to be studied without making contact. The distance between tip and sample is so small that accidental contact between them is quite possible in normal STM work, and should usually be avoided. However, the STM tip can also be used to create atomic point contact.

The principle of the STM-BJ is presented in Figure 2a. In a typical STM-BJ experiment, a nanometer-sized contact is created by pressing an STM tip into the substrate, followed by pulling it out, during which an atomic scale contact between the tip and substrate may form.

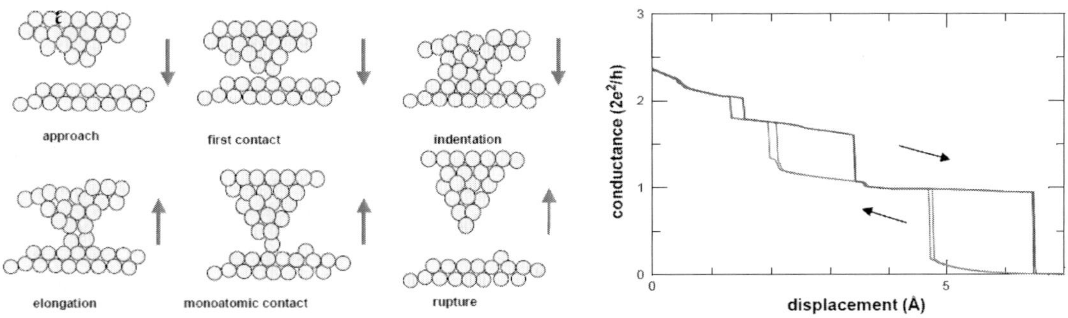

Figure 2. (a) Cartoon representation of contact fabrication using an STM. (b) Conductance curves of a gold nanocontact at low temperature (4.2 K) using a stable STM. Two complete consecutive cycles of approach-retraction are shown. The lower and higher curves correspond to approach and retraction, respectively.

The first report of the formation and study of a metallic contact of atomic dimensions with STM is that of Gimzewski and Möller [6]. The surface was gently touched and the transition from the tunneling regime to metallic contact was observed as an abrupt jump in the conductance. From the magnitude of the resistance at the jump (~ *10 kΩ*), the contact diameter was estimated to be 0.15 nm, which suggested that the contact should consist of one or two atoms. Further approaches of the STM tip to the surface and make a wide tip-sample contact. Next, the tip is pulled away from the substrate. As the tip retracts away elongation of the nanowire happens until a contact between the tip and the substrate is made by one atom. This process is followed by recording the variation of the current or the conductance as function of pulling distance (Figure 2b). In an STM experiment on metallic contacts, a bias voltage between the tip and sample is thus used and kept fixed (at a low value, 10 mV) and the current is recorded as the tip-sample distance is varied by ramping the z piezovoltage. The results are typically presented as a plot of the conductance (or current) versus stretching distance (or time). Figure 2b shows a typical STM conductance curve for a clean Au contact at low temperatures. Before contact the current depends exponentially on the distance with an apparent tunneling barrier of the order of the work function of the material. After this jump the conductance increases in a stepwise manner as the size of the contact increases. Reversing the motion of the tip shows that these steps are hysteretic. In the case of Au, the conductance of the first plateau is quite well defined with a value of approximately $2e^2/h=1G_0$ and corresponds to a one-atom contact. For other metals the conductance curves will look different, depending on the electronic structure and the valence of the metal.

Different groups have performed STM experiments on the conductance of atomic-sized contacts in different experimental conditions: at cryogenic temperatures [7, 8, 9, 10], at room temperature under ambient conditions [11, 12, 13, 14], and UHV [15, 16, 17]. (Note that the energy scales involved are so large that quantized conductances are easily visible at room temperature)

In order to form a large number of nanowires rapidly and to avoid the data selection, the procedure based on STM-BJ was fully automated with a feedback loop. The nanowires were created by repeatedly driving a gold STM tip into and out of contact with a gold substrate using a z-piezoelectric transducer. When the tip was pulled out of the contact, a nanowire was formed transiently between the STM tip and the substrate with a diameter decreasing all the way down to the atomic scale before breaking. The loop started by applying a linearly sweeping voltage to the *z*-piezoelectric transducer to drive the STM tip into contact with the substrate. When the contact was fully established as the conductance between the tip and substrate reached a preset value, the feedback loop activated the *z*-piezoelectric transducer to stretch the tip and form an atomically thin wire. The wire was stretched continuously until it broke completely, and the same procedure was then repeated to form the next nanowire [18]. Figure 3a shows a typical variation of the conduction as function of the pulling distance [19].

During the process, a metal neck is formed between the electrodes, due to the strong cohesive energy of metals, which is stretched into an atomically thin wire before breaking. The conductance during the stretching process changes in a stepwise fashion (see Figure 3a) and take an integer value of the quantum conductance. Figure 3b shows the conductance histogram of Au nanowires from more than 1000 unselected individual conductance traces. It shows the statistical occurrence of the conductance steps at various well defined peaks near integer multiples of G_0, with a major peak centered at $1G_0$. Similar observations were reported using various metals such as Cu, Co, Ni, and Ag. STM-BJ is thus an excellent method for the fabrication and characterization of atomic point contacts. However, the atomic point contact structures fabricated by STM-BJ are not very stable due to thermal or mechanical drift.

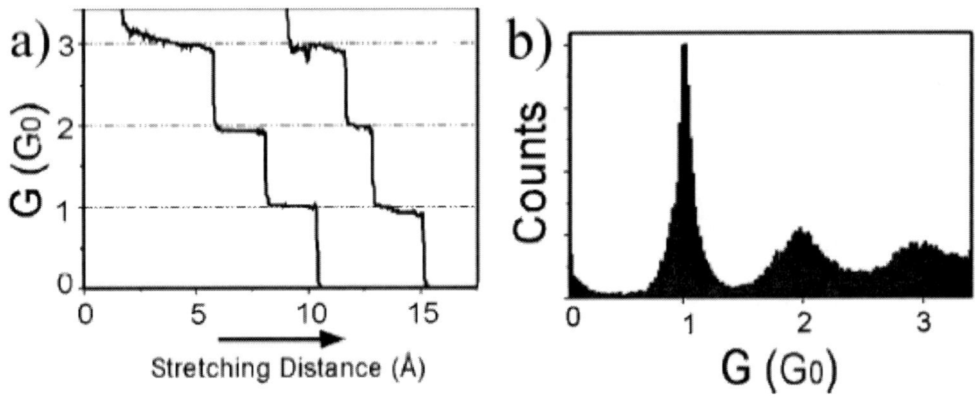

Figure 3. (a) Typical conductance vs. stretching distance traces that show the quantized variation in the nanowire conductance. (b) Conductance histograms of Au nanowires.

2.1.2. The Mechanically Controlled Break Junction (MCBJ)

In 1985 Moreland and Ekin [20] introduced "break" junctions for the study of the tunneling characteristics for superconductors. In this work an Nb-Sn filament mounted on a flexible glass beam was broken to form an electron tunneling junction between the fracture elements. They found a sharp superconducting energy gap in the *I/V* characteristics. Such behavior was attributed to the formation a tunneling barrier between the fracture elements. Several extensions to this concept have been introduced later, initially by Muller et al. [21, 22], who introduced the name Mechanically Controllable Break Junction (MCBJ). The technique is very fruitful for the study of atomic-sized metallic contacts and produce stable atomic point contacts due to its intrinsic stability.

The principle of the technique is illustrated in Figure 4. It shows a schematic top and side view of a MCBJ, where the studied metal has the form of a notched wire, typically 0.1 mm in diameter, which is fixed onto an insulated elastic substrate with two drops of epoxy adhesive. The notch is manually cut into the center of a piece of wire of the metal. The substrate is mounted in a three-point bending configuration between the top of a piezo-element and two fixed counter supports. This set-up was originally mounted inside a vacuum can and cooled down to liquid helium temperatures. Then the substrate is bent by moving the piezo-element forward. The bending causes the top surface of the substrate to expand and the wire to break at the notch. Two clean fracture surfaces are obtained, which remain clean when cryo-pumping and low-temperature vacuum are used. The fracture surfaces can be brought back into contact by relaxing the piezoelectric element. The roughness of the fracture surfaces usually results in a first contact at one point.

The method leads to a clean surface and the stability of the two electrodes with respect to each other is greatly enhanced compared to STM-BJ. From the noise in the current in the tunneling regime one obtains an estimate of the vibration amplitude of the vacuum distance, which is reported to be typically less than 10^{-4} nm.

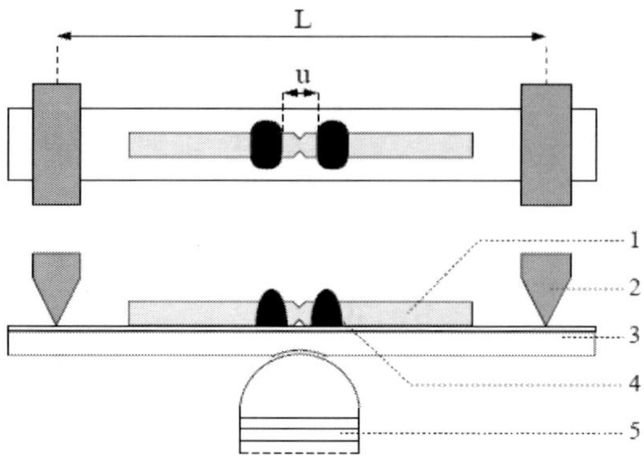

Reproduced from reference [3] Copyright (2003), with permission from Elsevier.

Figure 4. Schematic top and side view of the mounting of a MCBJ, with the notched wire (1), two fixed counter supports (2), bending beam (3), drops of epoxy adhesive (4) and the stacked piezo element (5).

In the first experiments using the MCBJ for a study of conductance in atomic-sized metallic contacts [21, 22], distinct steps were observed in the conductance of Pt and Nb contacts. Figure 5 shows two examples of recordings of the conductance as a function of the voltage *Vp* on the piezo-element, which controls the displacement of the two wire ends with respect to each other. In the experiment, when coming from the tunneling regime (left part), a contact is formed by moving the electrodes together. The contact is then slowly increased while recording the conductance. The conductance is observed to increase in a step-wise fashion, which is different each time a new contact is made.

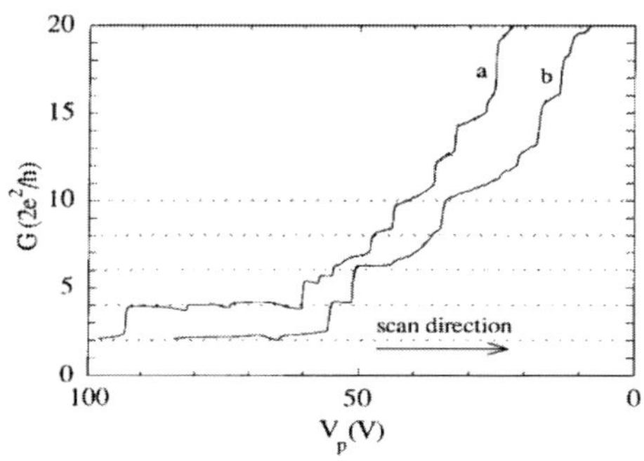

Reproduced from reference [22]. Copyright (1992), with permission from the American Physical Society.

Figure 5. Two examples of traces of the conductance, G, measured on Pt contacts using the MCBJ technique at a temperature of 1.2 K. The electrodes are pushed together by decreasing Vp. An estimate for the corresponding displacement is 10 V ~ 0.1 nm.

The principle of the break junction technique can be refined by employing micro-fabrication techniques to start from a metal bridge. The advantage is a further improved immunity to external vibrations and the possibility to control the electronic and electromagnetic environment of the junction. Figure 6 shows a lithographically fabricated MCBJ device on a silicon substrate, as developed by Zhou et al. [23].

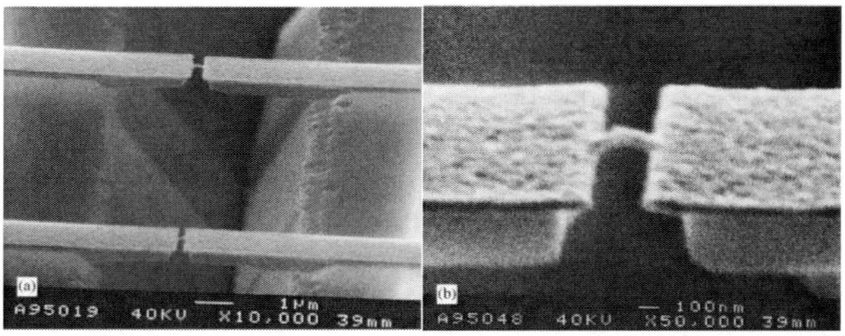

Reproduced from reference [23]. Copyright (1995), with permission from American Institute of Physics.

Figure 6. (a) Electron microscopy image of two micro-fabricated bridges suspended above a triangular pit in the silicon substrate. The close-up in (b) shows the two SiO2 cantilevers, which are about 700 nm apart. The cantilevers are covered by a gold layer from which the final conducting bridge of about 100 nm wide is formed, and which is broken by bending of the silicon substrate.

Metallic junctions of gold, silver and copper were carefully broken, using MCBJ method, and the conductance was shown to decrease stepwise until finally the contact has a cross-section equal to the diameter of only a single atom [24]. For the noble metals the conductance of single-atom contacts was found to be close to one quantum unit of conductance, G_0. For gold, the long plateaus were observed with conductances around *1* G_0 while monitoring the conductance of gold contacts as they are pulled apart. This observation was attributed to the fact that upon stretching a single atom contact, additional atoms can be pulled out of the electrodes to form a monatomic chain of up to seven atoms long [25, 26, 27]. Such long plateaus were not observed for silver and copper as can be seen in Figure 7 suggesting that silver and copper do not form long single atom wires like gold does.

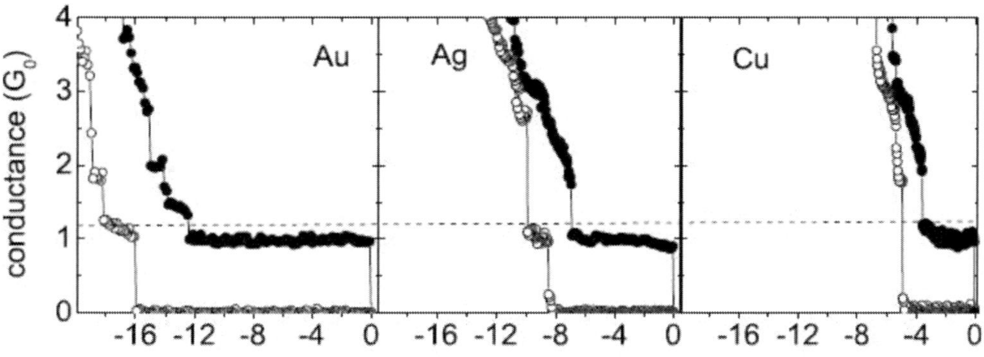

Reproduced from reference [24] Copyright (2008), with permission from IOP Publishing Ltd.

Figure 7. Typical conductance traces for gold, silver and copper performed using MCBJ method. The filled dots show the digitalized points measured while breaking the contact, while the open dots are measured when closing the junction again.

The smaller size of the device reduces effects such as thermal and mechanical instabilities. As a consequence, it enables the fabrication of much more stable atomic point contact or nanogap. The reduction ratio has been reduced to as low as 10^{-4} – 10^{-5} [28, 29, 30, 31] since the first introduction of micro-fabricated MCBJs in the mid-1990s [32, 33].

Patterned constriction of about 100–200 nm of the nanowire on a microchip by use of e-beam lithography gives rise to the possibility of sub-angstrom control of metallic separation size. However, such a large reduction ratio requires very sophisticated fabrication of the microchip which has to be patterned by expensive e-beam lithography equipment that is not commonly available. Besides, the e-beam lithographic experiment is time consuming and the metal material cannot be changed once the e-beam evaporation or sputtering process is finished. Moreover, the microchip must be rejected once a non-recoverable break of the metal wire occurs. These reasons have yielded the development of alternative methods to generate quantum point contact

2.2. Non-Mechanical Methods

To date, the most widely used method for creating a nanowire that exhibits the conductance quantization phenomenon is based on mechanically breaking a fine metal wire [34, 35] or separating two metal electrodes in contact [12, 36]. The breaking and separating are usually controlled by a break junction or a scanning tunnelling microscope (STM) that involves a stepping motor, or a piezoelectric transducer. A nanowire fabricated by the mechanical method cannot be removed from the apparatus and is, therefore, unsuitable for many applications.

2.2.1. Surface Diffusion Mediated Deposition (SDMD).

A new method, called "surface diffusion mediated deposition" (SDMD) has been recently reported. The principle of SDMD is presented in Figure 8a. Briefly, Au was evaporated through a shadow mask aligned perpendicular to a first electrode consisting of pyrolysed photoresist films (PPF) lines [37]. Vapor deposition was done using a tilt angle that renders direct gold deposition on the PPF electrode impossible. However, cold gold atoms can diffuse toward this electrode and ultimately can contact it.

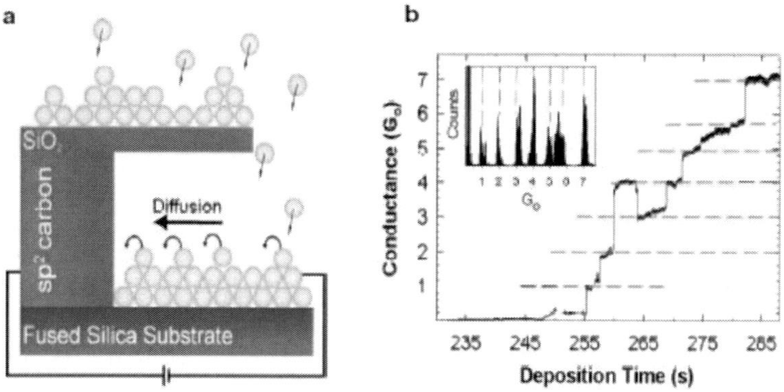

Figure 8. Schematic (a) and current vs time trace (b) for SDMD. Current is expressed as conductance, with one Go unit corresponding to 7.74 μA at the applied bias of 0.1 V. Inset in (b) is a histogram of the number of occurrences of each conductance value during the deposition, with full scale equaling 800 counts in histogram "bins" with widths of 0.05 Go units. Dashed lines indicate the conductance values corresponding to the histogram peaks.

In this study, the deposition rate was empirically selected with the aim to obtain at least 4000 conductance measurements during the observation of 4 to 10 conductance steps, with 4000 measurements chosen to ensure adequate sampling to construct the conductance histograms. Figure 8b shows a current-time trace using voltage of about 0.1 V. In this case, the first expected electrical contact is between the PPF layer and the first Au atom arriving from the diffusing Au "front". Discrete conductance steps are observed during metal diffusion. The quantized conductance is clearer in the histogram of the current readings, shown in the inset of Figure 8b and the spacing between peaks in the histogram is close to the expected value of one G_0 unit. These results clearly indicate that quantized conductance can be observed with the SDMD technique, with the value expected for a single atom contact. To date this method is too new to predict if it will be widely used in the fabrication of quantum point contact.

2.2.2. Electrochemical Deposition/Etching

The disadvantages of the mechanical methods can be overcome by using electrochemical methods. The first reported one was a simple etching/deposition method. By sequentially electrodepositing metal atoms from an electrolyte solution onto the working electrode surface, it is possible to narrow an initial separation between two defined electrode pairs until contact. The electrodeposition mode can be reversed to an electrodissolution mode to etch the atomic point contact to rebuild a separation. Switching between these two modes repeatedly is to some extent similar to the mechanical crashing and stretching process in STM-BJ and MCBJ experiments. The nanowires supported on a solid substrate may be removed from the fabrication set-up and used as a stand-alone device or for further investigation. The method was first demonstrated using a set-up with a thin (5– 25 μm) Cu wire attached to a glass substrate (Figure 9).

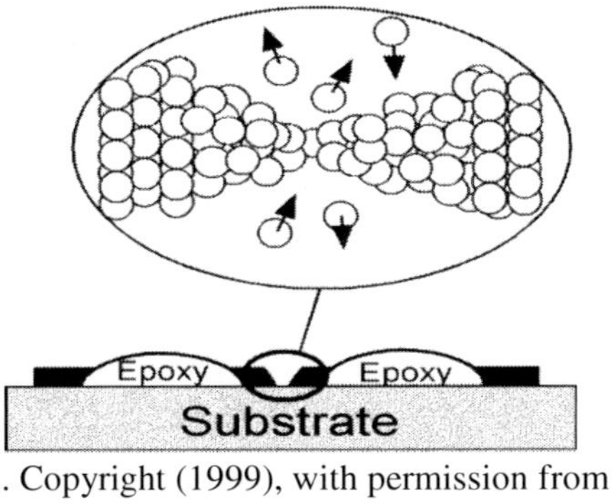

Figure 9. A schematic drawing of the experimental set-up. A single or an array of metal wires is supported on a glass substrate. Each wire is covered with a layer of insulator except the central portion which is then etched electrochemically until the conductance quantization appears.

The wire was coated with an insulation layer (wax or 5-minutes-epoxy) except a region smaller than 1 μm near the centre. The central region was then exposed to 1 mM $CuSO_4$ + 100 mM H_2SO_4 for electrochemical etching which was controlled by adjusting the electrochemical potential of the wire [38, 39]. The etching rate can be flexibly controlled with the electrochemical potential of the wire. In the beginning of the etching process, a relative large overpotential (~0.2 V) was typically used in order to quickly etch down the wire to the nanometre scale which was reflected by a decrease in the conductance to a few hundredths of

Siemens. The etching rate was slowed to trim the width of the nanowire down to the atomic scale. By carefully controlling the potential, the etching rate could be reduced enough so that the conductance stayed on a quantized plateau for many minutes which gave enough time to control the process. Figure 10 shows that the conductance of the nanowire stays at the quantized steps, $N = 9$, 8 and 7 for many minutes.

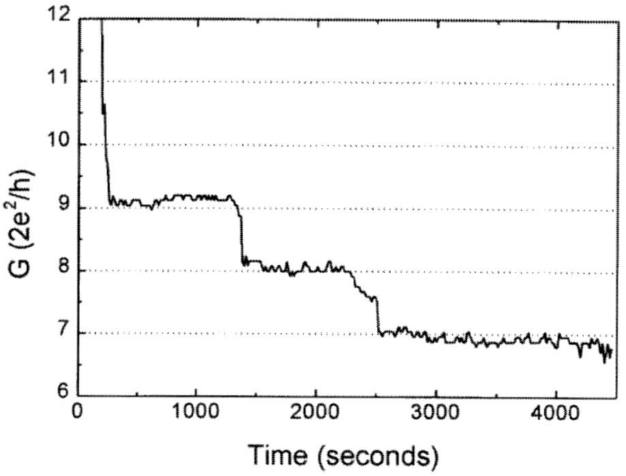

Reproduced from reference [39]. Copyright (1999), with permission from IOP Publishing Ltd.

Figure 10. The conductance of a Cu wire as its width is being slowly decreased by electrochemical etching. The stepwise change is due to conductance quantization.

Once the conductance dropped down to the order *10* G_0, an automatic switch was activated in order to fabricate a nanowire with a preset quantized conductance. When the conductance reached the preset value, the switch turned the etching off by cutting off the electrochemical etching current that flows between the nanowire and the counter electrode.

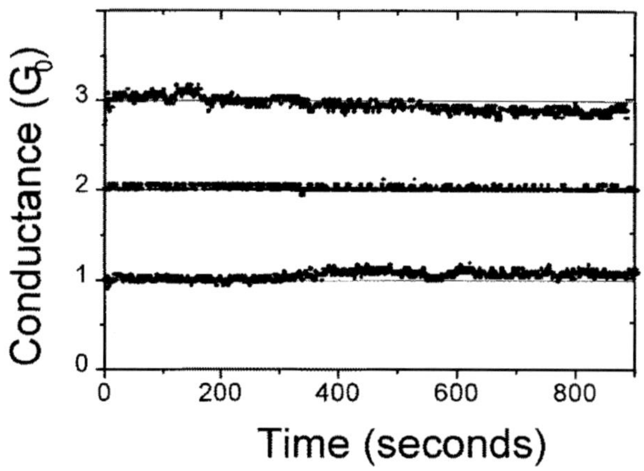

Reproduced from reference [39]. Copyright (1999), with permission from IOP Publishing Ltd.

Figure 11. Cu nanowires with conductance, 1 G_0, 2 G_0 and 3 G_0 fabricated by stopping the etching process at the selected steps with an automatic switch.

Figure 11 shows the conductance plots of three nanowires fabricated using the described procedure. The conductance of each nanowire was preset at *1*, *2* and *3* G_0, respectively. This experiment shows the possibility to control the number of atoms making the contact while keeping large time stability.

2.3. Self-Terminated Method

Apart from the previously reported electrochemical methods [39, 40], Tao et. al. described a new method which is not only simpler but also has a built-in self-termination mechanism [41]. The principle of this method is sketched in Figure 12a.

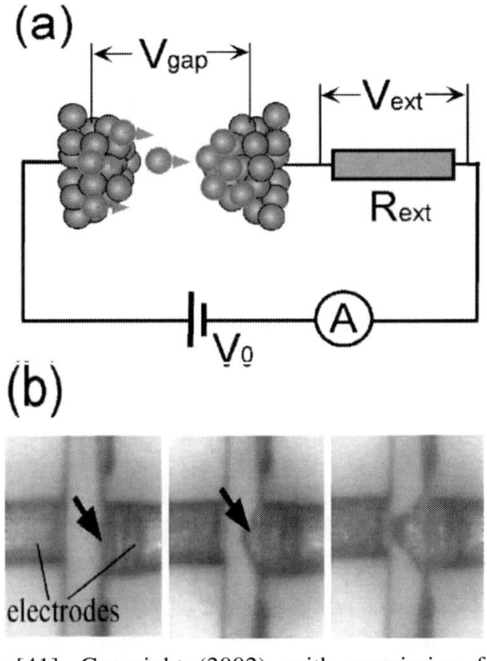

Figure 12. (a) Schematic drawing of the setup. Metal atoms etched off the left electrode are guided and deposited onto the right electrode by applying a voltage between the electrodes with an external resistor in series. As the gap between the two electrodes shrinks, the gap resistance decreases, which results in a drop in etching/deposition voltage and eventually terminates the etching/deposition processes. (b) Snapshots of an experiment that started with a pair of 25 μm Cu electrodes separated with a 20 μm gap.

It starts with a pair of electrodes separated with a relative large gap in an electrolyte or even pure water. When a bias voltage is applied between the two electrodes, metal atoms are etched off the anode and dissolved into the electrolyte as metal ions, which are then deposited onto the cathode. It was found experimentally, that the etching takes place all over the anode surface, but the dissolved metal ions are guided by the electric field and deposited onto the sharpest point of the cathode. Consequently, the gap decreases to the atomic scale and then completely closes as the two electrodes form a contact.

In order to fabricate an atomic gap or contact in a controlled fashion, the etching and deposition processes must be terminated promptly once a desired gap or contact is formed. A simple self-termination mechanism by connecting one electrode to an external resistor (R_{ext}) was proposed. The effective voltage applied to the electrochemical two electrode cell for etching and deposition is given by

$$V_{gap}= [R_{gap}/(R_{gap}+R_{ext})]\ V_0$$

where R_{gap} is the gap resistance between the two electrodes ($V_{gap}= R_{gap}I$), and V_0 is the total applied bias voltage ($V_0=(R_{gap}+R_{ext})I$). R_{gap} is determined by electron tunneling across the gap and by ionic conduction (leakage current) between the electrodes. The gap is initially large and the electron tunneling is negligible, so $R_{gap}>>R_{ext}$, and $V_{gap}\sim V_0$, or the entire applied voltage is used for etching and deposition. Consequently, the etching and deposition processes take place at the maximum rates. As the gap narrows, the tunneling current rises exponentially and R_{gap} decreases, which results in a decrease in V_{gap}, and, therefore, a slowdown in the etching and deposition rates. Eventually, when $R_{gap}<<R_{ext}$, $V_{gap}\sim 0$, which terminates the etching and deposition and results in a gap whose width depends on R_{ext}. If R_{ext} is equal or smaller than $\sim 12.7\ k\Omega$, a contact between the electrodes is formed and the tunneling is replaced with ballistic transport.

In the original work of Tao, the starting electrodes were prepared from a 25 mm Cu, Au, or Ag fibers coated with epoxy on a glass substrate. The initial gap between the electrodes was 10–20 mm, created by carefully cutting the metal. The electrodes were immediately covered by the etching/deposition solutions (e.g., pure water, KCl, HCl, and H_2SO_4). Figure 13a shows the conductance (normalized against G_0) between two Cu electrodes during electrochemical etching and deposition in water with R_{ext} preset at $3\ k\Omega$ ($<12.7\ k\Omega$).

The initial conductance due to ionic conduction is negligibly small compared to G_0. A few minutes after applying a 1.2 V voltage, the conductance suddenly jumps to ~ $2\ G_0$ and the deposition terminates itself as a contact is formed between the electrodes. In the presented data, the initial contact breaks within seconds and the conductance drops back to zero. The drop in the conductance reactivates the deposition process, and the contact with conductance near $1\ G_0$ is reformed a few seconds later. While the conductance tends to stabilize, a large noise (sharp spikes) in the conductance is clearly visible. Zooming in, the noise reveals stepwise fluctuations in the conductance between ~ $1\ G_0$ and ~ $0\ G_0$ [inset of Figure 13a], corresponding to a constant breakdown and reformation of the contact. The conductance fluctuates violently when keeping the voltage at 1.2 V, but the fluctuations are significantly smaller after reducing the bias voltage to 0.2 V (Figure 13b). During the formation of the contact, the conductance varies in a stepwise fashion that can be recorded when the variation is slow enough (Figure 13c). Finally, Figure 13d is an example of the *I–V* curves for a contact with conductance near $1\ G_0$. The current is linearly proportional to the bias voltage, and a simple ohmic behavior was observed.

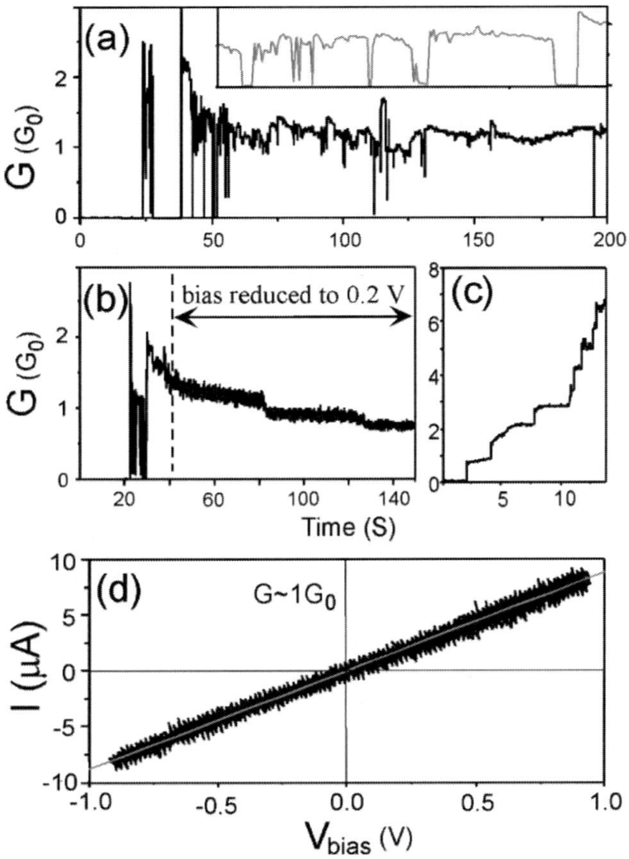

Figure 13. Choosing a R_{ext}<12.7 kΩ, an atomic-scale contact with quantized conductance can be fabricated. (a) is an example that shows the conductance change during the formation of a contact (R_{ext} = 3 kΩ and V_0=1.2 V). (b) shows a similar experiment except that the bias was reduced to 0.2 V from 1.2 V, 20 s after the formation of a contact. The reduction of the bias largely eliminates the on–off fluctuations. A smaller R_{ext} (e.g., 1 kΩ) results in a higher conductance contact and the conductance often increases in a stepwise fashion before reaching a final value (c). The current through the contact is linearly dependent on the voltage across the contact (d).

2.4. Scanning Electrochemical Microscopy SECM

More recently, the generation of metallic nanowires starting from a Scanning Electrochemical microscope (SECM) setup was investigated. The method is clearly derived from the self-terminated method introduced by Tao but also presents some aspects of STM-BJ methods since SECM has a piezo element that can be used to create the initial gap between a tip and a substrate or to stretch a generated nanowire. Figure 14 summarizes the principle and the procedure followed for the fabrication of copper nanowires.

A first study was carried out using the SECM setup operating in a two-electrode (anode and cathode) configuration (Figure 14a) [42]. Briefly, the Pt tip was positioned near the substrate to make a micrometric gap. In a next step, a bias voltage was applied between the two electrodes leading to etching of metal atoms from the anode (Cu substrate) and dissolution into the electrolyte as metal ions followed by deposition on the cathode (Pt ultramicroelectrode). Following this, a contact was established, and the copper junctions generated, in water, were characterized by following the variation of the current versus time (which was plotted as conductance versus time) and by recording the *I–V* curve.

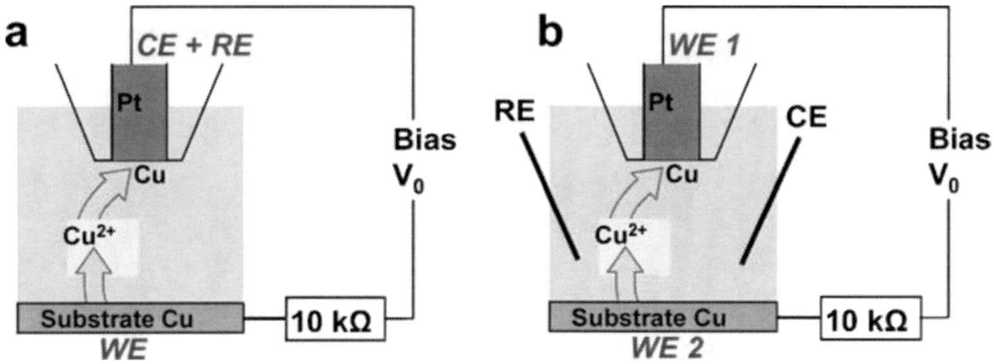

Figure 14. Schematic drawing of SECM setup for the fabrication of copper nanowires in MilliQ water using (a) two-electrode configuration, (b) four-electrode configuration

Figure 15 shows the variation of the normalized conductance (G/G_0) versus time before and after contact in two different experiments. Figure 15a shows that the initial conductance due to ionic conduction is in the *nS* range and is negligible compared to the $77\ \mu S$ corresponding to the conductance quantum. After contact the electrochemical deposition process terminates itself, thanks to the control by the external resistance, and current flows between the tip and the substrate through the copper junction. The conductance stabilizes at a value close to $2\ G_0$ for more than 100 s, then jumps to $3\ G_0$ then $4\ G_0$ and finally back to $3\ G_0$. It varies in stepwise fashion similar to that observed when other techniques are used [43, 44]. Another proof that an atomic-scale junction is formed is its ohmic behavior (see inset in Figure 15a).

All the results given above are based on a SECM setup using a two-electrode anode and cathode configuration. The method was adapted to a four-electrode configuration as shown in Figure 14b [42]. It has two working electrodes, a Pt UME tip and a Cu substrate, the latter being connected to an external 10 kΩ resistance to self-terminate the process after contact, a quasi-reference electrode, and a Pt counter-electrode. The electrochemical process was performed in MilliQ water by applying a bias voltage between the tip and the substrate. In these experiments, the substrate was polarized at +0.5 V so that copper dissolution can occur whereas the Pt tip was polarized at −0.5 V so that copper deposition is possible.

Figure 16a shows the variation of the conductance with time during the growth of Cu nanowires and after contact is established. During the first 50 s corresponding to the growth of Cu nanowires the conductance is close to zero, and then jumps abruptly when contact is established between the two electrodes. It is stable for more than 500 s at value around $1\ G_0$, suggesting the formation of a monoatomic contact. Another behavior characteristic of the formation of atomic-scale junctions is the stepwise conductance variation as shown in the inset in Figure 16a. Figure 16b shows a linear variation of the current versus the applied bias voltage demonstrating ohmic behavior. The slope of this curve is close to *1* G_0, similar to the conductance obtained in Figure 16a. These observations confirm that it is possible to generate atomic-scale contacts using the four-electrode SECM setup.

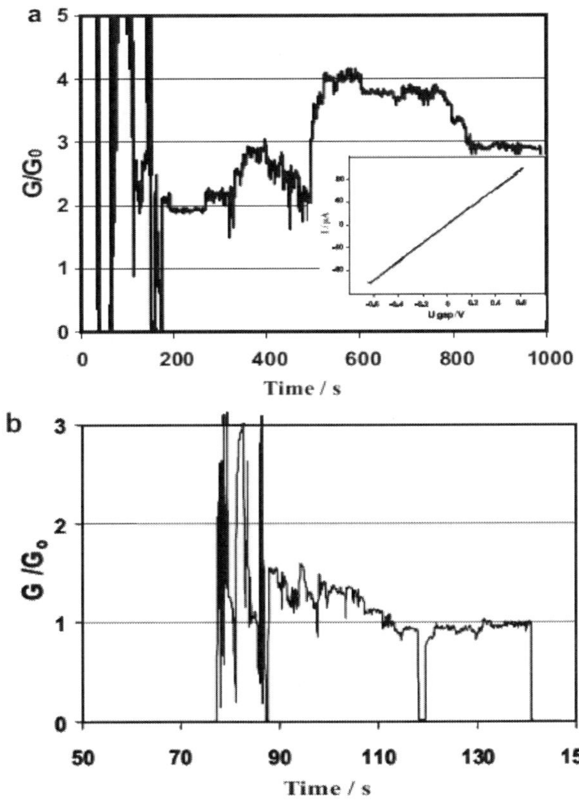

Reproduced from reference [42]. Copyright (2012), with permission from Elsevier.

Figure 15. (a) and (b) two examples of the conductance vs. time plot of copper nanowire growth in MilliQ water. The inset of (a) shows the I–V characteristic of the contact.

2.5. STM-BJ and MCBJ Coupled with Electrochemistry

Electrochemistry has also been used to extend the capability of STM-BJ to create nanowires of various metals. The principle of the approach is based on the electrochemical STM tip-induced nanostructuring [45, 46, 47, 48, 49] and is illustrated in Figure 17 [50].

The tip and substrate are made of the same metal (M_1). A metal of interest (M_2), which is does not form a surface alloy at room temperature with M_1, is being continuously electrodeposited onto the tip (Figure 17a). Usually, atoms of M_2 on the tip transfer to the surface to create a nanoconstriction of M_2 (Figure 17b). Hence, the nanoconstriction is gradually stretched to atomic size and eventually breaks, during which the current trace is recorded simultaneously. By repeating the whole process always at new positions of the surface to ensure the well-controlled jump-to-contact mechanism, thousands of current traces can be recorded, from which conductance histogram can be constructed. Such an electrochemically assisted and jump to-contact facilitated strategy extends the capability of STM-BJ to form a chemically well-defined contact. The feasibility of the present approach was demonstrated by taking Cu as a model system.

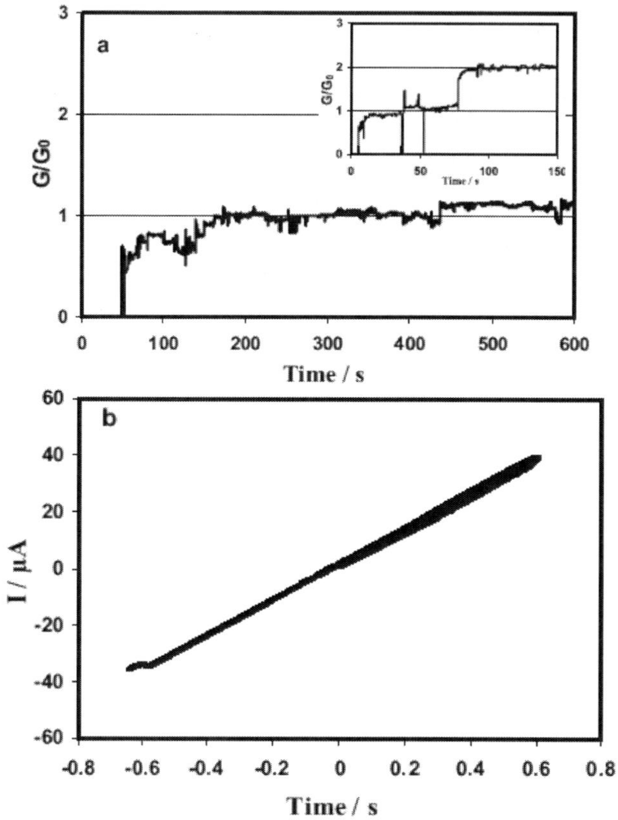

Reproduced from reference [50]. Copyright (2002), with permission from Elsevier.

Figure 16. (a) Conductance vs. time plot of copper nanowire generated in MilliQ water using four-electrode SECM setup (the substrate was polarized at +0.5 V vs. SCE whereas the Pt tip was polarized at −0.5 V), (b) I–V characteristic of the generated contact.

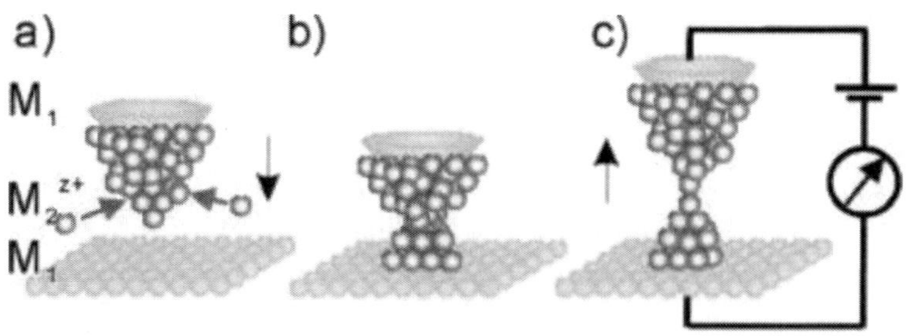

Reproduced from reference [50]. Copyright (2008), with permission from American Chemical Society.

Figure 17. Schematic illustration of the electrochemically assisted and jump to contact facilitated strategy for STM-BJ.

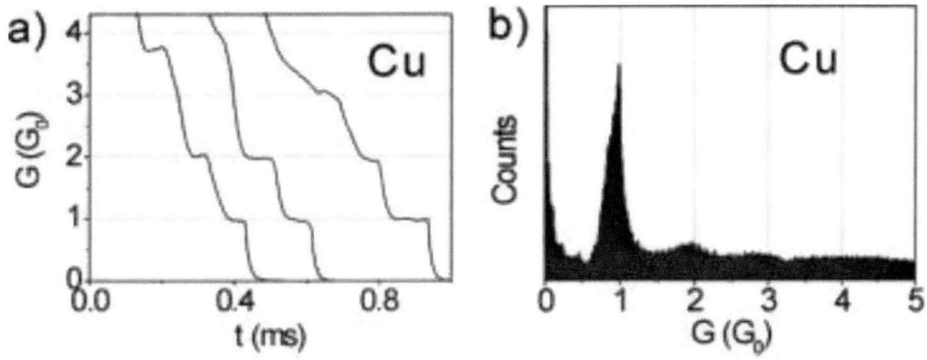

Figure 18. (a) Typical conductance traces and (b) the conductance histogram of Cu from 1600 traces without data selection. Bias (tip vs substrate) = -25 mV.

The conductance measurements were carried out in the aqueous solution of 1 mM $CuSO_4$ + 50 mM H_2SO_4. The Au tip was controlled at -23 mV vs Cu wire for Cu bulk deposition, while the Au(111) substrate was controlled at potentials slightly positive of the Nernst potential where Cu bulk deposition cannot proceed. Figure 18a gives typical conductance traces of Cu. The conductance has a clear stepwise feature in the last two plateaus at near multiples of $1\ G_0$ and drops sharply from the last plateau into the tunnelling current regime. The histogram constructed from 1600 traces without data selection is given by Figure 18b, and the most preferential conductance is found at $\sim 1\ G_0$. Although peaks at multiples of $1\ G_0$ are rather weak, a small yet distinguishable peak can be found at $2\ G_0$.

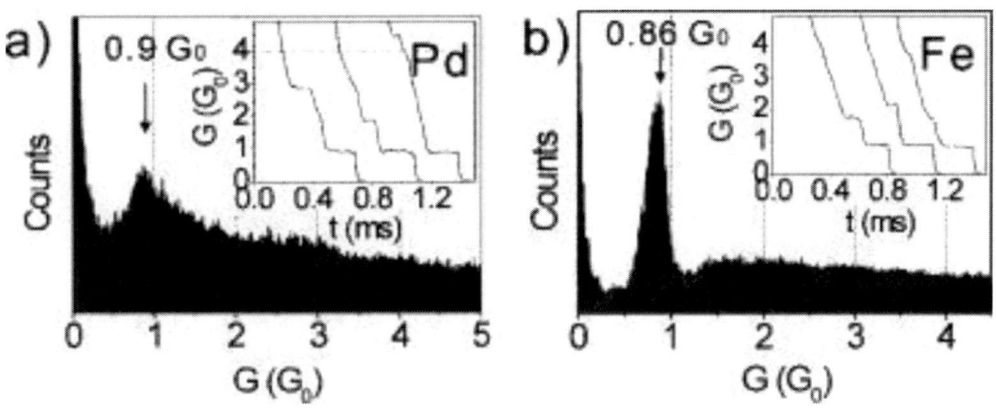

Figure 19. Conductance histograms of (a) Pd and (b) Fe. Insets of (a) and (b): typical conductance traces of Pd and Fe obtained in $BMIBF_4$ ionic liquid, respectively.

The above-proven approach is advantageous for conductance measurement of transition and magnetic metals, such as Pd and Fe for which the conductance has been found difficult to measure by other approaches, especially at room temperature. For Pd, by using the conventional STM-BJ technique, Murakoshi and co-workers obtained a point-contact conductance of $0.9\ G_0$ in solution, but only with the assistance of hydrogen evolution reaction (HER) [51]. When using the above approach, a hydrogen evolution reaction (HER) can be avoided and the complexity due to an absorbed hydrogen influence is removed. As shown in the inset of Figure 19a, well-defined final plateaus at near *1* G_0, corresponding to single-atom contacts, are distinguished despite the complicated stepwise decrease of conductance at high values of some traces. The conductance histogram constructed from 1600 of such traces gives a statistic preference of conductance at near *1* G_0 (i.e., *0.9* G_0) as shown in Figure 19a. Similarly, in the case of Fe the mean conductance of the atomic wire was found around *0.86* G_0 (Figure 19b). Similarly, combining of MCBJ method with electrochemistry was recently presented [52]. An interesting conductance histogram evolution was demonstrated upon repeatedly connecting and disconnecting the two electrodes. Moreover, it was found that the EC–MCBJ method offers high quality data and a high success rate for atomic contact formation, comparable to STM-BJ and conventional MCBJ, but it is cost effective, fast, and environmentally friendly. The authors concluded that the method is expected to extend and complement the conventional e-beam lithographic MCBJ method.

3. Quantized Conductance Atomic Switch

Making stable quantum point contacts, and controlling its conductance require a feedback control system such as that used in a STM [53]. However, in view of introducing QPCs into actual devices, it is not feasible for each atomic contact, namely, each 'bit,' to employ its own feedback control system containing large circuits. Using a mechanical positioning system such as a piezodevice is also undesirable from the viewpoint of integrating devices.

A new type of QPC was developed by Aono et al., which is formed by a solid electrochemical reaction occurring in a nanogap. Since it works without a mechanical positioning system and its conductance is controlled simply by applying a certain bias to it, this QPC is easily introduced into actual devices. Here, the new QPC, particularly its function as a switching device has been investigated [54, 55].

To do so, a silver wire covered by silver sulfide (Ag_2S) crystal, which is a mixed ionic and electronic conductor, was used as one of two electrodes. The other electrode (Pt) was set at a distance of 1 nm from the Ag_2S electrode.

In the original work [56], the switching mechanism was explained as follows. When a negative bias is applied to the Pt electrode with respect to the Ag_2S electrode, silver ions in Ag_2S are reduced to silver atoms by electrons flowing from the Pt electrode, resulting in the precipitation of the silver atoms at the surface of Ag_2S, as shown in Figure 20a. Between the two electrodes, the silver atoms form an atomic bridge whose conductance can be quantized. When the opposite bias is applied, silver atoms in the bridge are oxidized and dissolved into the bulk of Ag_2S, as shown in Figure 20b, which results in the thinning and breaking of the bridge. In this manner, the formation and annihilation of the QPC can be controlled by applying a bias between the electrodes.

The switching cycles was demonstrated between ON and OFF states. Figure 21 shows switching between an ON state with a *2* G_0 QCP and an OFF state with no contact in between the two electrodes. All contacts exhibiting the same conductance value of *2* G_0 were obtained by applying pulse bias voltages with switching frequencies of 10 kHz Figure 21a. The switching speed up to *40 kHz* was confirmed as shown in Figure 21b. In this case, however, the maximum number of cycles of successive ON/OFF transitions obtained was around ten.

Further cycles of bias pulses were reported at that time to deviate the particle from a regular cyclic motion, which ended up in either ON or OFF state [57].

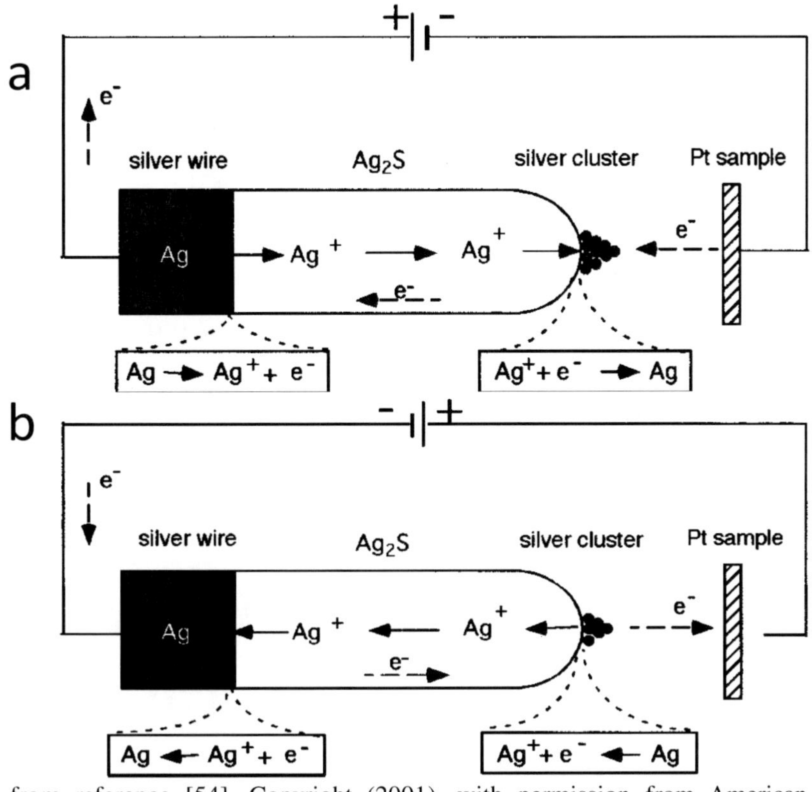

Figure 20. Principle of a QPC switch formed by a solid electrochemical reaction. (a) the growth of the Ag cluster on the Ag_2S tip at a negative sample bias, and (b) the shrinkage of the Ag cluster at a positive sample bias.

Since these seminal experiments, several integrated devices using microfabricated technics such as *12×12* cross bar switch arrays, logic gates, or non-volatile *1kb* memories were presented in collaboration with NEC. As an important result, it was demonstrated that atomic switch can be microfabricated easily. The best performance reported to date for these systems are a scalability below 10 nm, a retention time of 10 years, a switching time of few nanoseconds and a cycling endurance above 10^5 times [58, 59]. New directions, such as the use of polymer electrolyte based atomic switch [60], photosensitive atomic switch [61] are currently investigated. Extending these systems using three terminal configuration is also studied [62]. Finally, the synaptic characteristics of atomic switch have also been recently reported [63, 64, 65].

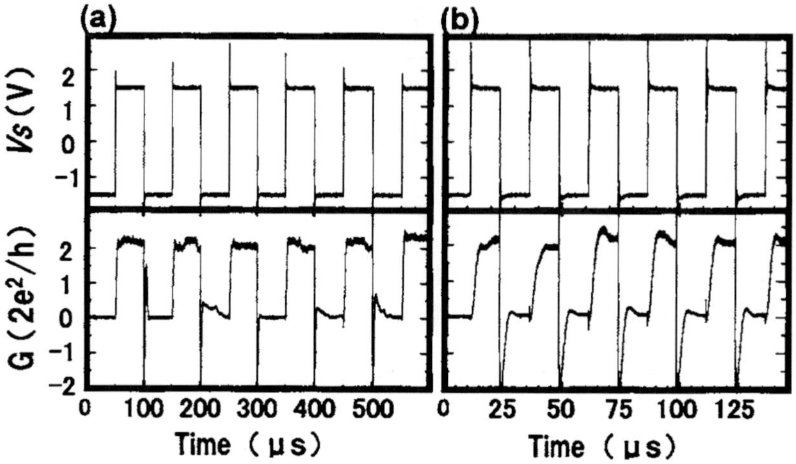

Reproduced from reference [57]. Copyright (2006), with permission from American Institute of Physics.

Figure 21. Demonstration of switching with the alternate bias voltage pulse of Vs of −1.5 V and 1.5 V at the frequencies of (a) 10 kHz and (b) 40 kHz. The dull response of the signal is due to the delay time of the preamplifier used.

4. Deviation from Integer Values of G_0 in Quantized Point Contact

The electrical conductance of a quantum coherent structure accommodating *N* channels of transmission Ti is given by the Landauer formula $G = 2e^2/h\sum_i T_i$. From this formula the conductance can deviate from integer values of G_0 since the $\sum_i T_i$ is not quantized. Several phenomenon affects $\sum_i T_i$ such as the nature of the metal, the contact roughness and the chemical environment.

4.1. Nature of the Metal

For some elements the conductance deviates from the simplest case, $G = NG_0$ where *N* represents the number of atoms making the contact. Indeed, the conductance of given metal is determined by the number of available valence orbitals, and so strongly differs for metallic elements in different series of the periodic table. Scheer and co-workers [66] were the first to show the signature of chemical valence in the electrical conduction through a single atom contact. The quantized conductance for different metallic elements (Pb, Al, Nb and Au), covering a broad range of valences and orbital structure were studied and compared (Figure 22.

Figure 22 shows the conductance *G* for the four metals as the contacts are stretched. The conductance decreases by successive steps until the contact breaks. The step heights are in the order of G_0, while the series of ‘‘plateaus’’ is different for each individual opening of the contact. It was then noted that the shape of the curve (typical value of *G* at which the contact breaks, the lengths of the plateaus, and the behavior within the plateau) appears to be characteristic of each material. Pb shows plateaus with negative slope, and an almost continuous decrease of G between 3 G_0 and 1 G_0, where the contact usually breaks. Al has plateaus with positive slope, and breaks at around 1 G_0. Nb displays negatively inclined plateaus, smallest contacts with conductances between $2\ G_0$ and $3\ G_0$. For Au the last contact before break typically has a conductance between $0.3\ G_0$ and $0.8\ G_0$ and is thus below 1 G_0.

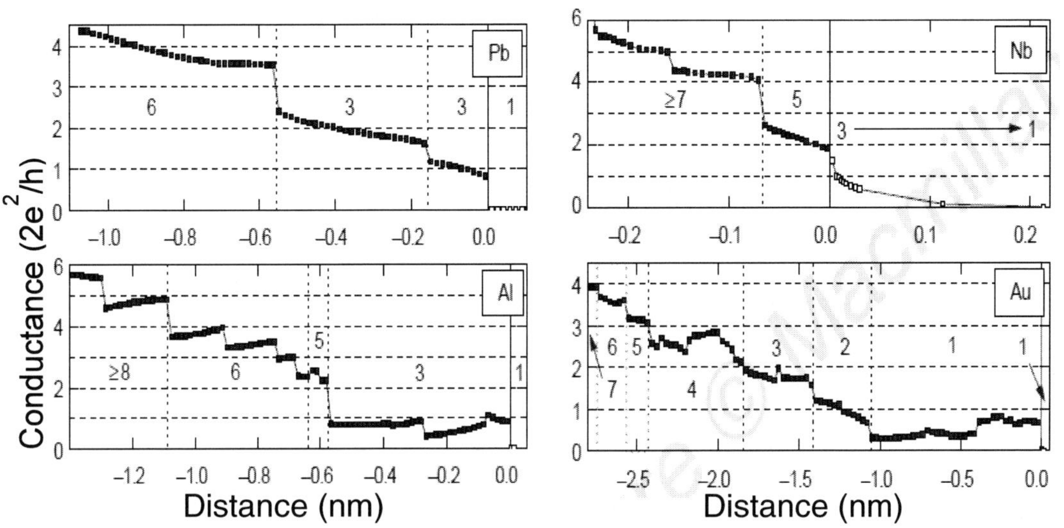

Reproduced from reference [66]. Copyright (1998), with permission from Nature Publishing Group.

Figure 22. Typical conductance G as a function of distance, recorded during a continuous opening of the samples, for four different metals Pb, Al, Nb and Au.

Measuring the conductance is equivalent to measuring the sum $\sum_i T_i$ and gives no information about the individual T_i. The authors developed a technique based on the nonlinearities in the current–voltage characteristics of superconducting constrictions and providing the latter and missing information. These nonlinearities, called 'subgap structure' [67], are sensitive to the exact values of the individual transmissions and can therefore be used as a tool to determine the channel ensemble, $\{Ti\}$. The variation of the conductance as fraction of G_0, observed in Figure 22, was thus linked to valence orbitals of the metal and to the number of conducting channels in one-atom contacts. For Pb it turns out that three or four channels contribute to the total current for the smallest contacts. Despite the fact that more than two channels are available, G can be well below $2\ G_0$. For larger contacts with $G > 3\ G_0$, six or more channels were found. Applying a similar analysis to different metals, they demonstrated the important conclusion that the maximum number of channels N_{max} for the smallest contacts is characteristic for a given metal. In the case of Al smallest contacts found to be the contributions of three channels, whereas for Nb five channels were detected. These results show unambiguously that the conduction channels in an atomic contact are determined by the chemical nature of the central atom. As a crude rule of thumb, the number N of active channels corresponds to the number of valence electrons.

In the Landauer formula, the factor 2 in the quantum of conductance, $G_0 = 2e^2/h$, is due to spin degeneracy. As a consequence, in contacts involving magnetic electrodes the spin degeneracy of transport channels may be lifted. Each spin-polarized channel then may contribute up to $G_0/2$ to the total conductance. Néel and co-workers [68] have studied a single

Co atom adsorbed on Cu(111) or on ferromagnetic Co islands contacted with nonmagnetic W or ferromagnetic Ni tips operating in a STM configuration. A single magnetic atom on a ferromagnetic island or a nonmagnetic substrate was contacted with nonmagnetic and ferromagnetic tips. The obtained results are presented in Figure 23. The conductance of a single-Co atom is *1* G_0 when two nonmagnetic electrodes are used. With ferromagnetic electrodes the conductance is *0.5* G_0, while, conductances of *0.9* G_0 are observed for a combination of a nonmagnetic and a ferromagnetic electrode.

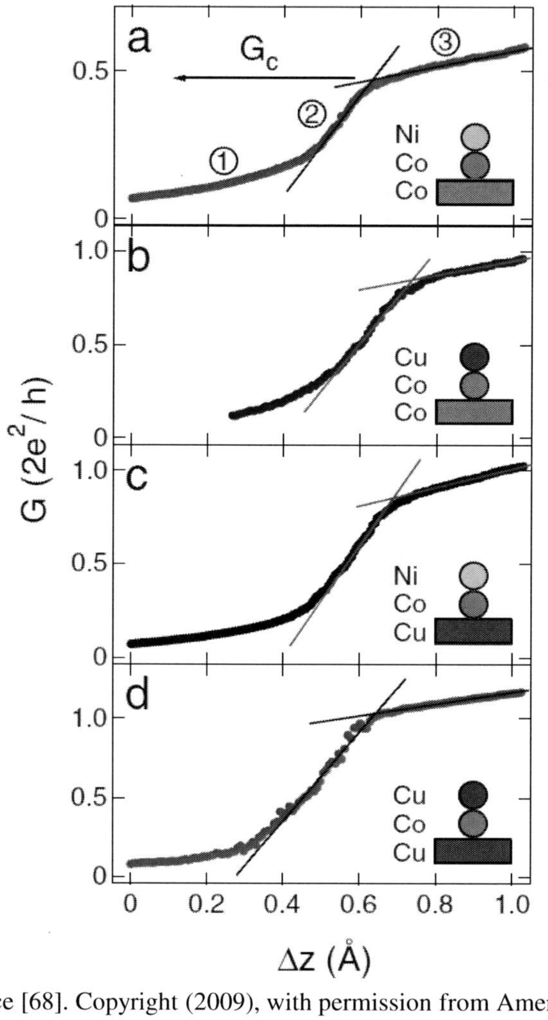

Figure 23. Conductance G versus tip displacement Δz recorded from a single Co atom with different electrode combinations. (a) Ni–Co–Co. (b) Cu–Co–Co. (c) Ni–Co–Cu. (d) Cu–Co–Cu. Tunneling (1), transition (2) and contact (3).

Moreover, the switching of quantum conductance associated with a change in nanoscale ferromagnetic domain structure was also reported. The conductance of the Ni nanowire was studied and found to be clearly quantized in unit of G_0 in a zero magnetic field, but was switched to *0.5* G_0 upon the application of magnetic fields above 60 Oe. This effect was presented as the possibility of generating a new device, a nanoscale colossal magneto-resistive sensor [69].

4.2. Effects of Molecular Adsorption on the Conductance of Metallic Quantum Wires

The principle is based on the phenomenon that a conductor changes its conductance upon molecular adsorption onto its surface [70, 71]. This phenomenon is known in the classical regime in which diffusive scattering of electrons determines the conductance. Zhang et al. [72] have for instance shown that the self-assembly of thiol compounds onto a Au film can be detected by measuring the conductance of the Au film. Several groups have studied the conductance changes of thin-film electrodes due to electrochemically controlled molecular and anion adsorptions [73, 74, 75]. In such experiments, the adsorbate-induced conductance change is due to the scattering of the conduction electrons by the adsorbates as the electrons impinge on the surface [76] but a reduction in the conduction electron density by adsorbates is also believed to be involved [74]. Since different molecules have different density of states, the conductance change should be specific for each adsorbate, which has been confirmed, to some extent, for classical conductors. Moreover, it was shown that the adsorbate-induced conductance change is inversely proportional to the thickness of the film. As a consequence, thinner films mean higher sensitivity in terms of sensor applications. Based on the above argument, it is thus expected that the atomically thin nanowires will give the highest sensitivity and quantum point contact have been proposed as the basis of single molecule sensors.

A number of theoretical attempts have been made to calculate the effects of defects in/on metallic nanowires on the quantized conductance [77, 78]. Bradbyge et al. [79] have shown that the scattering by defects lowers the nanowire conductance, as found in the classical theories. Consequently the conductance steps shift away from the integer multiples of G_0 to fractional values. Landman et al. have also simulated the adsorption of a methyl thiol onto a gold nanowire made of a chain of four Au atoms between two well-defined electrodes [80]. They found that the molecule binds strongly to the gold atoms and becomes a part of the nanowire. The final conductance of the molecule-gold nanowire can be even greater than that of the gold nanowire. Lang has calculated the conductance through a chain of three Al atoms connected to Jellium electrodes [81]. Substituting one of the Al atoms with a sulfur atom was predicted to decrease the conductance. The theoretical works mentioned above predict an adsorbate-induced conductance change, which is the basis of the molecular detection applications of nanowires.

To this end, the electrochemical methods are well adapted to study molecular adsorption effects, since the generated contact exhibits large time stability. In general, two experimental approaches were used. The first one consists on the fabrication of nanowire at a chosen quantum step, next sample molecules were injected into the solution and the subsequent conductance variation was recorded. The second method is based on the generation of the metallic nanowires in the presence of molecules in solution.

As an example, Figure 24a shows the conductance of Cu nanowires quantized at various steps upon introduction of dopamine into the solution cell [82].

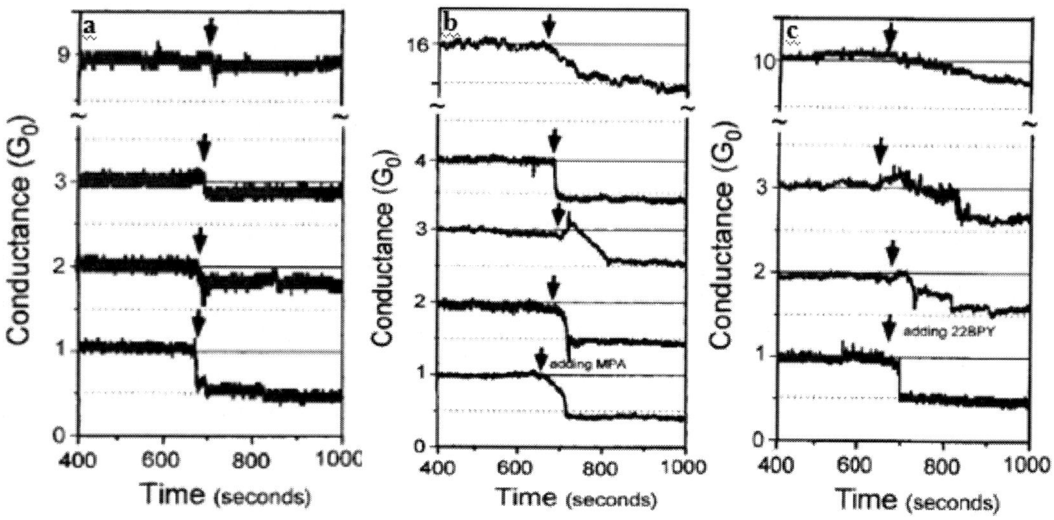

Figure 24. (a) Conductance change of Cu nanowires with conductance quantized at 1 G_0, 2 G_0, 3 G_0, ..., 9 G_0 upon dopamine adsorption from the electrolyte. (b) Conductance change of Cu nanowires with conductance quantized at 1 G_0, 2 G_0, 3 G_0, 4 G_0, ..., 16 G_0 upon MPA adsorption from the electrolyte. (c) Conductance change of Cu nanowires with conductance quantized at 1 G_0, 2 G_0, 3 G_0, ..., 10 G_0 upon 22BPY adsorption from the electrolyte.

For the nanowire with conductance at the lowest quantum step (*1 G_0*), the conductance decreased drastically a few seconds after the introduction of a drop of dopamine, and it then stabilized at around *0.5 G_0*. Adding another drop of dopamine caused no further change, which indicates no further adsorption once the nanowires are covered with the molecules. Nanowires with conductance stabilized at higher quantum steps were also investigated in a similar way. For a nanowire at the second step (*2 G_0*), dopamine adsorption caused also a sharp decrease in the conductance, but the net change is about *0.25 G_0*, much smaller than that of the first quantum step. For a nanowire at the third step (*3 G_0*), the decrease was found to be even smaller. Figure 24b shows also the conductance change due to the adsorption of mercaptopropionic acid, MPA, onto Cu nanowires stabilized at various conductance steps. At the first quantum step (*1 G_0*), the adsorption induces a decrease in the conductance by *0.6 G_0*, which is somewhat greater than that of dopamine. At higher quantum steps, the MPA-induced conductance change stays more or less the same. This observation is in sharp contrast to the dopamine-induced change that diminishes quickly as the conductance increases to higher quantum steps. The effect of MPA adsorption on the quantized conductance appeared thus much greater than that of dopamine. This was correlated with the relative binding strengths of the two molecules to Cu electrodes. In another experiment, upon the adsorption of 2,2'-bipyridine, BPY, the conductance of Cu nanowires also shifted to lower values Figure 24c). Similar to the case with dopamine, the BPY-induced conductance change diminishes as the conductance increases to higher steps, but it diminishes more slowly than the dopamine-induced change.

Absorption of small anions on quantum point contact has also been studied [83, 84]. Some ions such as ClO_4^- and F^- absorb only weakly whereas others like Cl^-, Br^-, and I^- absorb more strongly onto gold or silver with binding strength in the order of Cl^- < Br^- < I^- [85]. Furthermore, the absorption of such ions is known to change with the electrochemical potential. Modulation of the electrochemical potential was shown to induce conductance changes with phase shift of 180°. Since adsorption strength of the anions increases with the potential, the phase shift observed shows that anionic absorption induces a decrease in the conductance. Comparing anions with different adsorption strength also indicated that the $\Delta G/G$ increases when absorption strength increases.

In another example, quantum conductance variation in the presence of various surfactants was investigated [86]. Contacts between two copper microelectrodes were generated using self-terminated electrochemical methods in aqueous media containing various types of surfactants. It was first shown that atomic contacts with conductance close to the conductance quantum (G_0) can be generated in these media.

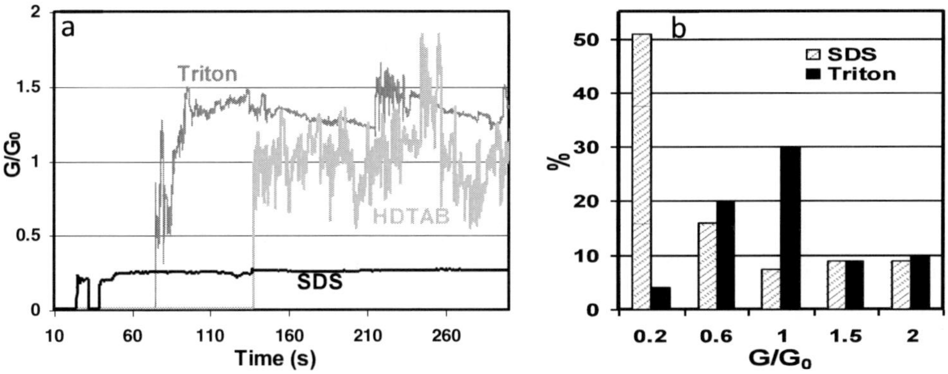

Reproduced from reference [86]). Copyright (2011), with permission from American Chemical Society.

Figure 25. (a) Conductance of copper atomic contacts grown in presence of different surfactants media: neutral (Triton), cationic (HDTAB), and anionic (SDS). (b) Statistics obtained for the formation of atomic contacts in SDS (in grey) and in Triton (in black).

Figure 25a shows that for the three media under investigation, an abrupt increase of the conductance is observed. Such behavior indicates that whatever the used surfactants, contact between the two electrodes was obtained. Comparison in detail of the three curves clearly showed three distinct but reproducible behaviors. First, in the case of cationic surfactant (HDTAB) conductances close to *1* G_0 were observed but the contacts were found to be highly unstable with large fluctuations. Second, in the case of neutral surfactant (Triton) the contact exhibited smaller conductance fluctuations but conductance values were still close to 1 G_0. For both media the conductances were mainly around *1* G_0 but deviated from an integer value of the quantum conductance suggesting random molecular adsorption onto the contact. On the contrary, copper atomic contacts grown in anionic surfactant solution (SDS) showed a more specific behavior. In the example shown in Figure 25a, the recorded data shows that conductance increases abruptly after 100 s and reaches a value around *0.25* G_0, which indicates the occurrence of contact between the two electrodes. A strong decrease of the conductance was thus observed and values below *1* G_0 were reproducibly obtained. Moreover, the fabricated contacts were stable for more than 400 s and showed less fluctuation when compared to those obtained using cationic and neutral surfactant. The (*I/V*) characterization showed an ohmic behavior supporting the formation of metallic contact with quantized conductance. Figure 25b shows a histogram distribution of copper atomic contact generated in the presence of Triton and SDS surfactant. The two distributions are clearly different with a pronounced peak at conductance centered at $0.2\ G_0$ corresponding to more than 50% of the counts when SDS is used. The low conductance value ($<1\ G_0$) was attributed to the occurrence of molecular adsorption on the contact. This specific effect was also linked to the critical micellar concentration (CMC). Indeed, when anionic surfactant concentration was increased above the CMC, the generated contacts had conductance close to that obtained in neutral or cationic surfactant solution.

Similarly, copper metallic contacts in between two microelectrodes were generated in aqueous media containing ODN, nitrogenous nucleobases Poly A and Poly C, at various concentrations using the self-terminated electrochemical method [87]. The obtained results are summarized in the histogram presented in Figure 26.

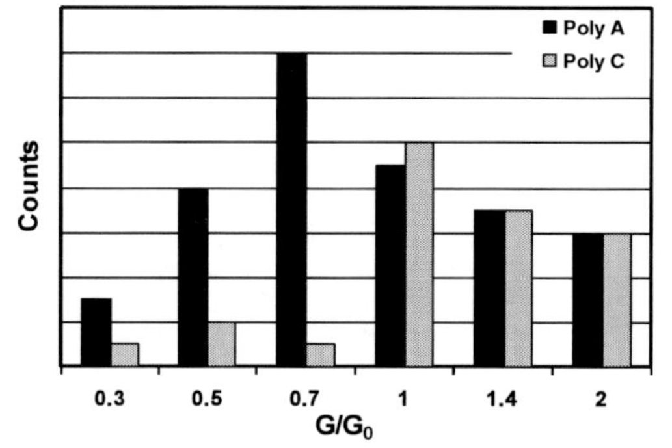

Reproduced from reference [87]. Copyright (2011), with permission from Elsevier.

Figure 26. Statistics for the formation of atomic contacts in aqueous ODN media at 10^{-6} mol.L^{-1} (back distribution) Poly A and (grey distribution) Poly C. As a result the two terminal devices generated are connected by a single metallic nanowire suitable for conductance measurements. The contact can be generated in 10^{-6} mol.L^{-1} PBS solution independently of the ODN nitrogenous nucleobases. The analysis of the data by mean of histogram shows that the conductance of the generated contacts depends strongly on the used ODN. Indeed, those obtained in Poly A have below $1\ G_0$ conductance in marked contrast with those generated in Poly C. These suggest that Poly-A tends to absorb specifically and more strongly than Poly C on the quantum wire.

4.3. Protected Atomic Point Contacts

In another study, Cu atomic point contacts have been generated via electrochemistry in water/β-cyclodextrine media [88]. Cyclodextrins are naturally occurring molecular tubes composed of several α-1,4-linked glucopyranose units. The rim-to-rim tube length is 0.87 nm but this dimension varies with the conformation, and cyclodextrins often pack more closely than this in the solid state through hydrogen-bonding interactions to form channel structures. They have a hydrophilic external surface and a sub-nanometric hydrophobic cavity. As a consequence, CDs form inclusion complexes with a wide variety of organic guests in aqueous solution. In molecular electronics, they have been used to generate encapsulated molecular wires [89]. Some of these encapsulated organic wires were produced from aromatic monomers using oxidative electropolymerization [90, 91].

In a similar way, copper electroreduction in water/β-cyclodextrin media were studied [88]. The authors showed that in this media two types of copper atomic point contacts are obtained. The first is shown to be similar to atomic contacts generated in water. The second has unusual properties. Contacts of this type have below-one G_0 conductance, low conductance fluctuation with time and appear to be protected or partially protected from external molecular probes (Figure 27). Such contacts are not obtained in water, in water/glucose or when β -CD is added after 1 G_0 contacts have been generated in water. These results indicate strong, permanent and specific adsorption of β -cyclodextrin on the atomic contacts. Given the particular shape of β -CD, these results have been explained by assuming that encapsulated atomic contacts are obtained. However, direct independent structural evidence is still needed to confirm or infirm this interpretation.

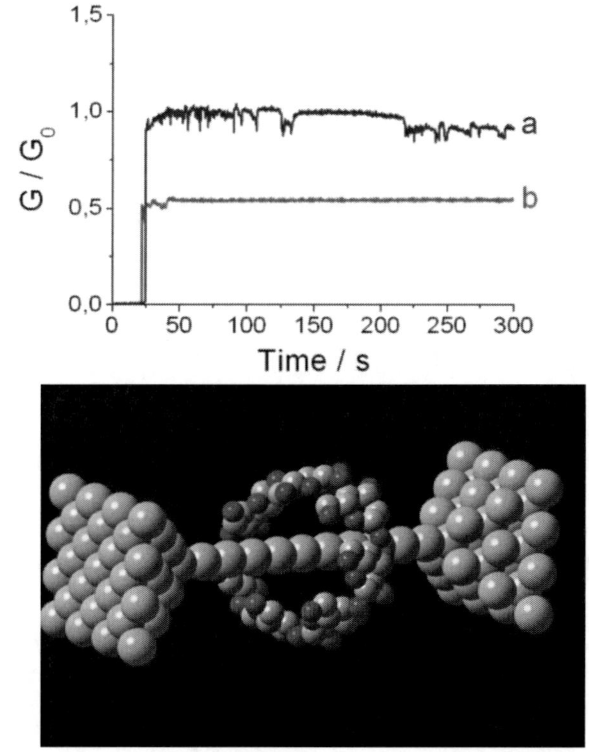

Figure 27. Left: Conductance vs time during the electrochemical process for contacts generated in: a) water; b) water + 10^{-4} M β-CD. Right: Metallic atomic contact encapsulated by one β-cyclodextrin. (Note that the numbers of Cu atoms and β -CDs involved are unknown).

Conclusion

Quantum point contacts or atomic scale metal wires show conductance quantization as a specific property. They can be easily generated using mechanical and electrochemical methods which have now been extensively used for this purpose. New methods combining the advantages of both (SECM and EC-STM) and new methods such as surface diffusion mediated deposition are currently reported. Important applications seem to be reachable. Single-atom switches in nanoelectronic circuits have been demonstrated and have now been pushed towards integrated devices and the next decade may witness the use of quantum point contact in non-volatile and volatile memories and/or in on/off switching circuits or logic gates. Conductance variation upon molecular adsorption has also been proposed as the basis for chemical sensor applications. However, no specificity has yet been demonstrated, since all the molecules surrounding the contact may adsorb or leave the contact at any time, causing fluctuations in the conductance versus time curve. First steps towards this application have been reported and even though an experimental knowledge has now been reached concerning the interplay between molecules and atomic wires, many important fundamental questions remain to be addressed with new experimental and theoretical developments.

REFERENCES

[1] Landauer, R. *IBM J. Res Dev*. 1957, *1*, 223.
[2] Büttiker, M. *IBM J. Res Dev*. 1988, *32*, 63.
[3] Agraït, N.; Yeyati, A. L.; van Ruitenbeek, J. M. *Phys. Rep*. 2003, *377*, 81.
[4] Berndt, R.; Kröger, J.; Néel, N.; Schull, G. *Phys. Chem. Chem. Phys*. 2010, *12*, 1022.
[5] Wiesendanger, R. Scanning Probe Microscopy and Spectroscopy, Cambridge University Press, Cambridge, 1994.
[6] Gimzewski, J. K.; Möller, R. *Physica B* 1987, *36*, 1284.
[7] Agraït, N. ; Rodrigo, J. G. ; Vieira, S. *Phys. Rev. B* 1993, *47*, 12345.
[8] Agraït, N.; Rodrigo, J. G.; Sirvent, C.; S. Vieira, *Phys. Rev. B* 1993, *48*, 8499.
[9] Sirvent, C.; Rodrigo, J. G.; Agraït, N.; Vieira, S. *Physica B* 1996, *218*, 238.
[10] Sirvent, C.; Ridrigo, J. G.; Vieira, S.; Jurczyszyn, L.; Mingo, N.; Flores, F. *Phys. Rev. B* 1996, *53*, 16086.
[11] Pascual, J. I.; Méndez, J.; Gómez-Herrero, J.; Baró, A. M.; García, N.; Thien Binh, V. *Phys. Rev. Lett*. 1993, *71*, 1852.
[12] Pascual, J. I.; Méndez, J.; Gómez-Herrero, J.; Baró, A. M.; García, N.; Landman, U.; Luedtke, W. D.; Bogachek, E. N.; Cheng, H.-P. *Science* 1995, *267*, 1793.
[13] Dremov, V. V.; Shapoval, S. Y. *JETP Lett*. 1995, *61*, 336.
[14] Landman, U.; Luedtke, W. D.; Salisbury, B. E.; Whetten, R. L. *Phys. Rev. Lett*. 1996, *77*, 1362.
[15] Olesen, L.; Legsgaard, E.; Stensgaard, I.; Besenbacher, F.; Schiøtz, J.; Stoltze, P.; Jacobsen, K. W.; Nørskov, J. K. *Phys. Rev. Lett*. 1994, *72*, 2251.
[16] Costa-Krämer, J. L.; García, N.; García-Mochales, P. ; Serena, P. A. *Surf. Sci*. 1995, *342*, L1144.
[17] Brandbyge, M.; Schiøtz, J.; Sørensen, M. R.; Stoltze, P.; Jacobsen, K. W.; Nørskov, J. K.; Olesen, L.; Legsgaard, E.; Stensgaard, I.; Besenbacher, F. *Phys. Rev. B* 1995, *52*, 8499.
[18] He, H.; Tao, N. J. *Encyclopedia of nanoscience and nanotechnology*, volume Electrochemical fabrication of metal nanowires. American Scientific Publishers, 1998.
[19] He, H. X.; Shu, C.; Li, C. Z.; Tao, N. J. *J. Electroanal. Chem*. 2002, *522*, 26.
[20] Moreland, J.; Ekin, J. W. *J. Appl. Phys*. 1985, *58*, 3888.

[21] Muller, C. J.; van Ruitenbeek, J. M.; de Jongh, L. J. *Physica C* 1992, *191*, 485.

[22] Muller, C. J.; van Ruitenbeek, J. M.; de Jongh, L. J. *Phys. Rev. Lett.* 1992, *69*, 140.

[23] Zhou, C.; Muller, C. J.; Deshpande, M. R.; Sleight, J. W.; Reed, M. A. *Appl. Phys. Lett.* 1995, *67*, 1160.

[24] Thijssen, W. H. A.; Strange, M.; Aan de Brugh, J. M. J.; van Ruitenbeek J. M. *New Journal of Physics* 2008, *10*, 033005.

[25] Yanson, A. I.; Rubio Bollinger, G.; van den Brom, H. E., Agraït, N.; van Ruitenbeek, J. M. *Nature* 1998, *395*, 783.

[26] Hakkinen, H.; Barnett, R. N.; Scherbakov, A. G.; Landman, U. *J. Phys. Chem. B* 2000, *104*, 9063.

[27] Bahn, S. R.; Jacobsen, K. W. *Phys. Rev. Lett.* 2001, *87*, 266101.

[28] Gonzalez, M. T.; Wu, S. M.; Huber, R.; van der Molen, S. J.; Schonenberger, C.; Calame, M. *NanoLett.* 2006, *6*, 2238.

[29] Martin, C. A.; Ding, D.; van der Zant, H. S. J.; van Ruitenbeek, J. M. *New J. Phys.* 2008, *10*, 065008.

[30] Huisman, E. H.; Trouwborst, M. L.; Bakker, F. L.; de Boer, B.; van Wees, B. J.; van der Molen, S. J. *Nano Lett.* 2008, *8*, 3381.

[31] Yang, Y.; Liu, J-Y.; Chen, Z.-B.; Tian, J.;-H. Jin, X.; Liu, B.; Li, X.; Luo, Z-Z.; Lu, M.; Yang, F-Z.; Tao, N.; Tian, Z.-*Nanotechnology* 2011, *22*, 275313.

[32] Zhou, C.; Muller, C. J.; Deshpande, M. R.; Sleight, J. W. Reed, M. A. *Appl. Phys. Lett.* 1995, *67*, 1160.

[33] van Ruitenbeek, J. M.; Alvarez, A.; Pineyro, I.; Grahmann, C.; Joyez, P.; Devoret, M. H.; Esteve, D.; Urbina, C. *Rev. Sci. Instrum.* 1996, *67*, 108.

[34] Muller, C. J.; Krans, J. M.; Todorv, T. N.; Reed, M. A. *Phys. Rev. B* 1996, *53*, 1022.

[35] Krans, J. M.; van Ruitenbeek, J. M.; Fisun, V. V.; Yanson, I. K. de Jongh, L. J. *Nature* 1995, *375*, 767.

[36] Landman, U.; Luedtke, W. D.; Salisbury, B. E.; Whetten, R. L. *Phys. Rev. Lett.* 1996, *77*, 1362.

[37] Bonifas, A. P.; McCreery R. L. *Nano Lett.* **2011**, *11*, 4725.

[38] Li, C. Z.; Tao, N. J. *Appl. Phys. Lett.* 1998, *72*, 894.

[39] Li, C. Z.; Bogozi, A.; Huang, W.; Tao, N. J. *Nanotechnology* 1999, *10*, 221.

[40] Morpurgo, A. F.; Marcus, C. M.; Robinson, D. B. *Appl. Phys. Lett.* 1999, *14*, 2082.

[41] Boussaad, S.; Tao N. J. *Appl. Phys. Lett* 2002, *80*, 2398.

[42] Janin, M.; Ghilane, J.; Lacroix, J-C. *Electrochim. Acta* 2012, *83*, 7.

[43] He, H.; Tao, N. J. *Advanced Materials* 2002, *14*, 161.

[44] Pascual, J. I.; Mendez, J.; Gomez-Herrero, J.; Baro, A. M.; Garcia, N.; Binh, V. T.; *Physical Review Letters* 1993, *71*, 1852.

[45] Kolb, D. M.; Ullmann, R.; Will, T. *Science* 1997, *275*, 1097.

[46] Engelmann, G. E.; Ziegler, J. C.; Kolb, D. M. *J. Electrochem. Soc.* 1998, *145*, L33.

[47] Wang, J. G.; Tang, J.; Fu, Y. C.; Wei, Y. M.; Chen, Z. B.; Mao, B. W. *Electrochem. Commun.* 2007, *9*, 633.

[48] Wei, Y. M.; Zhou, X. S.; Wang, J. G.; Tang, J.; Mao, B. W.; Kolb, D. M. *Small* 2008, *4*, 1355.

[49] Kolb, D. M.; Engelmann, G. E.; Ziegler, J. C. *Angew. Chem., Int. Ed.* 2000, *39*, 1123.

[50] Zhou, X-S.; Wei, Y-M.; Liu, L.; Chen, Z-B.; Tang, J.; Mao B-W. *J. Am. Chem. Soc.* 2008, *130*, 13228.

[51] Kiguchi, M.; Murakoshi, K. *Appl. Phys. Lett.* 2006, *88*, 253112.
[52] Yang, Y.; Liu, J.-Y.; Chen, Z-B.; a Tian, J-H.; Jin, X.; Liu, B.; Li, X.; Luo, Z-Z.; Lu, M.; Yang, F-Z. ; Tao N.; Tian, Z-Q. *Nanotechnology* 2011, *22*, 275313.
[53] Smith D. P. E. *Science* 1995, *269*, 371.
[54] Terabe, K.; Nakayama, T.; Hasegawa, T.; Aono, M. *J. Appl. Phys.* 2001, *91*, 10110.
[55] Terabe, K.; Hasegawa, T.; Nakayama, T.; Aono M. *RIKEN Review 37*, 7.
[56] Terabe, K.; Hasegawa, T.; Nakayama, T.; Aono, M. *Nature*, 2005, *433*, 47.
[57] Akai-Kasaya, M.; Nishihara, K.; Saito, A.; Kuwahara Y.; Aono, M. *Appl. Phys. Lett.* 2006, *88*, 023107.
[58] Tsuruoka T.; Terabe, K.; Hasegawa, T.; Aono, M. *Nanotechnology* 2010, *21*, 425205.
[59] Banno, N.; Sakamoto, T.; Iguchi, N.; Matsumoto, M.; Imai, H.; Ichihashi, T.; Fujieda, S.; Tanaka, K.; Watanabe, S.; Yamaguchi, S.; Hasegawa, T.; Aono, M. *Appl. Phys. Lett.* 2010, *97*, 113507.
[60] Wu, S.; Tsuruoka, T.; Terabe, K.; Hasegawa, T.; Hill, J. P., Ariga, K.; Aono, M. *Adv. Funct. Mater.* 2011, *21*, 93.
[61] Hino, T.; Tanaka, H.; Hasegawa, T.; Aono, M.; Ogawa, T. *Small* 2010, *6*, 1745.
[62] Hasegawa, Y.; Itoh, Y.; Tanaka, H.; Hino, T.; Tsuruoka, T.; Terabe, K.; Miyazaki, H.; Tsukagoshi, K.; Ogawa, T.; Yamaguchi, S.; Aono, M. *Appl. Phys. Express 4.* 2011, 015204.
[63] Ohno, T.; Hasegawa, T.; Tsuruoka, T.; Terabe, K.; Gimzewski J. K.; Aono, M. *Nature Mater.* 2011, *10*, 591.
[64] Stieg, A. Z.; Avizienis, A. V.; Sillin, H. O.; Martin-Olmos, C.; Aono, M.; Gimzewski, J. K. *Adv. Mater.* 2012, *24*, 286.
[65] Avizienis, A. V.; Sillin, H. O.; Martin-Olmos, C.; Shieh, H. H.; Aono, M.; Stieg, A. Z.; Gimzewski J. K. *Plos ONE 7* 2012, e42772.
[66] Scheer, E.; Agraït, N.; Cuevas, J. C.; Yeyati, A. L.; Ludoph, B.; Martín-Rodero, A.; Bollinger, G. R.; van Ruitenbeek, J. M.; Urbina, C. *Nature*, 1998, *394*, 154.
[67] Octavio, M.; Tinkham, M.; Blonder, G. E.; Klapwijk, T. M. *Phys. Rev. B* 1983, *27*, 6739.
[68] Néel, N.; Kröger, J.; Berndt, R. *Phys. Rev. Lett.* 2009, *102*, 086805.
[69] Ono, T.; Ooka, Y.; Miyajima, H. *Appl. Phys. Lett.* 1999, *75*, 1622.
[70] Schumacher, D. In *Surface Scattering Experiments with Conduction Electrons*; Hohler, G., Ed.; Springer: New York, 1993; Vol. 128.
[71] Rauh, M.; Heping, B.; Wissmann, P. *Appl. Phys. A* 1995, *61*, 587.
[72] Zhang, Y.; Terrill, R. H.; Bohn, P. W. *J. Am. Chem. Soc.* 1998, *120*, 9969.
[73] Tucceri, R. I.; Posadas, D. J. *Electroanal. Chem.* 1985, *191*, 387.
[74] Tucceri, R. I.; Posadas, D. J. *Electroanal. Chem.* 1990, *283*, 159.
[75] Hanewinkel, C.; Otto, A.; Wandlowski, T. *Surf. Sci.* 1999, *429*, 255.
[76] Ishida, H. *Phys. Rev. B* 1995, *52*, 10819.
[77] Nikolic, K.; MacKinno, A. *Phys. Rev. B* 1994, *50*, 11008.
[78] Garcia-Mochales, P.; Serena, P. A. *Phys. Rev. Lett.* 1997, *79*, 2316.
[79] Bradbyge, M.; Jacobsen, K. W.; Norskov, J. K. *Phys. Rev. B* 1997, *55*, 2637.
[80] Hakkinen, H.; Barnett, R. N.; Landman, U. *J. Phys. Chem.* 1999, *103*, 8814.
[81] Lang, N. D. *Phys. Rev. B* 1995, *52*, 5335.
[82] Bogozi, A.; Lam, O.; He, H.; Li, C.; Tao, N. J.; Nagahara, L. A.; Amlani, I.; Tsui, R. *J. Am. Chem. Soc.* 2001, *123*, 4585.

[83] Xu, B.; He, H.; Boussaad, S.; Tao, N. J. *Electrochim. Acta* 2003, *48*, 3085.
[84] Xu, B.; He, H.; Tao N. J., *J. Am. Chem. Soc.* 2002, *124*, 13568.
[85] Lipkowski, J.; Shi, Z. C.; Chen, A. C.; Pettinger, B.; Bilger, C.; *Electrochim. Acta*, 1998, 43, 2875.
[86] Janin, M.; Ghilane, J.; Randriamahazaka, H.; Lacroix J-C. *J. Phys. Chem. C*, 2011, *115*, 549.
[87] Molet, P.; Ghilane, J.; Martin, P.; Noel, V.; Randriamahazaka, H.; Pham, M. C.; Lacroix, J-C. *Electrochem. Comm.* 2011, 13, 272.
[88] Leroux, Y. R.; Fave, C.; Zigah, D.; Trippe, G.; Randriamahazaka, H.; Lacroix, J.-C.; *J. Am. Chem. Soc.* 2008, 130, 13465.
[89] Taylor, P. N.; O'Connell, M. J.; McNeill, L. A.; Hall, M. J.; Aplin, R. T.; Anderson, H. L., *Angew. Chem. Int. Ed.* 2000, *39*, 3456.
[90] Lagrost, C.; Chane-Ching, K. I.; Lacroix, J. C.; Aeiyach, S.; Jouini, M.; Lacaze, P. C.; Tanguy, J., *J. Mat. Chem.* 1999, *9*, 2351.
[91] Lagrost, C.; Lacroix, J. C.; Aeiyach, S.; Jouini, M.; Chane-Ching, K. I.; Lacaze, P. C., *Chemical Commun.* 1998, 489.

In: Applied Surface Chemistry of Nanomaterials
Editors: Mohamed M. Chehimi and Jean Pinson
ISBN: 978-1-62808-351-4

Chapter 7

PREPARATION, SURFACE MODIFICATION AND FUNCTIONALIZATION OF ANISOTROPIC FERROMAGNETIC METAL NANOPARTICLES: TOWARDS MULTIFUNCTIONAL MATERIALS

Jean-Yves Piquemal[1,*], Kahina Aït Atmane[1], Guillaume Viau[2], Mona Ibrahim[2], Philippe Decorse[1], Aazdine Lamouri[1] and Mohamed M. Chehimi[1]

[1]Université Paris Diderot, Sorbonne Paris Cité, ITODYS, CNRS UMR 7086, Paris Cedex 13, France

[2]Université de Toulouse, INSA, LPCNO, Toulouse Cedex 4, France

ABBREVIATIONS AND SYMBOLS

1D	Mono-dimensional
2D	Bi-dimensional
3D	Tri-dimensional
AAO	Anodic aluminum oxide
B	Magnetic induction (SI: T; cgs: G)
bcc	Body-centered cubic
CMS	Cubic mesoporous silica
CNT	Carbon nanotube
CVD	Chemical vapor deposition
fcc	Face-centered cubic
H_C	coercivity (SI: $A \cdot m^{-1}$; cgs: Oe)
hcp	Hexagonal close-packed
HDA	Hexadecylamine
HMS	Hexagonal mesoporous silica

* Corresponding author: jean-yves.piquemal@univ-paris-diderot.fr.

M_R	Remanent magnetization (SI: $A \cdot m^{-1}$; cgs: $emu \cdot g^{-1}$ or $emu \cdot cm^{-3}$)
M_S	Saturation magnetization (SI: $A \cdot m^{-1}$; cgs: $emu \cdot g^{-1}$ or $emu \cdot cm^{-3}$)
OA	Oleic acid
ODE	Octadecene
TEPC	Track-etched polycarbonate
(HR)TEM	(High resolution) transmission electron microscopy
ND	Nanodisk
NP	Nanoparticle
NR	Nanorod
NT	Nanotube
NW	Nanowire
TOP	Trioctylphosphine
TOPO	Trioctylphosphine oxide
VLS	Vapor liquid solid
VS	Vapor solid

1. INTRODUCTION

Anisotropic or one-dimensional (1D) nanoparticles are a very important class of nanomaterials which have attracted considerable attention in recent years, since they display unique properties compared to either bulk materials or their isotropic zero-dimensionnal (OD) counterparts. The effect of the elongated shape on physical properties (optical, magnetic, catalytic, mechanical) has been reviewed recently for metal, semi-conductor and oxide nanoparticles.[1] Ferromagnetic anisotropic metal particles combine magnetic and conductive properties which make them good candidates for various important technological applications (see Scheme 1). Because there are elongated, with 1D nanocatalysts, different selectivities can be observed as a result of the privileged development of specific facets, in contrast to isotropic particles.[2] For elongated magnetic particles, the shape anisotropy can make a very important contribution of the total magnetic anisotropy, leading to high coercivities. 1D magnetic particles can be used for the elaboration of polymer-based composites or for high-density magnetic recording devices provided it is possible to control their organization on a substrate. The possibility of obtaining stable dispersions in a liquid due to their very small diameter allows to consider the development of new kinds of ferrofluids.[3,4] Last, but not least, 1D nanostructures correspond also to model materials which can be used for theoretical studies related to magnetism or electron transport, for example.

1D particles encompass rods, wires and tubes as well as ribbons and belts, and are characterized by their aspect ratio, which is the ratio between the mean length and the mean diameter. Even if the distinction between wires and rods is not always clearly established, the former have higher aspect ratios than the latter. In this review, we will focus mainly on 1D anisotropic metal nanostructures based on iron, cobalt and nickel. Table 1 shows that iron and nickel crystallize in the isotropic system while cobalt possess several allotropes with the hexagonal-close packed, the face-centered cubic and the ε[5,6] (isomorphous to β-Mn) varieties. For magnetic applications, iron, which is also the cheapest ferromagnetic metal, is a good candidate since it displays a very high saturation magnetization. On the other hand, hcp cobalt exhibits the highest magnetocrystalline anisotropy constant. For anisotropic Co particles with their long axis along the crystallographic c axis, it is thus possible to obtain a very high magnetic anisotropy, combining magnetocrystalline anisotropy with shape anisotropy.

Several excellent reviews have been published on the elaboration of 1D nanostructures.[8,9] Xia *et al.* have inventoried six strategies:[9] i) anisotropic growth governed by the crystallographic system; ii) use of a solid template with a uniaxial pore system; iii) confinement of growth by a liquid droplet similar to the Vapor-Liquid-Solid (VLS) process; iv) use of capping agents which are preferentially adsorbed on some crystallographic faces of a seed; v) self-assembly of discrete nanoparticles;[10] vi) size reduction of 1D microstructures (a top-down approach). Although the elaboration of ferromagnetic nanowires using a uniaxial porous solid template has been reviewed recently,[11] reviews dealing with the solution–based preparations of magnetic nanowires are very scarce. Section 2 is related to the preparation of magnetic nanowires and nanorods *via* bottom-up approaches either by vapor- or liquid-phase processes.

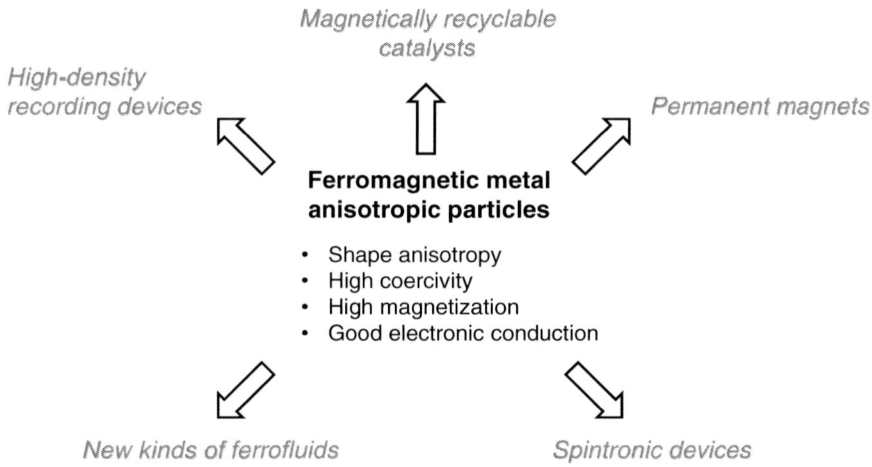

Scheme 1. Potential applications for ferromagnetic metal anisotropic particles.

Table 1. Structural and magnetic properties of Fe, Co and Ni

	Fe	**Co**	**Ni**
Crystal structure	bcc	hcp, fcc, ε	fcc
M_S[a], kA m^{-1} (emu·g^{-1})	1735 (220)	1445 (162)	509 (57)
T_C[b], K	1043	1388	627
Magnetocrystalline anisotropy constant K_1, kJ m^{-3}	48	530	– 4.5
D_c[c], nm	15	15 (hcp) 7 (fcc)	55

[a] M_S: saturation magnetization at 300 K.
[b] T_C: Curie Temperature.
[c] d_c: estimated critical diameter for a spherical particle.[7]

In Section 3, different strategies for the functionalization and the surface modification of 1D particles, directly during the synthesis, or after the preparation of the anisotropic particles, are presented. These surface modification and functionalization are efficient tools to: i) endowing the final material with a new physicochemical property, ii) protecting metallic particles from core oxidation and, in some cases, iii) driving the anisotropic growth when appropriate ligands or additives are added during the synthesis. Finally, we will conclude this review giving some general remarks on the way future research could be directed.

2. Preparation of Anisotropic Ferromagnetic Particles

In this section, we describe different synthetic strategies used for the preparation of anisotropic metal nanoparticles. We only focus on bottom-up approaches since they generally allow for a better control of the mean size and size distribution of the resulting materials, compared to top-down routes. These bottom-up methods can be classified in two categories: vapor-phase and liquid-phase processes. The objective, here, is not to give an exhaustive view of the preparation methods but to show the great diversity of methods which can be used for the preparation of 1D ferromagnetic metal NPs. Several excellent reviews, published a decade ago, are related to the preparation of anisotropic inorganic particles to which the interested reader can refer.[8,9]

We first begin by describing briefly vapor-phase processes before to focus on liquid-phase methods. These latter generally allow to obtain a higher diversity of morphologies and can be easily up-scaled, allowing to obtain larger amounts of nanoparticles than those obtained with the vapor-phase methods.

2.1. Vapor Phase Processes

As regards vapor phase methods, two main types have been used for the elaboration of 1D nanostructures: i) vapor-liquid-solid (VLS) processes and ii) vapor-solid (VS) strategies which often consist in the template-directed synthesis under ultra-high vacuum using nanostructured surfaces. In the VLS process, the anisotropic particle is grown from a molten (or semi-molten) metal nucleus which has been first deposited on a support. Gold has been extensively used as the seed, but numerous other metals have been found to catalyze the nanowire growth, such as Fe, Co, Ni or Ti.[12] The growth is achieved through the incorporation of vapor species of the material to be grown into the droplet leading to supersaturation, nucleation at the liquid-solid interface and finally growth perpendicular to the support. Since the growth is limited to the liquid-solid interface, the droplet diameter dictates that of the 1D particle. The control of the catalyst size is thus a way to control the diameter of the grown nanowires.

Although the VLS method has been widely used for the production of, *e.g.* semi-conducting metal oxides or III-V materials,[9,13] to the best of our knowledge, pure metallic Fe, Co or Ni nanowires were not prepared following this approach. Indeed, Xia *et al.* indicated that it seems impossible to employ the VLS method for the production of metallic 1D particles.[9] Magnetic cobalt-containing anisotropic particles have nevertheless been obtained. For instance, Shihui *et al.* reported on the preparation of $Co_xSn_{1-x}O_2$ (x = 0.013-0.22) nanowires prepared using the VLS method.[14] These diluted semiconductor nanowires were ferromagnetic at room temperature, and the magnetic moment decreased sharply with the Co content. This was explained by superexchange antiferromagnetic coupling *via* oxo ions.

A vapor-solid-solid mechanism has also been proposed to account for the formation of single-crystal ferromagnetic Ni_3Co nanowires vertically aligned on a sapphire substrate (see Figure 1a).[15] The syntheses were performed in a hot wall dual-zone furnace and the growth of the wires was catalyzed by Ni_3Co_2 particles at their tip, as evidenced by HRTEM. The density of the wires can be controlled by the temperature difference between the two heating zones in the furnace and by the distance between the sapphire substrate and the metal precursors. Morevover, by adjusting the ratio of the Ni(II) and Co(II) chloride precursors, Ni and NiCo anisotropic particles, exhibiting a diameter of about 100 nm, were also prepared.[15]

A very useful VS strategy for the preparation of metallic ferromagnetic nanowires is to use well-defined metal surfaces which can be considered as two-dimensional templates. The use of metal and oxide surfaces for the production of metal, oxide, semi-conductor and organic nanostructures has been recently reviewed.[16] The main difference with VLS methods (*vide supra*) is that, in this case, the growth is achieved parallel to the surface. Two types of surfaces have been used for the deposition of ferromagnetic particles: reconstructed and vicinal surfaces.[16] Due to a lower coordination number for atoms at the surface than for those in the bulk, surface atoms can rearrange in order to minimize surface energy. Reconstructed or relaxed surfaces are thus obtained, depending if the symmetry of the top layer has been modified or not. Vicinal surfaces are generated by cutting a crystal at an angle slightly off from a low-index lattice plane (the so-called miscut angle). Gold has been extensively studied because of its high stability. Depending on the value of the miscut angle, several vicinal surfaces can be obtained (see ref.[17] for Au as a representative example) which results in different terrace widths. For instance, with Au(111), a miscut angle of 3.8 ± 0.2° leads to a stepped reconstructed surface with a terrace width of 3.5 nm which can serve for the templated growth of Co nanowires (see Figure 1b).[18] In Table 2 are listed several vicinal surfaces used for the preparation of Fe, Co and Ni nanowires.

Table 2. Some vicinal surfaces used for the preparation of Fe, Co and Ni 1D nanoparticles

Metal	Surface	Miscut angle, °	Ref.
Fe	Cu(111)	1.2	[20]
	Au(111)	5.8	[21]
	Au(111)	3.5	[22]
Co	Au(111)	3.8	[18]
	Cu(111)	8.5	[23]
	Au(111)	3.5	[22]
	Pt(111)	6.5	[24]
Ni	Au(111)	3.5	[22]

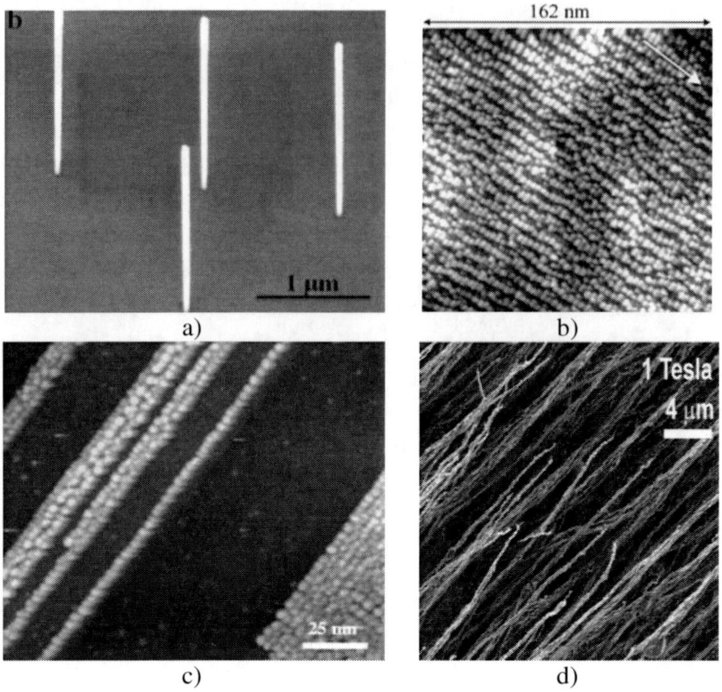

Figure 1. Monodimensional ferromagnetic metal particles prepared using vapor-phase processes: a) Ni_3Co obtained *via* a vapor-solid-solid mechanism with a sapphire substrate,[15] b) Co nanowires on a Au(111) vicinal surface,[18] c) Co nanoclusters aligned on a Al_2O_3/NiAl(100) template[19] and d) Co necklace-type nanowires prepared without a solid template by reducing flame synthesis under the application of an external magnetic field (B = 1 T).[25] Reproduced with the authorization of the American Chemical Society,[15] Elsevier[18] and IOP Publishing Ltd.[19,25]

Apart from the use of vicinal surfaces, it is also possible to take benefit from the patterning of a template surface under peculiar conditions. This was done with a Al_2O_3 thin film which was grown on a NiAl(100) substrate.[19] Thermal treatments performed at 1000 K under exposure of 1000 L O_2 (1 L corresponds to an exposure of $133 \cdot 10^{-6}$ Pa for 1 s) yielded complex crystalline Al_2O_3 protrusion networks on the NiAl substrate which patterned the growth of very regular Co nanoclusters.

As regards vapor-phase processes and magnetic nanostructures, the use of a solid template is not always a prerequisite to obtain 1D nanostructures. Athanassiou *et al.* have reported on the formation of necklace-type nanowires prepared by reducing flame synthesis under the application of an external magnetic field.[25] In the absence of the magnetic field, only a random growth of 20-50 nm-sized Co nanoclusters could be obtained. Depending on the magnetic field strength applied, either loose nanofibers (with B = 0.05 T) or thick self-aligned bundles of nanowires (with B = 1 T) can be generated (see Figure 1d). Beside the fact that is is a template-free method, another advantage is that the process can be easily scaled up to the multigram or even kilogram level.

2.2. Solution-Based Methods

Although vapor-phase methods are generally highly reproducible and allow obtaining a very good control of the size and shape, liquid-phase syntheses appear to be more versatile in terms of control of the structure, the composition and the morphology of the desired product. Moreover, with solution-based methods, it is also possible to obtain, in some favorable cases, 2D or 3D particles assemblies (*vide infra*). Their moderate cost is also a clear advantage compared to that of vapor-phase methods since it allows for the production of large quantities of materials. In the following, synthetic methods relying on the use of a solid template, either organic, inorganic or biologic, are first presented, before to focus on template-free syntheses, which are undoubtedly more challenging.

2.2.1. Using a Template

2.2.1.a. Organic and Inorganic Templates

Uniaxial Porous Membranes

As with vapor-phase methods, templates can be used for solution-based syntheses. Numerous efforts have been directed towards the use of uniaxial porous membranes for the directed growth of 1D nanomaterials. The most used templates are commercially available anodic aluminum oxide (AAO) and radiation track-etched polycarbonate (TEPC) membranes, but other types have also been successfully utilized, such as radiation track-etched mica membranes.[26,27] Indeed, the first report on the use of porous templates for the preparation of nanometer-sized anisotropic particles was reported by Possin as early as 1970, with a track-etched mica template for the preparation of tin, zinc and indium metal nanowires.[28]

AAO displays very uniform pores ranging from tens to hundreds of nanometers which self-arrange during anodization into hexagonal symmetry, while with TEPC membranes pores as small as 10 nm can be obtained, but due to the preparation procedure, pores are randomly distributed within the TEPC material. AAO and TEPC membranes have very high pore densities, about 10^{11} and 10^{9} pores·cm^{-2}, respectively. Apart from the common planar geometry, cylindrical porous alumina was recently reported: it was obtained after the two-step anodization of a 100 μm diameter Al wire.[29]

About a decade ago, the different synthetic strategies for the preparation of metallic and non-metallic nanomaterials inside the porosity of these solid templates were reviewed.[30] These methods are: i) electrodeposition, which is considered to be the most employed strategy, ii) chemical deposition which requires a reducing agent, iii) chemical polymerization for the preparation of organic polymers such as polypyrrole or polyaniline, iv) sol-gel deposition which is a way to prepare anisotropic metal oxides and v) chemical vapor deposition requiring the use of volatile precursors.

The elaboration of 1D metal magnetic particles is very often accomplished *via* electrodeposition.[26,31,32] To the best of the authors' knowledge, the first report for the preparation of ferromagnetic metal nanowires was published in 1975 with AAO as the template.[33] Pure Fe, Co and Ni wires were obtained, as well as Fe-Co, Fe-Ni and Co-Ni alloys.[33] Since then, an impressive number of studies have been devoted to the preparation of Fe, Co and Ni (and their alloys) nanowires and nanotubes following this approach (see Table 3 and references [34-40] for Fe, [35,39-44] for Co and [39-45] for Ni as representative examples). This has been motivated, *inter alia*, by the fact that such parallel assemblies of ferromagnetic NWs can find apppplications for high-density recording technologies, *i.e.* as a perpendicular magnetic recording medium. The maximum coercivity measured at room temperature for Fe, Co and Ni nanowires parallel assemblies in host matrixes were 3.0, 2.6 and 0.95 kOe, respectively.[31] The observed coercivity is the sum of two contributions: i) shape anisotropy and ii) magnetocrystalline anisotropy. While for Fe and Ni particles, the shape anisotropy contribution overcomes the magnetocrystalline one and tends to align the magnetization along the NW axis, the situation is a little more complex for Co. Indeed, for this metal, the two contributions can be of the same order of magnitude since hcp Co exhibits the highest magnetocystalline anisotropy constant of the three ferromagnetic metals (see Table 1). Depending on the experimental conditions such as the electrodeposition parameters (DC or AC, voltage, current density, pH of the electrolytic bath, etc.) and the diameter of the template pores, it was found that not only a different crystallographic phase of Co (cubic, hexagonal or a mixture of both allotropic varieties) can be obtained but also different textures were observed.[35,42,46] For instance, with hcp Co, the c axis (which corresponds to the easy magnetization axis) was found to range from an orientation in the substrate plane (perpendicular to the wire axis) to an orientation perpendicular to the substrate plane, *i.e.* parallel to the wire axis. Note, however, that for hcp Co, the easy magnetization axis is very often found to lie perpendicular to the wire axis.[32] In such a case, the shape anisotropy and magnetocrystalline anisotropy contributions are competing, which results in reduced coercivity and squareness of the hysteresis cycle.[47] It has been shown that it is possible to control, to some extent, the orientation of the easy magnetization axis relative to the NW axis by using an external magnetic field during electrodeposition: with μ_0H_{ext} = 5 T, the c axis moved from 90 to 70°.[48]

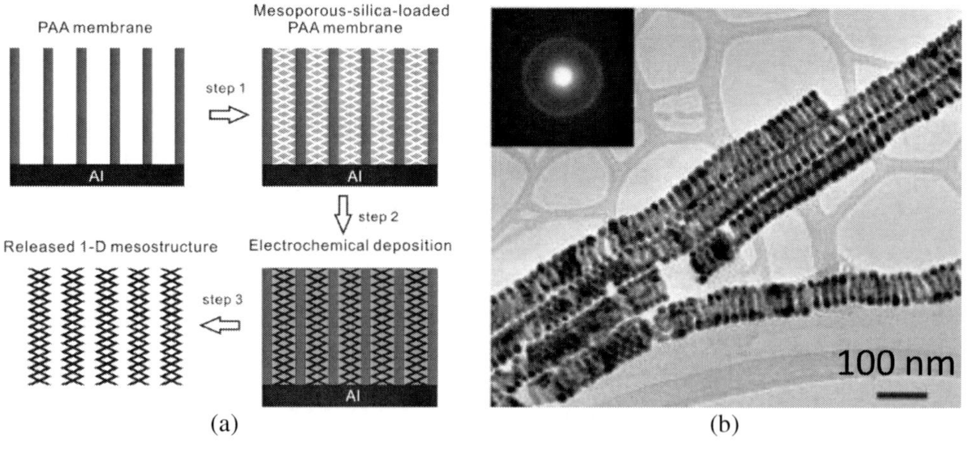

Figure 2. Schematic view of the preparation of mesostructured NWs in a mesoporous-silica-loaded porous anodized membrane (PAA) by electrodeposition (a) and TEM image of Ni NWs prepared using this hierarchical template (b). Reproduced with permission from ref. 64. Copyright 2004 American Chemical Society.

Mesostructured Silicas

Hexagonal or cubic mesoporous silicas, resulting from the cooperative assembly of inorganic species and surfactants, are another class of solid template which can be used for the preparation of anisotropic metal ferromagnetic particles by electrodeposition.[49-51] Depending on the mesoporous silica used, either aligned nanowires[50,51] or 3D ferromagnetic networks[51] were obtained.

Apart from electrodeposition, the incorporation of well-suited precursors in the porosity of the solid and their subsequent treatment to yield the metal particles is another strategy. Eliseev *et al.* have reported on the incorporation of pentacarbonyliron(0) in the hydrophobic pores of the hexagonal silica matrix, followed by UV-assisted decomposition of the molecular complexes in vacuum and, finally, annealing in a H_2 flow at a temperature ranging from 260 to 400°C.[52-56] The Fe/SiO_2 composite material annealed at 375°C, prepared with a silica having a mean pore diameter of 2.9 nm, exhibited a coercivity of 222 Oe at room temperature.[53] The formation of ferromagnetic nanowires can also been achieved *via* electroless deposition as showed by Zhang *et al.*[57,58] The multi-step synthesis consisted in i) grafting trimethylsilyl groups on the external surface of a SBA-15[59] silica to obtain a hydrophobic external surface, ii) incorporation of Pd(II) ions and their subsequent reduction to yield Pd nanoparticles inside the porosity which allowed iii) the catalyzed electroless deposition of Ni nanowires inside the pores. Increasing the electroless deposition time was found to increase the density of Ni nanowires inside the composite material although a complete filling of the porosity was not reported. An improvement of this method was reported by Li *et al.*[60] In their work, the mesostructured silica was prepared with a mixture of (3-aminopropyl)triethoxysilane and tetraethoxysilane (TEOS) instead of TEOS alone. The amine functions located inside the pores led to a higher affinity of the the silica matrix towards the Pd(II) ions and, consequently, to a very high filling of the porosity with nickel: after removing of the silica template, 3D ordered Ni nanowires were obtained. Very high filling levels have also been reported with the inclusion of Si nanowires into hexagonal mesoporous silica using a supercritical fluid inclusion technique.[61] Indeed, ^{29}Si MAS NMR indicated that about 80% of the mesopores were filled with the Si nanowires. This method was also applied to the formation of Co nanowires but no information was given on the filling in this case.[62] An interesting result is that the reaction conditions (applied pressure and temperature) influenced directly the nature of the obtained crystallographic phases. It is noteworthy that Co has 3 different allotropes: the cfc α-Co phase, the hcp β-Co phase and ε-Co,[5,6] which is isomorphous to β-Mn. Using supercritical fluid inclusion techniques, pure cfc β-Co as well as mixtures of α-, β- and ε-Co were obtained as a result of the applied pressure and temperature.[62]

It is generally argued that the interest in using mesoporous silica templates, instead of AAO, is that lower diameters are attainable for the former, since AAO often contain pore diameters greater than 30 nm.[50] It is however possible to combine mesoporous silica with AAO templates to obtain very interesting hierarchical morphologies. Stucky *et al.* reported the preparation of composite mesostructures resulting from the development of mesostructured silica inside the porosity of an AAO membrane (see Figure 2a).[63] The composite material was prepared by sol-gel dip-coating an AAO membrane with an acidic alcohol-water solution containing the silica precursor, TEOS, and the P123 non ionic triblock copolymer surfactant. The removing of the surfactant by thermal treatment yielded a template with a porous hierarchical porous structure. Subsequent electrodeposition allowed to obtain monocrystalline Ni nanowires with an unprecedented mesoscale texture (see Figure 2b).[64] It would have been interesting to evaluate the magnetic properties of these Ni NWs but no magnetic characterization was reported in this study.

Carbon Nanotubes

Since the discovery of carbon nanotubes (CNT) by Iijima in 1991,[65] extensive efforts have been directed towards the incorporation of ferromagnetic metals into the internal space of these nanotubes. One obvious advantage of this method is that the metallic core is protected from oxidation by the carbon coating. A first strategy consists in using a solid template to generate M@C (M = Co,[66] Ni[67,68]) hybrid anisotropic nanoparticles (Scheme 2). The carbon layer can be generated inside the cylindrical pores of the matrix, generally AAO, either before[66,67] or after[68] the metal nanowire growth. Apart from graphitic carbon nanotubes, tubes made of C60 and C70 fullerenes have also been used for the growth of fcc Ni nanowires.[67]

The thermolysis of aerosols of a metal precursor, generally a metallocene, with a possible additional carbon source such as acetylene,[69] benzene,[70] toluene,[71] fullerene[72] or trichlorobenzene[73,74] is a second route, template-free, for the preparation of carbon-encapsulated metal nanowires. A two-heating zone furnace is generally used with a cooler zone for the sublimation of the metal precursor and a hotter zone for the pyrolysis of the gaseous material. It has been shown that experimental conditions including the sublimation temperature, the pyrolysis temperature, the gaseous flow rate as well as the working pressure had to be finely controlled in order to obtain the desired phase with optimum magnetic properties.[69,71,63] However the metal deposition yield inside CNTs is generally low, in the order of 60%.[69,72,63]

The filling of CNT by capillarity is another strategy to obtain Metal@CNT.[75,76, 66] It was shown that only substances having surface tensions lower than 200 $mN \cdot m^{-1}$ will fill spontaneously CNTs by capillarity.[77] Liquid metallic iron, cobalt and nickel have high surface tensions, respectively 1650, 1779 and 1810 $mN \cdot m^{-1}$,[78] which *a priori* precludes their incorporation inside the hollow space of the CNTs by capillarity but it is possible to use aqueous solutions of the metal precursor to fill the internal space then to perform a thermal treatment to obtain the desired metallic phase. Iron nanowires encapsulated into double wall CNTs have been prepared following this approach with an aqueous solution of $FeCl_3$.[76] Remarkably, only bcc α-Fe was evidenced by electron diffraction while very often, mixtures consisting in variable amounts of α-Fe, γ-Fe, and cementite (Fe_3C) are obtained, whatever the synthesis procedure adopted.[71,79]

Mixtures of hcp, fcc and disordered polytypic structure have also been observed for Co@CNT hybrid particles prepared using a 2 wt. % Co Y zeolite catalyst, but no cobalt carbide phase was detected.[80] Even if polycrystalline materials are very often obtained, the preparation of monocrystalline $Fe_{65}Ni_{35}$[70] and FeCo[81] nanowires embedded in CNT has been achieved successfully *via* the thermolysis of metallocene precursors in an inert atmosphere.

Table 3. Some representative examples of Fe, Co and Ni anisotropic particles grown into host templates and their magnetic properties

Metal	Substrate[a]	Particle type[b]	Mean diameter, nm	Mean length, nm	Electro-deposition pH	$H_C(\perp)$,[c,d] kOe	$H_C(//)$,[c,d] kOe	$(M_R/M_S)_\perp$[c]	$(M_R/M_S)_{//}$[c]	Ref.
Fe	AAO	NW	20-180	25-965	~ 3.0	0.23-0.65	0.75-2.10	0.08-0.16	0.09-0.98	[34]
	AAO	NW	15	4 800	2.7-3.0	0.40	2.21	0.006	0.98	[36]
	AAO	NT	-[e]	40 000	-[e]	0.15	0.15	0.22	0.05	[40]
	TEPC	NT	~ 600	6 000	-[e]	0.16	0.36	0.35	0.09	[40,82]
	TEPC	NW	50	2 000	-[e]	0.23	0.15	0.07	0.11	[83]
	TEPC	NR	100	-[f]	-[e]	0.29	0.47	-[e]	-[e]	[84]
Co	AAO	NW	~ 10	-[g]	-[e]	-[e]	2.6	-[e]	> 0.9	[43]
	AAO	NW	~ 40	~ 40 000	2.0-4.0	0.48-1.08	0.09-0.15	-[e]	-[e]	[42]
	AAO	NT	-[e]	40 000	-[e]	0.075	0.12	0.05	0.03	[40]
	TEPC	NT	~ 600	6 000	-[e]	0.20	0.16	0.37	0.11	[40,82]
	CMS[h]	3D network	5-6	-[e]	-[e]	0.23[i]	0.23[i]	-[e]	-[e]	[50,51]
	HMS[j]	NW	8	-[e]	-[e]	0.89[i]	0.39[i]	~ 0.1	~ 0.4	[50]
	TEPC	NR	100	-[f]	-[e]	0.32	0.48	-[e]	-[e]	[84]
Ni	TEPC	NW	50	2 000	-[e]	0.24	0.35	0.20	0.44	[83]
	TEPC	NR	100	-[f]	-[e]	0.36	0.69	-[e]	-[e]	[84]
	TEPC	NT	~ 600	6 000	-[e]	0.13	0.08	0.31	0.06	[40,82]

Table 3. (Continued)

Metal	Substrate[a]	Particle type[b]	Mean diameter, nm	Mean length, nm	Electro-deposition pH	$H_C(\perp)$,[c,d] kOe	$H_C(//)$,[c,d] kOe	$(M_R/M_S)_\perp$[c]	$(M_R/M_S)_{//}$[c]	Ref.
	AAO	NT		40 000	-[e]	0.10	0.11	0.24	0.16	[40]
	AAO	NW	280	4 500	4.0	0.055	0.53	~ 0	~ 0.6	[39]
	TEM	NW	120	5 000	-[e]	~ 0.09	~ 0.24	~ 0	0.89	[85]

[a] AAO: anodic aluminum oxide; TEPC: track-etched polycarbonate; CMS: cubic mesoporous silica; HMS: hexagonal mesoporous silica; TEM: track-etched mica.

[b] NW: nanowire; NT: nanotube; NR: nanorod.

[c] unless otherwise stated, // and ⊥ stand for parallel and perpendicular to the wire axis.

[d] values given at room temperature.

[e] not indicated.

[f] Length in the range 4000 – 9000 nm.

[g] the mean aspect ratio was higher than 20.

[h] the cubic mesoporous silica was prepared with the non-ionic Brij-58 surfactant and deposited on ITO.

[i] // and ⊥ correspond to orientations parallel and perpendicular relative to the substrate.

[j] the hexagonal mesoporous silica was prepared with the pluronic P123 triblock copolymer template. The pores channels lie parallel to the planar ITO substrate.

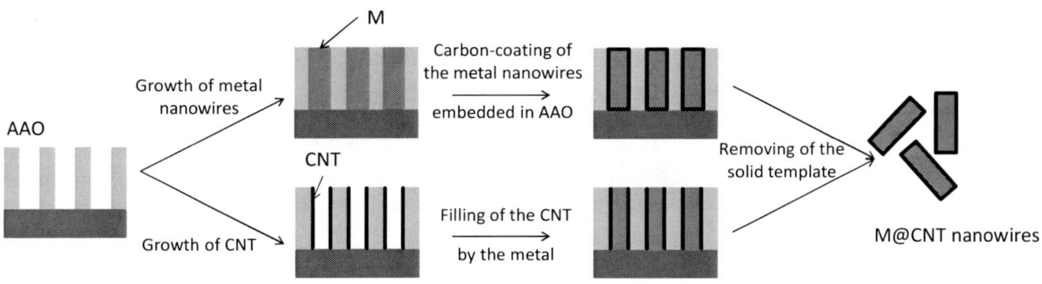

Scheme 2. Preparation strategies for core-shell metal (M) nanoparticles coated by carbon nanotubes (CNT) using an aluminum anodic oxide membrane (AAO).

2.2.1.b. Biological Templates

Several biological systems have been used as templates for the preparation of anisotropic ferromagnetic metal particles: microtubules,[86] DNA[87,88] and viruses.[89-93]

Microtubules, which are tubular protein structures of the cytoskeleton of eukaryotic cells with a 25 nm outer diameter and a length of several micrometers, have been advantageously utilized for the electroless deposition of Co and Ni.[86] In this process, the microtubules had to be first activated by the adsorption of Pd or Pt nanoparticles onto their surface. Otherwise, no metal plating is observed for the nonactivated microtubules. The outer diameter of the hybrid 1D particles ranged between 50 and 60 nm indicating an about 15 nm metal layer. The magnetic properties of the metallized nanotubes were not evaluated but the particles responded readily to an external magnetic field.

Following a similar synthetic strategy, Knez *et al.* have reported the use of the Tobacco Mosaic Virus (TMV) to obtain anisotropic hybrid biological-inorganic nanostructures (see Figure 3).[89,90] TMV can be seen as a nanocylinder with a length of 300 nm, a central channel with a diameter of 4 nm and an external diameter of 18 nm. Depending on the experimental conditions, it was possible to spatially direct the growth of anisotropic Co or Ni wires inside the central channel or to deposit nearly spherical Ni or Co clusters (> 50 nm in diameter) on the outer surface of the TMV by electroless deposition.[90] In the last case, necklace-type 1D nanostructures were generated. As in the work of Mertig *et al.*,[86] the Ni and Co deposition could be realized only if TMV has been first activated with Pd(II) or Pt(II) tetrachloro complexes. The metallic nature and the crystallographic phase of the inorganic part of the hybrid material were not detailed.

While fcc CoPt and FePt are soft magnetic phases, $L1_0$-ordered CoPt and FePt alloys exhibit a large magnetocrystalline anisotropy and are thus potential candidates for high density recording media.[1] Using TMV, bimetallic anisotropic ferromagnetic particles can also be obtained. CoPt and $FePt_3$ nanoalloys were indeed prepared using a multistep ultrasonically-assisted synthesis with $NaBH_4$ as the reducing agent.[92] The authors underlined that the single reduction of Co(II) species cannot be realized, while with Pt(II) alone, only irregular particles were formed outside the central channel of the TMV. This suggests that the presence of the two cations is thus necessary to observe the formation of the wire.

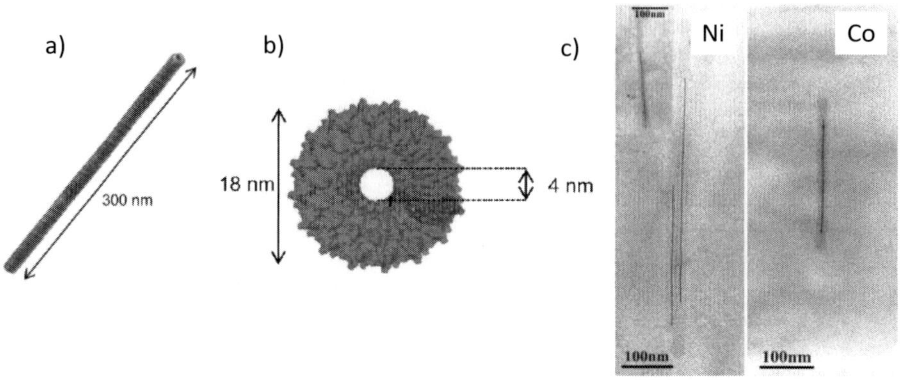

Figure 3. a) Schematic illustration of the tobacco mosaic virus (TMV),[93] b) top view showing the central channel of TMV[93] and c) Ni and Co nanowires grown inside the central channel of TMV by electroless deposition after Pd(II) activation (see text for details).[89] Reproduced with the authorization of the American Chemical Society.

Moreover, some viruses can be genetically modified in order to obtain a control of the particle phase and composition. This was the case with the M13 bacteriophage: by modifying the nature of the peptides in the linear self-assembled filamentous construct of the virus, either single-crystal anisotropic semiconducting ZnS and CdS or the $L1_0$ phase of magnetic CoPt and FePt nanowires were obtained.[91] This shows the versatility of this method since it is possible to direct the synthesis towards the formation of a desired material by controlling the genetic material of the biological template.

DNA is another biological template which can be used to initiate 1D growth. This approach was followed by Gu *et al.*, who described the electroless elaboration of cobalt-metallized 1D DNA nanostructures (see Figure 4).[87] Again, without preliminary activation of DNA with Pd(II) species and their subsequent reduction to Pd(0) nanoparticles (using dimethylamine borane), no deposition of metallic Co nanoclusters could be observed. It is very similar to what was observed with the TMV.[89,90] No evidence for the metallic nature of cobalt or its crystallographic structure was given. However, the composite material appeared to be ferromagnetic with a saturation magnetization of about 160 $emu{\cdot}g^{-1}$, very close to that of bulk Co (see Table 1) and a low coercivity of 214 Oe, as compared to materials prepared with uniaxial porous membranes.[31]

2.2.2. Template-Free Methods

Even if considerable progress have been made these last years, the formation of 1D particles with a control of particle size and distribution without the use of a template still correspond to a real challenge.

Essentially, two methods have been devised for the preparation of ferromagnetic nanorods and nanowires: the organometallic pathway and the polyol process. They are briefly presented in the following sections.

2.2.2.a. The Organometallic Pathway

Organometallic compounds can be advantageously used for the production of anisotropic metal particles. Some representative examples are given in Table 4. Contrary to Co with an anisotropic hexagonal structure, it is much more challenging to obtain, without the use of a solid template, anisotropic Fe and Ni particles which crystallize in the cubic system (see Table 1). Nevertheless, Park *et al.* have reported the formation of uniform 2 nm × 11 nm Fe nanorods (see Figure 5).[94] Firstly, 2 nm-sized iron NPs are synthesized by reduction of [$Fe(CO)_5$] in the presence of a mixture of trioctylphosphine (TOP) and trioctylphosphine oxide (TOPO) at 340°C in an Ar atmosphere. These particles are then dispersed in a solution of didodecylmethylammonium bromide (DDAB) in pyridine and the resulting mixture is heated at reflux for 12 h yielding Fe nanorods, probably *via* the coalescence of the isotropic NPs. To the best of the authors' knowledge it is the only example of anisotropic iron nanoparticles obtained *via* a solid template-free route. The anisotropic growth was tentatively explained by considering the irreversible adsorption of DDAB on the central region of the growing nanoparticle.[94]

Chaudret *et al.* have developped an organometallic route for the preparation of (anisotropic) magnetic metallic NPs in a H_2 atmosphere. Olefinic complexes are preferably used instead of carbonyl compounds since CO ligands can readily adsorb at the particle surface and lead to a decrease in the magnetic properties.[95] At the applied hydrogen pressure, olefinic ligands give coordinatively inert alkanes yielding particles with non-contaminated surfaces. Ni nanoparticles were prepared following this approach with the [$Ni(COD)_2$] complex (COD: cycloocta-1,5-diene) using either hexadecylamine (HDA) or TOPO in tetrahydrofuran at 70°C in 3 bar H_2 pressure for 12 h. With 10 equiv. HDA relative to Ni, 4 nm × 15 nm nanorods were mostly obtained while with TOPO, a mixture of small monodisperse spherical particles and tear-drop-shaped nanorods was evidenced.[96]

Table 4. Selected examples of Fe, Co, Ni anisotropic nanoparticles prepared by the organometallic pathway

Metal	Precursor	morphology	dimensions	Structure[a]	H_C, kOe (T, K)	M_S, emu·g^{-1}	Ref.
Fe	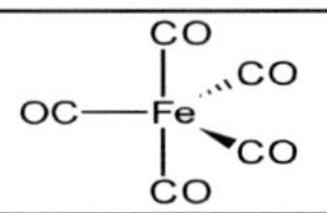	NR	2 nm × 11 nm	bcc	-[b]	-[b]	94
Co	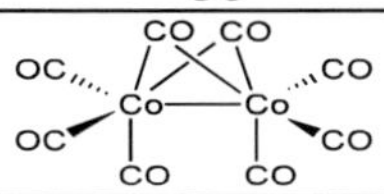	ND	4 nm × 35 nm	hcp	0 (200)	-[b]	97
	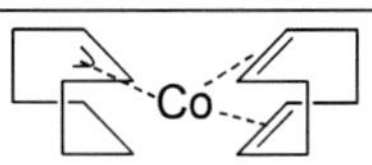	NR	9 nm × 40 nm	hcp	8.9 (300)	160	98
	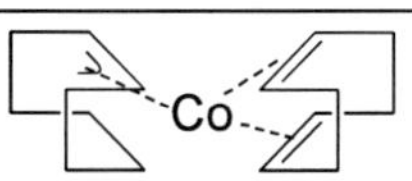	ND	12.5 nm × 21.5 nm	hcp	1.0 (300)	156	99
Ni	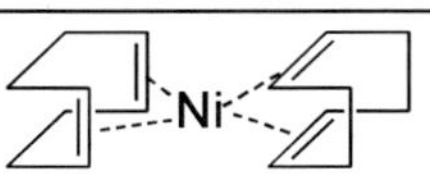	NR	4 nm × 15 nm	fcc	~ 0.5 (2)	55	96

[a] bcc: body-centered cubic, hcp : hexagonal close-packed, fcc : face-centered cubic.
[b] not indicated.

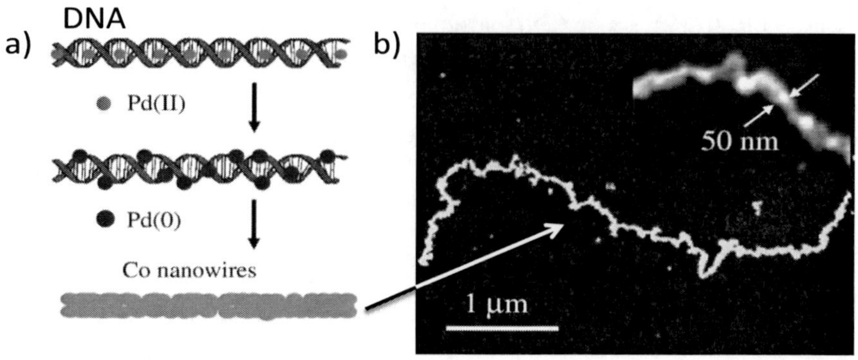

Figure 4. a) Schematic illustration of Pd nanoparticle-controlled Co nanowire growth and b) SEM image of a single continuous Co nanowire. Inset: magnified view. Adapted with permission from ref. 88. Copyright 2008 Elsevier.

At 3 bar H_2 the reaction of $[Co(\eta^3\text{-}C_8H_{13})(\eta^4\text{-}C_8H_{12})]$ in anisole, $C_6H_5OCH_3$, with an equimolar mixture of octadecylamine and oleic acid led after 48 h to the formation of 9 nm × 40 nm NRs.[98] Depending on the number of carbon in the alkyl chain of the amine, anisotropic particles with mean aspect ratios ranging from 1.7 to 22 were obtained. Remarkably, these Co NRs are self-organized in solution into 2D or 3D superlattices.[100] Using the $[Co\{N(SiMe_3)_2\}_2]$ complex, similar superlattices were generated which presented very high coercivities, $\mu_0H = 1.2$ T at 2 K.[101] Apart from nanorods, other anisotropic shapes are also possible: Co nanodisks (NDs) were synthesized with the $[Co(\eta^3\text{-}C_8H_{13})(\eta^4\text{-}C_8H_{12})]$ complex using HDA and Rhodamine B.[99]

Alivisatos *et al.* also proposed an organometallic route for the preparation of hcp Co NDs.[97] In this work, the dicobaltoctacarbonyl precursor is decomposed in dichlorobenzene (DCB) with oleic acid (OA) and octadecylamine (ODA). The organometallic compound is injected in a refluxing solution of OA and ODA in DCB in an Ar atmosphere yielding 4 nm × 30 nm Co NDs. However, the reaction should be under kinetic control since after tens of minutes, ε-Co nanospheres are eventually obtained. To explain the formation of such particles, it is tentatively proposed that the long-chain amines are selectively adsorbed on the (0001) facets thus inhibiting its growth. These Co NDs self-arranged into long 1D nano-ribbons, probably *via* dipolar interactions.

Recently, based on different results obtained with systems including $[Co_2(CO)_8]$ in 1,2-dichlorobenzene and several additives (either cetyltrimethylammonium bromide, benzylpyridine, or mixtures of OA and oleylamine), Dreyer *et al.* suggested that the disk habitus observed in the work of Alivisatos *et al.*[103] could actually be due to the application of a weak external magnetic field during the synthesis since, without the application of an external magnetic field, only cube-like ε-Co particles were obtained in their hands.[104]

Anisotropic magnetic nanoalloys can also be prepared following the organometallic pathway. As an example, FePt NRs and NWs were synthesized using pentacarbonyliron(0) and bis(acetylacetonate)platinum(II) in a mixture of oleylamine and octadecene at 160°C (see Figure 5).[102] Depending on the oleylamine/octadecene volume ratio, it was possible to control the length of the anisotropic particles: with pure oleylamine, more than 200 nm-sized nanowires were observed, a volume ratio of 3/1 gave nanowires with a length of 100 nm while FePt NRs with a mean length of 20 nm were isolated with a 1/1 ratio.

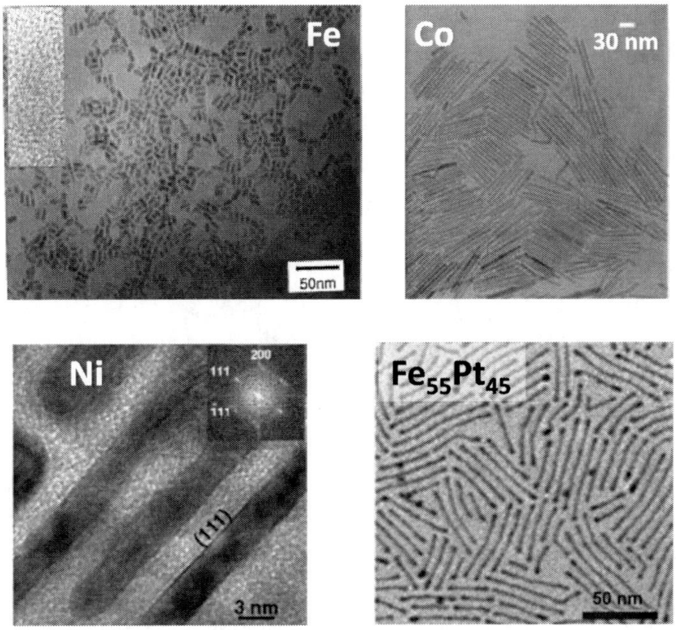

Figure 5. Preparation of anisotropic ferromagnetic metal particles by the organometallic route: Fe,[94] Co,[100] Ni[96] and $Fe_{55}Pt_{45}$.[102] Reproduced with the authorization of the American Chemical Society[94,96] and Wiley-VCH.[100,102].

These selected examples show that the thermolysis of organometallic complexes is thus an efficient way to obtain anisotropic particles with a good control of the shape and of the mean sizes. Carbonyl complexes have been widely used because the metal is zerovalent (the decomposition of the complex leads directly to Fe(0) monomers) and because of their commercial availability. However, as already indicated, carbonyl ligands can lead to a decrease of the magnetic properties *via* adsorption at the particle surface. Other complexes can be utilized, such as olefinic compounds, but in this case, one has generally to prepare these air-sensitive products. Moreover the choice of the ligands used as additives in the synthesis is of paramount importance.[97,104] Until recently, as far as "hot injection" techniques[105] are concerned, the reaction time corresponding to the addition of the metal precursor in the reaction mixture was also considered to be an important parameter since the sudden injection at high temperature was believed to initiate the nucleation step. However Timonen *et al.* have shown that "heat-up" procedures can be substituted to "hot injection" syntheses if the heating up of the solution is accurately controlled.[106] These authors have indeed shown that the number of nuclei in the "hot injection" method depends more on temperature kinetics after the injection than on the injection itself. Moreover, unexpected factors such as the rate of addition of the metal precursor to a ligand stock solution can have a marked influence on the final shape: 4.6 nm-sized spherical Co particles are obtained after a slow addition ($0.05\ mol{\cdot}min^{-1}$) of the $[Co\{N(SiMe_3)_2\}_2(thf)]$ complex in a lauric acid-hexadecylamine mixture while sea-urchin like particles are observed for a higher addition rate ($1.40\ mol{\cdot}min^{-1}$).[107] This was related to the reaction of a part of lauric acid with $HN(SiMe_3)_2$ yielding a silyl ester. Even if this reaction is of course specific to this system, this raises the question of the becoming of the stabilizing ligands used in the reaction, especially at high temperature.[107]

2.2.2.b. The polyol process

Initially developped for the production of fine powders of easily reducible (Cu, Pt, Ag, etc.) or less-easy reducible (Fe, Co, Ni and their alloys) metals,[108] the polyol process has quickly been extended to the preparation of oxides,[109] alcoxides,[110] hydroxycarboxylates,[111] phosphates[112] and sulfides.[113] As concerns metals, the polyol acts as a reducing agent, a solvent for the metal precursor, a growth medium and, in certain cases as a chelating ligand.[114] Polyols can be roughly classified into two classes, α-diols such as ethylene glycol and ether glycols such as diethyleneglycol. Some of the most used polyols for the synthesis of nanomaterials are presented in Figure 6, together with relevant physicochemical data. Even if the polyol is a mild reducing agent, to some extent, its nature is a way to control its reducing ability. Because of its shorter hydrocarbon chain, ethyleneglycol is thought to have a higher reducing ability than diethleneglycol.[115,116]

Several studies related to the formation of magnetic particles of Co, Ni and their alloys have been reported in liquid polyols. Monodisperse spherical Co_xNi_{100-x} and $Fe_z[Co_xNi_{100-x}]$ particles with a mean diameter ranging from about 25 to 500 nm were prepared in basic 1,2-propanediol at reflux for a couple of hours.[117] A heterogeneous nucleation approach was followed with Pt(II) or Ag(I) compounds. These noble divalent metals are easily reduced in these conditions and allowed to initiate nucleation of the less easily reducible Fe(II), Co(II) and Ni(II) precursors. Apart from Pt and Ag, Ru can also been used as a nucleating agent since Ru(III) is also easily reduced in polyols.[118]

As regards the formation of ferromagnetic metals in liquid polyols, a solid intermediate phase is evidenced in these systems, its formation being favored by the basicity *via* addition of sodium hydroxide. This solid intermediate phase acts as a cation reservoir which controls the precursor(s) concentration in polyol. This ensures a good separation of the nucleation and growth steps, which is admitted to be of major importance to obtain monodisperse particles, according to the model of LaMer and Dinegar.[119] Several non-reduced phases have been reported for Ni(II) and/or Co(II) acetate species. For high hydrolysis ratio h = $[H_2O]/[M]$ and moderate reaction temperatures, hydroxyacetates lamellar compounds with the general formula $M(OH)_{1.5}(CH_3COO)_{0.5}\cdot xH_2O$ have been isolated[111] while for higher temperatures and lower h values, metal alkoxides, such as $M(OCH_2CH_2O)$ in 1,2-ethanediol (M = Co,[120] Ni[121,122]), have been obtained.

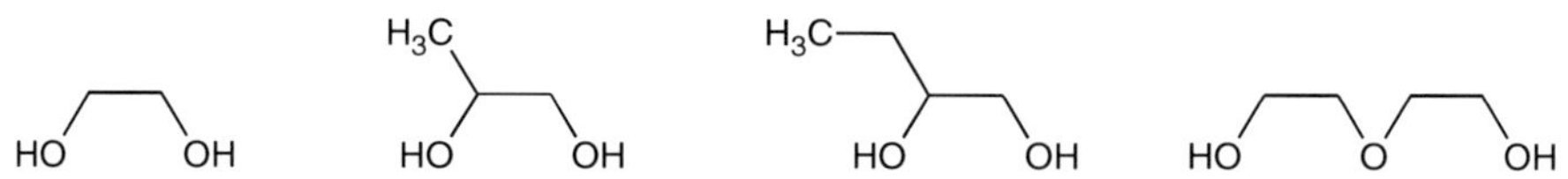

	ethyleneglycol	1,2-propanediol	1,2-butanediol	diethyleneglycol
$T_{boiling}$(°C)	198	187	192	245
ε_r	38	32	22	32

Figure 6. Some polyols used for the preparation of nanomaterials with selected physicochemical properties. ε_r: relative permittivity.

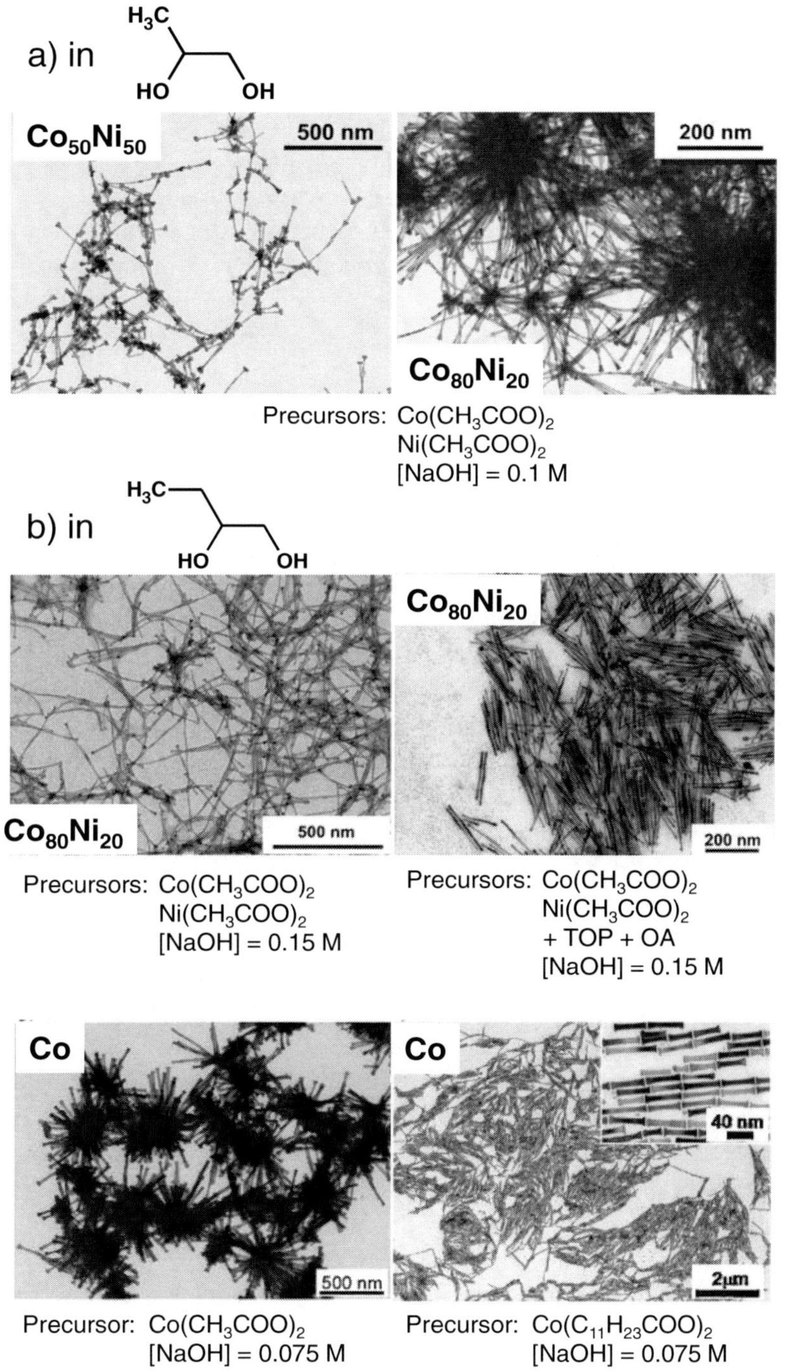

Figure 7. Co_xNi_{100-x} (x = 20, 50, 100) particles prepared in a) 1,2-propanediol and b) 1,2-butanediol *via* heterogeneous nucleation with Ru. TEM images reproduced with the authorization of the American Chemical Society,[123] the Royal Chemical Society[124] and Wiley-VCH.[125]

As regards syntheses of Co_xNi_{100-x} ($0 < x < 100$) particles performed in 1,2-propanediol, Ung *et al.* have evidenced two main parameters which have a marked influence on the shape of the resulting objects: i) the basicity of the medium and ii) the Co/Ni composition.[123] Several shapes were observed such as nanoplates, (short and long) dumbbells, diabolos and wires (see Figure 7). The introduction of Ni in the system was found to favor the formation of isolated particles and well isolated $Co_{50}Ni_{50}$ nanowires were obtained; the TEM images showed conical heads at the tips of the wires. For richer Co compositions such as $Co_{80}Ni_{20}$ or pure Co, anisotropic particles were generated but were grown from a metallic core.[123,126] The proportion of the core relative to the wires (or rods) can be important leading to sea-urchin morphologies, as was observed for pure Co when Co(II) acetate precursor is used (see Figure 7).[123]

Replacing 1,2-propanediol to 1,2-butanediol allowed to obtain well-isolated $Co_{80}Ni_{20}$ nanowires even if some multipods were also detected on the TEM images.[124] The wires crystallized in the hexagonal system with the c axis parallel to the long axis of the wire with a mean length in the range 200-300 nm (standard deviation equal to 25 % of the mean length) and a quasi monodisperse diameter of 7-10 nm (standard deviation equal to 10 % of the mean diameter). These wires are ferromagnetic at room temperature with a coercivity of 3.5 kOe. For a parallel assembly of wires in 1,2-butanediol frozen at 140 K, the coercivity is significantly increased in agreement with the theoretical model of Stoner-Wohlfarth, with a value of 6.5 kOe and a very high remanent to saturation magnetization ratio of 0.97.[124] It is noteworthy that the time-controlled addition of a mixture of one equivalent of TOP and OA led to much more rigid wires than those prepared without surfactant, with no tip at their extremities (see Figure 7, see also section 3.1).

Because of the intrinsic magnetic properties of Co, higher coercivities are expected for pure Co anisotropic particles than for Co_xNi_{100-x} alloys. However, as already indicated above, with Co(II) acetate as the metal precursor in 1,2-propanediol, only agglomerated sea-urchin-like particles were obtained which displayed only a coercivity of 2.2 kOe at 300 K. The influence of the cobalt precursor was evaluated recently. Replacing $Co(CH_3COO)_2$ by $Co(C_nH_{2n+1}COO)_2$ with $n \geq 7$ resulted in the formation of well-isolated nanorods (see Figure 7).[125] Moreover, it was found that the mean diameter and mean length could be varied in the range 10-30 nm and 100-250 nm, respectively, depending on the metal carboxylate precursor, the basicity and the temperature ramping rate. At room temperature Co nanorods exhibited a coercivity of 4.5 kOe and a saturation magnetization of 113 $emu \cdot g^{-1}$, corresponding to 70 % of the bulk value. This reduced magnetization can be explained by the presence of a thin CoO layer as well as by the adsorption of the carboxylate species at the particle surface.[127,128] At 140 K, a very high coercivity of 9.0 kOe and a remanent to saturation magnetization ratio of 0.95 were measured for a parallel assembly of Co nanorods.

Based on an experimental procedure adapted from the work of Ung *et al.*, Liu *et al.* have prepared Co nanorods in 1,2-propanediol using stearic acid *via* a hydrothermal approach.[129] The morphology of the Co particles was found to depend strongly on the basicity of the medium and the stearic acid concentration. No magnetic data was presented in this study but the Co nanowires were found active for the hydrogenation of citral (3,7-dimethyl-2,6-octadienal) to citric alcohol and hydrogenolysis of glycerol to 1,2-propanediol. In the last case, the Co nanowires were even found to give higher yields (51 %) than noble metals such as Ru, Rh and Pt (with a yield of 35 %).[129]

Hierarchical superstructures have also been reported with Co in liquid polyols. Using the "$Co(CH_3COO)_2{\cdot}4H_2O/CH_3COONa$/ethylenediamine/diethyleneglycol/ polyvinylpyrrolidone" system, pine-tree-leaf-like architectures were formed.[130] The morphology can be described as central stems from which have grown Co nanorods with a mean length of 5-10 μm and a mean diameter in the range 150-200 nm. The addition of ethylenediamine was found to be crucial for the formation of these complex architectures: without this chelating ligand, no cobalt product was obtained. The solids are ferromagnetic at room temperature with a moderate coercivity, about 192 Oe and a saturation magnetization of 137 $emu{\cdot}g^{-1}$, about 85% of the bulk value.[130]

2.3. Other Methods

Other strategies have been devised for the production of anisotropic ferromagnetic particles *via* liquid-phase processes. It is possible to take advantage of the natural anisotropy of goethite, α-FeO(OH), which crystallizes in the orthorhombic system (space group Pnma), to prepare needles of this oxyhydroxide by precipitation-based methods. A subsequent dehydration of the material will result in the formation of maghemite, γ-Fe_2O_3, which, after a final reducing step in H_2 at high temperature will give metallic bcc iron.[131] Experimental conditions have to be adjusted in order to keep the acicular shape during the different preparation steps. It has to be noted that these particles have been industrially manufactured as magnetic recording media.[131]

Sodium hypophosphite, NaH_2PO_2, can be advantageously used for the reduction of hcp Co(II) species in basic aqueous solution. Qian *et al.* have developed synthetic strategies with this reducing agent, commonly used in electroless processes, for the preparation of Co particles with several morphologies: nanowires,[132] nanobelts[133] and hierarchical dendritic structures.[134] The growth orientation for the nanowires and the nanobelts was found to be [001] in the two cases, and the origin of shape anisotropy was explained by the adsorption of citrate ions for the wires and of sodium dodecyl benzenesulfonate for the belts. Without these two additives, the morphology was dramatically modified. Although Co nanowires and nanobelts were monocrystalline in nature, low coercivities were measured, respectively 167[132] and 411[133] Oe at 300 K, significantly lower than that of Co nanorods and nanowires prepared via the organometallic pathway or the polyol process (*vide supra*).

3. Surface Modification and Functionalization

Because of their elongated shape, anisotropic magnetic metal particles generally have high coercivities which make them good candidates for several important applications such as high-density magnetic recording, biosensing or the elaboration of magnetic composites. However, because of their metallic nature, these particles are prone to oxidation, which can, in favorable cases, be limited to the surface. Surface modification appears as a tool to inhibit, to some extent, the particle oxidation. Moreover, surface functionalization is also a way to add a supplementary function to the material, yielding multifunctional materials. An increasing number of studies can indeed be found in the literature with materials displaying several functionalities such as fluorescent, magnetic and plasmonic properties.[135] In the following, methods corresponding to the modification of the surface during the synthesis are first presented before to give some selected examples related to the post-functionalization of magnetic anisotropic materials. The surface modification is generally performed through adsorption or coordination of ligands or additives, either molecular or polymeric.

3.1. During the Synthesis

This section has more to do with surface modification than surface functionalization, even if some relevant examples are given below for the latter case. The addition of well-chosen additives such as long-chain ligands or polymeric species may lead, in certain cases, to the anisotropic growth. Actually, almost all the syntheses described in section 2.2.2 have been performed in the presence of additives which act as growth-directing ligands and can be seen as surface modifiers. For instance, with hcp Co, the interfacial energy of the (0001) facet is lower than that of the $\{10\bar{1}0\}$ surface,[78] which indicates that, in the absence of a ligand, the compact (0001) is the more stable facet: hcp Co platelets should normally be obtained. Ligands such as long-chain amines and carboxylates, which preferentially adsorb on certain crystallographic planes, allow obtaining the observed anisotropic morphology. Similarly, some of the M@CNT prepared with metallocene precursors are obtained in a one-step procedure (see section 2.2.1.a).

Other synthetic procedures can involve the supplementary addition of ligands or additives, which lead to a modified surface state or even different morphologies. In section 2.2.2.b, the effect of the addition of a mixture of surfactants, namely TOP and OA, on the morphology of $Co_{80}Ni_{20}$ nanowires prepared using metal acetates *via* the polyol process was clearly evidenced on the TEM images (Figure 7): the wires are much more rigid and do not present tips at their extremities in contrast to materials performed without surfactant. This might seem anecdotal but micromagnetic simulations showed that it is highly desirable to reduce these conical heads which play a key role in the magnetization reversal mechanism.[136] The ellipsoidal shape was found to give the highest coercivity for a given mean aspect ratio and theoretical calculations showed that aspect ratios of at least 20 have to be obtained in order to reach the maximum theoretical shape anisotropy.[136]

Using a liquid crystalline co-polymer possessing carboxylic acid functions, the *in situ* synthesis of Co NPs embedded in the polymer matrix has been described recently.[137] The carboxylic acid functions of the polymer can bind to the metal surface, in the same manner as the long-chain carboxylic acids, which have been used for the preparation of Co anisotropic particles alone (see section 2.2.2.a). The addition of HDA appeared to be necessary in order to obtain Co nanorods within the polymer matrix since without HDA only spherical and cubic particles (which can arrange in necklace type aggregates) were obtained. However, increasing further the HDA amount reduced strongly the NRs yield. The composite materials were found to be ferromagnetic at room temperature and can find applications as magnetic composites.

Wang *et al.* have reported the formation of amorphous silica-sheathed Ni nanowires *via* a vapor-solid approach by reacting $NiCl_2$ and silicon powder in a tubular furnace at 700°C in a nitrogen atmosphere.[138] TEM analyses revealed 1D structures with an anisotropic Ni core of 50 nm in diameter and an outer diameter of about 200 nm. To explain the formation of such anisotropic materials, the authors proposed that: i) SiO_2 amorphous nanoparticles were first produced due to the presence of O_2 and/or H_2O traces in N_2 and ii) these SiO_2 nanoparticles adsorbed preferentially on some crystallographic Ni facets, leading to the observed anisotropic morphology. Moreover, depending on the reaction conditions, assemblies of Ni nanoparticles embedded in amorphous SiO_2 1D structures were also obtained. Even if this aspect has not been discussed in this work, the amorphous SiO_2 shell could protect the metal core from oxidation.

Another strategy to protect the 1D ferromagnetic materials from oxidation consists in adsorbing organic species onto the metal surface, provided that the outer shell formed by these organic species is sufficiently dense to prevent O_2 diffusion. In recent studies devoted to the evaluation of microwave absorption properties of iron nanowires prepared from $FeCl_3 \cdot 6H_2O$ and sodium borohydride as the reducing agent, polyvinylpyrrolidone[139] or dextran[140] was added in the reaction mixture to minimize surface oxidation. Anisotropic growth was obtained as a result of the application of an external magnetic field during the synthesis. The nanowires crystallized mainly with the bcc structure even if Fe_3O_4 was also detected by XRD as a minor phase. Even if no evidence is given for the presence of adsorbed organic species in these studies, it was shown that iron anisotropic particles prepared in the presence of organic species displayed better microwave absorption performances than either isotropic particles[139] or anisotropic particles prepared without organic moieties.[140]

3.2. Post-Synthesis Methods

This section deals with surface functionalization of existing magnetic 1D particles. In this case, a new functionality is added to the material. Some procedures which aim to achieve the deposition of a given material at the particle surface in order to enhance the resistance of the material towards oxidation are also described.

Grafting of Molecules of Biomedical Interest

Tanase *et al.* have reported the preparation of fluorescent magnetic Ni nanowires by a two-step procedure.[141] First, the Ni nanowires are electrochemically grown in an AAO template, then extracted from the host matrix and further reacted with Hematoporphyrin IX, a fluorescent porphyrin. This fluorescent compound binds to the native metal oxide layer which developed at the Ni surface *via* two carboxylic acid groups. The composite particles were highly fluorescent, thus optically detectable, and can be easily manipulated or oriented using an external magnetic field, which make them potential candidates for *e.g.* biomedical applications.

Indeed, these last years, considerable efforts have been directed towards the development of magnetic nanoparticles for biomedical applications. These particles can be used for various applications such as therapy, medical imaging or cell separation.[142] Because of their biocompatibility and biodegradability, iron oxide colloids, magnetite (Fe_3O_4) and maghemite (γ-Fe_2O_3), have been extensively studied.[142] Metal nanoparticles are also highly desirable since they possess higher saturation magnetizations than their corresponding oxides, but have to be protected in order to prevent oxidation, which in turn can improve their biocompatibility. Recently, Gao *et al.* reported the surface functionalization of Ni nanowires by an antibody, biotin anti-CD31.[143] The nanowires were prepared by electrodeposition in a uniaxial porous AAO membrane and further modified by pimelic acid. This dicarboxylic acid adsorbed at the particle surface and allowed the binding of streptavidin. Finally, the antibody anti-CD31 was bound to the Ni nanowire *via* streptavidin-biotin coupling. The resulting materials showed performances toward separation of mouse endothelial cells equivalent to or superior than commercial micrometer-sized magnetic epoxy-coated beads. Note however, that too high a concentration of Ni nanowires was found to be cytotoxic.[143]

Biocompatible and bioactive Ni nanowires were prepared by direct deposition of polysaccharides using the layer-by-layer (lbl) technique on an array of vertically aligned Ni nanowires obtained by electrodeposition in the pores of an AAO template.[144] An interesting point is that the deposition of the biopolymer was performed directly on the AAO-supported nanowires assembly and not on the free-standing nanowires, eliminating the need for several centrifugation, rinsing and redispersion steps. The organic layer was about 6-8 nm thick as evidenced by TEM. In a second step the glucose oxidase enzyme was immobilized onto polysaccharide-coated Ni NWs.

Carbon-coating

The direct, *i.e.* one-step, formation of M@CNT composite materials has already been described in section 2.2.1.a. Here, we focus on post-synthetic strategies for the deposition of carbon-based materials onto the particle surface.

Graphene nanotubes have been grown using Ni NWs prepared first by electrodeposition in a AAO template.[145] The graphene NTs were prepared by CVD using ethylene at about 750°C. It was found that the number of graphene layers depended on the growth time but was independent on the tube diameter and length. In a second step, the Ni NW could be removed by chemical etching to give the free-standing graphene NTs. Moreover, molecular dynamics simulations have shown that already-formed graphene sheets (GS) can spontaneously wrap around Fe NWs.[146] It is believed that the GS/Fe NWs as well as the π-π stacking interactions could be responsible for this phenomenon.

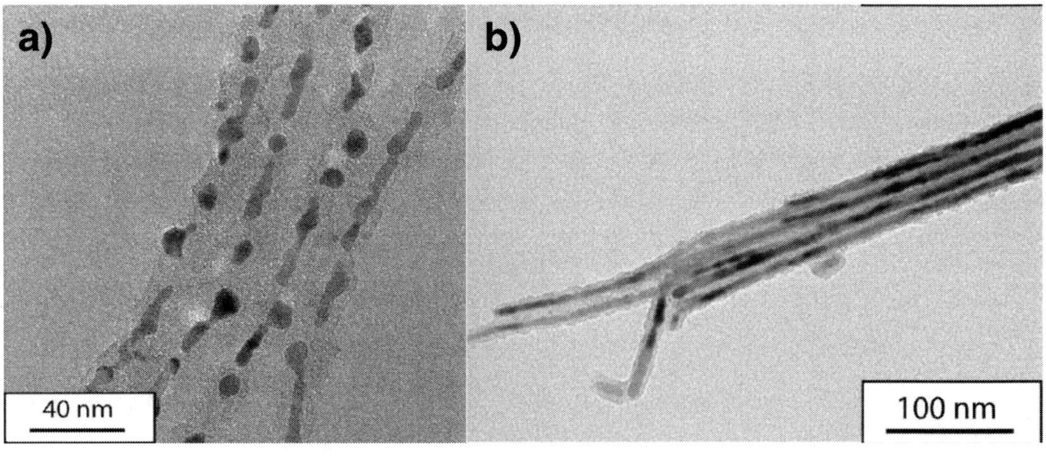

Figure 8. TEM images of Co NWs thermally treated at 350°C for 30 min in a static Ar atmosphere with two temperature rates: a) 10°C·min^{-1} and b) 5°C·min^{-1}. Reproduced with permission from ref. 147. Copyright 2009 American Chemical Society.

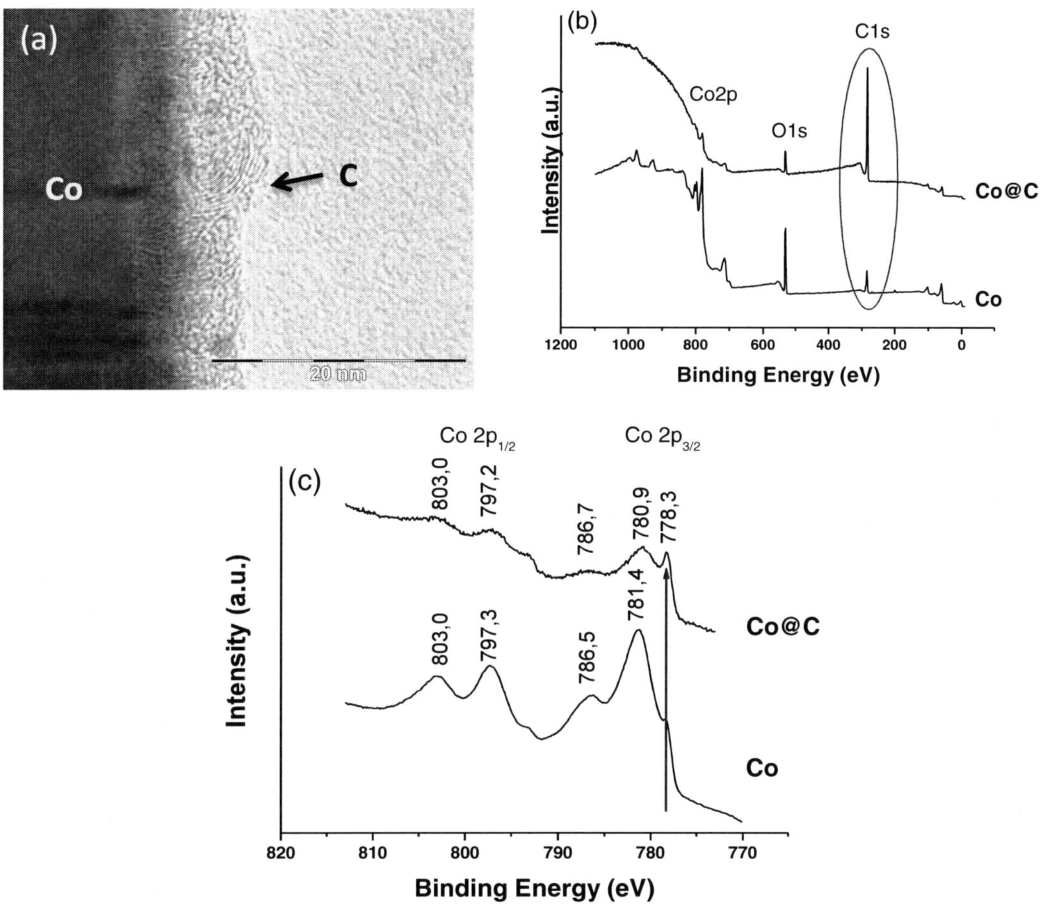

Figure 9. a) TEM image of a detail of a carbon-coated Co NR, b) XPS survey scans of Co NRs prepared by the polyol process and carbon-coated Co NRs after treatement in ODE at 300°C for 3h and c) high resolution XPS spectra of the Co 2p peaks.

As already indicated, the deposition of a carbon-based material onto metal anisotropic particles is a way to protect them from oxidation. Given the very interesting magnetic properties of metal NWs and NRs prepared by solution-based methods, several efforts were directed to their coating by carbon-based materials. The carbon-coating of Co NWs prepared by the organometallic pathway was realized by annealing the NWs at 350°C for 30 min in a static Ar atmosphere.[147] It has to be noted that in this study, the carbon source corresponds to the adsorbed ligands used for the synthesis of the Co particles. Two different heating rates were used: 5 and 10°C·min^{-1}. With the higher ramping rate, 10°C·min^{-1}, the NW is fragmented into chain of particles inside a rod-like carbon matrix as a result of the so-called Rayleigh-Plateau instability (see Figure 8a). With the lower temperature rate, the NW morphology was maintained (Figure 8b).

The decomposition of high-boiling-point organic solvents is also a way to generate carbon coatings. This was realized at 300°C for 3h with octadecene (ODE), oleylamine or a mixture of these two solvents in the presence of Co NRs.[148] When ODE was used alone, a disordered carbon shell developed at the particle surface (see Figure 9a), without alteration of the particle shape or the chemical nature of the rod, since pure hcp Co was evidenced by

XRD. However, treatments in ODE/oleylamine and prolonged treatments in oleylamine at 300°C led to a partial to total carburization of the metallic core (depending on the time elapsed at the final temperature). Note that only Co_2C (not Co_3C) was detected by XRD. The XPS survey scans of the Co NRs, before and after treatment in ODE at 300°C for 3h are presented in Figure 9b. The most striking feature is that the C1s peak centered at 285 eV has strongly increased for the carbon-coated material, denoted Co@C, compared to the non-coated Co NRs. This result is in agreement with the deposition of a carbon shell onto Co NRs. The high-resolution XPS spectra corresponding to the Co $2p_{1/2}$ and $2p_{3/2}$ components and their shake-up peaks are given in Figure 9c. The binding energy Co $2p_{3/2}$ of the untreated cobalt rods is 781.4 eV corresponding to cobalt in the 2+ valence state and a lower energy shoulder is also detected; it is attributed to metallic Co. The main contribution associated with Co(II) species is easily explained since previous studies have shown the presence of a thin polycrystalline CoO layer[128] which screens the metallic cobalt. For the carbon-coated Co NRs, the peak corresponding to metallic Co, at 778.3 eV has a much higher intensity than that of the non-treated sample. These results show clearly that during the deposition of the carbon shell, there was also a partial reduction of the native oxide layer.

Since the pioneering work of Delamar *et al.* in 1992,[149] who described the electrografting of aryl radicals on a carbon surface, numerous studies have been related to the reduction of diazonium salts on various substrates such as metal, semi-conducting, carbon and even oxide surfaces.[150,151] One clear advantage of this method is that the organic moieties are bonded covalently to the target surface. Electrochemical reduction is the most used method for the grafting of aryl radicals but the activation of diazonium salts by micro-wave[152] or ultrasonic irradiation[153,154] as well as by photochemistry[155,156] have been reported. These surface functionalization pathways among others have been recently critically reviewed.[150,157-159]

We investigated the grafting of aryl radicals using the $[N_2\text{-}C_6H_4\text{-}N(CH_3)_2]^+Cl^-$ salt under ultrasonic irradiation at room temperature on polyol-made Co NRs with different initial surface states: i) Co NRs with a very limited surface oxidation (corresponding to particles kept in the reaction medium), ii) Co NRs with a CoO thin surface layer (correspondind to samples dried in air) and carbon-coated Co NRs prepared as indicated previously.[160] If the grafting of the $C_6H_4\text{-}N(CH_3)_2$ radicals is efficient, it would allow to subsequent graft a polymer from the particle surface by photochemistry as was described recently for gold surfaces.[161]

Preliminary XPS results have clearly shown the grafting of the aryl groups for the metallic surface as well as the carbon-coated one, without alteration of the particle shape nor the crystal structure, but not for the Co NRs with a thin CoO layer.[160] These first results are an incentive to use these hybrid materials for the grafting of polymers from the particle surface and, in a second step, to use these polymer-grafted magnetic particles as building blocks for the formation of magnetic composite materials.

Formation of Hybrid Multifunctional Particles

Hybrid multifunctional nanomaterials consist in the association of two types of materials combining at least two different physicochemical properties. This emerging field in nanosciences opens the way to fundamental issues such as the influence of one type of property on the other and *vice-versa*. In this respect, an increasing number of studies deal with the elaboration of nanocomposites combining magnetic and plasmonic properties which can be used for biomedical applications, such as photothermal heating. Janus particles as well as core@shell materials can be prepared but core@shell particles with a metallic magnetic core and a plasmonic shell are attractive since the metallic core is protected in this case by the noble metal (Ag or Au) of the shell.

Gold-decorated Co NRs have been prepared by reacting [AuCl(PPh_3)] with a suspension of Co NRs synthesized *via* the organometallic pathway, without any reducing agent added.[162] Depending on the reaction conditions (temperature, Co/Au(I) ratio, addition of HDA, lauric acid or both), gold particles were localized at the tips or on the whole surface of the NRs (see Figure 10). It is assumed that Au(I) is reduced by the organic ligands adsorbed on the nanorod surface acting as mild reductants. Note that using [AuCl(tht)] (tht: tetrahydrothiophene) instead of [AuCl(PPh_3)] led to the total collapse of the Co NRs, due to galvanic displacement of Co by Au. These results show that Co NRs served as seeds for the nucleation of the Au particles. Based on HREM analyses, it was found that the Au NPs have grown epitaxially on the lateral sides of the Co NRs while the large mismatch on the lattice parameter for Au(111) and Co(0002) prevents simple epitaxial growth.[162]

It would be highly desirable to obtain materials with a continuous noble metal shell. Indeed, based on model calculations with a cobalt core diameter of 6 nm and a length of 80 nm and a gold shell thickness of 5 nm, it has been shown recently that such hybrid materials would be ideal nanoprobes for real-time immunodiagnostics.[163] To address this important issue, it is possible to benefit from Au NPs immobilized first at the particle surface which act as nucleation for gold deposition, as was reported for Fe_xO@Au magnetic-plasmonic core shell nanoparticles.[164] The preliminary deposition of gold NPs onto the particle surface is not always a prerequisite: Ni@Au core shell nanowires were prepared by electroless gold plating of Ni NWs prepared with an AAO membrane by electrodeposition.[165] A continuous shell could be obtained depending on the plating time (see Figure 11), but the authors indicated that the degree of cleanliness of the Ni surface was critical since the coating process could be obtained only on the clean metallic surface.

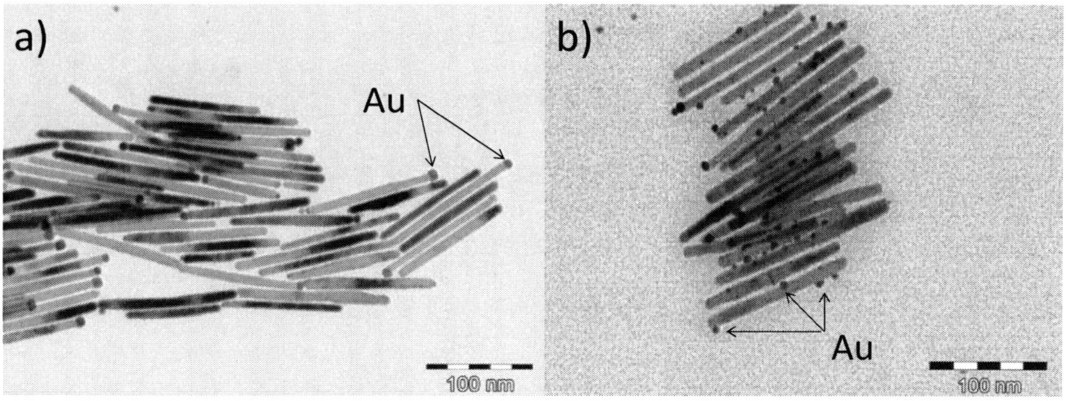

Figure 10. TEM images of a) gold-tipped Co NRs prepared at 40°C and b) gold-decorated Co NRs prepared at 20°C. Reaction time: 48 h. Reproduced with the authorization of Wiley-VCH.[162]

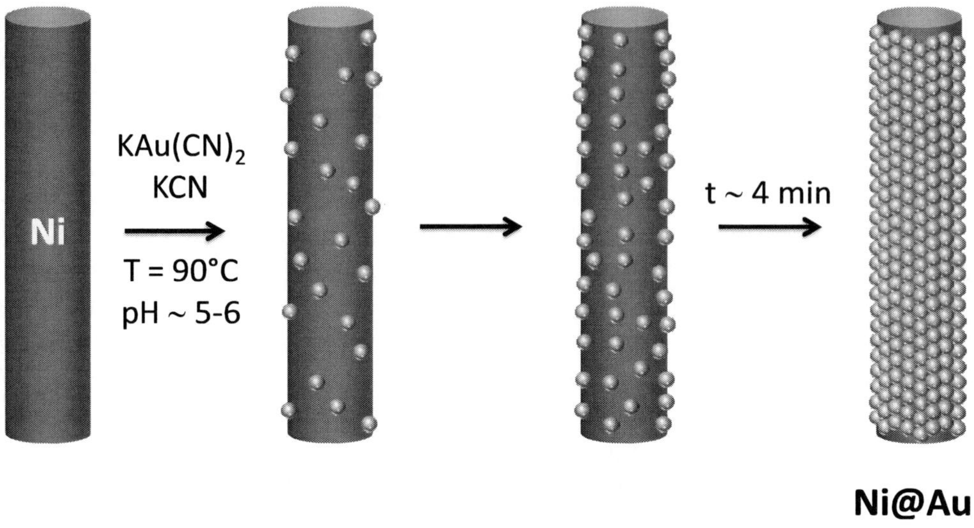

Figure 11. Schematic formation of the hybrid core shell nanowires by electroless plating of Ni nanowires with gold. After ref. 165.

Conclusion and Final Remarks

Ferromagnetic metal anisotropic nanoparticles are a very important class of materials with key applications in the domains of high-density magnetic recording, spintronics, catalysis or the elaboration of polymer-based magnetic composites. They can be prepared using vapor- or liquid-based strategies with or without templates. Vapor-phase methods are generally highly reproducible and allow obtaining a very good control of the size and shape, but liquid-phase syntheses appear to be more versatile in terms of control of the structure, the composition and the morphology of the desired product. Moreover, the latter are more appropriate for the production of large quantities of materials.

As regards 1D particles obtained in solution without solid inorganic or biological templates, it is largely accepted that the anisotropic growth is due to the privileged adsorption of long-chain ligands such as carboxylic acids, phosphines or amines on certain crystallographic facets ot the particle. However, the explicit role of these ligands and the reaction mechanisms accounting for the anisotropic growth are actually poorly understood. It would be highly desirable to study, at a molecular scale the origin of the shape preference. This should permit a better understanding of the nucleation and growth steps and should lead to a better control of structure and properties of the resulting materials.

Surface modification and functionalization of anisotropic magnetic metal particles are efficient tools for: i) consolidating the mechanical properties of these materials, ii) improving their resistance towards oxidation and iii) adding an interesting supplementary functionality to the material (other than magnetic). Multifunctional materials, which have several potential important technological applications such as catalysis and biomedicine, are interesting not only because they display several physicochemical functions into a single object, but also because they are model materials allowing the study of the possible interactions between these different functions. It is believed that this somewhat new research field will find a constantly growing interest in the near future.

ACKNOWLEDGMENTS

The authors would like to thank Dr. John Lomas and Prof. Fernand Fiévet for scientific discussions and for correcting the manuscript. ANR (Agence Nationale de la Recherche) and CGI (Commissariat à l'Investissement d'Avenir) are gratefully acknowledged for the partial financial support of this work through Labex SEAM (Science and Engineering for Advanced Materials and devices).

REFERENCES

[1] Krahne, R., Morello, G., Figuerola, A., George, C., Deka, S. & Manna, L. (2011). *Phys. Rep.*, *501*, 75.

[2] Uzio, D. & Berhault, G. (2010). *Catal. Rev.*, *52*, 106.

[3] Gómez-Ramírez, A., López-López, M. T., Durán, J. D. G. & González-Caballero, F. (2009). *Soft Matter*, *5*, 3888.

[4] Birringer, R., Wolf, H., Lang, C., Tschöpe, A. & Michels, A. (2008). *Z. Phys. Chem.*, *222*, 229.

[5] Dinega, D. P. & Bawendi, M. G. (1998). *Angew. Chem., Int. Ed.*, *38*, 1788.

[6] Sun, S. & Murray, C. B. (1999). *J. Appl. Phys.*, *85*, 4325.

[7] Lu, A. H., Salabas, E. L. & Schuth, F. (2007). *Angew. Chem. Int. Ed.*, *46*, 1222.

[8] Rao, C. N. R., Deepak, F. L., Gundiah, G. & Govindaraj, A. (2003). *Progress Solid State Chem.*, *31*, 5.

[9] Xia, Y., Yang, P., Sun, Y., Wu, Y., Mayers, B., Gates, B., Yin, Y., Kim, F. & Yan, H. (2003). *Adv. Mater.*, *15*, 353.

[10] Tang, Z. & Kotov, N. A. (2005). *Adv. Mater.*, *17*, 951.

[11] Schlörb, H., Haehnel, V., Khatri, M. S., Srivastav, A., Kumar, A., Schultz, L. & Fähler, S. (2010). *Phys. Status Solid. (B)*, *247*, 2364.

[12] Nguyen, P., Ng, H. T. & Meyyappan, M. (2005). *Adv. Mater.*, *17*, 1773.

[13] Mohammad, S. N. (2010). *J. Appl. Phys.*, *107*, 114304.

[14] Shihui, G., Jinglei, Y. & Huaxin, Z. (2008). *J. Appl. Phys.*, *104*, 063906.

[15] Bagkar, N., Seo, K., Yoon, H., In, J., Jo, Y. & Kim, B. (2010). *Chem. Mater.*, *22*, 1831.

[16] Becker, C. & Wandelt, K. (2009). *Top. Curr. Chem.*, *287*, 45.

[17] Rousset, S., Pourmir, F., Berroir, J. M., Klein, J., Lecoeur, J., Hecquet, P. & Salanon, B. (1999). *Surf. Sci.*, *422*, 33.

[18] Repain, V., Berroir, J.-M., Rousset, S. & Lecoeur, J. (2000). *Surf. Sci.*, *447*, L152.

[19] Luo, M. F., Chiang, C. I., Shiu, H. W., Sartale, S. D. & Kuo, C. C. (2006). *Nanotech.*, *17*, 360.

[20] Shen, J., Skomski, R., Klaua, M., Jenniches, H., Manoharan, S. S. & Kirschner, J. (1997). *Phys. Rev. B*, *56*, 2340.

[21] Shiraki, S., Fujisawa, H., Nantoh, M. & Kawai, M. (2004). *Phys. Rev. Lett.*, *92*.

[22] Shiraki, S., Fujisawa, H., Nantoh, M. & Kawai, M. (2004). *Appl. Surf. Sci.*, *237*, 284.

[23] Zaki, N., Potapenko, D., Johnson, P. & Osgood, R. (2009). *Phys. Rev. B*, *80*.

[24] Gambardella, P., Blanc, M., Bürgi, L., Kuhnke, K. & Kern, K. (2000). *Surf. Sci.*, *449*, 93.

[25] Athanassiou, E. K., Grossmann, P., Grass, R. N. & Stark, W. J. (2007). *Nanotech.*, *18*, 165606.

[26] Aranda, P. & García, J. M. (2002). *J. Magn. Magn. Mater.*, *249*, 214.

[27] Martin, C. R. (1996). *Chem. Mater.*, *8*, 1739.

[28] Possin, G. E. (1970). *Review of Scientific Instruments*, *41*, 772.

[29] Sanz, R., Hernandez-Velez, M., Pirota, K. R., Baldonedo, J. L. & Vazquez, M. (2007). *Small* 2007, *3*, 434.

[30] Huczko, A. (2000). *Appl Phys A*, *70*, 365.

[31] Sellmyer, D. J., Zheng, M. & Skomski, R. (2001). *J. Phys.: Condens. Matter*, *13*, R433.

[32] Hernández-Vélez, M. (2006). *Thin Solid Films*, *495*, 51.

[33] Kawai, S. & Ishiguro, I. (1976). *J. Electrochem. Soc.*, *123*, 1047.

[34] AlMawlawi, D., Coombs, N. & Moskovits, M. (1991). *J. Appl. Phys.*, *70*, 4421.

[35] Kröll, M., Blau, W. J., Grandjean, D., Benfield, R. E., Luis, F., Paulus, P. M. & de Jongh, L. J. (2002). *J. Magn. Magn. Mater.*, *249*, 241.

[36] Peng, Y., Zhang, H.-L., Pan, S.-L. & Li, H.-L. (2000). *J. Appl. Phys.*, *87*, 7405.

[37] Yang, S., Zhu, H., Yu, D., Jin, Z., Tang, S. & Du, Y. (2000). *J. Magn. Magn. Mater.*, *222*, 97.

[38] Menon, L., Zheng, M., Zeng, H., Bandyopadhyay, S. & Sellmyer, D. J. (2000). *J. Electronic Mater.*, *29*, 510.

[39] Qin, J., Nogués, J., Mikhaylova, M., Roig, A., Muñoz, J. S. & Muhammed, M. (2005). *Chem. Mater.*, *17*, 1829.

[40] Han, X.-F., Shamaila, S., Sharif, R., Chen, J.-Y., Liu, H.-R. & Liu, D.-P. (2009). *Adv. Mater.*, *21*, 4619.

[41] Sanz, R., Vazquez, M., Pirota, K. R. & Hernández-Velez, M. (2007). *J. Appl. Phys.*, *101*, 114325.

[42] Huang, X., Li, L., Luo, X., Zhu, X. & Li, G. (2008). *J. Phys. Chem. C*, *112*, 1468.

[43] Zeng, H., Zheng, M., Skomski, R., Sellmyer, D. J., Liu, Y., Menon, L. & Bandyopadhyay, S. (2000). *J. Appl. Phys.*, *87*, 4718.

[44] Schönenberger, C., van der Zande, B. M. I., Fokkink, L. G. J., Henny, M., Schmid, C., Krüger, M., Bachtold, A., Huber, R., Birk, H. & Staufer, U. (1997). *J. Phys. Chem. B*, *101*, 5497.

[45] Li, D., Thompson, R. S., Bergmann, G. & Lu, J. G. (2008). *Adv. Mater.*, *20*, 4575.

[46] Caffarena, V. R., Guimarães, A. P., Folly, W. S. D., Silva, E. M. & Capitaneo, J. L. (2008). *Mater. Chem. Phys.*, *107*, 297.

[47] Fert, A. & Piraux, L. (1999). *J. Magn. Magn. Mater.*, *200*, 338.

[48] Chaure, N. B., Stamenov, P., Rhen, F. M. F. & Coey, J. M. D. (2005). *J. Magn. Magn. Mater.*, *290-291*, 1210.

[49] Campbell, R., Bakker, M. G., Havrilla, G., Montoya, V., Kenik, E. A. & Shamsuzzoha, M. (2006). *Micropor. Mesopor. Mater.*, *97*, 114.

[50] Luo, H., Wang, D., He, J. & Lu, Y. (2005). *J. Phys. Chem. B*, *109*, 1919.

[51] Wang, D., Luo, H., Kou, R., Gil, M. P., Xiao, S., Golub, V. O., Yang, Z., Brinker, C. J. & Lu, Y. (2004). *Angew. Chem. Int. Ed.*, *43*, 6169.
[52] Chernysheva, M. V., Eliseev, A. A., Napolskii, K. S., Lukashin, A. V., Tretyakov, Y. D., Grigoryeva, N. A., Grigoryev, S. V. & Wolff, M. (2006). *Thin Solid Films*, *495*, 73.
[53] Eliseev, A. A., Kolesnik, I. V., Lukashin, A. V. & Tretyakov, Y. D. (2005). *Adv. Eng. Mater.*, *7*, 213.
[54] Eliseev, A. A., Kolesnik, I. V., Lukashin, A. V., Tretyakov, Y. D. & Görnert, P. (2008). *Physica E: Low-dimensional Systems and Nanostructures*, *40*, 2531.
[55] Eliseev, A. A., Napolskii, K. S., Lukashin, A. V. & Tretyakov, Y. D. (2004). *J. Magn. Magn. Mater.* 2004, *272-276*, 1609.
[56] Grigorieva, N. A., Grigoriev, S. V., Eckerlebe, H., Eliseev, A. A., Napolskii, K. S., Lukashin, A. V. & Tretyakov, Y. D. (2006). *J. Magn. Magn. Mater.*, *300*, e342.
[57] Zhang, Z., Dai, S., Blom, D. A. & Shen, J. (2002). *Chem. Mater.*, *14*, 965.
[58] Zhang, Z., Pan, Z., Mahurin, S. M. & Dai, S. (2003). *Chem. Commun.*, 2584.
[59] Zhao, D., Feng, J., Huo, Q., Melosh, N., Fredrickson, G. H., Chmelka, B. F. & Stucky, G. D. (1998). *Science*, *279*, 548.
[60] Li, H., Lin, H., Xie, S., Dai, W., Qiao, M., Lu, Y. & Li, H. (2008). *Chem. Mater.*, *20*, 3936.
[61] Coleman, N. R. B., Morris, M. A., Spalding, T. R. & Holmes, J. D. (2001). *J. Am. Chem. Soc.*, *123*, 187.
[62] Crowley, T. A., Ziegler, K. J., Lyons, D. M., Erts, D., Olin, H., Morris, M. A. & Holmes, J. D. (2003). *Chem. Mater.*, *15*, 3518.
[63] Wu, Y., Cheng, G., Katsov, K., Sides, S. W., Wang, J., Tang, J., Fredrickson, G. H., Moskovits, M. & Stucky, G. D. (2004). *Nat. Mater.*, *3*, 816.
[64] Wu, Y., Livneh, T., Zhang, Y. X., Cheng, G., Wang, J., Tang, J., Moskovits, M. & Stucky, G. D. (2004). *Nano Lett.*, *4*, 2337.
[65] Iijima, S. (1991). *Nature*, *354*, 56.
[66] Bao, J., Tie, C., Xu, Z., Suo, Z., Zhou, Q. & Hong, J. (2002). *Adv. Mater.*, *14*, 1483.
[67] Tao, F., Liang, Y., Yin, G., Xu, D., Jiang, Z., Li, H., Han, M., Song, Y., Xie, Z., Xue, Z., Zhu, J., Xu, Z., Zheng, L., Wei, X. & Ni, Y. (2007). *Adv. Func. Mater.*, *17*, 1124.
[68] Zeeshan, M. A., Shou, K., Pane, S., Pellicer, E., Sort, J., Sivaraman, K. M., Baro, M. D. & Nelson, B. J. (2011). *Nanotech.*, *22*, 275713.
[69] Rao, C. N. R. & Sen, R. (1998). *Chem. Commun.*, 1525.
[70] Grobert, N., Mayne, M., Terrones, M., Sloan, J., Dunin-Borkowski, R. E., Kamalakaran, R., Seeger, T., Terrones, H., Ruhle, M., Walton, D. R. M., Kroto, H. W. & Hutchison, J. L. (2001). *Chem. Commun.*, 471.
[71] Morelos-Gómez, A., López-Urías, F., Muñoz-Sandoval, E., Dennis, C. L., Shull, R. D., Terrones, H. & Terrones, M. (2010). *J. Mater. Chem.*, *20*, 5906.
[72] Grobert, N., Hsu, W. K., Zhu, Y. Q., Hare, J. P., Kroto, H. W., Walton, D. R. M., Terrones, M., Terrones, H., Redlich, P., Rühle, M., Escudero, R. & Morales, F. (1999). *Appl. Phys. Lett.*, *75*, 3363.
[73] Lv, R., Kang, F., Gu, J., Gui, X., Wei, J., Wang, K. & Wu, D. (2008). *Appl. Phys. Lett.*, *93*, 223105.
[74] Lv, R., Kang, F., Zhu, D., Zhu, Y., Gui, X., Wei, J., Gu, J., Li, D., Wang, K. & Wu, D. (2009). *Carbon*, *47*, 2709.

[75] Borowiak-Palen, E., Mendoza, E., Bachmatiuk, A., Rummeli, M. H., Gemming, T., Nogues, J., Skumryev, V., Kalenczuk, R. J., Pichler, T. & Silva, S. R. P. (2006). *Chem. Phys. Lett.*, *421*, 129.
[76] Jorge, J., Flahaut, E., Gonzalez-Jimenez, F., Gonzalez, G., Gonzalez, J., Belandria, E., Broto, J. M. & Raquet, B. (2008). *Chem. Phys. Lett.*, *457*, 347.
[77] Dujardin, E., Ebbesen, T. W., Hiura, H. & Tanigaki, K. (1994). *Science*, *265*, 1850.
[78] Jiang, Q. & Lu, H. M. (2008). *Surf. Sci. Rep.*, *63*, 427.
[79] Shpak, A., Kolesnik, S., Mogilny, G., Petrov, Y., Sokhatsky, V., Trophimova, L., Shanina, B. & Gavriljuk, V. (2007). *Acta Mater.*, *55*, 1769.
[80] Blank, V. D., Kulnitskiy, B. A., Perezhogin, I. A., Polyakov, E. V. & Batov, D. V. (2010). *Acta Mater.*, *58*, 1293.
[81] Elías, A. L., Rodríguez-Manzo, J. A., McCartney, M. R., Golberg, D., Zamudio, A., Baltazar, S. E., López-Urías, F., Muñoz-Sandoval, E., Gu, L., Tang, C. C., Smith, D. J., Bando, Y., Terrones, H. & Terrones, M. (2005). *Nano Lett.*, *5*, 467.
[82] Sharif, R., Shamaila, S., Ma, M., Yao, L. D., Yu, R. C., Han, X. F. & Khaleeq-ur-Rahman, M. (2008). *Appl. Phys. Lett.*, *92*, 032505.
[83] Xu, X. & Zangari, G. (2005). *J. Appl. Phys.*, *97*, 10A306.
[84] Xue, S. H. & Wang, Z. D. (2006). *Mater. Sci. Eng.: B*, *135*, 74.
[85] Sun, L., Searson, P. C. & Chien, C. L. (2001). *Appl. Phys. Lett.*, *79*, 4429.
[86] Mertig, M., Kirsch, R. & Pompe, W. (1998). *Appl Phys A*, *66*, S723.
[87] Gu, Q., Cheng, C., Haynie, D. T. (2005). *Nanotech.*, *16*, 1358.
[88] Gu, Q. & Haynie, D. T. (2008). *Mater. Lett.*, *62*, 3047.
[89] Knez, M., Bittner, A. M., Boes, F., Wege, C., Jeske, H., Maib, E. & Kern, K. (2003). *Nano Lett.*, *3*, 1079.
[90] Knez, M., Sumser, M., Bittner, A. M., Wege, C., Jeske, H., Martin, T. P. & Kern, K. (2004). *Adv. Func. Mater.*, *14*, 116.
[91] Mao, C., Solis, D. J., Reiss, B. D., Kottmann, S. T., Sweeney, R. Y., Hayhurst, A., Georgiou, G., Iverson, B. & Belcher, A. M. (2004). *Science*, *303*, 213.
[92] Tsukamoto, R., Muraoka, M., Seki, M., Tabata, H. & Yamashita, I. (2007). *Chem. Mater.*, *19*, 2389.
[93] Kobayashi, M., Seki, M., Tabata, H., Watanabe, Y. & Yamashita, I. (2010). *Nano Lett.*, *10*, 773.
[94] Park, S.-J., Kim, S., Lee, S., Khim, Z. G., Char, K. & Hyeon, T. (2000). *J. Am. Chem. Soc.*, *122*, 8581.
[95] Respaud, M., Broto, J. M., Rakoto, H., Fert, A. R., Thomas, L., Barbara, B., Verelst, M., Snoeck, E., Lecante, P., Mosset, A., Osuna, J., Ely, T. O., Amiens, C. & Chaudret, B. (1998). *Phys. Rev. B*, *57*, 2925.
[96] Cordente, N., Respaud, M., Senocq, F., Casanove, M.-J., Amiens, C. & Chaudret, B. (2001). *Nano Lett.*, *1*, 565.
[97] Puntes, V., Zanchet, D., Erdonmez, C. K. & Alivisatos, A. P. (2002). *J. Am. Chem. Soc.*, *124*, 12874.
[98] Dumestre, F., Chaudret, B., Amiens, C., Fromen, M.-C., Casanove, M.-J., Renaud, P. & Zurcher, P. (2002). *Angew. Chem. Int. Ed.*, *41*, 4286.
[99] Comesaña-Hermo, M., Ciuculescu, D., Li, Z.-A., Stienen, S., Spasova, M., Farle, M. & Amiens, C. (2012). *J. Mater. Chem.*, *22*, 8043.

[100] Dumestre, F., Chaudret, B., Amiens, C., Respaud, M., Fejes, P., Renaud, P. & Zurcher, P. (2003). *Angew. Chem.*, *115*, 5371.
[101] Wetz, F., Soulantica, K., Respaud, M., Falqui, A. & Chaudret, B. (2007). *Mater. Sci. Eng.: C*, *27*, 1162.
[102] Wang, C., Hou, Y., Kim, J. & Sun, S. (2007). *Angew. Chem. Int. Ed.*, *46*, 6333.
[103] Puntes, V. F., Zanchet, D., Erdonmez, C. K. & Alivisatos, A. P. (2002). *J. Am. Chem. Soc.*, *124*, 12874.
[104] Dreyer, A., Peter, M., Mattay, J., Eckstädt, K., Hütten, A. & Jutzi, P. (2012). *Eur. J. Inorg. Chem.*, *2012*, 198.
[105] Kwon, S. G. & Hyeon, T. (2011). *Small*, *7*, 2685.
[106] Timonen, J. V., Seppala, E. T., Ikkala, O. & Ras, R. H. (2011). *Angew. Chem. Int. Ed.*, *50*, 2080.
[107] Liakakos, N., Cormary, B., Li, X., Lecante, P., Respaud, M., Maron, L., Falqui, A., Genovese, A., Vendier, L., Koinis, S., Chaudret, B. & Soulantica, K. (2012). *J. Am. Chem. Soc.*, *134*, 17922.
[108] Fiévet, F. & Brayner, R. (2013). *Nanomaterials: A Danger or a Promise? A Chemical and Biological Perspective*; Springer.
[109] Feldmann, C. & Jungk, H.-O. (2001). *Angew. Chem. Int. Ed.*, *40*, 359.
[110] Chakroune, N., Viau, G., Ammar, S., Jouini, N., Gredin, P., Vaulay, M. J. & Fiévet, F. (2005). *New J. Chem.*, *29*, 355.
[111] Poul, L., Jouini, N. & Fiévet, F. (2000). *Chem. Mater.*, *12*, 3123.
[112] Azib, T., Ammar, S., Nowak, S., Lau-Truing, S., Groult, H., Zaghib, K., Mauger, A. & Julien, C. M. (2012). *J. Power Sources*, *217*, 220.
[113] Feldmann, C. & Metzmacher, C. (2001). *J. Mater. Chem.*, *11*, 2603.
[114] Rivers, J. H., Carroll, K. J., Jones, R. A. & Carpenter, E. E. (2010). *Acta Crystallogr. C*, *66*, m83.
[115] Wang, Y., Zheng, Y., Huang, C. Z. & Xia, Y. (2013). *J. Am. Chem. Soc.*, *135*, 1941.
[116] Zheng, Y., Tao, J., Liu, H., Zeng, J., Yu, T., Ma, Y., Moran, C., Wu, L., Zhu, Y., Liu, J. & Xia, Y. (2011). *Small, 116,* 2307.
[117] Toneguzzo, P., Viau, G., Acher, O., Fiévet-Vincent, F. & Fiévet, F. (1998). *Adv. Mater.*, *10*, 1032.
[118] Viau, G., Brayner, R., Poul, L., Chakroune, N., Lacaze, E., Fiévet-Vincent, F. & Fiévet, F. (2002). *Chem. Mater.*, *15*, 486.
[119] LaMer, V. K. & Dinegar, R. H. (1950). *J. Am. Chem. Soc.*, *72*, 4847.
[120] Chakroune, N., Viau, G., Ammar, S., Jouini, N., Gredin, P., Vaulay, M. J. & Fiévet, F. (2005). *New J. Chem.*, *29*, 355.
[121] Tekaia-Elhsissen, K., Delahaye-Vidal, A., Nowogrocki, G. & Figlarz, M. (1989). *C. R. Acad. Sci., Ser. 2*, *309*, 349.
[122] Tekaia-Elhsissen, K., Delahaye-Vidal, A., Nowogrocki, G. & Figlarz, M. (1989). *C. R. Acad. Sci., Ser. 2*, *309*, 469.
[123] Ung, D., Soumare, Y., Chakroune, N., Viau, G., Vaulay, M.-J., Richard, V. & Fiévet, F. (2007). *Chem. Mater.*, *19*, 2084.
[124] Soumare, Y., Piquemal, J.-Y., Maurer, T., Ott, F., Chaboussant, G., Falqui, A. & Viau, G. (2008). *J. Mater. Chem.*, *18*, 5696.
[125] Soumare, Y., Garcia, C., Maurer, T., Chaboussant, G., Ott, F., Fiévet, F., Piquemal, J.-Y. & Viau, G. (2009). *Adv. Func. Mater.*, *19*, 1971.

[126] Ung, D., Viau, G., Ricolleau, C., Warmont, F., Gredin, P. & Fiévet, F. (2005). *Adv. Mater.*, *17*, 338.

[127] Aït Atmane, K., Zighem, F., Soumare, Y., Ibrahim, M., Boubekri, R., Maurer, T., Margueritat, J., Piquemal, J.-Y., Ott, F., Chaboussant, G., Schoenstein, F., Jouini, N. & Viau, G. (2013). *J. Solid State Chem.*, *197*, 297.

[128] Maurer, T., Zighem, F., Ott, F., Chaboussant, G., André, G., Soumare, Y., Piquemal, J.-Y., Viau, G. & Gatel, C. (2009). *Phys. Rev. B*, *80*.

[129] Liu, Q., Guo, X., Chen, J., Li, J., Song, W. & Shen, W. (2008). *Nanotech.*, *19*, 365608.

[130] Li, C. C. & Zeng, H. C. (2010). *J. Mater. Chem.*, *20*, 9187.

[131] Hisano, S. & Saito, K. (1998). *J. Magn. Magn. Mater.*, *190*, 371.

[132] Xie, B.-Q., Qian, Y., Zhang, S., Fu, S. & Yu, W. (2006). *Eur. J. Inorg. Chem.*, *2006*, 2454.

[133] Xie, Q., Dai, Z., Huang, W., Liang, J., Jiang, C. & Qian, Y. (2005). *Nanotech.*, *16*, 2958.

[134] Fan, N., Yu, H., Ju, Z., Ma, Q., Hu, Y., Xu, L. & Qian, Y. (2009). *Mater. Lett.*, *63*, 551.

[135] Bigall, N. C., Parak, W. J. & Dorfs, D. (2012). *Nano Today*, *7*, 282.

[136] Ott, F., Maurer, T., Chaboussant, G., Soumare, Y., Piquemal, J. Y. & Viau, G. (2009). *J. Appl. Phys.*, *105*, 013915.

[137] Zadoina, L., Soulantica, K., Ferrere, S., Lonetti, B., Respaud, M., Mingotaud, A. F., Falqui, A., Genovese, A., Chaudret, B. & Mauzac, M. (2011). *J. Mater. Chem.*, *21*, 6988.

[138] Wang, J., Zhang, H., Ge, J. & Li, Y. (2006). *J. Phys. Chem. B*, *110*, 807.

[139] Yang, R.-B., Liang, W.-F., Lin, W.-S., Lin, H.-M., Tsay, C.-Y. & Lin, C.-K. (2011). *J. Appl. Phys.*, *109*, 07B527.

[140] Liang, W. F., Yang, R. B., Lin, W. S., Jian, Z. J., Tsay, C. Y., Wu, S. H., Lin, H. M., Choi, S. T. & Lin, C. K. (2012). *J. Appl. Phys.*, *111*, 07B545.

[141] Tanase, M., Bauer, L. A., Hultgren, A., Silevitch, D. M., Sun, L., Reich, D. H., Searson, P. C. & Meyer, G. J. (2001). *Nano Lett.*, *1*, 155.

[142] Reddy, L. H., Arias, J. L., Nicolas, J. & Couvreur, P. (2012). *Chem. Rev.*, *112*, 5818.

[143] Gao, N., Wang, H. & Yang, E. H. (2010). *Nanotech.*, *21*, 105107.

[144] Magnin, D., Callegari, V., Matéfi-Tempfli, S., Glinel, K., Jonas, A. M. & Demoustier-Champagne, S. (2008). *Biomacromolecules*, *9*, 2517.

[145] Wang, R., Hao, Y., Wang, Z., Gong, H. & Thong, J. T. (2010). *Nano Lett.*

[146] Li, Y., Li, H., Zhang, K. & Liew, K. M. (2012). *Carbon*, *50*, 566.

[147] Ciuculescu, D., Dumestre, F., Comesaña-Hermo, M., Chaudret, B., Spasova, M., Farle, M. & Amiens, C. (2009). *Chem. Mater.*, *21*, 3987.

[148] Ibrahim, M., Garcia, C., Aït Atmane, K., Berrichi, E., Zwick, A., Warot-Fonrose, B., Lacroix, L.-M., Lachaize, S., Piquemal, J. Y. & Viau, G. *submitted for publication*.

[149] Delamar, M., Hitmi, R., Pinson, J. & Saveant, J. M. (1992). *J. Am. Chem. Soc.*, *114*, 5883.

[150] Belanger, D. & Pinson, J. (2011). *Chem. Soc. Rev.*, *40*, 3995.

[151] Griffete, N., Herbst, F., Pinson, J., Ammar, S. & Mangeney, C. (2011). *J. Am. Chem. Soc.* 2011, *133*, 1646.

[152] Liu, J., i Zubiri, M. R., Vigolo, B., Dossot, M., Humbert, B., Fort, Y. & McRae, E. (2007). *J. Nanosci. Nanotech.*, *7*, 3519.

[153] Mangeney, C., Qin, Z., Dahoumane, S. A., Adenier, A., Herbst, F., Boudou, J.-P., Pinson, J. & Chehimi, M. M. (2008). *Diamond Relat. Mater.*, *17*, 1881.

[154] Mirkhalaf, F., Mason, T. J., Morgan, D. J. & Saez, V. (2011). *Langmuir*, *27*, 1853.

[155] Pan, Y., Tong, B., Shi, J., Zhao, W., Shen, J., Zhi, J. & Dong, Y. (2010). *J. Phys. Chem. C*, *114*, 8040.

[156] Bouriga, M., Chehimi, M. M., Combellas, C., Decorse, P., Kanoufi, F., Deronzier, A. & Pinson, J. (2013). *Chem. Mater.*, *25*, 90.

[157] *Aryl Diazonium Salts: New Coupling Agents and Surface Science*; Chehimi, M. M., Ed.; Wiley-VCH Verlag: Weinheim, Germany, 2012.

[158] Mahouche-Chergui, S., Gam-Derouich, S., Mangeney, C. & Chehimi, M. M. (2011). *Chem. Soc. Rev.*, *40*, 4143.

[159] Salmi, Z., Gam-Derouich, S., Mahouche-Chergui, S., Turmine, M. & Chehimi, M. M. (2012). *Chem. Papers*, *66*, 369.

[160] Aït Atmane, K. (2012). Thesis of the University Paris Diderot.

[161] Gam-Derouich, S., Lamouri, A., Redeuilh, C., Decorse, P., Maurel, F., Carbonnier, B., Beyazit, S., Yilmaz, G., Yagci, Y. & Chehimi, M. M. (2012). *Langmuir*, *28*, 8035.

[162] Wetz, F., Soulantica, K., Falqui, A., Respaud, M., Snoeck, E. & Chaudret, B. (2007). *Angew. Chem. Int. Ed.*, *46*, 7079.

[163] Schrittwieser, S., Ludwig, F., Dieckhoff, J., Soulantica, K., Viau, G., Lacroix, L.-M., Lentijo, S. M., Boubekri, R., Maynadié, J., Huetten, A., Brueckl, H. & Schotter, J. (2011). *ACS Nano*, *6*, 791.

[164] Levin, C. S., Hofmann, C., Ali, T. A., Kelly, A. T., Morosan, E., Nordlander, P., Whitmire, K. H. & Halas, N. J. (2009). *ACS Nano*, *3*, 1379.

[165] Jeon, I. T., Cho, M. K., Cho, J. W., An, B. H., Wu, J. H., Kringel, R., Choi, D. S. & Kim, Y. K. (2011). *J. Mater. Chem.*, *21*, 12089.

In: Applied Surface Chemistry of Nanomaterials
Editors: Mohamed M. Chehimi and Jean Pinson
ISBN: 978-1-62808-351-4

Chapter 8

GOLD NANOPARTICLES AT POLYMERIC INTERFACES

Samia Mahouche-Chergui,[1] Sandrine Lépinay,[1] Mourad Lakhlifi,[1] Mohamed Guerrouache,[1] Mohamed M. Chehimi[2] and Benjamin Carbonnier[1,*]

[1]Institut de Chimie et des Matériaux Paris-Est – Equipe Systèmes Polymères Complexes, UMR CNRS 7182 *and* Université Paris-Est, Thiais, France
[2]Université Paris Diderot, Sorbonne Paris Cité, ITODYS, UMR CNRS 7086, Paris, France

1. INTRODUCTION

The combination of structures with characteristic length scales in the nano-sized regime with organic polymers opens fascinating pathways to designing smart nano-heterostructures with applications in varied scientific and technological fields. Indeed polymer-supported/ embedded nanoparticles (NPs) heterostructures have found applications as composite systems with reinforced mechanical properties, [1] hydrogels for biotechnological applications, [2] light emitting materials, [3] nano-catalysts for C-C coupling [4] or hydrogenation [5] reactions, sensors based on infra-red [6] or electrochemical [7] responses…, to name but a few.

Among the large variety of polymer-based nanocomposites developed to date, materials incorporating metal NPs (MNPs) have been the subject of numerous investigations. This intense interest derives from the unique physical and chemical properties – *undoubtedly different from both atoms and bulk materials* – of metallic NPs arising from high surface-area-to-volume ratios.

Among the large variety of metal NPs investigated to date, gold NPs (GNPs) have attracted a lot of interest for a long time. [8] Nano gold exhibits remarkable properties not only from a fundamental point of view but also for practical uses and numerous reports have demonstrated their ease of preparation, [9] specificity of interaction with thiol ligands and

* Corresponding author: Benjamin Carbonnier. Institut de Chimie et des Matériaux Paris-Est – Equipe Systèmes Polymères Complexes, UMR CNRS 7182 and Université Paris-Est, 2-8 rue Henri Dunant, 94320 Thiais, France. E-mail: carbonnier@icmpe.cnrs.fr; Tel: +331 4978 1154; Fax: +331 4978 1208.

thus versatility of bioconjugation, [10] pronounced biosafety and biocompatibility, [11] as well as enhanced catalytic [12] and optical properties [13].

The use of polymers as supports for GNPs, and MNPs in general, can be easily rationalized on the one hand, by the commercial availability of a large variety of polymerizable monomers bearing functional side groups acting as chelating groups for reduced or precursors metal species. On the other hand, the versatility of macromolecular synthesis tools allows tuning the macromolecular topology, chemical composition and functional groups distribution along the polymer chains according to the targeted application. In addition to size, shape and size uniformity, control of dispersion of NPs is a key parameter to minimize the loss of properties such as surface chemical activity and electronic states related to the nanosize regime. Although there are many examples where the polymer solely plays the role of passive matrix, the study of nanoparticle-polymer matrix interactions is of upmost importance and still presents several challenges. Indeed, incorporating *pre*-made GNPs or aqueous soluble gold precursors in a polymer matrix may be intrinsically challenging owing to unfavorable entropic interactions between both partners. The scope of this paper is not to provide thorough discussion about polymer-GNPs interaction, however it is worth mentioning in this introductive part that the use of functional polymers has been proposed to stabilize the GNPs and inhibit particle agglomeration, and further facilitate their compatibility with the polymer matrix. [14]

Other approaches reported on the synthesis of polymer-protected Au nanoparticles from metal precursors in the presence of a protective polymer with mercapto [15] or amino [16] groups. Herein, we discuss mainly the case where GNPs are deposited on the surface of polymer supports but we have also extended the scope of this study to the case where GNPs are incorporated within polymer matrix playing an active role in the function of the composite materials. Preparation methods involving the polymerization of reactive mixture with a passive role around preformed nanoparticles along with investigation of the effect of incorporated NPs on mechanical properties and glassy transition of polymer bulk will not be considered in this chapter. This contribution does not intend to be exhaustive in the field of gold nanoparticles at polymeric interfaces, but instead to provide comprehensive discussion on critically selected papers relevant for the considered field. Although important, studies on conductive polymers and gold nanoparticle are not included in this paper, the reader is referred to the following references. [17, 18]

2. Preparation Methods of Gold Nanoparticles at Polymeric Interface

Various methods of designing nano-assemblies made of gold nanoparticles at polymeric interface have been reported in the literature based on which one may distinguish two main strategies depending on performing the mixing step with either (i) preformed objects or (ii) precursors (i.e., metal ions or monomers). Figure 1 displays general pathways for making polymer-GNP hybrids through one step or multistep mechanism.

The approach (*i*) consisted largely in the adsorption of *pre*-synthesized GNPs on *pre*-existing polymeric interface containing chemical groups allowing specific chelation of the NPs.

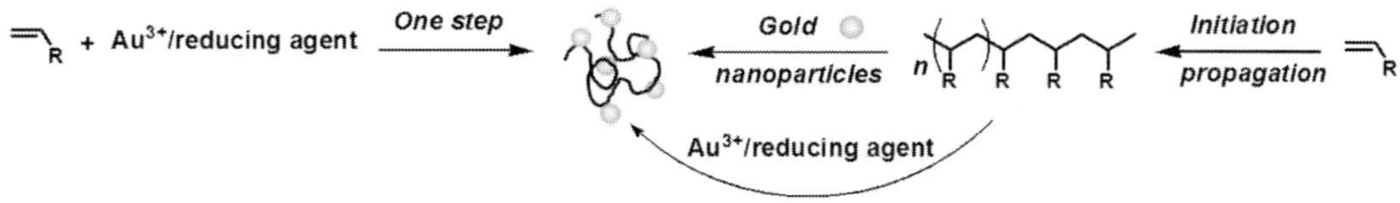

Figure 1. Schematic representation of pathways for the preparation of polymer-gold nanoparticles hybrid systems.

Functional groups, such as mercapto (-SH), nitrile (-CN), carboxylic (-COOH) and amine ($-NH_2$) are well-suited for controlling the adsorption behaviors of metal nanoparticles and gold nanoparticles in particular. By this method, synthetic control over each partners morphology, topology and chemical composition is easily attainable enabling a nearly limitless range of composite materials to be synthesized depending on the imagination of the designer. This strategy is particularly well-adapted for ultra-thin polymer films and porous polymers.

The strategy (*ii*) implies the preparation of one of the nanocomposite components in the presence of the other one and thus, implementation of successive step is generally required. For instance, pre-synthesized polymer systems with well-defined chemical functionality can be put into contact with a solution of metal precursors and the reduction-induced *in-situ* formation of the MNPs is performed in a further step. Alternatively, combination of thiol-gold affinity and thiol-ene curing process was successively proposed to prepare polymer coating containing well-dispersed nanogold particles from trifunctional thiol and ene monomers and nanometer-sized preformed GNPs. [19]

As a third strategy, it is also possible to prepare gold nanoparticle-containing polymer composites starting with monomers and metal salts precursors. In this case, both the GNPs and the polymer are formed at the same time. As a representative example one may mention the study by Yagci and co-workers dealing with the preparation of gold-cross-linked poly(ethylene glycol) nanocomposites by photoinduced electron transfer and free radical polymerization processes. [20] The use of a suitable photoinitiator, 1-[4-(2-hydroxyethoxy) phenyl]-2-hydroxy-2-methyl-1-propane-1-one, leads to the generation of radicals allowing starting the polyaddition reaction of acrylic resin, and, simultaneously, reducing Au^{3+} to Au^0. As a result, gold nanoparticles are formed *in situ* during the polymer network formation. This one pot strategy also proved efficient for conductive polymers. [21]

3. Gold Nanoparticles at Polymeric Interface for Separation Applications

As mentioned above, GNPs have gained importance in separation science because they can provide not only a large surface area-to-volume ratio giving rise to a significantly increased interaction area with the analytes but also an easy tunable surface chemistry via functionalization with thiol-containing molecules. In the field of chromatographic separation GNPs have been incorporated to polymer-based monoliths to prepare a large variety of stationary phases. The basic idea related to designing polymer monoliths relies on the thermally or photochemically driven free radical copolymerization of a reactive mixture containing functional monomers, cross-linkers, initiator and porogens. [22]

After completion of the polymerization and removal of the porogens, a crosslinked polymer is obtained. The structure consists of small globuli fused together into spheres of about 1 micron in size and forming a highly porous scaffold with, in-between, tortuous porous channel network. Major advantages of monolithic supports, are easy in-situ preparation and tuning of surface functionality and of up-most importance the fast mass transport achievable between the monolith polymer and the surrounding mobile phase. These transport phenomena have been well-described by various authors. [23] Preparation can be performed through a single step by copolymerizing a monomer bearing a chromatographic selector as side group or by a multi-step process involving the post-polymerization modification of the polymer monolith incorporating reactive monomers. [24] The latter approach has been largely experienced for preparing GNPs-decorated monoliths.

In a communication published in 2010, Connolly and co-workers reported for the first time a method to prepare high-capacity gold nanoparticle functionalized polymer monoliths. [25] Although no application was reported the study highlighted the importance of the chemical nature of the functional group on the monolith pore surface to provide attachment of 20 nm GNPs. The preparation pathway involves first, the synthesis of poly(butyl methacrylate-*co*-ethylene dimethacrylate) monolith through free radical polymerization of the respective monomers; second, the photografting of vinyl azlactone to cover the pore surface of with reactive moieties; and third, the grafting of either cysteamine or ethylenediamine to deliver NH_2 or SH surface attachment sites for GNPs. Demonstration was given that the amine groups on the monolith surface allow a full, homogenous and dense covalent attachment of GNPs as compared to the case where thiol groups were attached.

In another implementation, amino groups were grafted on the surface of porous monoliths *via* a photochemically-driven thiol-yne click strategy with the objective to spatially control the deposition of GNPs on the monolith surface. [26] The synthetic route started with the preparation of reactive macroporous monolith within th confines of fused-silica capillary by UV-initiated free radical polymerization of N-acryloxysuccinimide and ethylene dimethacrylate in toluene, as unique porogen (Figure 2). Subsequent post-functionalization with propargylamine gave rise to monolith with thiol-sensitive surface moieties through standard electrophile-nucleophile chemistry. Monolith with surface-grafted alkyne groups was then reacted with cysteamine through radical-mediated thiol–yne addition. Photochemical initiation permitted spatial control over the reaction site and further chelating character of the monolith interface. Site-specific adsorption of 20 nm size GNPs was possible upon simple dynamic loading with colloidal solution of commercially available GNPs. The chelating behavior of the amino-containing monolith areas was assigned to their hydrated state, in other words highly hydrophilic nature.

In several articles published in the 2010-2012 period, the group of Svec from Berkeley, the major contributor to innovation in the field of monolithic stationary phases for chromatographic separation purposes, has discussed the application of monolithic supports containing GNPs to the separation of small and large molecules. [27] Thiol-modified monolithic columns were prepared from a generic poly(glycidyl methacrylate-*co*-ethylene dimethacrylate) monolith through reaction of pendant epoxide moieties with cysteamine.

Two strategies were applied for the immobilization of GNPs. Attachment of GNPs on the monolith surface was possible *via* either *in situ* (within the monolithic column) reduction of chloroauric acid or flowing of a solution of colloidal GNPs through the pore channels of the monoliths.

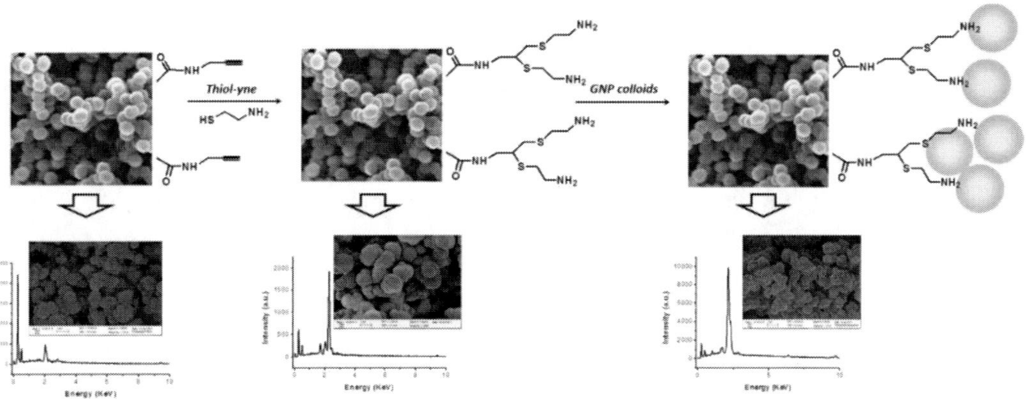

The figure is adapted with permission from reference [26].

Figure 2. Schematic representation of the successive synthesis steps for the chemical modification of alkyne-functionalized monolith surface; cysteamine is grafted *via* thiol-yne click chemistry leading to amino functionalized monolith surface allowing the subsequent immobilization of gold nanoparticles *via* adsorption. In the scheme bis-cysteamine addition on alkyne units is assumed to be the major reaction mechanism.

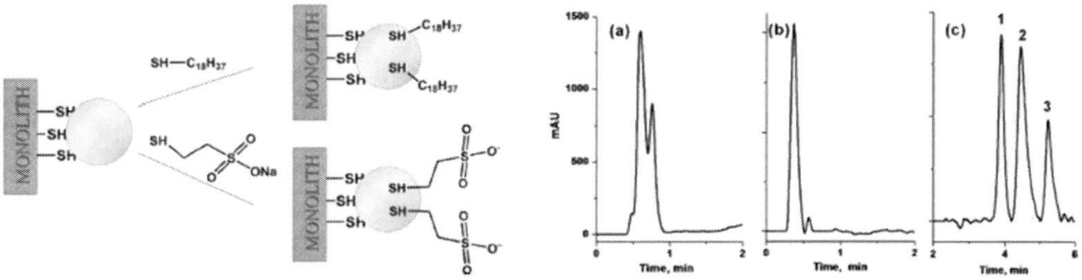

The two figures are reproduced with permission from reference [27b].

Figure 3. (Left hand side) schematic showing the surface modifications of a monolith containing gold nanoparticles through a process of exchangeable chemistry based on gold-thiol affinity. The use of octadecylthiol and sodium-2-mercaptoethanesulfonate afforded monolithic columns for reversed phase and ion exchange chromatography, respectively. (Right hand side) chromatograms obtained for the injection of a protein mixture made of (1) ribonuclease A, (2) cytochrome c and (3) myoglobin on the monolith column after the various synthesis step detailed here-above. Briefly, (a) generic poly(glycidyl methacrylate-*co*-ethylene dimethacrylate) monolithic, (b) gold nanoparticles functionalized poly(glycidyl methacrylate-*co*-ethylene dimethacrylate) monolith and (c) after surface functionalization of the gold nanoparticles with 1-octadecanethiol.

The authors concluded that although the first approach is faster, the resulting coverage of nanoparticles is lower than the one achievable through the second method. Affinity of gold for thiol-containing molecules was exploited successfully for the *in situ* surface functionalization of the GNPs with a series of chromatographic selectors, and for the capture of cystein containing peptides (Figure 3).

The very attractive feature of the study is that the type of GNPs-attached grafts can be easily varied due to the exchangeable nature of the thiol-gold surface interaction.

In this respect, different types of monolithic stationary phases incorporating GNPs with specific surface groups such as carboxylic acid, hydroxyl, amine, sulfonated and aliphatic units were designed for ion exchange and reversed-phase (electro)chromatographic modes.

Similarly, reversibility of the capture process of cysteine-containing peptide was possible in the presence of an excess of 2-mercaptoethanol acting as a competiting agent. A loading capacity of 2.58 μmol/m as determined using frontal elution was reported for L-cysteine. With the aim to enhance the surface coverage of the monolith pore surface with gold nanoparticles, poly(GMA-*co*-EDMA) monoliths were first reacted with cystamine, followed by the reduction of the disulfide bonds with tris(2-carboxylethyl)phosphine producing SH functionality. [28]

By this method, the surface coverage reached a value of 51.6wt% for gold nanoparticles with a diameter of 20 nm that is more than a twofold increase as compared to the value reported by Conolly.

Cysteamine-functionalized poly(glycidyl methacrylate-*co*-ethylene dimethacrylate) served also as GNPs-support for preparing solid phase microextractor. [29] Here again, the specificity of the formation of Au–S bonds was beneficial for the concentration of thiol-containing drugs such as captopril in plasma.

4. Gold Nanoparticles at Polymeric Interface for Sensing Applications

In the context of sensing, incorporation of GNPs within polymeric matrices was found to enhance significantly the sensors response thanks to the unique opto-electronic properties of the dispersed nano-metals. Nanoparticle-enhanced surface plasmon resonance (SPR) for ultra-sensitive detection has been reported over the last few years for the detection of large and small molecules. [30] This approach was extensively applied to the case of the detection of low molecular weight molecules using molecularly imprinted polymer, MIP in short, films where the *in-situ* and *during polymerization* generated nanocavities serve as recognition elements for the target molecules. [31] Tokareva and co-workers described the combined use of ultrathin molecularly imprinted polymer and enhanced transmission surface plasmon resonance spectroscopy for the detection of cholesterol. [32] The authors have shown that the presence of cholesterol into the MIP was accompanied with an easily measureable shift (56 nm) of the absorption maximum of the interacting gold nanoislands and nanoparticles which sandwich the MIP. Enhancement of the electrochemical response of MIP sensors for dopamine was also reported thanks to the presence of embedded GNPs as compared to related MIP films sensors without GNPs. [7,33] The preparation strategy relied on a two-step electrografting process (Figure 4). (*i*) Thiol-functionalized gold plates were prepared using diazonium salt and were used as immobilization platform for GNPs. (*ii*) Photoinitiator derived from 4-benzoylbenzenediazonium salt was immobilized on top of the GNPs allowing the further covalent anchoring of thin molecularly imprinted polymer film through surface initiated photo-driven free radical polymerization of methacrylic acid crosslinked by ethyleneglycol dimethacrylate in presence of dimethylaniline serving as co-initiator. For the as-synthesized sensing layers, specific and selective recognition ability towards dopamine was obtained with a limit of detection (LOD), as measured by square wave voltammetry, of 0.35 $nmol.L^{-1}$. This value is almost 3 times lower than the LOD achieved without any embedded gold NPs. Covalent attachment of the MIP film on the GNPs resulted in high stability and reproducibility to the prepared sensor.

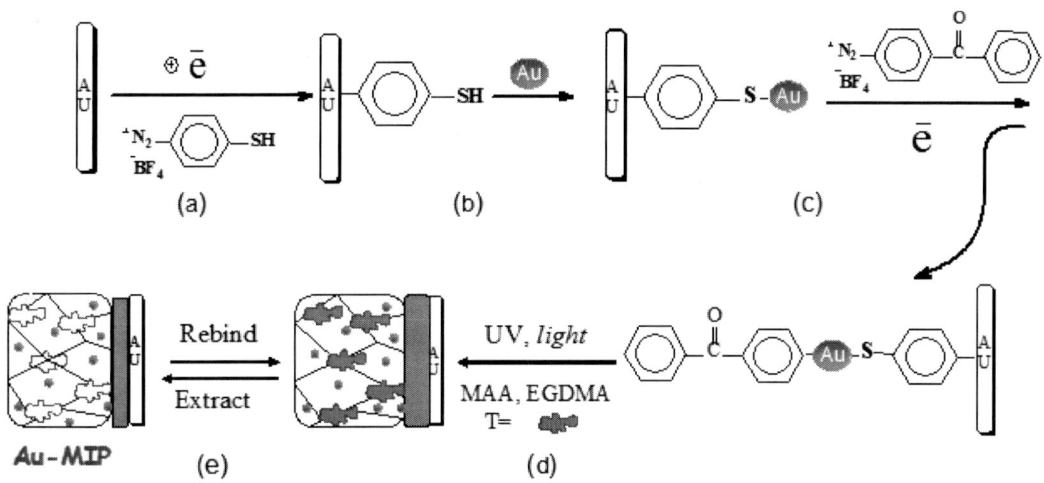

The figure is reproduced with permission from reference [7].

Figure 4. Illustration presenting the preparation protocol of molecularly imprinted polymer grafted films with embedded gold nanoparticles on gold electrodes. The successive steps are (a) modification of gold electrode surface with mercaptobenzene diazonium salt; (b) attachment of gold nanoparticles to the modified Au electrode *via* the sulfhydryl moiety of the aryl layer; (c) electrografting of 4-benzoyl benzene diazonium salt serving in (d) as photoinitiator; (d) photopolymerization of the MIP formulation consisting of methacrylic acid, ethylene glycol dimethacrylate and template molecule; (e) principle of the sensor based on reversible redinding and extraction processes.

A sensitive molecularly imprinted electrochemical sensor incorporating Au nanoparticles was created for the selective detection of imipramine, a tricyclic antidepressant. [34] The thin molecularly imprinted film-based sensor was fabricated onto the indium tin oxide electrode through stepwise modification of GNPs by self-assembly and a thin film of MIPs via sol–gel technology. The authors observed that the molecularly imprinted film displayed excellent selectivity towards the target molecule imipramine. Meanwhile, the introduced GNPs exhibited noticeable catalytic activities towards imipramine oxidation. This effect permitted to remarkably enhance the sensitivity of the imprinted film. Due to such combination, quick response to imipramine was obtained for the as-prepared sensor within only 1 min of incubation. It was also found that the differential voltammetric anodic peak current exhibited linear variation as a function of the logarithm of imipramine concentration in the concentration range 5.0×10^{-6} to 1.0×10^{-3} $mol.L^{-1}$. Moreover, detection limit of 1.0×10^{-9} $mol.L^{-1}$ was achieved. This proposed method was successfully extended to the determination of imipramine in drug tablets, and proven to be reliable compared with conventional UV method. These obtained results reveal that such GNPs-combined polymer sensor fulfills the selectivity, sensitivity, speed and simplicity requirements for clinical application in physiological fluids.

An elegant pathway to take advantage of both properties of polymeric supports and GNP relies on the combination of responsive polymers with nanosized gold particles. Polymer brushes consisting of pH-responsive polymers were produced *via* atom-transfer radical polymerization (ATRP) initiated from the surface of GNPs. [35] For this purpose, citrate stabilized gold nanoparticles were chemically modified by a disulfide ATRP initiator prior to proceed to the growth of 4-vinylpyridine (4VP) chains from the surface of gold particles.

The Au@P4VP assembled nanocomposites demonstrated pH-responsive behavior in relation to the protonation/deprotonation properties of the pyridyl groups. Indeed, P4VP is prone to protonation at low pH.

As a result of the inherent positive charge in acidic medium (pH <3.2) the polymer chains grafted on GNPs are expanded by electrostatic forces, and the polymer layer is loosely swelled. In contrast, at pH above 3.2, the polymer chains shrink and wrap around the gold particle surface, leading to the aggregation of the latter having around a thin shrunken polymer layer as observed by TEM observation. Such pH-induced conformation changes were well-documented by 3-D AFM images (Figure 5).

The concept of responsive polymer brushes@GNP hybrids for sensing application has been extended by the group of Stamm to the elaboration of nano-assemblies in the form of flat surface tethered polymer brushes and GNP. [36] Polystyrene and poly(N-isopropyl acrylamide) (PNIPAM) brushes were used as solvent and temperature responsive polymers, respectively. In the former study, immobilization of gold nanoparticles was achieved on end-functionalized polystyrene brushes grafted on an underlying substrate. In the latter report, the fabrication process involved the grafting of carboxyl terminated PNIPAAm chains on an underlying substrate in a brush conformation followed by the immobilization of surface functionalized Au NPs through physical interaction based on hydrogen bonding.

The presence of gold nanoparticles on both types of polymer brushes was confirmed by a combination of experimental methods (atomic force microscopy and X-ray photoelectron spectroscopy). The resulting polystyrene (PNIPAAm)–Au nanoassemblies have been used as nanosensors for the detection of a variety of organic solvents (temperature changes) in surrounding media. The sensing mechanism is based upon the change in the proximity of the immobilized gold nanoparticles as a consequence of the solvent (temperature) induced reversible swelling–deswelling of polystyrene (PNIPAAm) chains (Figure 6).

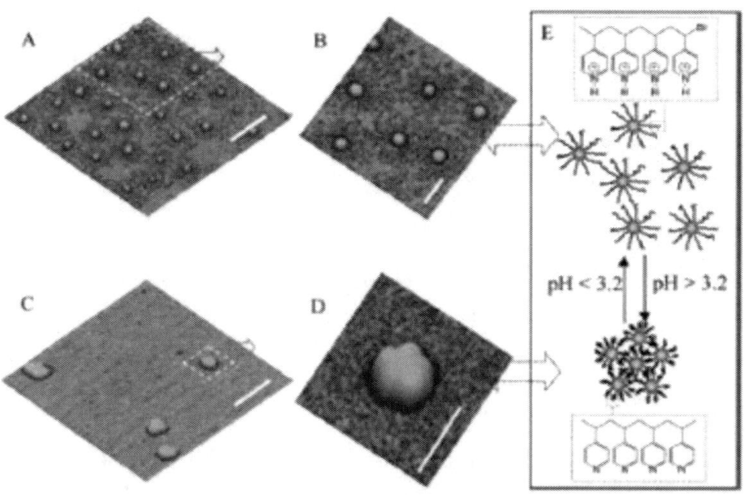

Figure 5. 3D-AFM images obtained for Au@P4VP nano-assembled particles at different pH values are presented in regard to the assumed conformation-induced mechanism of pH response of Au@P4VP nanocomposites. (A, B) pH = 1.2, and (C and D) pH = 5.0. The white bars in AFM images present size of (A, C) 500 nm and (B and D) 200 nm.

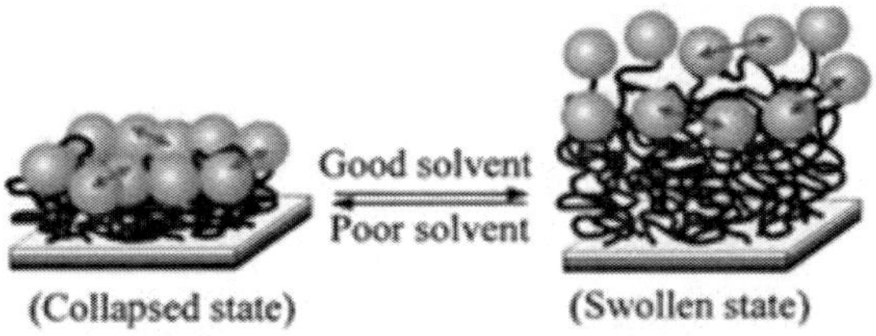

Figure 6. Scheme showing the change in chains conformation observed for surface-attached responsive polymer brushes end-decorated with gold nanoparticles upon change in solvent. Depending on the solvent quality, the polymer brushes expanse towards the solution, i.e. outwards the surface. The arrows in the scheme indicate the resulting local changes in distance between gold nanoparticles. Such effect can induce a variation in a measurable signal at the origin of the nanosensing process. The figure is reproduced with permission from reference [36b]. Such swelling/shrinking process can also be observed for thermoresponsive polymer brushes upon change in temperature.

More specifically, the sensing ability of PS-GNPs systems was easily demonstrated by UV−vis spectroscopy measurements, through monitoring the significant blue shift of 32 nm in the plasmon resonance band of immobilized Au nanoparticles as changing the surrounding media from air to the toluene. PNIPAM brushes have been found to suppress the aggregation of NPs and, hence, facilitate the complete surface coverage. The described approaches are facile and versatile in nature, i.e. of a large interest for fabrication of a broad variety of nanosensors based on responsive polymer brushes−nanoparticle assemblies.

PNIPAM chains grafted on gold surface and terminated by GNPs were also investigated for surface enhanced Raman scattering (SERS) applications by the group of Chehimi. [37] Plasmonic nanostructures were prepared using a smart synthesis strategy relying on a combination of diazonium salt electrografting, surface-initiated atom transfer radical polymerization and (SI-ATRP), and click chemistry. For thin polymer layer of about 20 nm, temperature-driven phase transition observed through crossing over the PNIPAM low critical solution temperature (~37°C) was accompanied by a significant change in the SERS signal assigned to the presence of adsorbed methylene blue molecules. Importantly, these variations were found to be fully reversible, making the hybrid layered structure a thermal switch for near-field coupling regime.

5. Gold Nanoparticles at Polymeric Interface for Catalysis Applications

In the catalysis domain, polymers have been used as supports because of their ability to strongly bind and disperse metallic nanoparticles. Ballauff's research group described the use of spherical polyelectrolyte brushes of linear chains morphology as templates for the immobilization of GNPs. [38]

In their study, poly(2-methylpropenoyloxyethyl) trimethylammonium chloride served for immobilizing $AuCl_4^-$ ions and generating, upon addition of a reducing agent, namely $NaBH_4$, GNPs with a diameter of about 1.25 nm. Cryo-TEM analyses clearly evidenced the presence of the nanoparticles near the surface of the PS core particles. Catalytic activity of the supported GNPs was investigated towards the reduction of nitrophenol.

In addition to their potential as pH nanosensors, the assembled Au@P4VP nanocomposites described in Reference [35] are well-suited as smart supports for entrapping transition metal ions. Subsequently, the metal ions can be reduced *in situ* allowing designing novel bimetallic nanocomposites, which are regarded as intelligent catalysts with environmental stimuli activity. Inspired by our previous work on designing nanostructured Pd-based catalysts we have recently extended our ATRP-click chemistry tandem approach to the design of CNT@polymer brushes@GNPs hybrids, where CNT stands for carbon nanotubes. [39] For the specific case of gold, the nature of the chemical functionality on the polymer brushes was changed to amine groups as compared to carboxylic units previously used for Pd NPs (Figure 7). In short, carbon nanotube-grafted P4VP (CNT@P4VP) were prepared by Huisgen click reaction between alkynylated CNT*s* (CNT-C≡CH) and azide-terminated P4VP chains (P4VP-N_3). Chemical modification of the CNTs was possible using alkynylbenzenediazonium salt as a surface modifier whilst P4VP was synthesized by ATRP and post-modified using NaN_3. The CNT@P4VP and reference materials were characterized by a combination of complementary experimental techniques. All the collected data confirmed the proposed structures. Of particular interest was the demonstration of the pH-responsive behaviour on the basis of a quantitative analysis of the XPS signal assigned to the protonated fraction of nitrogen atoms in the pyridinyl groups of P4VP. The CNT@P4VP nanocomposite was further shown to be efficient nanoplatform for immobilizing GNPs leading to complex, while well-defined, CNT@P4VP@Au heterostructure.

From the above, it is clear that the tandem diazonium surface chemistry and "click" reactions is a simple and efficient process for the design of *hairy* carbon nanotubes for a variety of purposes. Gold nanoparticles-cored dendrimers were successfully prepared through simultaneous reduction of of $HAuCl_4$ and diazo-functionalized Dendron. [40] These materials consisted of a gold core of approximately 5 nm in diameter to which the dendrons of generations 1–4 are connected radially by means of gold–carbon bonds. Although the authors did not mention any application, catalytic activity could be envisioned as it was previously shown for their Pd-counterparts. [41] These studies stress also the key role of diazonium compounds in modern surface and interface materials chemistry. [42, 43]

CONCLUSION

Designing gold nanoparticles-decorated polymeric interfaces for a broad rand of application domains has become a major research field in nanoscience. Indeed, nanostructured systems with unprecedently reported functions can be prepared thanks not only to the intrinsic properties of nano-gold and polymers but also due to a synergic effect resulting from properties combination of each partner.

Choice of the macromolecular architecture, typically topology, chemical composition and distribution of the functional groups along the polymer chains of the polymeric supports is driven by the applications. Herein, after a brief description of strategies for making the discussed materials, we focused on three applications namely separation, detection and catalysis. In this respect, polymer platforms in the form of thin polymeric films made of either surface-tethered chains, in other words polymer brushes, or covalently attached crosslinked polymer, porous monoliths and polymer particles are presented.

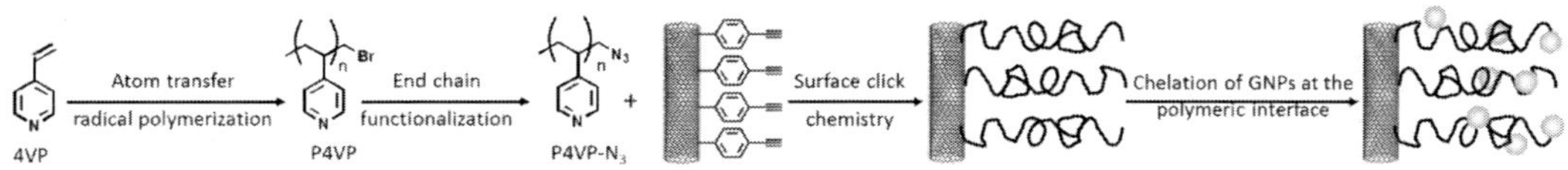

The figure is adapted from Reference [39b].

Figure 7. Schematic representation of the whole elaboration pathway for designing nanohetero-structured materials made of a combination of carbon nanotubes, polymer brushes and gold nano-particles.

As a relevant example of synergic effect observed for polymer-GNPs composites, one may cite the case of nanosensors based of responsive polymer brushes. In this case, external stimuli-induced changes in chain conformation is easily detected with a high sensitivity through GNPs-induced signal enhancement.

REFERENCES

[1] (a) Haraguchi, K., Li, H. J. *Macromolecules* 2006, 39, 1898. (b) Hemmasi, A. H., Khademi-Eslam, H., Talaiepoor, M., Kord, B., Ghasemi, I., *J. Reinf. Plast. Comp.* 2010, 29, 964.

[2] Schexnailder, P., Schmidt, G. *Colloid. Polym. Sci.* 2009, 287, 1-11,

[3] Park, J. H., Lim, Y. T., Park, O. O., Kim, J. K., Yu, J. W., Kim, Y. C., *Chem. Mater.* 2004, 16, 688.

[4] Mahouche-Chergui, S., Ledebt, A., Mammeri, F., Herbst, F., Carbonnier, B., Ben Romdhane, H., Delamar, M., Chehimi, M. M., *Langmuir* 2010, 26, 16115.

[5] Xing, R., Liu, Y., Wu, H., Li, X., He, M., Wu, P. *Chem. Commun.* 2008, 47, 6297.

[6] Pradhan, B., Setyowati, K., Liu, H., David, H. Waldeck and Jian Chen *Nano. Lett.* 2008, 8, 1142.

[7] Gam-Derouich, S., Mahouche-Chergui, S., Truong, S., Ben Hassen-Chehimi, D., Chehimi, M. M. *Polymer* 2011, 52, 4463.

[8] Hayat, M. A. (ed.), *Colloidal Gold: Principles, Methods and Applications*, San Diego, CA: Academic Press, Vols. 1 and 2, 1989.

[9] Grzelczak, M., Perez-Juste, J., Mulvaney, P., Liz-Marzan, L. M. *Chem. Soc. Rev.* 2008, 37, 1783.

[10] Li, N., Larson, T., Nguyen, H. H., Sokolovb, K. V., Ellington, A. D. *Chem Commun.* 2010, 46, 392.

[11] Connor, E. E., Mwamuka, J., Gole, A., Murphy, C. J., Wyatt, M. D. *Small*, 2005, 1, 325.

[12] Han, J., Liu, Y., Guo, R. *J. Am. Chem. Soc.* 2009, 131, 2060.

[13] Dreaden, E. C., Alkilany, A. M., Huang, X., Murphy, C. J., El-Sayed, M. A. *Chem. Soc. Rev.* 2012, 41, 2740.

[14] Corbierre, M. K., Cameron, N. S., Sutton, M., Mochrie, S. G. J., Lurio, L. B., Rühm, A., Lennox, R. B. *J. Am. Chem. Soc.*, 2001, 123, 10411.

[15] (a) Teranishi, T., Kiyokawa, I., Miyake, M. *Adv. Mater* 1998, 10, 596. (b) Shimizu, T., Teranishi, T., Hasegawa, S., Miyake, M. *J. Phys. Chem. B* 2003, 107, 2719.

[16] Youk, J. H. *Polymer* 2003, 44, 5053.

[17] (a) Qiu, L., Peng, Y., Liu, B., Lin, B., Peng, Y., Malik, M. J., Yan, F. *Appl. Catal. A General* 2012, 413-414, 230. (b) Selvan, S. T., Spatz, J. P., Klok, H.-A., Möller, M. *Adv. Mater.*, 1998, 10, 132. (c) Fujii, S., Mouri, M., Aichi, A., Akamatsu, K., Nawafune, H., Nakamura, Y. *J. Adh. Soc. Jpn.*, 2008, 44, 12.

[18] Mangeney, C., Bousalem, S., Connan, C., Vaulay, M.-J., Bernard, S., Chehimi, M. M. *Langmuir* 2006, 22, 10163.

[19] Phillips, J. P., Mackey, N. M., Confait, B. S., Heaps, D. T., Deng, X., Todd, M. L., Stevenson, S., Zhou, H., Hoyle, C. E. *Chem. Mater.* 2008, 20, 5240.

[20] Yagci, Y., Sangermano, M., R. Giancarlo. *Chem. Commun.* 2008, 2771.

[21] Breimer, M. A., Yevgeny, G., Sy, S., Sadik, O. A. *Nano Lett.,* 2001, 1, 305.

[22] Buchmeiser, M. R. *Polymer* 2007, 48, 2187.

[23] (a) Rodrigues, A. E., Lopes, J. C., Lu, Z. P., Loureiro, J. M., Dias, M. M. *J. Chromatogr.* 1992, 590, 93. (b) Rodrigues, A. E., Loureiro, J. M., Chenou, C., De la Vega, M. R. *J. Chromatogr. A* 1995, 664, 233. (c) Rodrigues, A. E., *J. Chromatogr. B* 1997, 699, 47. (d) Siouffi, A-M., *J. Chromatogr. A* 2006, 1126, 86. (d) Meyers, J. J., Liapis, A. I. *J. Chromatogr. A* 1999, 852, 3.

[24] (a) Peters, E. C., Petro, M., Svec, F., Fréchet, J. M. J. *Anal. Chem.* 1997, 69, 3646. (b) Peters, E. C., Petro, M., Svec, F., Fréchet, J. M. J. *Anal. Chem.* 1998, 70, 2288.

[25] Connolly, D., Twamley, B., Paull, B., *Chem. Commun.* 2010, 46, 2109.

[26] Guerrouache, M., Mahouche-Chergui, S., Chehimi, M. M., Carbonnier. B. *Chem. Commun.*, 2012, 48, 7486.

[27] (a) Xu, Y., Cao, Q., Svec, F., Fréchet, J. M. J. *Anal. Chem.* 2010, 82, 3352. (b) Cao, Q. Xu, Y. Liu, F. Svec, F. Fréchet, J. M. J. *Anal. Chem.* 2010, 82, 7416.

[28] Lv, Y., Alejandro, F. M., Fréchet, J. M. J., Svec, F., *J. Chromatogr. A* 2012, 1261, 121-128.

[29] Zhang, A-Z., Ye, F-G., Lu, J-Y., Wei, Z., Peng, Y., Zhao, S-L., *Chin. J. Anal. Chem.* 2011, 39, 1247.

[30] (a) Kim, S., Lee, J., Lee, S. J., Lee, H. J., *Talanta*, 2010, 81, 1755. (b) Fang, S., Lee, H. J., Wark, A. W., Corn, R. M. *J. Am. Chem. Soc.* 2006, 128, 14044.

[31] Lépinay, S., Kham, K., Millot, M-C., Carbonnier, B. *Chem. Pap.* 2012, 66, 340.

[32] Tokareva, I., Tokarev, I., Minko, S., Hutter, E., Fendler Janos, H. *Chem. Commun.,* 2006, 3343.

[33] Gam-Derouich, S., Jouini, M., Ben Hassen-Chehimi, D., Chehimi, M. M. *Electrochim. Acta* 2012, 73, 45.

[34] Xu, X., Zhou, G., Li, H., Liu, Q., Zhang, S., Kong, *J. Talanta* 2009, 78, 26.

[35] Li, D., He, Q., Cui, Y., Li, *J. Chem. Mater.* 2007, 19, 412-417.

[36] (a) Gupta, S., Agrawal, M., Uhlmann, P., Simon, F., Stamm, M., *Chem. Mater.* 2010, 22, 504. (b) Gupta, S., Agrawal, M., Uhlmann, P., Simon, F., Oertel, U., Stamm, M., *Macromolecules* 2008, 41, 8152-8158.

[37] Gehan, H., Fillaud, L., Chehimi, M. M., Aubard, J., Hohenau, A., Felidj, N., Mangeney, C. *ACS Nano* 2010, 4, 6491.

[38] Yu, M., Lu, Y., Schrinner, M., Polzer, F., Ballauff, M. *Macromol. Symp.* 2007, 254, 42.

[39] (a) Mahouche Chergui, S., Ledebt, A., Mammeri, F., Herbst, F., Carbonnier, B., Ben Romdhane, H., Delamar, M., Chehimi, M. M. *Langmuir* 2010, 26, 16115. (b) Salmi, Z., Epape, C., Mahouche-Chergui, S., Carbonnier, B., Omastová, M., Chehimi, M. M. *J. Colloid Sci. Biotechnol.* 2013, 2, 53-61.

[40] Kumar, V. K. R., Gopidas, K. R. *Chem. Asian J.* 2010, 5, 887.
[41] Kumar, V. K. R., Gopidas, K. R. *Tetrahedron Letters* 2011, 52, 3102.
[42] Bélanger, D., Pinson, *J. Chem. Soc. Rev.* 2011, 40, 3995.
[43] *Aryl Diazonium Salts. New Coupling Agents in Polymer and Surface Science*, M. M. Chehimi, ed., Wiley-VCH, Weinheim, 2012).

POLYPYRROLE NANOSTRUCTURES

In: Applied Surface Chemistry of Nanomaterials
Editors: Mohamed M. Chehimi and Jean Pinson
ISBN: 978-1-62808-351-4

Chapter 9

NANOSTRUCTURED POLYPYRROLE MATERIALS: FOCUS ON TEMPLATELESS SYNTHETIC METHODS AND ON SOME APPLICATIONS

Catherine Debiemme-Chouvy***
Laboratoire Interfaces et Systèmes Electrochimiques
UPR15 du CNRS, Université P. et M. Curie, Paris - France

INTRODUCTION

Among the intrinsically conducting polymers, organic polymers with π-conjugated chains, polypyrrole (PPy) is one of the most important because of its high electronic conductivity, environmental stability, ion exchange capacity and biocompatibility. [1] Polypyrrole is usually synthesized by oxidative polymerization of pyrrole monomers. It could be prepared either by a chemical route, in this case pyrrole monomers are oxidized by an oxidizing agent such as iron(III) chloride ($FeCl_3$) or ammonium persulfate (= peroxydisulfate) (APS, $(NH_4)_2S_2O_8$) or by an electrochemical route [2] under potentiostatic, galvanostatic (intentiostatic) or potentiodynamic conditions. In general PPy is obtained under a powder form dispersed in the solution by the chemical route whereas by electrochemistry the polymer is deposited on the electrode surface as a compact film having a cauliflower-like morphology. The former permits the preparation of a high amount of polymer in bulk whereas electropolymerisation allows a deposition of the polymer as a film on the conductive substrate used as the working electrode. The pyrrole monomers can be dissolved in aqueous (monomer solubility: 0.9M) or in organic solvents whereas polypyrrole is not soluble in these solutions.

The (electro)chemical oxidation of pyrrole entities leads to the synthesis of polypyrrole in its oxidized form (doped form), following the reaction:

$$n\,Py + \gamma n\,A^- \rightarrow [(Py)_n^{(\gamma n)+}, \gamma n\,A^-] + 2(n-1)\,H^+ + ((2+\gamma)\,n-2)\,e^- \qquad (1)$$

* Corresponding author: Catherine Debiemme-Chouvy. LISE – UPR15 du CNRS, Université P. et M. Curie. 4 place Jussieu – 75005 Paris – France. Email : catherine.debiemme-chouvy@upmc.fr.

where γ stands for the doping level of the polymer, in general it is between 0.25 and 0.33. From this equation, it is interesting to note on the one hand that Py oxidation leads to the release of protons *i.e.* at the solution/polymer interface the pH decreases during Py polymerization, and on the other hand that the presence of anions (A^-) in the solution is essential. For electrochemical syntheses, it has been evidenced that the anodic current (Py polymerisation rate) is proportional to the anion concentration when this concentration is inferior to 1 M ($I = k\,[A^-]$). [3] Polypyrrole can be reversibly switched between conductive (p-doped state) and insulating (neutral state) states. In order to maintain charge neutrality during the dedoping process (reduction) either anions are expelled from the polymer backbone ('anionic' polymer) or cations are integrated into the polymer backbone ('cationic' polymer) and conversely during the polymer reoxidation:

$$[(Py)_n^{(\gamma n)+}, \gamma n\, A^-] + \gamma n\, e^- \leftrightarrows (Py)_n + \gamma n\, A^- \quad (2)$$

$$[(Py)_n^{(\gamma n)+}, \gamma n\, A^-] + \gamma n\, e^- + \gamma n\, C^+ \leftrightarrows [(Py)_n, \gamma n\, A^-, \gamma n\, C^+] \quad (3)$$

Actually, the reactions ((2) or/and (3)) which occur during the reduction/reoxidation of the polymer depend mainly on the size and the shape of the doping anions. [4-8] For example, using *ac*-electrogravimetry [9, 10], for polypyrrole doped with polyoxometallate species such as $SiMo_{12}O_{40}^{4-}$, it has been shown that in pH=0.5 solution only protons are involved in the charge compensation process (no cation and no solvent molecule are implied), [11, 12] whereas for films co-doped with monohydrogenosulfate and polyoxometallate anions both, cations and anions, are involved in the electrochemical processes. [13]

In order to produce polypyrrole nanostructures such as nanowires, nanotubes, nanoparticles, nanonetworks formed by the interconnection of nanowires, nanosheets, nanospheres, two categories of methods can be employed, the template-assisted methods that rely on external templates (hard template) or additives (soft template) to direct the polymer growth and the template-free methods. [14-18] The former processes can involve the use of a hard membrane template with nanopores of uniform diameter and length such as anodized aluminum oxide or a soft-template, typically consisting of a surfactant that can guide the growth of the polymer in or around self-assembled micelles but could also be gas bubbles. [19]

The main drawback of the use of template is the need to remove it after the growth of the nanostructures since this post-synthesis operation could lead to the alteration of the polymeric nanostructures. Chemical routes that do not use external templates typically exploit the predisposition of the material to self-assemble into well-defined morphologies. [20]

Nanostructured polymers have several potential advantages over their bulk counterparts. Indeed nanostructured materials provide enhanced surface-area-to-volume characteristics, electronic conductivity especially for fibers having diameter inferior to 150 nm [21] and light weight. Polypyrrole nanoparticles are very often prepared chemically in the presence of stabilizer [22] and unimodal spherical particles are generally obtained ranging from 50 to 400 nm in diameter, depending on the nature of the stabilizer and its concentration and molecular weight.

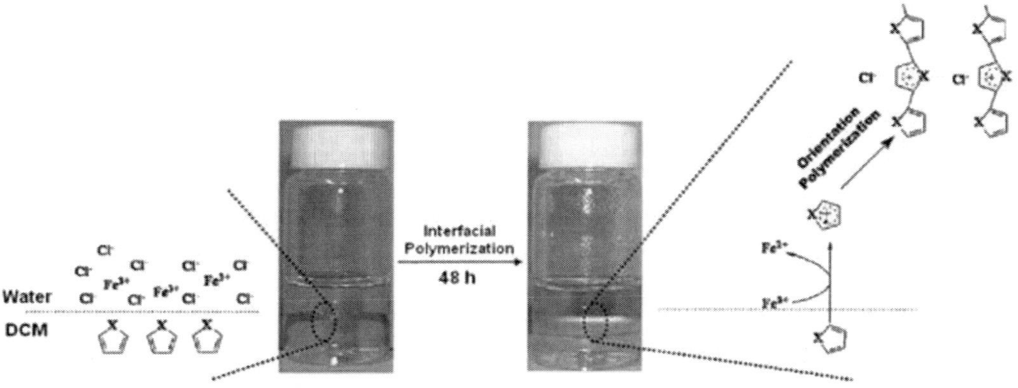

Adapted from Reference [26].

Figure 1. Interfacial polymerization of PPy (X = N) nanoneedles. Slow pyrrole polymerization through the dichloromethane (DCM) + 1 mg mL^{-1} Py / water + 0.1 mg mL^{-1} $FeCl_3$ interface.

Reprinted with permission from Reference [26].

Figure 2. (A) TEM image of PPy nanoneedles obtained by interfacial polymerization. (B) HR-TEM image of a PPy nanoneedle, scale bar = 3 nm. The arrow shows the direction of the long axis of 001.

This chapter focuses on the synthetic methods, recently reported in the literature, for the fabrication of nanostructured polypyrrole materials without the use of hard or soft templates. [23] Chemical routes are described in the first part, in the second part various electrochemical processes are presented. Finally some applications [20] of nanostructured polypyrrole materials are described.

1. Synthetic Methods

1.1. Chemical Route

In general, in solution the conducting polymer nanoparticles obtained by oxidation of the monomers are prevented from extensive agglomeration by steric stabilizers adsorbed or covalently bound to particle surface. [22, 24]

However polypyrrole nanostructures have been synthesized without stabilizer notably using an aqueous/organic or an aqueous/ ionic liquid [25] interfacial polymerization process. Using this method, Matsui *et al.* have been able to obtain single crystalline PPy nanoneedles. [26] In this process, the polymerization is performed in an immiscible aqueous/organic biphasic system. The monomer is dissolved in one phase and the oxidizing agent in the other, the principle of this method is summarized in Figure 1. In this case, the pyrrole monomers are dissolved in dichloromethane whereas the aqueous phase contains the oxidant, $FeCl_3$; pyrrole and oxidant are at low concentration, 1 mg/mL and 0.1 mg/mL, respectively. Needle-shape polypyrrole is obtained after 48 hours. The average length and diameter of these nanostructures are 70 and 20 nm, respectively (Figure 2). It was shown that the conductance switching of this crystalline polymer is fast in the time scale of milliseconds.

Besides, it has been pointed out that the oxidant has a significant role on the size of the polymer structure. [27] Notably it has been shown that the use of persulfate instead of iron(III) as the oxidizing agent permits to reduce the size of the PPy particles by about one order of magnitude.

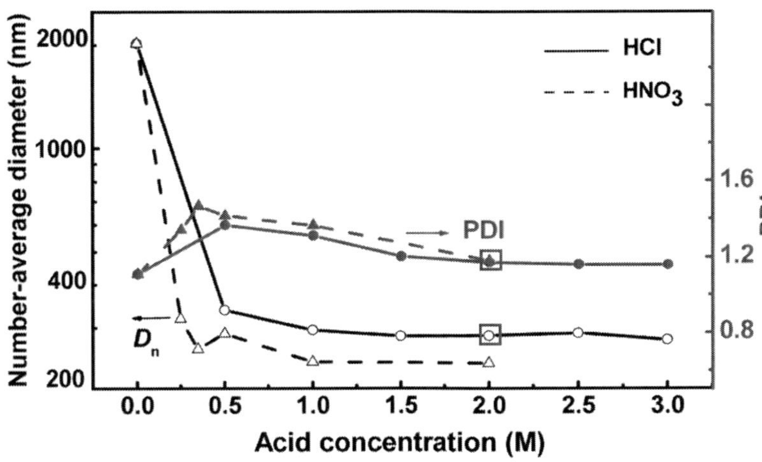

A

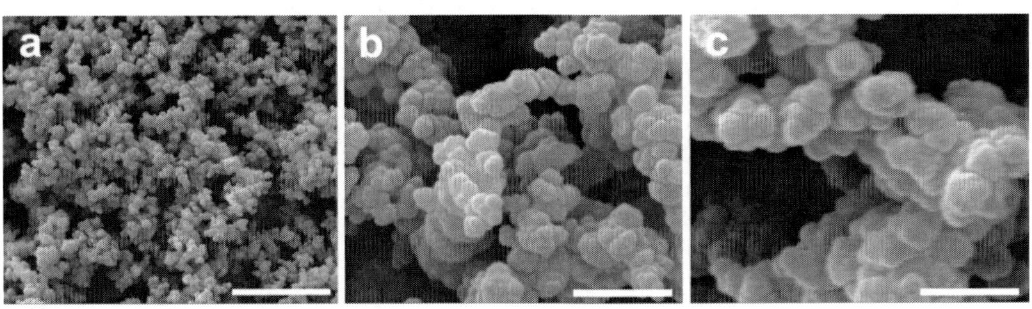

B

Adapted from Reference [28].

Figure 3. A) Influence of HCl and HNO_3 concentrations on average diameter (D_n) and polydispersity index (PDI=D_w/D_n) of the PPy particles synthesized in aqueous solutions with a fixed $(NH_4)_2S_2O_8$/ pyrrole molar ratio of 1 for 6 h, at 0 °C. B) SEM images of PPy nanoparticles obtained using a 2.0 M HCl solution (framed scatters in A). Scale bars: (a) 5 μm, (b) 1 μm, (c) 500 nm.

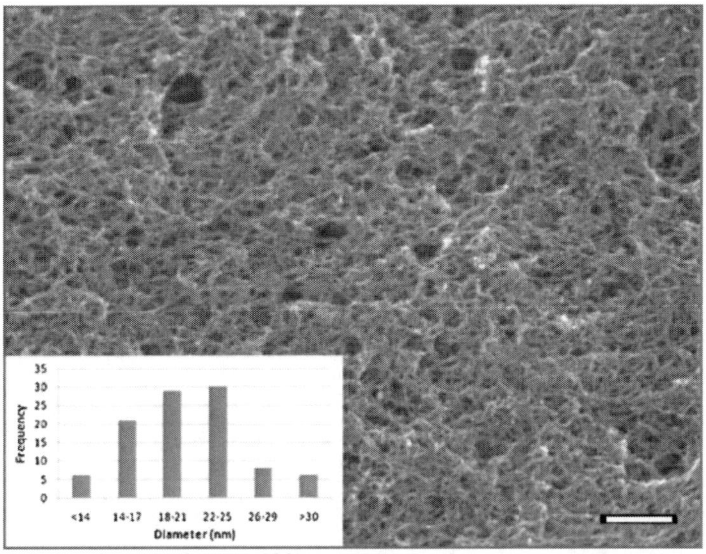

Reprinted with permission from Reference [32].

Figure 4. SEM image of PPy nanofibers synthesized by chemical route using 5% bipyrrole added as an initiator. Inset: Histogram showing the diameter distribution of the polypyrrole nanofibers. Scale bar: 1 μm.

Li and coworkers have synthesized pure polypyrrole nanoparticles without any stabilizer using persulfate as oxidizing agent in quiescent aqueous acid (HCl or HNO_3), at 0°C. [28] They have studied the influence of the nature and concentration of the acid on the nanoparticle diameter, polymerization yield, electrical conductivity and polydispersity (Figure 3A). Figure 3B shows the SEM images of polypyrrole nanostructures obtained in 2M HCl with an oxidant/monomer molar ratio of 1. In these conditions, the polymerization yield was 72%, the average particle diameter was 285 nm and the electrical conductivity 8.5 μS cm^{-1}. This very low value can be explained by the formation of pyrrole trimers which lead to partially conjugated PPy. Indeed, at very low pH, the acid-catalyzed formation of a pyrrole trimer which has been identified as 2,2'-(2,5-pyrrolidinediyl)dipyrrole takes place. [29, 30, 31] This assumption is corroborated by the fact that using a less concentrated acid solution, 0.5M HCl, the nanoparticle conductivity is highly increased (37 mS cm^{-1}) but the polymerization yield is lower (52%). It is interesting to notice that these PPy nanoparticles could be a precursor of carbon nanoparticles having high conductivity.

In order to obtain polypyrrole nanostructures in bulk quantities chemically, Kaner *et al.* have shown that the addition of small amounts of bipyrrole (about 5%) into a solution of pyrrole in methanol enhances the polymerization rate in the presence of $FeCl_3$ (20 s *vs.* 3600 s). [32, 33] This increase is explained by the fact that the oxidation potential of bipyrrole is lower than the potential at which pyrrole is oxidized (0.55 *vs.* 1.2 V/SCE). [34] Actually, it has been proposed that the dimer addition promotes homogeneous nucleation which, in turn, organizes polypyrrole nanofiber formation. The resulting polymer is highly dispersible in water and possesses a morphology consisting of nanofibers with an average diameter of 20 nm (see Figure 4).

Another templateless chemical method used to synthesize unagglomerated PPy nanospheres has been developed by Gelling and coworkers. [35]

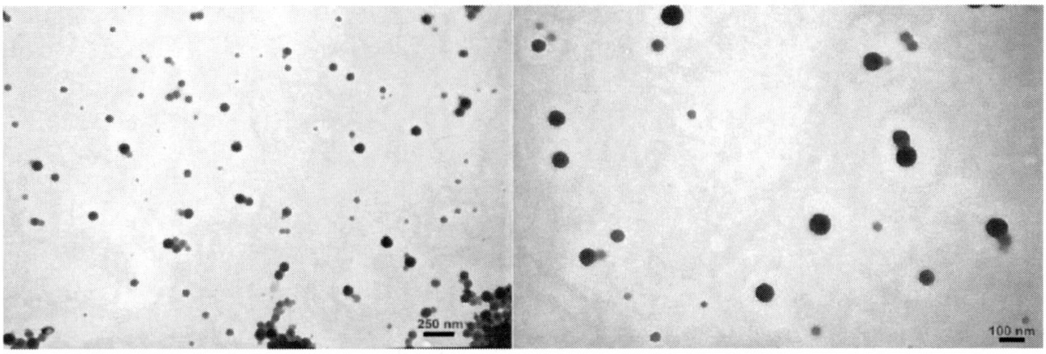

Adapted from Reference [35].

Figure 5. TEM images of PPy nanospheres synthesized in water containing 0.17 M Py and exposed to ozone during 60 s, at 4°C. Scale bars: 250 nm (left), 100 nm (right).

In this method, ozone is used as the oxidizing agent. During the synthesis this gas is bubbled for a few seconds (30 or 60 s) through an aqueous solution containing only the pyrrole monomers. After sealing, the flask is left at room temperature for several days before the analysis of the formed nanoparticles (Figure 5). Ozone acts as the oxidant of pyrrole monomers but it also leads to the overoxidation of at least the external layer of the PPy nanoparticles. This finding could explain that the nanoparticles do not agglomerate in the medium. Current sensing AFM (CS-AFM) experiments have shown that the nanospheres do not exhibit measurable conductivity, in good agreement with the overoxidation of the polymer (dedoped and non-conductive material). This overoxidation could be explained by the fact that in aqueous solution ozone leads to the formation of hydroxyl radicals which are responsible for the overoxidation of PPy. [36]

Finally, a method which combines chemical and physical routes to synthesize polymer nanostructures is the electrospinning. This method is a fiber-spinning technology used to produce very long fibers (up to a few kilometers) with diameters in the range of a few nanometers to a few microns (generally between 100 nm and 1 micron). This technique involves the use of a high voltage electrostatic field (Figure 6A). Obviously the interest of this method is the production of very high-volume nanofibers. Using this technique, pure PPy nanofibers (Figure 6B) have been prepared by Chronakis et al. [37] In order to avoid the addition of poly(ethylene oxide) (PEO) which is usually used as a carrier to improve PPy processability, they have used as the initial material a polypyrrole doped with di(2-ethylhexyl)sulfosuccinate (DEHS) ($[(Py_3)^+, DEHS^-]_n$, n being around 101); [38] this monomer was polymerized by oxidation of pyrrole by persulfate at 0 °C; it was soluble in various solvents and notably in DMF. Its solubility has been attributed to the nature of the dopant which contains three different types of oxygen atoms in addition to relatively long alkyl chains.

1.2. Electrochemical Way

Another route to generate polypyrrole is the electrochemical way which allows the anodic oxidation of pyrrole monomers.

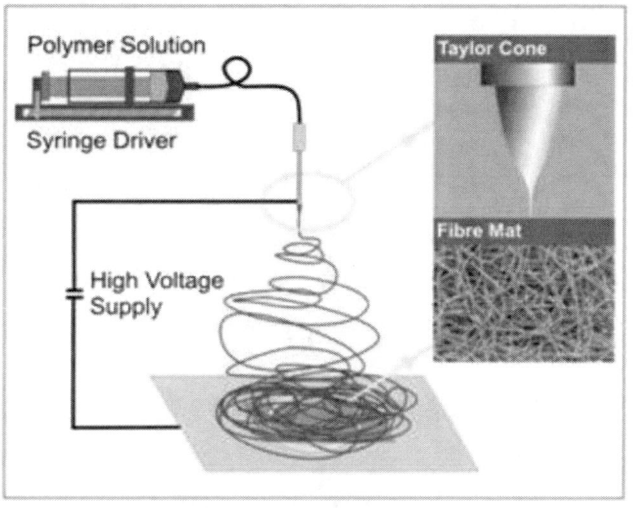

A

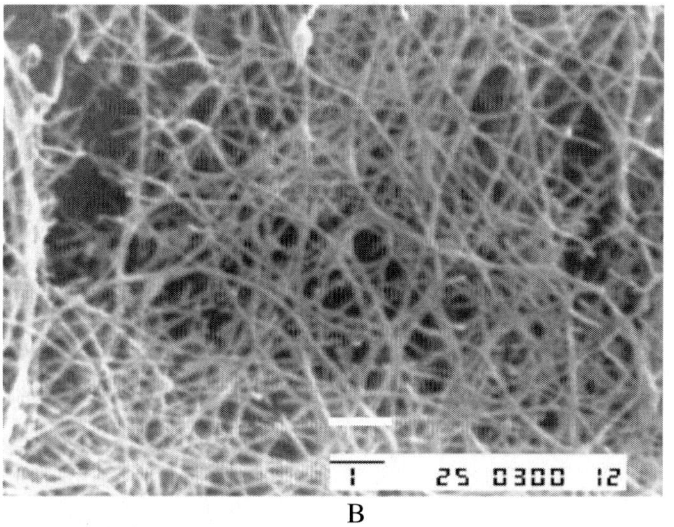

B

Adapted from Reference [37].

Figure 6. A) Schema of a conventional set-up for electrospinning. B) SEM image of electrospun PPy nanofibers from 7.5 wt% $((Py_3)^+(DEHS)^-)_x$ (x ≈100) solution in DMF. Scale bar: 1 μm.

Many methods could be used such as potentiodynamic (cyclic voltammetry or pulsed potential [39]), potentiostatic (constant potential) [40], galvanostatic (constant current) or galvanodynamic (pulsed current).

The solution in which the monomers are dissolved could be an aqueous solution, an organic solution or an ionic liquid such as 1-butyl-3-methylimidazolium tetrachloroferrate. [41]

In general, the PPy films synthesized by electrochemistry have surface with cauliflower-like morphology (Figure 7a). However, under specific experimental conditions the synthesis of nanostructured films could be achieved.

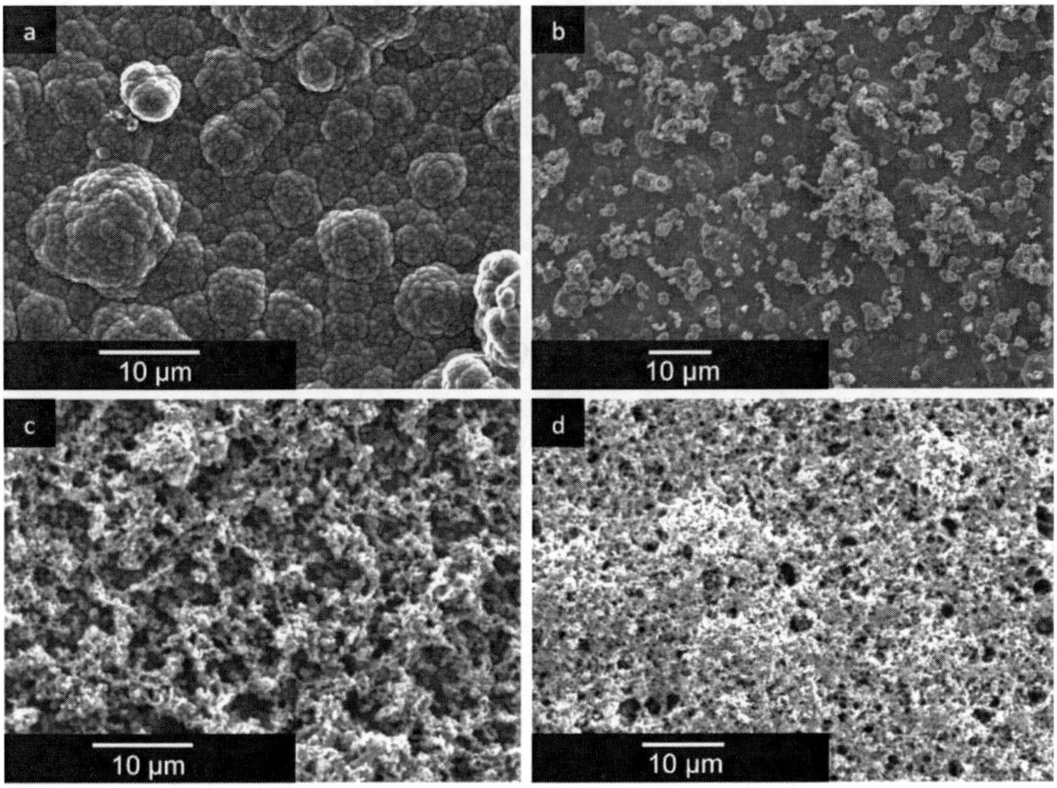

Adapted from Reference [42].

Figure 7. SEM images of PPy samples growth under galvanostatic conditions, with a current density of 0.15mA/cm^2, submitted to centrifugal accelerations of (a) 1g, (b) 50g, (c) 250g, (d) 2250g, for 90 min. The deposition solution contains 15 mM potassium perfluorooctanesulfonate + 0.1 M pyrrole in acetonitrile.

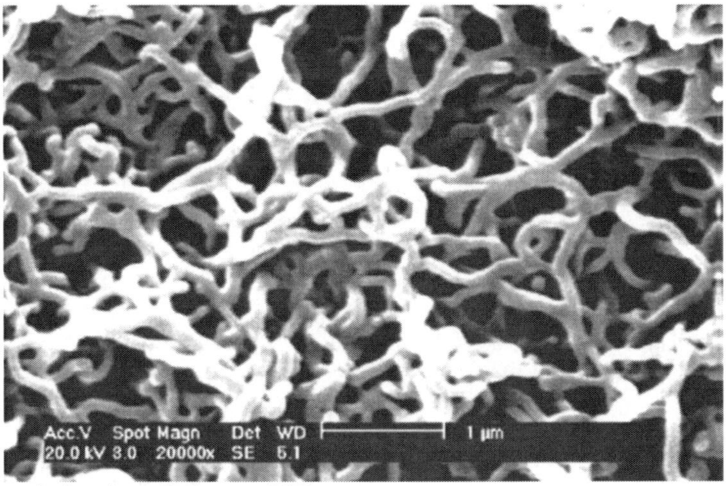

Reprinted with permission from Reference [43].

Figure 8. SEM image of the PPy modified electrode prepared potentiostatically (0.8 V/SCE) in aqueous solution containing 0.15 M pyrrole, 0.1 M $LiClO_4$ and 0.1 M carbonate.

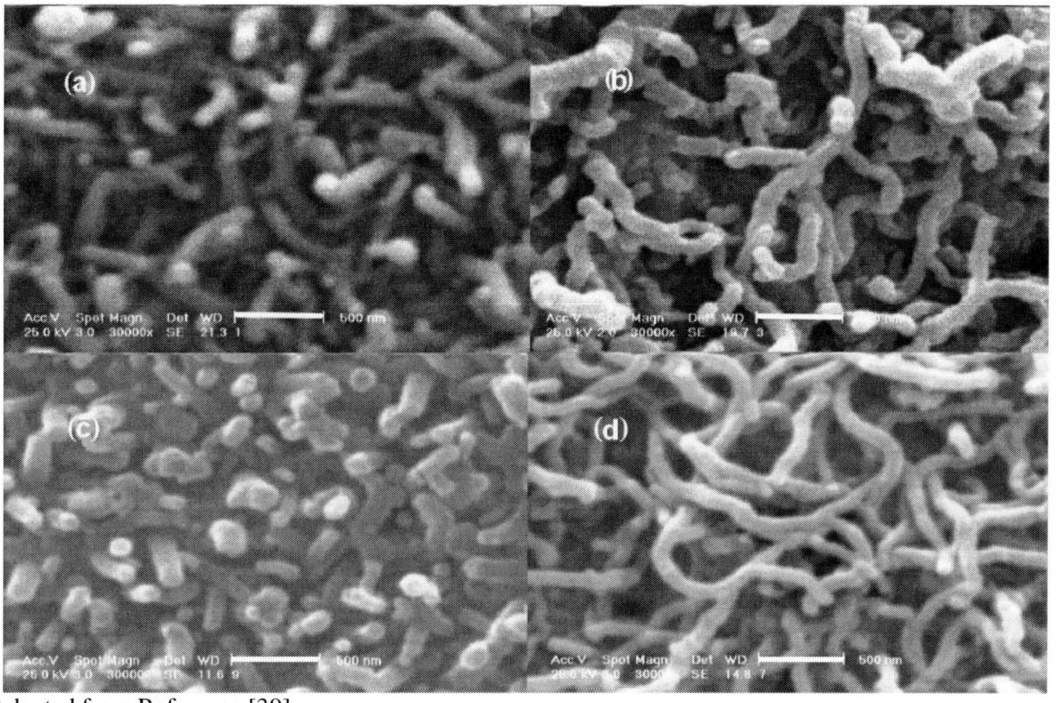

Adapted from Reference [39].

Figure 9. SEM images of PPy obtained by normal pulse voltammetry in 0.1 M Py + 0.1 M $LiClO_4$ 0.2 M phosphate buffer solution. Pulse duration and potential increment: (a) 0.4 s and 2 mV, (b) 0.4 s and 10 mV, (c) 0.2 s and 4 mV, (d) 0.4 s and 4 mV. Potential range: 0.0 to 0.8 V *vs.* Ag/AgCl. Scale bar: 500 nm.

For example, by applying a high centrifugal force on the electrodes during the PPy electrochemical deposition, Chang and coworkers have shown that the synthesized PPy film is nanostructured. [42] They have evidenced that under galvanostatic conditions (0.15 mA cm^{-2}, 90 min), as the applied g-force increased, the polymers progressed from having a classical morphology to nanostructured surface (Figure 7). The deposition solution contains 15 mM potassium perfluoro-octanesulfonate ($KC_8F_{17}SO_3$) and 0.1 M pyrrole in acetonitrile. It should be emphasized that the nature of the anions contained in the monomer solution is crucial. Indeed using a pyrrole solution that contains a salt that is not amphiphilic the polymer is not nanostructured.

Polypyrrole nanofibers have been electrogenerated under potentiostatic conditions (E = 0.8 V/SCE) in the presence of 0.15 M Py, 0.1 M $LiClO_4$ and 0.1 M carbonate (Figure 8). [43] Another method to electrosynthesize PPy nanofibers is the use of normal pulse voltammetry. Figure 9 shows that in these conditions PPy nanofibers 30-90 nm in diameter are obtained by electropolymerization of pyrrole. [39] The monomers were dissolved in 0.2 M phosphate buffer solution (PBS) containing 0.1 M $LiClO_4$. It should be noted that using classical cyclic voltammetry method (0 to 0.8 V/Ag/AgCl, 100 cycles, potential scan rate of 50 mV s^{-1}) PPy nanostructures were also obtained. Polypyrrole nanostructures were also produced by a stepwise electropolymerization process, multistep current strategy (0.3 mA cm^{-2} for 10 minutes then 0.2 mA cm^{-2} for one hour and finally 0.1 mA cm^{-2} for one hour). [44] The actual potential of the working electrode is not mentioned. The syntheses were performed in 40 mM

D-camphorsulfonic acid, 1 mM NaCl pyrrole aqueous solution (pH = 6). [44] In Figure 10 one observes that under these conditions conical nanocontainers were synthesized. The influence of the nature of the electrode was tested. The nanostructures were observed for platinum electrodes but not for gold coated platinum electrodes. The authors suggested the influence of micelles formed by the anions on the deposition mechanism.

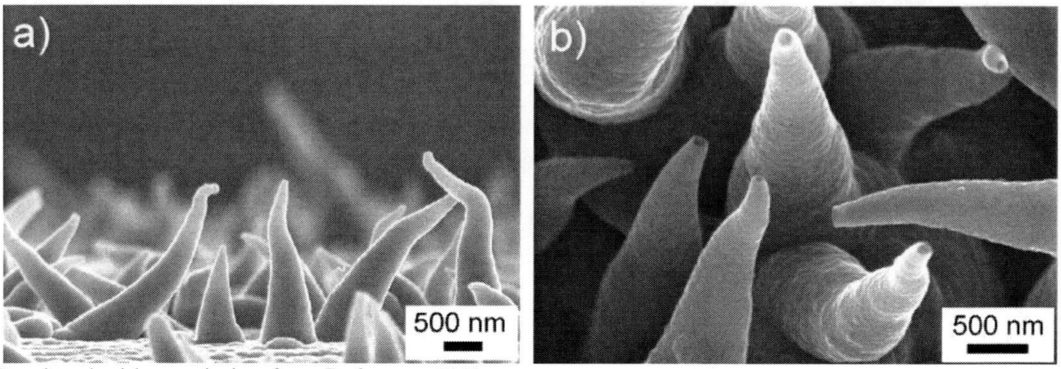

Reprinted with permission from Reference [44].

Figure 10. SEM images of PPy conical nanocontainers produced on a platinum substrate observed from a) side view with a tilt angle of 30° and b) top view.

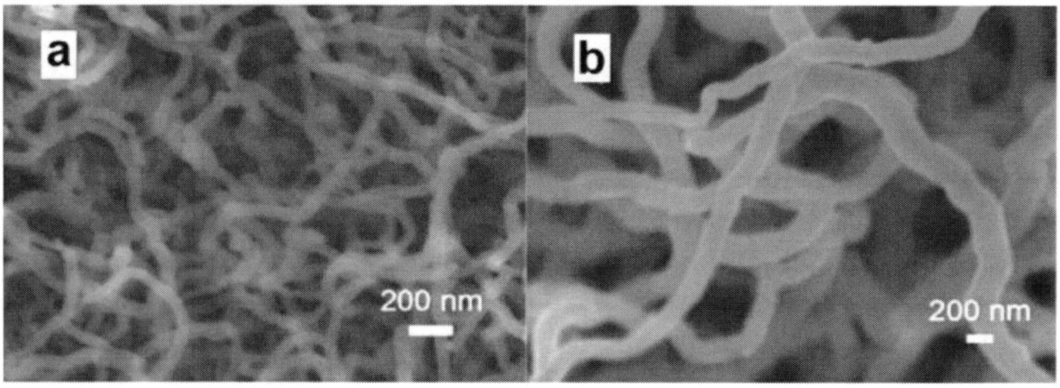

Adapted from Reference [46].

Figure 11. SEM images of PPy nanofiber networks formed at 0.85 V/(Ag/AgCl) for (a) 120 s and (b) 1 h in an aqueous solution containing 0.1 M Py, 0.07 M $LiClO_4$ and 0.2 M PBS (pH 6.8).

Polypyrrole films having hollow structures have also been prepared under galvanostatic conditions (5 mA cm^{-2}) in p-toluenesulfonate alkaline pyrrole solution (pH = 9). [45] Under the same experimental conditions but at pH 1.5, it has been observed that the PPy films synthesized have classical cauliflower-like morphology. The authors have proposed that the growth mechanism implies the formation of micelles which could act as a template with the assistance of hydroxyl ions.

Similar polypyrrole nanostructures have been obtained by Li *et al.* [46] They have developed a template-free electrochemical approach to fabricate a superhydrophilic PPy nanofiber network by using an aqueous solution containing 0.1 M Py, 0.07 M $LiClO_4$ and 0.2 M PBS (pH 6.8) and potentiostatic conditions (0.85 V/(Ag/AgCl)) (Figure 11).

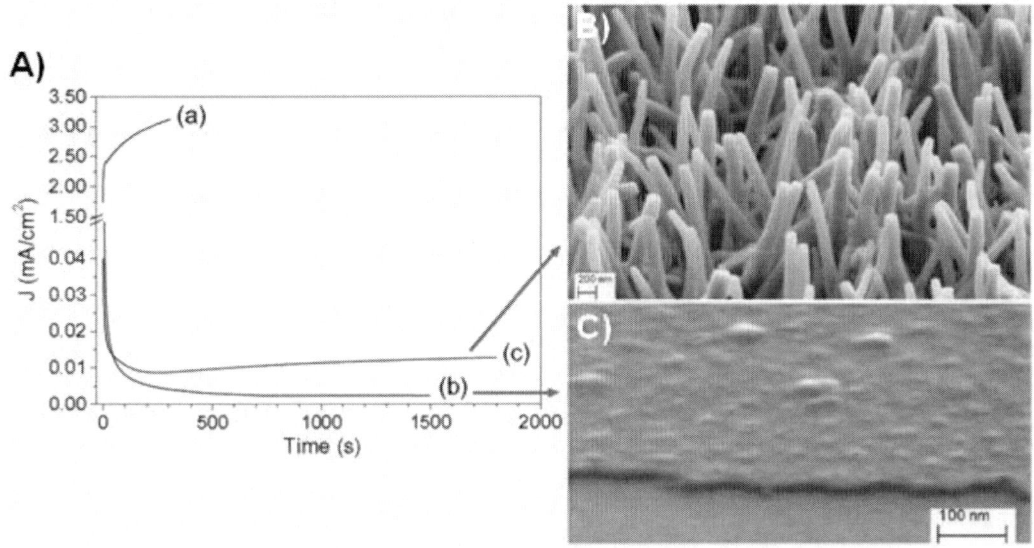

Adapted from Referece [40].

Figure 12. A) Current density vs. time of polarization at 0.8 V/SCE in 0.15 M Py + 0.2 M (a) ClO_4^-, (b) HPO_4^{2-}, (c) HPO_4^{2-} + 10^{-3} M ClO_4^-. B) and C) SEM images of cross-section (tilt angle 65°) of the polymer deposits obtained in conditions (b) and (c), respectively.

It was proposed that the presence of hydrogen bonding between phosphate anions and pyrrole oligomers is essential to produce the nanofiber structure. Systematic investigations have indicated that a relatively high concentration of PBS, high pH and low oxidation potential favor the formation of PPy nanofibers.

Besides, it has been reported that if only anions of weak acids (named weak-acid anions), such as monohydrogenophosphate, acetate, carbonate, monohydrogenocarbonate or tartrate, which confer a relatively high pH to the Py solution, are present in an aqueous pyrrole solution the electrooxidation of the monomers leads to a pinhole-free very thin (~ 10 nm) non-conductive overoxidized polypyrrole film (OPPy) under potentiostatic conditions (curve b in Figure 12A) or potentiodynamic conditions (see SEM image Figure 12C). [47, 48] It has been proposed that this overoxidation is indirectly due to the release of protons at the electrode/solution interface during Py oxidation (Equation 1).

Indeed, when the oxidation of the pyrrole monomers takes place the protons which are released are captured by the weak-acid anions, thus at a moment no or very few anions are available at the electrode/solution interface so that the Py oxidation cannot occur anymore. Since a high potential is applied at the working electrode (around 0.8 V/SCE), water oxidation takes place leading to the generation of hydroxyl radicals which react with the already deposited PPy film yielding its overoxidation with polymer dedoping *i.e.* release of the doping anions [36]:

$$H_2O \rightarrow OH^{\bullet} + H^+ + e^- \quad (4)$$

$$PPy(A^-)_n + OH^{\bullet} \xrightarrow{k_1} OPPy + n\,A^- + \ldots \quad (5)$$

It should be noted that the hydroxyl radicals can also dimerize leading to H_2O_2 that is further oxidized to O_2:

$$2\,OH^{\bullet} \xrightarrow{k_2} H_2O_2 \tag{6}$$

$$H_2O_2 \rightarrow O_2 + 2H^+ + 2e^- \tag{7}$$

Obviously, when the rate of formation of hydroxyl radicals is very slow, reaction (5) is the predominant one with respect to reaction (6) which is a second-order reaction with respect to the hydroxyl radical. It should be emphasized that the standard potential (E^0) of water oxidation reaction is function of the pH of the medium: $E^0 = 0.99 - 0.06$ pH V/SCE, at 25°C. Thus the higher the pH at the electrode/solution interface, the lower the potential at which the water oxidation occurs. Besides, we have shown that after addition of a small amount of perchlorate ions (1 mM) to a solution which yields electrodeposition of an ultra thin OPPy film (0.15 M Py + 0.2 M Na_2HPO_4) under potentiostatic conditions, the anodic current is always very low in comparison with the current obtained in the presence of only perchlorate anions (compare curves a and c in Figure 12A). Thus, the oxidation rate of the monomers is lower than in the 'classical' experimental conditions. However, the anodic current is higher than the one obtained when the anions present in the pyrrole solution are only monohydrohenophosphate (compare curves b and c in Figure 12A).

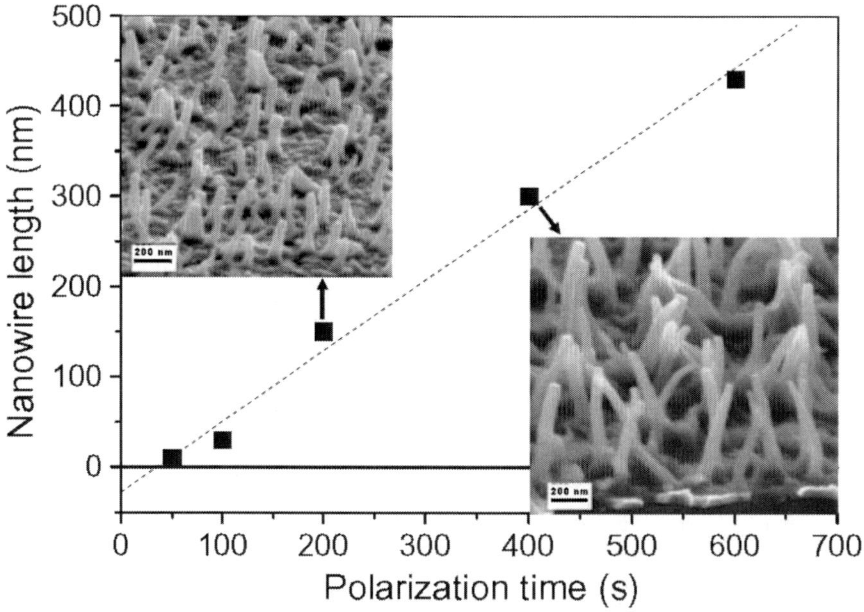

Reprinted with permission from Reference [40].

Figure 13. PPy nanowire height versus time of polarization at 0.75 V/SCE in 0.15 M Py + 0.2 M Na_2HPO_4 + 10^{-3} M $LiClO_4$ aqueous solution. Dashed line: linear regression line (slope: 0.78 nm/s; intercept: -28). The SEM images (side view) correspond to the third and the fourth points (200 and 400 s, respectively). Scale bar: 200 nm.

In this condition (anion concentration: 10^{-3} M ClO_4^- + 0.2 M HPO_4^{2-}), a polypyrrole oriented nanowire array has been electrogenerated (Figure 12B). It has also been reported that, under a given applied potential, the length of the nanowires is proportional to the duration of the anodic polarization (Figure 13), and the rate of the PPy nanowire growth is low, about 0.8 nm s^{-1}. [40] It was concluded that the thin polymeric film which surrounds the base of the nanowires is not conductive *i.e.* it is an OPPy film. The reaction mechanism proposed to explain the templateless formation of the nanowire array is based on the one hand on the variation of the interfacial pH and on the other hand on the anion concentration at the electrode/solution interface. Actually, this mechanism involves three steps: the main feature is that the initially formed PPy film is no more conductive except in some circular places where the current can continue to flow and, therefore, Py anodic polymerization can take place. Therefore, it has been proposed that in this process, the first step is the classical electrodeposition of a thin polypyrrole film (leading to proton release and therefore to a decrease of weak-acid anion concentration at the electrode/electrolyte interface). The second step which is very short concerns the oxidation of water leading in some places to the overoxidation of the polymer due to hydroxyl radicals (reaction 5) and in other ones to the formation of O_2 nanobubbles (reaction 7) that protect the already synthesized polymer against the action of hydroxyl radicals *i.e.* against overoxidation. Finally, when at the film/solution interface the anion concentration begins to increase due to anion diffusion from the electrolyte bulk and polymer dedoping, the last step concerns the slow growth (0.8 nm/s) of the PPy nanowires at places where O_2 has evolved. This reaction mechanism explains the disc shape of the nanostructure section. Actually, the electrode surface can be seen as shared into two domains, one corresponding to the nanostructures and the other domain which corresponds to the overoxidized film that surrounds the nanowires. The influence of the perchlorate concentration on the polypyrrole nanostructures (synthesized in the presence of 0.15 M pyrrole and 0.2 M K_2HPO_4 under potentiostatic conditions) has also been studied. It has been shown that for perchlorate concentrations lower than 0.7 mM, overoxidized polypyrrole films are formed whereas for higher perchlorate concentrations nanostructures are obtained and the films exhibit superhydrophilic properties. [49] The surface wettability of the films was tested by deposing a water droplet on the PPy surface; no contact angle could be measured because the water disappears inside the polypyrrole film which therefore is considered as superhydrophilic. The higher the perchlorate concentration, the higher the anodic charge (Q=It) passed through the electrode *i.e.* the amount of monomers oxidized during the imposition of the anodic potential (Figure 14A). Actually, two types of nanostructure are obtained either oriented nanowires for the lowest perchlorate concentrations or nanofiber network for the highest concentrations (see SEM images in Figure 14). Besides, the role of hydroxyl radicals in the growth mechanism described above has been confirmed by using a scavenger of these radicals. This scavenger is the dimethylthiourea (DMTU) that is largely used in biology in the studies of contribution of hydroxyl radical in tissue injury. [50, 51] Two concentrations of DMTU either 1 μM or 0.1 mM were added to the monomer solution which, without DMTU, leads to the formation of an overoxidized polypyrrole film. For the lowest concentration tested it has been shown that nanowires are formed whereas for the highest DMTU concentration, a thin film is deposited. [52] This last finding was explained by the fact that the DMTU reacts also with the pyrrole cation radicals formed during the first step of Py electropolymerisation ($Py \rightarrow Py^{+\bullet} + e^-$) [18, 53, 54, 55, 56] leading therefore to a decrease of the Py electropolymerization rate.

Finally biphasic electrochemical strategy has been used to synthesize PPy nanowires. [57] These polypyrrole nanostructures have been obtained by cyclic voltammetry. Py was dissolved in the organic phase and the dopant (L-camphorsulfonic acid) in the aqueous phase. Its concentration was 1 mM thus the anion concentration was very low and the resistivity of the solution high. Figure 15 shows the SEM image of the polymer obtained after 30 potential scan cycles. In this case the anionic dopant, L-camphorsulfonic acid (its structure is depicted in the left insert of Figure 15), was hydrophobic and polymer surface was superhydrophobic even if oriented nanowires are present on the surface. The contact angle is high, about 165° (right insert of Figure 15).

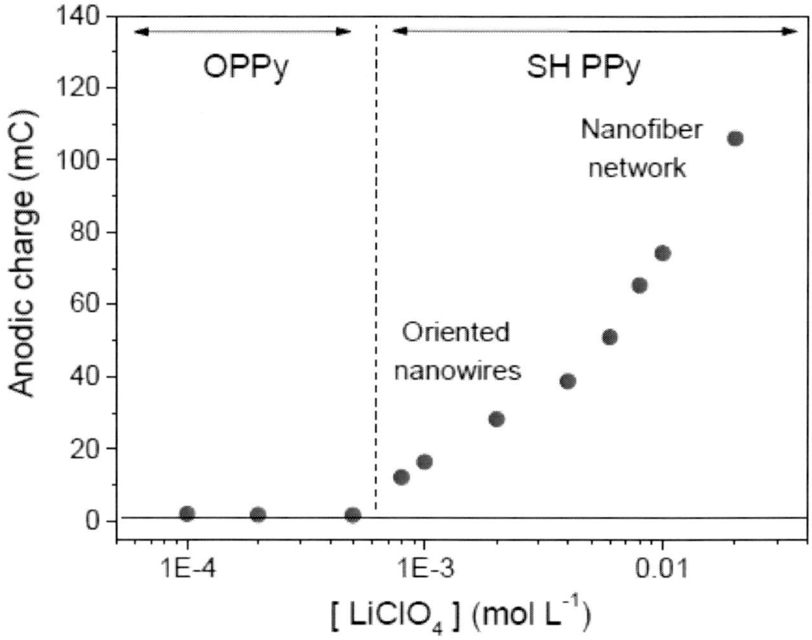

A

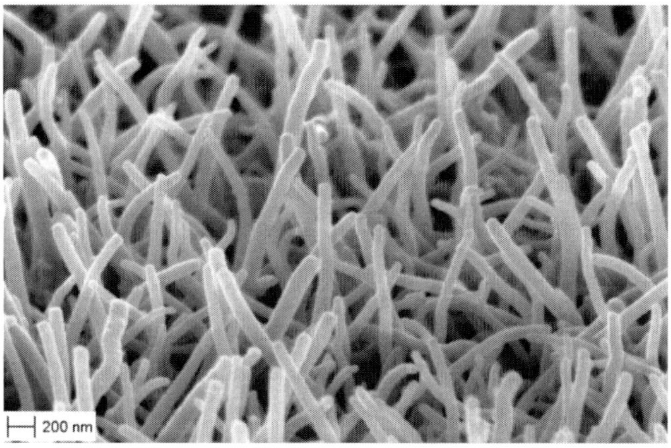

B

Figure 14 (Continued).

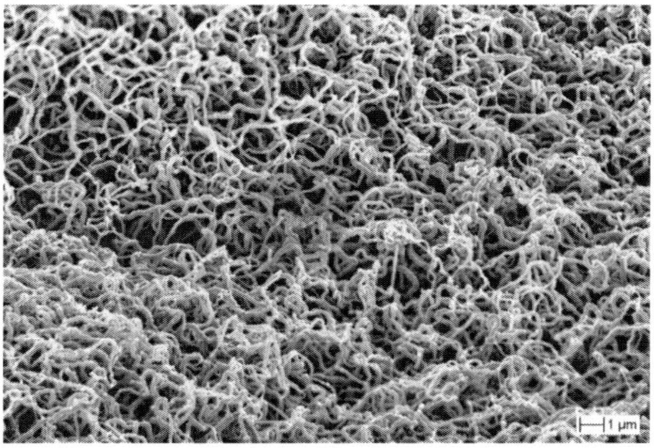

C

Adapted from Reference [49].

Figure 14. A) Anodic charge *vs.* perchlorate concentration of the pyrrole solution. Anodic charge passes through a Pt electrode (S = 0.07 cm^2) polarized at 0.78 V/SCE for 600 s. Electrolyte: $LiClO_4$ + 0.15 M Py + 0.2 M K_2HPO_4. OPPy: overoxidized PPy. SH PPy: superhydrophilic PPy. B) and C) SEM images of PPy nanowires and nanofibers obtained in the presence of 1 mM and 50 mM $LiClO_4$, respectively. Scale bars: (B) 200 nm, (C) 1 µm.

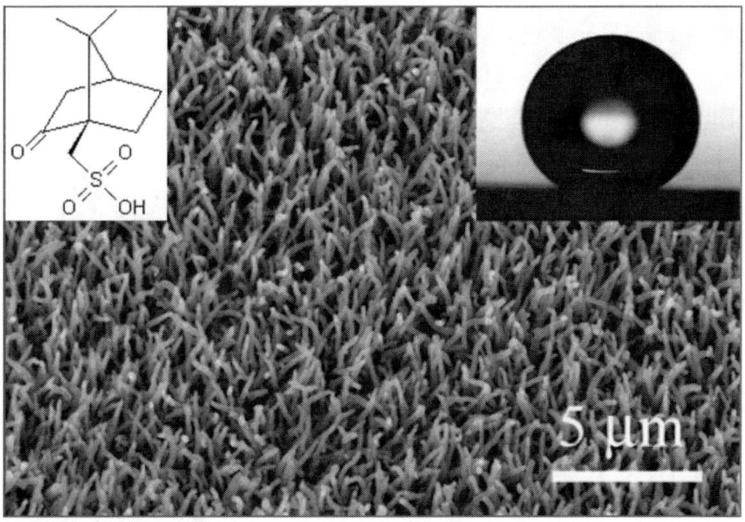

Adapted from Reference [57].

Figure 15. SEM image of the superhydrophobic aligned polypyrrole nanofibers film obtained after 30 potential scanning cycles using L-camphorsulfonic acid as dopant. Right inset: digital camera image of a water droplet on the nanofibers film, the contact angle is about 165°. Left inset: Structure of the L-camphorsulfonic acid (pK_a = 1.2).

To conclude this section, it is interesting to note that for all the examples presented herein, the templateless electrosynthesis of nanostructured polypyrrole is obtained when the pH of the monomer solution is relatively high and in the presence of either low anion concentration and/or high concentration of anions able to capture protons (CO_3^{2-}, HPO_4^-, $H_2PO_4^-/HPO_4^{2-}$ (PBS)) and therefore to limit the decrease of the interfacial pH which permits the water oxidation reaction to take place that leads to the production of hydroxyl radicals and O_2 nanobubbles.

2. Some Applications

Two domains of application of nanostructured polypyrrole are presented below: (i) (bio)sensors for detection and determination of analytes in solution [58] or gases and (ii) energy storage. [17, 59] For most of the applications, the conductive and redox properties of the polymer are exploited since the electrical and optical properties of polypyrrole can be reversibly changed by a doping/dedoping process. Nanostructured polypyrrole can yield a higher sensitivity and shorter response time than compact films for amperometric or conductometric sensors due to their very high surface area. One example concerns the determination of spermidine (NH_2-$(CH_2)_4$-NH-$(CH_2)_3$-NH_2 ; polyamine) for which a DNA/PPy electrochemical biosensor has been developed by Mousavi *et al.* [39] For this purpose, a PPy nanofiber film was electrogenerated using normal pulse voltammetry (process described in section I.2, Figure 9) and then double-stranded DNA was physisorbed onto the PPy film. Before the immobilization of DNA the nanofibrillar PPy film was oxidized at 0.5V/Ag/AgCl at pH 3 for 5 min. This amperometric biosensor is based on the oxidation of the guanine bases of DNA. Actually, the presence of spermidine in the medium leads to a decrease of the guanine oxidation because the binding of spermidine to the DNA guanine bases shields the oxidizable guanine group and therefore inhibits its oxidation. Therefore in the presence of spermidine the anodic current decreases. As one can notice from Figure 16, this decrease is proportional to the spermidine concentration with linear range for concentrations from 0.05 to 1.0 μM. For a compact PPy film obtained by classical cyclic voltammetry a linear decrease of the current was also observed but the linear range was observed for higher concentrations (1.0 to 15.0 μM).

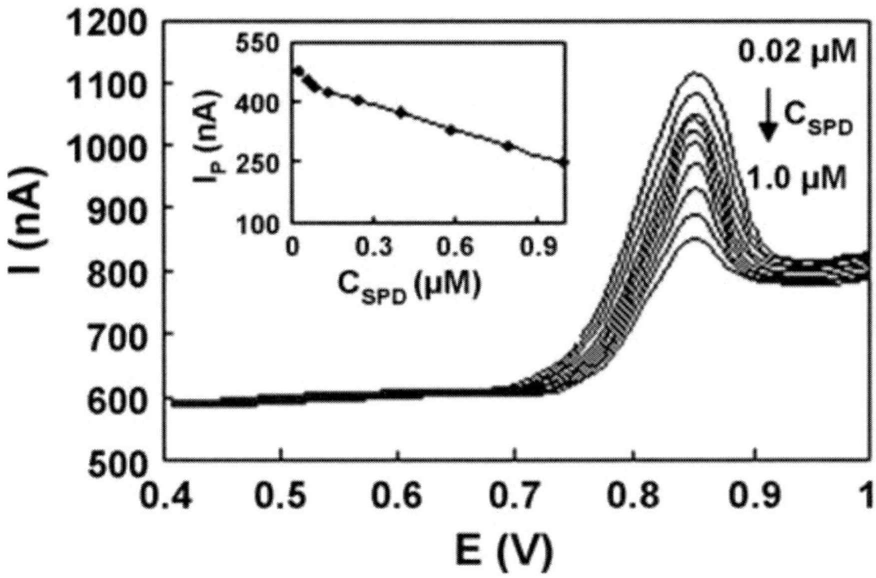

Adapted from Reference [39].

Figure 16. Differential pulse voltammetries at various concentrations of spermidine (SPD) (0.02 – 1.0 μM). Inset: calibration curve of Ip (I(at 0.84 V) – I(at 0.6 V)) versus SPD concentration. Electrode: DNA adsorbed on PPy nanofibers (SEM image shown in Figure 9d).

Some gas sensors are based on polypyrrole because in the presence of various chemical vapors its conductivity may change. [60] Indeed interacting gas molecules can donate or accept fractional electronic charge of the polymer leading to changes of its electronic properties. [61] It has been reported that the PPy conductivity increased in the presence of ammonia [32, 62] whereas it decreases in the presence of dihydrogen. [63-65]

Nanostructured polypyrrole-based chemical sensors such as surface acoustic wave (SAW) gas sensors [63] and conductometric hydrogen gas sensors [64, 65] have notably been developed. For these sensors, interdigitated electrodes (IDE) are used the polymer covering the electrodes and bridging the gaps.

With the water-dispersible PPy nanofibers (PPy obtained by chemical route), chemical sensors are fabricated by drop-casting a dispersion of such material directly onto the IDE. The porous features of the nanofibrillar film provide a high surface area that can enhance the sensitivity to the target molecules. In the case of SAW sensors and conductometric sensors, the active component was PPy nanofibers synthesized through a chemical route (bipyrrole was used as an initiator allowing the increase of the polymerization of monomers with $FeCl_3$ as oxidizing agent (see part I.1)).

After purification by centrifugation and washing with deionized water the purified PPy nanofibers were resuspended in deionized water and the solution was drop-casted onto the SAW or IDE transducer. The thickness of the resulting PPy nanofiber film was about 0.3 μm. The SAW sensor allows the detection of hydrogen and nitrogen dioxide. [63] The IDE transductor performance was analyzed during exposure to different concentrations of hydrogen gas in a synthetic air mixture. More recently, the same experiments have been conducted with polypyrrole nanofibers synthesized onto IDE by electrochemistry (Figure 17). In both cases (PPy nanofibers synthesized by chemical or electrochemical route) the response time for 10000 ppm of H_2 is similar (43s /54 s).

The sensitivity ($\Delta R/R_0$) obtained is better for the sensor prepared by drop casting, however the amount of PPy on the transducer is higher too.

In the case of synthesis performed by electrochemistry, it has been evidenced that the sensitivity of the IDE sensor is function of the amount of electrodeposited PPy (proportional to the anodic charge) on the IDE transductor (Figure 18). The higher the amount of polypyrrole deposited onto the IDE, the higher the sensitivity of the sensor towards hydrogen.

Another domain of applications concerned with nanostructured polypyrrole is energy storage, batteries and supercapacitors (charge-storage devices). [66-71] For polypyrrole-based supercapacitors, their overall capacitance is due to the double-layer capacitance but also to the pseudocapacitance involved in polymer bulk redox (Faraday) reactions. Obviously, the higher the surface area of the polymer the higher the double-layer capacitance.

Cyclic voltammetry (insert Figure 19A) is an effective tool to reveal the capacitive behavior of conducting polymers. Indeed a large current (I) proportional to the potential scan rate (v) and a rectangular form of the voltammogram, symmetric in anodic and cathodic directions are indications of a good capacitive nature ($C = dq/dV = I/v$).

Fan and Maier have obtained highly active polypyrrole electrodes for supercapacities by electrogenerating polypyrrole onto Ti foil by cyclic voltammetry between potential limits of -0.2 V/SCE and 0.8 V/SCE. The aqueous electrolyte was composed of 0.2 M oxalic acid ($pKa_1 = 1.2$ and $pKa_2 = 4.3$, at 25 °C) and 0.1 M Py.

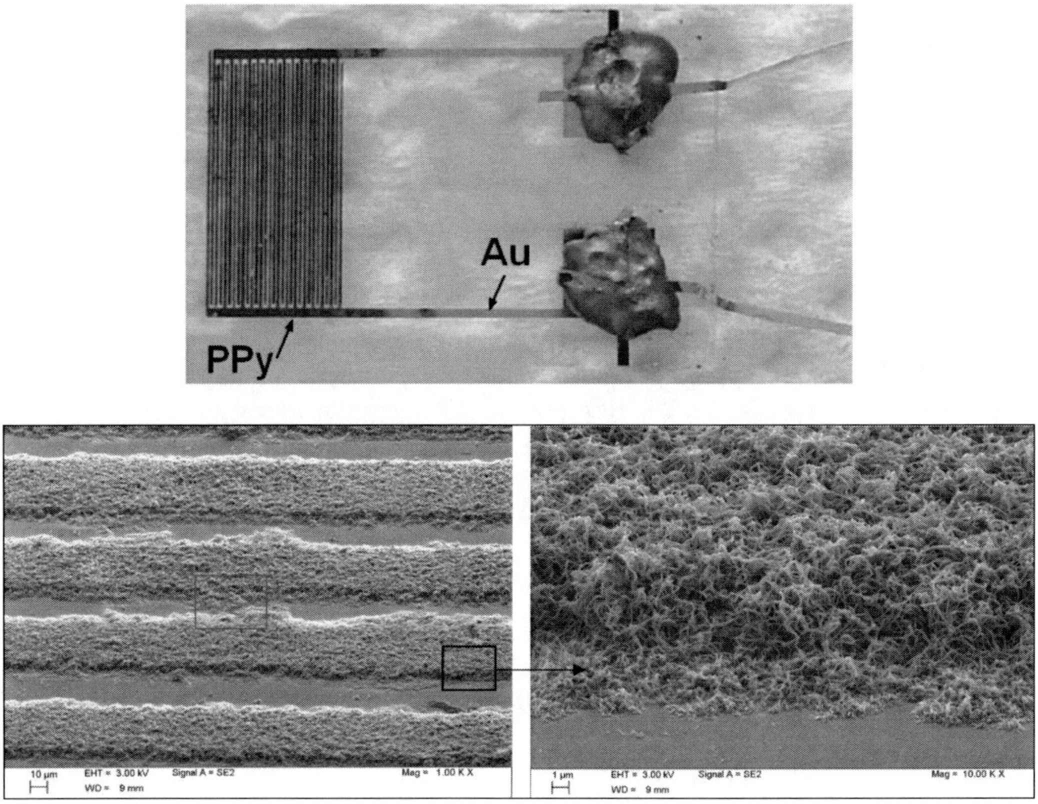

Adapted from Reference [65].

Figure 17. Optical image (top) and SEM images (bottom) of electrodeposited PPy nanofibers at 0.85 V/SCE in 0.15 M Py + 0.2M Na_2HPO_4 + 2 mM $LiClO_4$ aqueous solution, on two interdigital Au electrodes. During the pyrrole electropolymerisation the two interdigitated electrodes were connected. Scale bars: 10 μm (left), 1 μm (right).

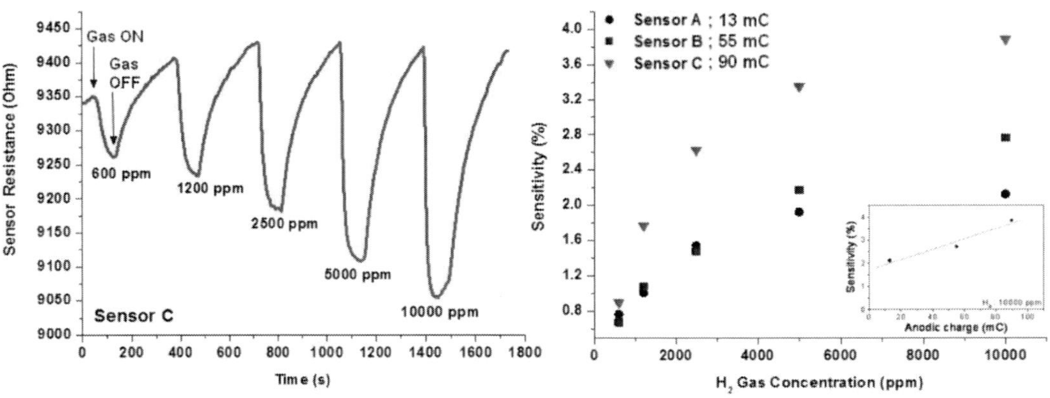

Adapted from Reference [65].

Figure 18. Dynamic response and sensitivities of hydrogen gas sensors obtained by electrodeposition of PPy nanofibers on a conductometric transducer's surface (see Figure 17). Insert: sensitivity *vs.* anodic charge (amount of PPy electrodeposited) in the presence of 10000 ppm of H_2.

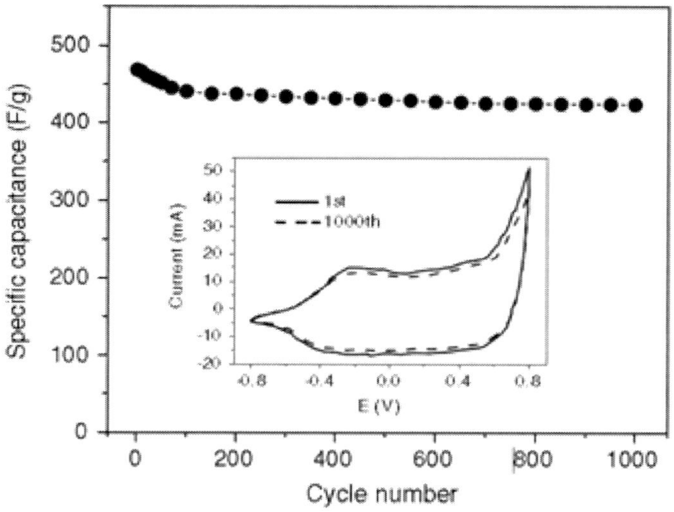

A

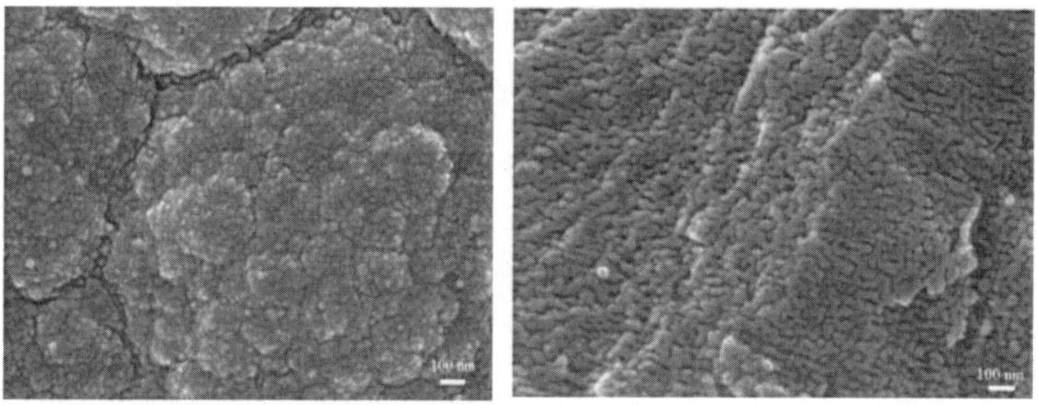

B

Adapted from Reference [66].

Figure 19. A) Cycle-life data of PPy electrode in 1 M KCl electrolyte. The inset shows the cyclic voltammograms of the 1st and the 1000th cycle of PPy electrode in 1M KCl solution at 50 mV/s. B) SEM images of PPy polymerized on Ti foil (top) surface, and (bottom) cross section. Scale bar: 100 nm. Specific mass of polypyrrole: 0.6 mg/cm^2. Polymerization conditions: 100 potential scans between 0.2 and 0.8 V/SCE at scan rate of 200 mV/s, in 0.2 M oxalic acid and 0.1 M pyrrole solution.

The PPy film has highly porous nanostructures which lead to high specific capacitance. It was shown that the electrode coated with the PPy film exhibits a high stability during cycle life test (see Figure 19). [66]

CONCLUSION

The templateless synthesis of nanostructured polypyrrole is possible using either a chemical or an electrochemical route.

The most promising route for chemical processes seems to be the addition of small amount of oligomer (dimer) to the monomer solution because it allows the synthesis of high amount of conductive nanostructures (20 nm in diameter).

In order to spatially control the polypyrrole deposition, the electrochemical way is the most appropriate route. To electrosynthesize polypyrrole nanostructures the key parameters are the pH and the anionic species composition of the monomer solution rather than the electrochemical method used. For some applications, it could be interesting to use substituted pyrrole in order to functionalize the polymer via covalent grafting. The monomer could be substituted by a functional group such as amino, carboxylic or affinity groups such as biotin. In this case, in order to control the steric hindrance, co-polymerisation with substituted and no-substituted monomers could be envisaged.

REFERENCES

[1] Wang, L. X., Li, X. G., Yang, Y. L. *React. Funct. Polym.* 2001, 47, 125.
[2] Diaz, A. F., Kanazawa, K. K., Gardini, G. P. *J. Chem. Soc. Chem. Comm.* 1979, 14, 635.
[3] Li, Y. F. *J. Electroanal. Chem.* 1997, 433, 181.
[4] Bidan, G., Lopez, C., Mendesviegas, F., Vieil, E., Gadelle, A. *Biosens. Bioelectron.* 1995, 10, 219.
[5] Kaneto, K., Bidan, G. *Thin Solid Films* 1998, 331, 272.
[6] Shimidzu, T. *React. Polym.* 1987, 6, 221.
[7] Lopez, C., Viegas, M. F. M., Bidan, G., Vieil, E. *Synthetic Met.* 1994, 63, 73.
[8] Levi, M. D., Lopez, C., Vieil, E., Vorotyntsev, M. A. *Electrochim. Acta* 1997, 42, 757.
[9] Gabrielli, C., Garcia-Jareno, J. J., Perrot, H. *Electrochim. Acta* 2001, 46, 4095.
[10] Gabrielli, C., Garcia-Jareno, J. J., Keddam, M., Perrot, H., Vicente, F. *J. Phys. Chem. B* 2002, 106, 3192.
[11] Debiemme-Chouvy, C., Rubin, A., Perrot, H., Deslouis, C., Cachet, H. *Electrochim. Acta* 2008, 53, 3836.
[12] Kim, L. T. T., Debiemme-Chouvy, C., Gabrielli, C., Perrot, H. *Langmuir* 2012, 28, 13746.
[13] Debiemme-Chouvy, C., Cachet, H., Deslouis, C. *Electrochim. Acta* 2006, 51, 3622.
[14] Malinauskas, A., Malinauskiene, J., Ramanavicius, A. *Nanotechnology* 2005, 16, R51.
[15] Tran, H. D., Li, D., Kaner, R. B. *Adv. Mater.* 2009, 21, 1487.
[16] Li, C., Bai, H., Shi, G. Q. *Chem. Soc. Rev.* 2009, 38, 2397.
[17] Yin, Z., Zheng, Q. *Adv. Energ. Mat.* 2012, 2, 179.
[18] Heinze, J., Frontana-Uribe, B. A., Ludwigs, S. *Chem. Rev.* 2010, 110, 4724.
[19] Qu, L. T., Shi, G. Q., Chen, F., Zhang, J. X. *Macromolecules* 2003, 36, 1063.
[20] Long, Y. Z., Li, M. M., Gu, C. Z., Wan, M. X., Duvail, J. L., Liu, Z. W., Fan, Z. Y. *Prog. Polym. Sci.* 2011, 36, 1415.
[21] Cai, Z. H., Martin, C. R. *J. Am. Chem. Soc.* 1989, 111, 4138.
[22] Pecher, J., Mecking, S., Nanoparticles of Conjugated Polymers. *Chem. Rev.* 2010, 110, 6260.
[23] Pan, L., Qiu, H., Dou, C., Li, Y., Pu, L., Xu, J., Shi, Y. *Int. J. Mol. Sci.* 2010, 11, 2636.

[24] Stejskal, J., Colloidal dispersions of conducting polymers. *J. Polym. Mater.* 2001, 18, 225.
[25] Pringle, J. M., Ngamna, O., Lynam, C., Wallace, G. G., Forsyth, M., MacFarlane, D. R. *Macromolecules* 2007, 40, 2702.
[26] Nuraje, N., Su, K., Yang, N. L., Matsui, H. *ACS Nano* 2008, 2, 502.
[27] Cawdery, N., Obey, T. M., Vincent, B. *J. Chem. Soc. Chem. Comm.* 1988, 17, 1189.
[28] Li, X. G., Li, A., Huang, M. R., Liao, Y. Z., Lu, Y. G. *J. Phys. Chem. C* 2010, 114, 19244.
[29] Potts, H. A., Smith, G. F. *J. Chem. Soc.* 1957, 797, 4018.
[30] Smith, G. F. *Adv. Heterocycl. Chem.* 1963, 18, 287.
[31] Lamb, B. S., Kovacic, P. *J. Polym. Sci. Pol. Chem.* 1980, 18, 1759.
[32] Tran, H. D., Shin, K., Hong, W. G., D'Arcy, J. M., Kojima, R. W., Weiller, B. H., Kaner, R. B. *Macromol. Rapid Comm.* 2007, 28, 2289.
[33] Tran, H. D., Wang, Y., D'Arcy, J. M., Kaner, R. B. *ACS Nano* 2008, 2, 1841.
[34] Gurunathan, K., Murugan, A. V., Marimuthu, R., Mulik, U. P., Amalnerkar, D. P. *Mater. Chem. Phys.* 1999, 61, 173.
[35] Vetter, C. A., Suryawanshi, A., Lamb, J. R., Law, B., Gelling, V. J. *Langmuir* 2011, 27, 13719.
[36] Debiemme-Chouvy, C., Tran, T. T. M. *Electrochem. Comm.* 2008, 10, 947.
[37] Chronakis, I. S., Grapenson, S., Jakob, A. *Polymer* 2006, 47, 1597.
[38] Oh, E. J., Jang, K. S., MacDiarmid, A. G. *Synthetic Met.* 2001, 125, 267.
[39] Ghanbari, K., Bathaie, S. Z., Mousavi, M. F. *Biosens. Bioelectron.* 2008, 23, 1825.
[40] Debiemme-Chouvy, C. *Electrochem. Comm.* 2009, 11, 298.
[41] Kim, J.-Y., Kim, J.-T., Song, E.-A., Min, Y.-K., Hamaguchi, H.-O. *Macromolecules* 2008, 41, 2886.
[42] Chang, J. H., de Leon, C. R. A., Hunter, I. W. *Langmuir* 2012, 28, 4805.
[43] Wang, J. S., Wang, J. X., Wang, Z., Wang, S. C. *Synthetic Met.* 2006, 156, 610.
[44] Huang, J. Y., Quan, B. G., Liu, M. J., Wei, Z. X., Jiang, L. *Macromol. Rapid Comm.* 2008, 29, 1335.
[45] Wang, J., Xu, Y. L., Yan, F., Zhu, J. B., Wang, J. P. *J. Power Sources* 2011, 196, 2373.
[46] Zang, J., Li, C. M., Bao, S.-J., Cui, X., Bao, Q., Sun, C. Q. *Macromolecules* 2008, 41, 7053.
[47] Debiemme-Chouvy, C. *Electrochem. Solid St. Lett.* 2007, 10, E24.
[48] Debiemme-Chouvy, C. *Biosens. Bioelectron.* 2010, 25, 2454.
[49] Fakhry, A., Debiemme-Chouvy, C. *Small,* In preparation.
[50] Fox, R. B., Harada, R. N., Tate, R. M., Repine, J. E. *J. Appl. Physiol.* 1983, 55, 1456.
[51] Fox, R. B. *J. Clin. Invest.* 1984, 74, 1456.
[52] Fakhry, A., Debiemme-Chouvy, *Unpublished results*
[53] Diaz, A. F., Castillo, J. I., Logan, J. A., Lee, W. Y. *J. Electroanal. Chem.* 1981, 129, 115.
[54] Genies, E. M., Bidan, G., Diaz, A. F. *J. Electroanall. Chem.* 1983, 149, 101.
[55] Waltman, R. J., Bargon, J. *Tetrahedron* 1984, 40, 3963.
[56] Sadki, S., Schottland, P., Brodie, N., Sabouraud, G. *Chem. Soc. Rev.* 2000, 29, 283.
[57] Li, M., Wei, Z. X., Jiang, L. *J. Mater. Chem.* 2008, 18, 2276.
[58] Ramanavicius, A., Ramanaviciene, A., Malinauskas, A. *Electrochim. Acta* 2006, 51, 6025.

[59] Khomenko, V., Frackowiak, E., Beguin, F. *Electrochim. Acta* 2005, 50, 2499.
[60] Bai, H., Shi, G. Q. *Sensors* 2007, 7, 267.
[61] Miasik, J. J., Hooper, A., Tofield, B. C. *J. Chem. Soc. Farad. T. I* 1986, 82, 1117.
[62] Kwon, O. S., Hong, J. Y., Park, S. J., Jang, Y., Jang, J. *J. Phys. Chem. C* 2010, 114, 18874.
[63] Al-Mashat, L., Tran, H. D., Wlodarski, W., Kaner, R. B., Kalantar-Zadeh, K. *Sensor Actuat. B-Chem.* 2008, 134, 826.
[64] Al-Mashat, L., Tran, H. D., Wlodarski, W., Kaner, R. B., Katantar-Zadeh, K. *IEEE Sens. J.* 2008, 8, 365.
[65] Al-Mashat, L., Debiemme-Chouvy, C., Borensztajn, S., Wlodarski, W. *J. Phys. Chem. C* 2012, 116, 13388.
[66] Fan, L. Z., Maier, J. *Electrochem. Comm.* 2006, 8, 937.
[67] Conway, B. E., Birss, V., Wojtowicz, J. *J. Power Sources* 1997, 66, 1.
[68] Sharma, R. K., Rastogi, A. C., Desu, S. B. *Electrochem. Comm.* 2008, 10, 268.
[69] Snook, G. A., Kao, P., Best, A. S. *J. Power Sources* 2011, 196, 1.
[70] Ramya, R., Sivasubramanian, R., Sangaranarayanan, M. V. *Electrochim. Acta* 2012, doi/10.1016/j.electacta.2012.09.116.
[71] Simon, P., Gogotsi, Y. *Nat. Mater.* 2008, 7, 845.

In: Applied Surface Chemistry of Nanomaterials
Editors: Mohamed M. Chehimi and Jean Pinson
ISBN: 978-1-62808-351-4

Chapter 10

CARBON NANOTUBES/POLYPYRROLE HYBRIDS

Matej Mičušík** and Mária Omastová
Polymer Institute, Slovak Academy of Sciences,
Bratislava, Slovakia

INTRODUCTION

The beginning of the carbon nanotubes (CNT) era is dated to 1991 when they were observed on the cathode of a carbon arc used to produce fullerenes [1], although nanotubes were commonly identified as contiguous hollow tubes in the core structure of carbon fibers in the 1970s [2]. Nanotubes are nowadays one of the most studied materials and their re-discovery in 1990s co-triggered the Nanotechnology Revolution [3].

CNT are graphene sheets rolled up into a cylinder of a few nanometers in diameter and up to some millimeters in length, which could give aspect ratio (length to diameter) more than 107. CNT could be prepared under well-defined conditions as singlewall carbon nanotubes (SWCNT) or multiwall carbon nanotubes (MWCNT). CNT consist of a single graphene sheet rolled in around itself or of multiple concentric cylinders of graphene.

CNT excite both fundamental scientists and engineers interested in applications due to their high electrical/thermal conductivity and extraordinary mechanical properties. These properties can be exploited in many applications as conductive and high strength nanocomposites [4], sensors [5, 6], field emission displays [7], nanoelectronics [8], energy storage and energy conversion devices and interconnectors, etc. [9]. CNT for their extraordinary electronic and mechanical properties offered also interesting applications in the combination with conducting polymers. CNT can be used as a high surface area conductive polymerization substrate for conducting polymers.

Conducting polymers have received considerable attention because of their superior physical and chemical properties. Extensive research has been carried out into conducting polymers over the last forty years. Of these, polypyrrole (PPy), polyaniline (PANI), and polythiophene (PTh) have been the most widely studied to date. In terms of the current

* Corresponding author: M. Mičušík, Polymer Institute, Slovak Academy of Sciences, Dúbravská cesta 9, 845 41 Bratislava 45, Slovakia. E-mail: matej.micusik@savba.sk; Tel.: +421 2 3229 4375.

knowledge of conducting polymers, research started in this area in the nineteenth century when J. Fritsche, described the preparation of a green product obtained by oxidation of aniline [10]. We now know that this product was a form of polyaniline. Aniline is produced from indigo, which forms part of a natural dye. In 1862 Letheby described the electrochemical oxidation of aniline on a platinum electrode resulting in the formation of a blue-black powder [11]. Polypyrrole was prepared in a similar manner in 1968 [12]. Real interest in conducting polymers began in 1977 following the discovery of polyacetylene doping [13]. Several subsequent studies substantially contributed to the understanding of the chemistry and physics of these materials. It was quickly established that neutral conjugated polymers were intrinsically semi-conducting materials in which the gap (in terms of width and position) was strongly dependent on the chemical constitution of the repeated unit. Neutral conjugated polymers can undergo oxidative (p-type) or reductive (n-type) doping which transforms the neutral chains into polycations or polyanions, respectively. At the same time, an equivalent number of counterions are incorporated into the polymer matrix to neutralise the charge imposed on the polymer chain during polymerization. A phenomenological manifestation of the doping reaction is an increase in the polymer electrical conductivity by several orders of magnitude. Thus, neutral semi-conducting materials are transformed into organic conductors in the course of the doping reaction. The type of doping depends on the position of the gap or, using molecular chemistry terms, on the position of the HOMO and LUMO levels (HOMO corresponds to the upper edge of the valence band and LUMO to the lower edge of the conduction band). Some polymers, such as polyacetylene or poly(p-phenylene), are amphoteric and can be p- and n-doped. Polypyrrole is distinguished from other conjugated polymers by the high positions of its HOMO and LUMO levels. This means that it can be easily p-doped but its n-doping is very difficult. As a consequence, under normal conditions, polypyrrole is much more stable in its doped (p-conducting) than in its undoped (neutral), form [14]. In 2000, Alan Heeger, Alan MacDiarmid, and Hideki Shirakawa were awarded the Nobel Prize in Chemistry for the discovery and development of electrically conductive polymers [15]. Progress in conducting polymer research has been aided by recent developments in nanoscience. In an environment with varied reaction conditions and in the presence of various surfactants [16] structured conducting polymers in the form of nanospheres, nanorods, nanowires, etc. can be prepared [17, 18].

Various methods can be used for PPy synthesis, electrochemical polymerization [19], chemical oxidative polymerization [20], photochemical polymerization [21], solid-state polymerization, plasma polymerization, etc.

However, the mechanism of pyrrole polymerization is still the subject of dispute; there is no single mechanism that is universally accepted. The controversy lies in the initiation step, as each mechanism proposes a different way of initiation of the reaction, varying between electron transfer, proton transfer and direct radical pyrrole formation [22].

The most accepted is pyrrole radical-cation creation in the initial step of the anodic polymerization of monomer. The pyrrole radical cation reacts with another monomer radical cation or monomer molecule, resulting in a dimer formation while the elimination of two protons occurs. The next step is oxidation of the pyrrole oligomers and their combination with another radical cation accompanied by deprotonation. The oxidation potentials for pyrrole oligomers become less positive with increasing chain length. Polypyrrole chain termination occurs most frequently when the reactive chain-end is blocked by further reaction, e.g. with water or when no further monomer is present for oxidation [20, 23]. The electrochemical oxidative polymerization of pyrrole produced PPy with conductivity about 100 S cm^{-1} with good mechanical properties and stability. For chemical oxidative polymerization of pyrrole many oxidants have been used, however iron chloride ($FeCl_3$) produced PPy with the highest conductivity [24, 25]. On the other hand, the poor mechanical properties and processability of PPy synthesized by chemical oxidative polymerization limited its commercial applications. The rapid increase in the number of methods for the preparation of PPy, as well as the range of their combinations with other materials, affords the possibility of various kinds of applications. Standard examples of the use of such composite materials are anti-corrosion coatings, radiation shields, electrochromic applications, but also a new category of sensors, prosthetic elements, and other human medicine applications are being developed. Recently new composite materials, based on various inorganic or organic substrates coated with a thin PPy layer have been developed as reported in review papers [26]. These composites combined the good physical and chemical properties of the individual components.

The key issue in the case of CNT/PPy is the interaction between conducting polymer and surface of CNT. Among the many studies and approaches related to this issue polymerization of pyrrole in the presence of CNT proved to be an easy and interesting way of preparing material, which combine the properties of both components involved, i.e. conducting polymer and carbon nanotubes. In this chapter we will try to cover all possible methods of CNT/PPy hybrids synthesis known nowadays and also summarize their possible applications.

PREPARATION OF CNT/PPY HYBRIDS

By coating of inorganic or organic particles and surfaces by means of in situ chemical oxidative polymerization of pyrrole new kinds of composites are created with many possibilities of application. Eventhough, there can be a problem associated with the homogeneous coverage of the surface of particles by surface of particles by polypyrrole layer because of the mutual incompatibility. Since CNT are electron acceptor and PPy are electron donor CNT/PPy particles can create charge-transfer complexes. The better surface interaction of PPy and CNT can be provided via surface active substances as ionic surfactants, functionalization of CNT surface with proper chemical groups or direct covalent binding of PPy onto the CNT surface via intermediate chemical linkage.

Fan et al. [27] described in-situ chemical oxidative polymerization of pyrrole monomer on CNT in 0.1 M solution of HCl with $(NH_4)_2S_2O_8$ as oxidant at 0-5°C. The resulting CNT/PPy composites showed a uniform tubular morphology with diameter less than 100 nm. The pristine CNT have diameter about 30 nm. The thickness of PPy layer was tuned by varying the initial amount of pyrrole monomer. The conductivity of CNT/PPy was 16 S cm^{-1}, in between the electrical conductivity of pure PPy (~3 S cm^{-1}) and pristine CNT (~40 S cm^{-1}). Han et al. [28] prepared PPy/MWCNT nanocomposites with core–shell structure by a novel ternary phase interfacial polymerization. The utilizations of both ternary system and 1-butyl-3-methylimidazolium tetrachloroferrate (Bmim[$FeCl_4$]) for the interfacial polymerization favoured the self-assembly of PPy on the surface of MWCNT. Increasing the pyrrole content from 50 to 90 wt.%, the thickness of the PPy layer increased from about 30 nm to 80 nm. Saba et al. [29] were able to get PPy coating as thin as ~ 3 nm by the electrodeposition method (Figure 1b). Their aim was to prevent the dispersion of MWCNT after sonication in ethanol by the protecting biocompatible PPy layer, which is important for safety issue. Figure 1a displays the resulting pictures of the solutions after sonication. It is clearly visible that the thicker the PPy layer, the more transparent are the solutions.

The structural characterization of CNT/PPy did not show any significant interaction between CNT and polypyrrole [27], i.e. CNT served just as a scaffold for the development of the PPy layer, but Chen et al. showed that CNT can act as anionic dopant for PPy [30]. The absence of chemical interaction in CNT/PPy was also confirmed by Chang et al. [31], but ESR study showed some decrease of the density of polarons and the increase of orientation of PPy in CNT/PPy compared with PPy. Wu et al. [32] observed slight shift of the absorption peak assigned to the polaron–polaron transition at 484 nm to the smaller wavelength with increasing the MWCNT content, indicating the interaction between the quinoid rings of PPy and MWCNT. An increased interaction between CNT and PPy can be achieved by surface functionalization of CNT. One of the possible ways is to introduce COOH group onto the surface of CNT by purification with aqueous solution of HNO_3. Qu et al. [33] studied MWCNT-COOH/PPy composites by FTIR and detected "blue shift" of the asymmetric stretching vibration of pyrrole ring at 1552 cm^{-1} compared with pure PPy (1546 cm^{-1}), which is also lower than that of C–C skeleton vibration of carbon ring in MWCNT (1579 cm^{-1}). This indicated strong interaction between the carboxyl groups on the MWCNT surface and PPy, which could be induced by two factors, one could be the hydrogen bonds between –COOH groups of chemically modified MWCNT and N–H groups of PPy, the other is that a donor– receptor complex was formed between PPy and MWCNT, resulting in an evenly distributed electron cloud.

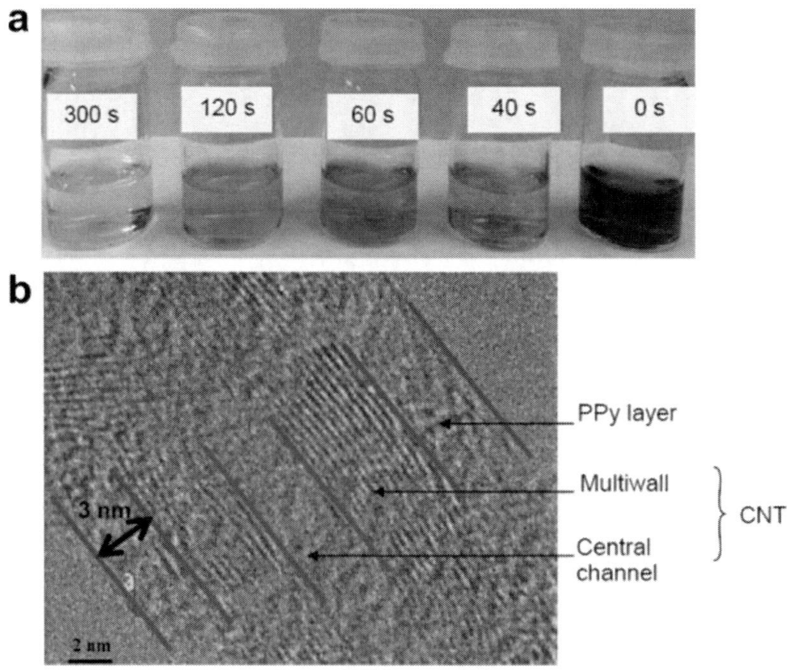

Figure 1. Dispersion tests of PPy coated CNT–CF by ultrasonication in ethanol: (a) pictures of the solution after sonication for various polymerization times; (b) HRTEM picture of a PPy coated CNT found in the ethanol solution after sonication. Reproduced from [29] with permission of the publisher.

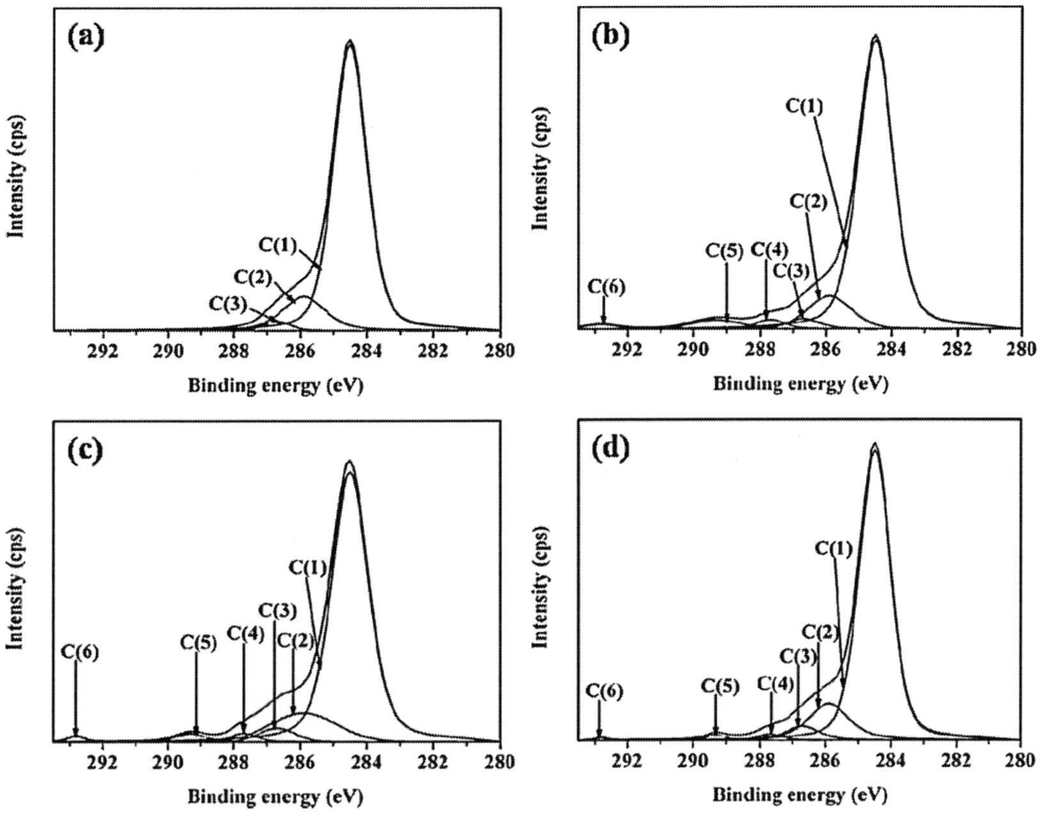

Figure 2. C 1s deconvolution of (a) the pristine MWCNT and the MWCNT oxyfluorinated with oxygen to fluorine ratio of (b) 2:8, (c) 5:5, and (d) 8:2. C(1) = C-C, C(2) = C-O, C(3) = C=O, C(4) = semi-ionic C-F bond, C(5) = covalent CF_2 bond, C(6) = perfluorinated CF_3 bond. Reproduced from [34] with permission of the publisher.

The oxyfluorination treatment on MWCNT introduced the hydrophilic functional groups resulting in a homogeneous distribution and higher interfacial affinity between PPy and MWCNT [34]. The high resolution XPS spectra of C1s are showed in Figure 2. In the case of oxygen to fluorine ratio 8/2, the total content of C–O and C=O increased more than 40% compared with that of 2/8 sample. These carbon–oxygen bonds contributed to the improved hydrophilicity of MWCNT. The variation in the F 1s region (not shown here) is in good agreement with the result of C 1s region (Figure 2) variation after the oxyfluorination. The covalent C–F bonds were produced in larger amounts by oxyfluorination with increasing fluorine content in the reactive gases. As observed by SEM (not shown here) the thickness of PPy on the surface of MWCNT decreased as the hydrophilic groups on MWCNT produced by the oxyfluorination treatment increased and thus PPy could be uniformly coated on MWCNT by increasing the oxygen content in oxyfluorination condition.

Another approach to improve the coating of PPy on the CNT surface is to use surface active substances. Zhang et al. [35] observed a formation of CNT/PPy nanocables when cationic cetyltrimethylammonium bromide (CTAB) or nonionic polyethylene glycol mono-p-nonylphenyl ether (O π-10) were used. They stated that an anionic surfactant such as sodium dodecyl sulphate (SDS) did not produce nanocables, because doped PPy has counterions and thus anionic surfactants serve as counterions for the polymer chains and permanently disrupt the structure of the columnar microregion, SDS proved to be a good supporting electrolyte for electropolymerization of PPy and alignment of disordered CNT within conducting polymer/CNT composite films was achieved [36]. Electrical measurement showed that electrical contacts between PPy and CNT were improved and that nanocable structure favours the charge transport between PPy and CNT [35]. Cationic CTAB proved to be better to the synthesis of nanocables than nonionic Oπ-10 (see Figure 3).

The diameter of nanocables ranges from 29 to 50 nm (pristine CNT had diameter ~16 nm) varying the pyrrole/CNT mass ratio from 10/3 to 30/3. At such high PPy amount (77-90 %) in the reaction solution it is very likely that polymerization took place also outside the CNT surface and then the bulk PPy could be adsorbed to the CNT as depicted in Figure 4 [37]. Often this critical issue not addressed regarding the making of optimally tailored CNT/PPy dual phase composite, i.e. the supposed incompatibility that might develop during the polymerization reaction. This may result in composite that will contain discrete non-interacting CNT and PPy phases [38]. Goldman and Lellouche [39] addressed this issue and designed a new methodological concept to prepare dual phase CNT/PPy composite. Their concept combined two key steps: (i) first, covalently coupling/grafting oxidized polycarboxylated MWCNT (COOH-MWCNT) with hydroxylated or animated Pyrrole(Py)/Carbazoyl (Cz) containing linkers affording Py- and Cz-decorated MWCNT (MWCNT-Py/Cz), and (ii), second, in situ oxidatively polymerizing various functional monocarboxylated Py- and Cz-monomers (Liquid phase polymerization (LPP) conditions) in the presence of former MWCNT-Py/Cz. This intermediate MWCNT-Py/Cz acted as a nucleophilic nanosized phase able to trap the cation/radical electrophilic polyPy- and polyCz-conducting polymer chains generated in the bulk medium during LPP experiments. They used statistically relevant design of experiments (DOE) to optimize multiparametric LPP process. DOE methods are powerful statistical tools that enable the discovery of global and not local minima of optimized process/reaction sets of conditions. This DOE mediated study led to two main conclusions. First, specific experimental sets of LPP conditions have been determined which provided optimal depositions of PPy polymer in a 17.7–18.1 % weight range. Second, only one influential LPP parameter, i.e., the amount of oxidizable Py- monomer has been identified. Pumera et al. [40] showed that only 4.6 wt.% of PPy in MWCNT-COOH/PPy composite is enough to create ultrathin (~7 nm) coating of the whole MWCNT surface (Figure 5). This ultrathin PPy coating effectively prevented electron transfer between the individual carbon nanotubes. Mi et al [41] presented a microwave hydrothermal strategy to functionalize CNT-COOH with a thornlike methyl orange-iron chloride complex. This complex then served as a template to directionally grow PPy chains out of the CNT surface. This 1-D nanostructured PPy perpendicular-modified CNT can found application as electrode for supercapacitors. Rana et al. used Cu(I)-catalyzed azide–alkyne cycloaddition click chemistry method to covalently attach the uniform film of PPy onto the MWCNT [42]. They prepared core–shell nanowires having MWCNT as a core and PPy as a shell. TEM examination indicated uniform PPy layers with thickness of 10–20 nm formed on the MWCNT surface. In particular 'grafting from' click coupling was more effective in obtaining uniform and stable core–shell nanowires as well as in the reaction yield (see Figure 6), compared to 'grafting to' click coupling. This is due to chemical bond formation between PPy and MWCNT at equal intervals along the longitudinal axis of the MWCNT, achieved by 'grafting from' click coupling. As a result, the core–shell nanowires were very stable even in the sonication of nanowires and showed an enhanced electrical conductivity of 80 S cm^{-1}, due to the synergetic interaction between MWCNT and PPy, which is higher than the conductivity of pristine MWCNT and pure PPy.

Bhatia et al. [43] prepared MWCNT/PPy coaxial fibrils by template free in-situ electrochemical deposition of PPy over MWCNT. They used propylene carbonate as the solvent and ionic liquid tetrabuthylammonium hexafluorophosphate in the electrochemical polymerization (see Figure 7). Interesting result, beside the non-linear I-V characteristics, was the Y-junctions of MWCNT/PPy fibrils (Figure 8), which may find some application in organic electronics.

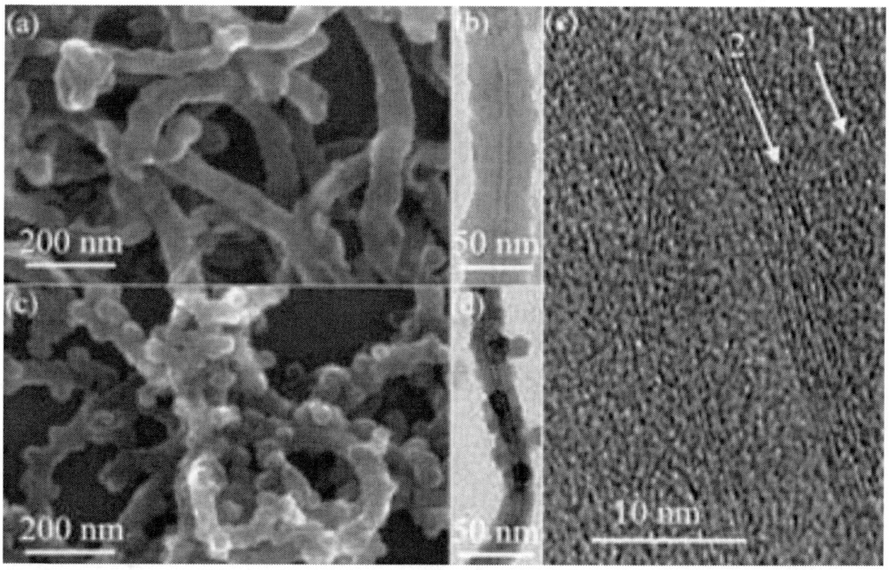

Figure 3. Surfactant-directed CNT/PPy nanocables: SEM image of CTAB-directed nanocables (a), TEM image of an individual CTAB-directed nanocable (b), SEM image of Oπ10-directed nanocables (c); TEM image of an individual Oπ10-directed nanocable (d), and HRTEM image of an individual CTAB-directed nanocable where arrow 1 indicates outer amorphous layer, and arrow 2 inner, crystalline layer (e). Reproduced from [35] with permission of the publisher.

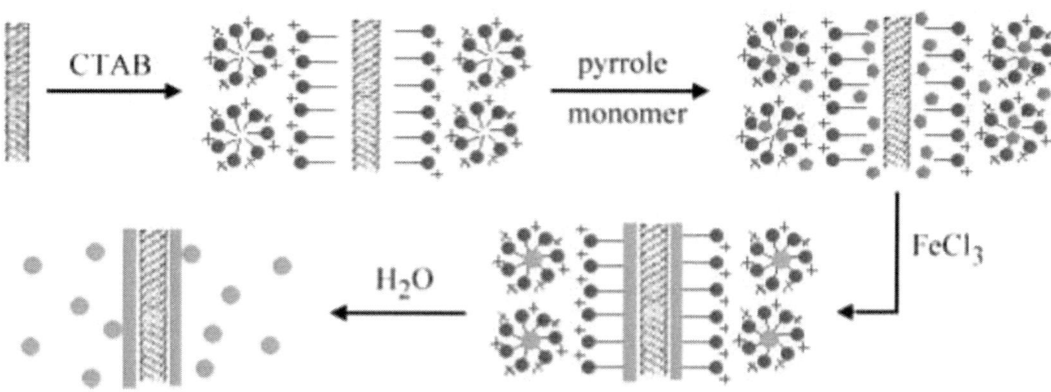

Figure 4. Schematic drawing of the mechanism governing the formation of MWCNT/PPy composites. Reproduced from [37] with permission of the publisher.

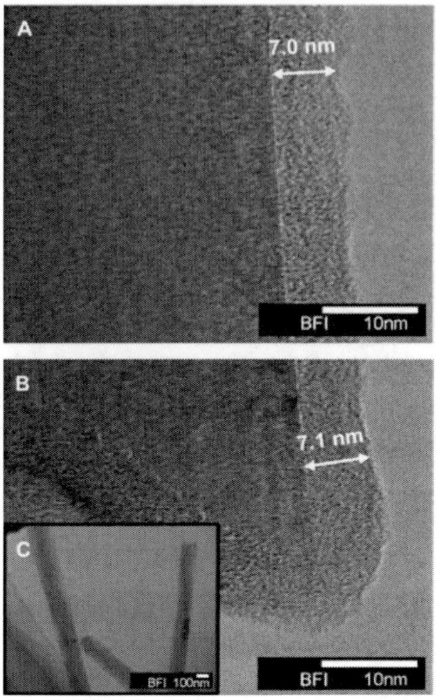

Figure 5. TEM images of PPy-coated MWCNT. Detailed view of PPy A) at the side, and B) the end of the PPy-coated MWCNT. Inset C shows a TEM image of several PPy-coated MWCNT at lower magnification. Coating conditions: concentration of pyrrole 10 mM; coating time, 24 h. Reproduced from [40] with permission of the publisher.

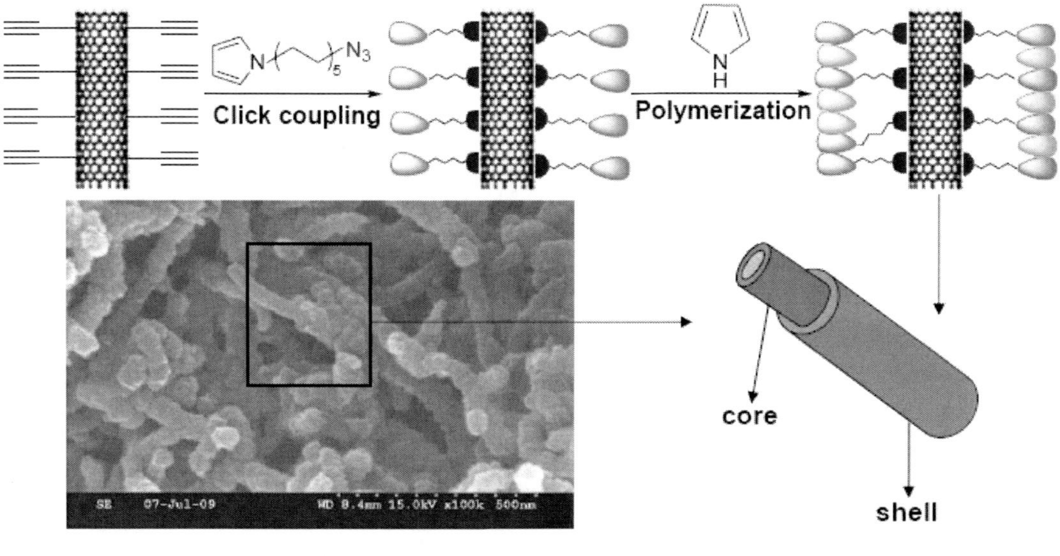

Figure 6. Schematic illustration of 'grafting from' click coupled core–shell nanowires of MWCNT/PPy. Reproduced from [42] with permission of the publisher.

Wu et al. [44] prepared ferromagnetic CNT/PPy using the decoration of CNT surface with magnetite prior to in-situ polymerization of pyrrole. Different procedures of decoration by inorganic nanoparticles was reported by Qu et al. [45]. They used CNT/PPy composite as support for Pt nanoparticles to prepare catalyst for methanol oxidation.

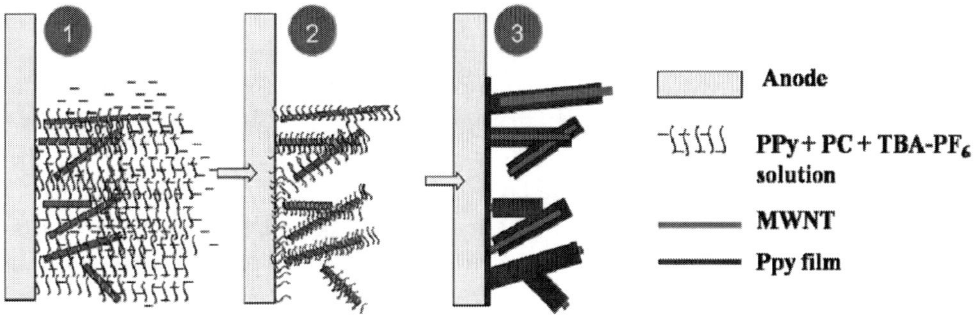

Figure 7. Schematic of MWNT/PPy fibrils formation. Reproduced from [43] with permission of the publisher.

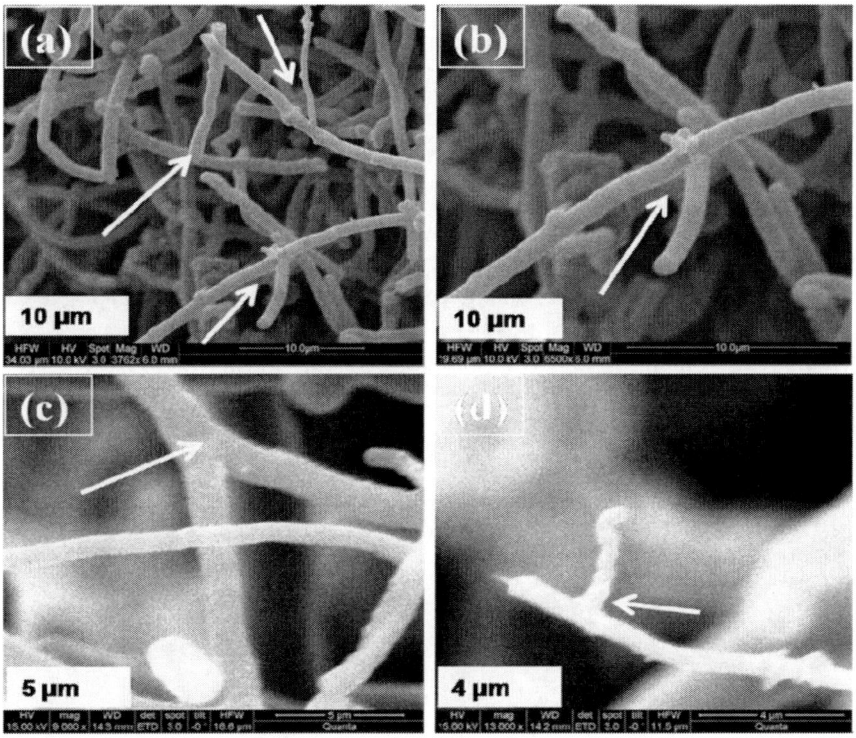

Figure 8. a) and b) TEM images of MWCNT/PPy fibrils and c) and d) SEM images of Y- junctions in MWCNT/PPy fibrils. Reproduced from [43] with permission of the publisher.

APPLICATIONS OF CNT/PPY HYBRIDS

Supercapacitors

One of the possible applications of CNT/PPy hybrids is in devices such as fuel cells, consumer electronics, and batteries, where supercapacitors are needed. Supercapacitors (SC) so called ultracapacitors or electrochemical capacitors, are devices with very high capacitance and high specific power, which show negligible deterioration under hundreds of thousands charge and discharge cycles. SC have been historically proposed in applications as memory back-up in consumer electronics and storage systems for microsolar power generators and now they are proposed for high power/energy applications, such as hybrid and electric vehicles, power quality systems and smart grids [46]. CNT/PPy hybrids are composite SC, where the storage of electrical charge beside the charge storage by the double layer between electrode and electrolyte (Electric double-layer capacitor - EDLC), is assured due to faradaic mechanisms as electrosorption, redox reactions and intercalation. This latter case also called pseudocapacitance is able to double the energy density when compared to EDLC. Conducting polymers such as PPy and also some metal oxides are promising candidates for supercapacitor applications owing to their high pseudocapacitance [47, 48, 49]. The advantages of conducting polymers over metal oxides are their ease and versatility in synthesis, high conductivity and low cost [50, 51]. PPy is an ideal electroactive material which demonstrates good stability and relatively high mass specific capacitance. However, virgin PPy shows poor cycling performance due to the volume change caused by the doping–dedoping of ions during charging and discharging [52, 53]. One of the solutions to overcome the stability problem of the PPy cells is to synthesize the composites with CNT.

The rectangular shape of voltammetry characteristics in the Figure 9 represents an ideal double layer capacitance, in this type of energy storage is purely electrostatic and current is independent on potential [54]. On the other hand, pseudocapacitors show a deviation from the rectangular shape and reversible redox peaks connected with pseudofaradaic reactions are remarkable (see Figure 9), where the accumulated charge in the capacitor is strongly dependent on the electrode potential.

Many papers deal with the preparation and characterization of CNT/PPy hybrids SC [55, 56, 57, 58, 59, 60, 61]. Mass specific capacitance of CNT/PPy SC ranges from 144 [61] up to 890 F/g [62]. There are many factors affecting the CNT/PPy final capacitor behaviour.

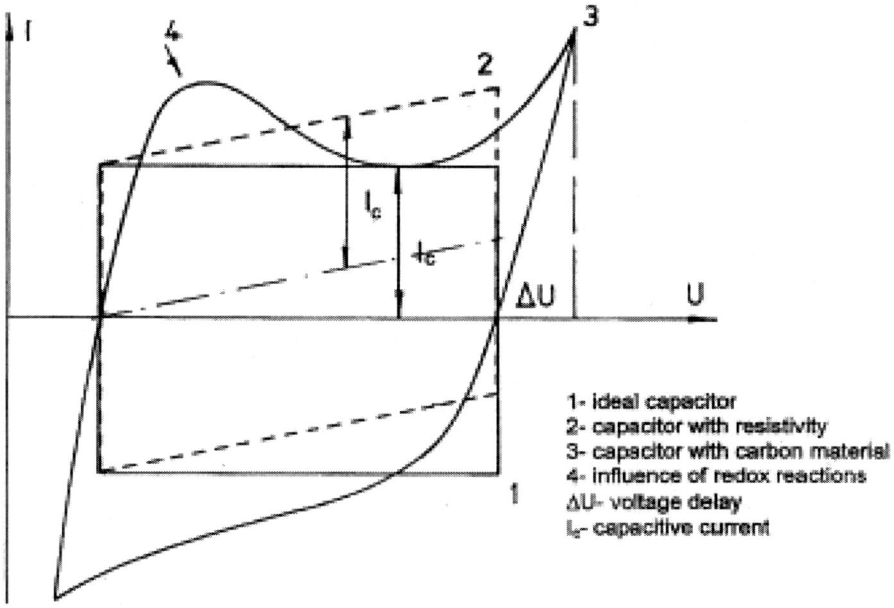

Figure 9. Typical charge / discharge voltammetry characteristics of an electrochemical capacitor. Reproduced from [54] with permission of the publisher.

Wang et al. [63] studied different types of anion dopants and their capacitive characteristics. They electrodeposited SWCNT functionalized by sulfuric and nitric acid (SWCNT-COOH) and conducting PPy doped by SWCNT-COOH, Cl^-, toluenesulfonate (T_OS^-), and dodecylbenzenesulfonate (DBS^-). Mass specific capacitance of SWCNT-COOH/PPy-T_OS was slightly lower than that of SWCNT-COOH/ PPy-Cl at low scanning rates due to the greater mass of its dopant T_OS^-, but greater in the condition of fast charging–discharging process (scanning rate >100 mVs^{-1}), which satisfies a crucial requirement of application in supercapacitors. SWCNT-COOH/ PPy-DBS nanocomposites had very low electrochemical activity mainly because DBS^-, with its overlarge size, was embedded in PPy matrix and consequently, overcome the anions exchange during charging–discharging process. Paul et al. [64] recently reported the importance of electrolyte of the system on the capacitive characteristics of CNT/PPy used as SC. They found that MWCNT/ PPy composite materials in terms of stability and therefore, superior SC performance, preferred a neutral electrolyte system over acidic and alkaline media. A relatively high mass specific capacitance ~165 F/g with excellent stability (1000 cycles) of the SC in a 1.0 M KCl electrolyte was achieved. The linear shape of the charge–discharge curves under a high current density demonstrates the ideal supercapacitor behaviour [64]. How it is important to have homogeneous coating of PPy in CNT/PPy SC has been shown by Fu et al. [55]. In Figure 10 is the comparison of capacitance cycle life of sulfonated MWCNT and non sulfonated MWCNT used for preparation of CNT/PPy hybrids. The prepared MWCNT-SO_3H/PPy composite exhibited good rate ability and specific capacitance loss was only 3 % after 1000 cycles. This good stability could be attributed to the uniform morphology which enhanced the effective electrochemical accessibility and MWCNT-SO_3H serving as a large molecular dopant integrated into the PPy chains ensures stability of the desired electrochemical properties. Also the alignment of the CNT in the CNT/PPy SC can lead to the better charge storage capacities. This improved performance of aligned arrays of MWCNT coated with PPy is coming from the combination of electrolyte accessibility, reduced diffusion distances, and increased conductivity in a redox pseudocapacitive composite structure [65].

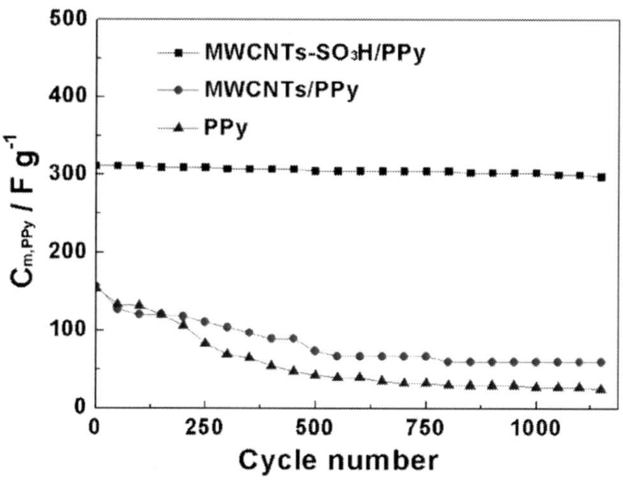

Figure 10. Cycle-life of PPy, MWCNT/PPy and MWCNT/C-SO_3H/PPy at a constant current density of 1.0 Ag^{-1}. Reproduced from [55] with permission of the publisher.

Sensors

Since polypyrrole is very susceptible to chemical or physical doping influencing its electrical properties, it is a good candidate for sensor materials [66, 67]. To improve the sensing properties of PPy the composites with various inorganic and organic particles resulting in some synergistic effects were prepared [68, 69]. Particularly, the CNT/PPy hybrids has shown significant improvements in the sensing performance of PPy [70, 71, 72, 73]. The impact is considered to arise from the interfacial interaction between PPy and CNT. Huyen et al. [74] prepared core-shell composite SWCNT/PPy by in situ chemical polymerization of pyrrole, where the shell had a thickness of 15-20 nm. This composite has shown synergistic effect in the sensitivity of gas sensor of O_2 (see Figure 11) and NH_3. The sensing mechanism is about the change of conductivity after adsorption and desorption of gases, which could be oxidizing (electron acceptor) or reducing (electron donor) the sensor material. Upon exposure of oxygen the electrons are extracted from semiconducting PPy and SWCNT resulting in the conversion of the material into p-type semiconductor. The more oxygen is adsorbed the more electrons are extracted resulting in the increase of conductivity. Huyen et al. [74] stated that due to the core-shell morphology, the PPy shell likely take the most share in the sensitivity. The diffusion of electrons from the SWCNT into PPy region could strongly affect the PPy band gap along the 15–20 nm width [75], the work function of electrons on the surface of PPy shell then could be reduced and easier to be donated [74]. The high sensitivity to O_2 as well as to NH_3 with inverse nature observed in SWCNT/PPy material indicates that the adsorbed O_2 molecules could be a reason to promote the adsorption of NH_3 [74, 76].

To improve the selectivity of the sensors based on CNT/PPy hybrids some papers reported synthesis of ternary systems involving some specific substance to assure some specific sensing properties [77, 78, 79, 80]. This specific specie improving the selectivity and sensitivity of the sensor could be directly the dopant of PPy. Good example is nitrazine yellow as anion dopant of PPy, which improved the sensitivity of glassy carbon/MWCNT-COOH/PPy electrode on Naltrexone. This sensor was then applied for the accurate determination of trace amounts of Naltrexone in pharmaceutical and clinical preparations [81]. Immunosensor can be prepared when the antibodies are used as dopant in electropolymerization of CNT/PPy film on microelectrodes [78]. Jin et al. [77] added Prussian blue ($K_3[Fe(CN)_6]$) which is known for its electroreduction of H_2O_2 in the CNT/PPy hybrids to prepare H_2O_2 sensor. The glassy carbon electrode modified with MWCNT/PPy/$K_3[Fe(CN)_6]$ exhibited good sensitivity (345 $\mu A\ mM^{-1}\ cm^{-2}$), wide linear range (0.004–0.517 mM) and low detection limit (0.08 μM) to detection of H_2O_2. Another group has prepared CNT/PPy sensor by layer-by-layer deposition. First they electropolymerized PPy onto indium tin oxide (ITO) electrode and then MWCNT-COOH dispersion in methanol was spread on PPY thin film by simple casting [80]. To prepare glucose sensitive sensor they covalently immobilized glucose oxidase (GOx) together with horseradish peroxidase (HRP) onto MWCNT-COOH/PPy/ITO electrode using glutaraldehyde chemistry. The biochemical reaction going on the electrode is depicted in Figure 12. This nanobiocomposite electrode can be used to estimate glucose in blood serum as was shown in [80]. Similar approach of this group was applied for preparation of biosensor for cholesterol, where cholesterol oxidase and cholesterol esterase were immobilized on MWCNT-COOH/PPy/ITO electrode [79].

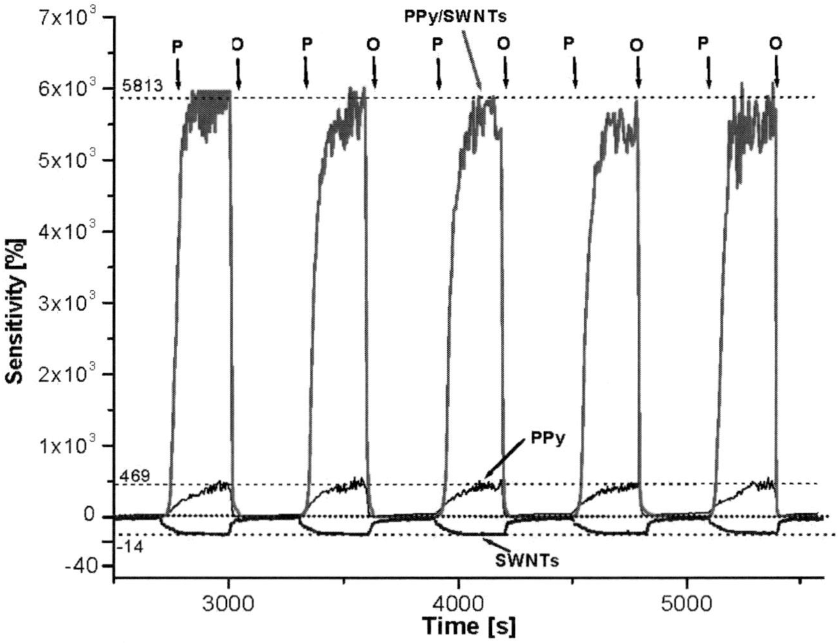

Figure 11. Oxygen sensitivity of PPy, SWCNT, and SWCNT/PPy upon exposure to open air (P: low air pressure as pump on, O: open air). Reproduced from [74] with permission of the publisher.

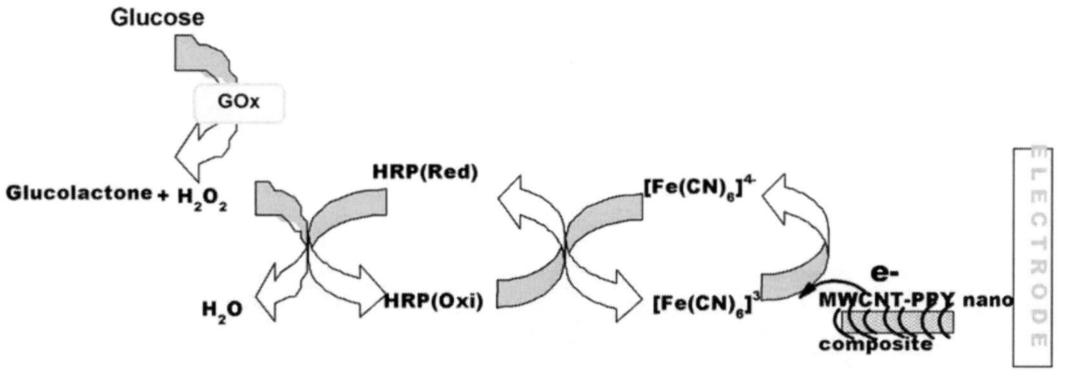

Figure 12. Proposed biochemical reaction on the GOx-HRP/MWCNT/PPY/ITO (in phosphate buffer (50 mM, pH 7.0, 0.9% NaCl) containing 5 mM $[Fe(CN)_6]^{3-/4-}$). Reproduced from [80] with permission of the publisher.

Other Applications

CNT/PPy hybrids could find also applications in emerging field of novel lightweight actuators operating at low voltage, mainly used in biomedical devices [82, 83]. Ionic actuators as CNT/PPy hybrids appear to be a good candidate for this kind of application. The resulting devices benefit from the actuation behaviour of the actuating properties of both CNT and PPy resulting in materials suitable for soft actuation application [84]. Generally, the addition of CNT lowers the actuation strain, but substantially improved the actuation under load [85].

Mukai et al. [86] prepared double layered SWCNT/PPy bending-type actuator. SWCNT layer and the PPy layer were constructed through the electropolymerization of pyrrole on the SWCNT film surface. They considered that actuation of the SWCNT/PPy double layer films was induced by volume changes of both SWCNT and PPy layers. The actuator is bended to the PPy side at negative voltage (Figure 13a) and to the SWCNT side at positive voltage (Figure 13b).

The requirements for actuators can vary depending on the end use. For example, an actuation strain of 5% at a 5MPa isotonic stress is necessary for a medical "rehabilitation glove", while a fast response rate to electrical stimulus is required for a valve or switch used as a component in micro electromechanical systems [85].

Another possible application of CNT/PPy hybrids is in the electromagnetic shielding materials. Electromagnetic shielding is used to avoid interference of electronic devices, which emitting the electromagnetic energy in the same frequency bands as some nearby other device resulting in the malfunction. Electromagnetic interference could be also harmful to human body. That is why it is important to limit and shield the electronic equipments against all sources of interference [87, 88]. Yun et al. [89] prepared MWCNT-COOH/PPy by in situ chemical oxidative polymerization and the new electromagnetic materials reached the EMI SE number around 28 dB, which is well above the commercial application SE requirement (more than 20 dB) [90].

Also, in the last years, some other possible applications of CNT/PPy hybrids were reported. MWCNT supported PPy in alkyd paint coatings has improved the corrosion protection of steel [91]. Lu et al. [92] have demonstrated the feasibility and advantages of co-deposited SWCNT/PPy films on Pt for improving the electrode-neural tissue interface for possible improving long-term performance of neuroprosthetic devices. In all possible application the most important thing is ability to control the PPy layer creating on supported CNT particles to ensure the development of the homogeneous film of conducting polymer. This is especially important in the case for organic electronic applications, where device quality film is a necessity [93].

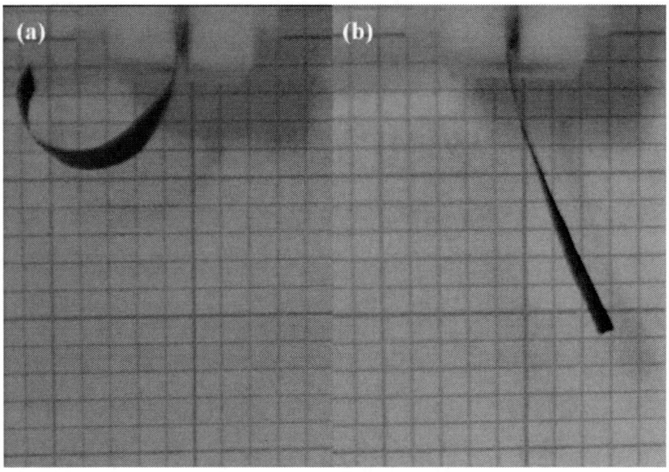

Figure 13. Photographs of deformed shapes of the SWCNT/PPy double layer film when applied at (a) -0.85 V and (b) +0.6 V (sample size:12 mm×2 mm). Reproduced from [86] with permission of the publisher.

Conclusion

CNT/PPy hybrids are an interesting group of materials, which have experienced an enormous increase in published papers in the last decade. The basis for all applications is the control of PPy layer growing on CNT surface and thus the control of the interphase. The methods of CNT/PPy hybrids preparation are generally very well described in details and provide a basis for possible applications of these materials. A rapid development of new applications of these hybrids can be expected in the near future, where they can replace materials containing noble metals, which are much more expensive and their sources are limited. Supercapacitors and sensors based on new kind of hybrid structures show dynamic development with real applications in the near future.

Acknowledgment

This work was financially supported by the Scientific Grant Agency of the Ministry of Education of the Slovak Republic and the Slovak Academy of Sciences (project No. 2/0064/10), and by the Slovak Research and Development Agency under contract No. APVV-0593-11.

References

[1] Iijima, S. (1991). Helical microtubules of graphitic carbon. *Nature*, *354*, 56-58.

[2] Oberlin, A., Endo, M. & Koyama, T. (1976). Filamentous growth of carbon through benzene decomposition. *J. Cryst. Growth*, *32*, 335.

[3] Ascheron, Claus E. (2008). Topics in Applied Physics: Carbon nanotubes advanced topics in the synthesis, structure, properties and applications. *Springer-Verlag Berlin Heidelberg*, 111.

[4] Pottish, N. (2005). Nanoparticles, nanotubes and nanocomposites. *Comp. Tech.*, *11*, 42–47.

[5] Vichchulada, P., Zhang, Q. & Lay, M. D. (2007). Recent progress in chemical detection with single- walled carbon nanotube networks. *Analyst*, *132*, 719–723.

[6] Moulton, S. E., Minett, A. I. & Wallace, G. G. (2005). Carbon nanotube based electronic and electrochemical sensors. *Sensor Lett.*, *3*, 183–193.

[7] Xu, N. S. & Huq, S. E. (2005). Novel cold cathode materials and applications. *Mat. Sci. Eng. R.*, *48*, 47–189.

[8] Bandaru, P. R. (2007). Electrical properties and applications of carbon nanotube structures. *J. Nanosci. Nanotechnol.*, *7*, 1239–1267.

[9] Baughman, R. H., Zakhidov, A. A. & De Heer, W. A. (2002). Carbon nanotubes - The route toward applications. *Science*, *297*, 787–792.

[10] Fritsche, J. (1840). Ueber das Anilin, ein neues Ersetzungsproduct des Indigo. *Journal fur praktische chemie,* 1840, *20,* 453–459.

[11] Letheby, H. (1862). XXIX. –On the production of a blue substance by the electrolysis of sulphate of aniline. *J. Chem. Soc.*, *15,* 161–163.

[12] Dall'Olio, A., Dascola, G., Varacca, V. & Biochi, V. (1968). Resonance paramagnetique electronique et conductivité d'un noir d'oxypyrrol electrolytique. *Comptes Rendus Acad. Sci. C (Paris), 267C*, 433–435.

[13] Shirakawa, H., Louis, E. J., MacDiarmid, A. G., Chiang, C. K. & Heeger, A. J. (1977). Synthesis of electrically conducting organic polymers: halogen derivatives of polyacetylene, (CH)x. *J. Chem. Soc. Chem. Commun.*, 578–580.

[14] Pron, A. & Rannou, P. (2002). Processible conjugated polymers: from organic semiconductors to organic metals and superconductors. *Prog. Polym. Sci.*, *27*, 135–190.

[15] http://www.nobel.se/chemistry/laureates/2000.

[16] Omastová, M., Trchová, M., Kovářová, J. & Stejskal, J. (2003). Synthesis and structural study of polypyrroles prepared in the presence of surfactants. *Synth. Metals.*, *138*, 447–455.

[17] Pecher, J. & Mecking, S. (2010). Nanoparticles of conjugated polymers. *Chem. Rev.*, *110*, 6260–6279.

[18] Trchova, M., Sedenkova, I., Konyushenko, E. N., Stejskal, J. Holler, P. & Ciric-Marjanovic, G. (2006). Evolution of polyaniline nanotubes: The oxidation of aniline in water. *J. Phys. Chem. B.*, *110,* 9461–9468.

[19] Diaz, A. F., Kanazawa, K. K. & Gardini, G. P. (1979). Electrochemical polymerization of pyrrole. *J. Chem. Soc. Chem. Commun.*, 635–636.

[20] Street, G. B. (1986). Polypyrrole from powers to plastics. In T. A. Skotheim (Ed.), Handbook of conducting polymers New York, NY, USA: Marcel Dekker., Vol. *1*, 265–291.

[21] Attia, M. F., Azib, T., Salmi, Z., Singh, A., Decorse, P., Battaglini, N., Lecoq, H., Omastova, M., Higazy, A. A., Elshafei, A. M., Hashem, M. M. & Chehimi, M. M. (2013). One-step UV-induced modification of cellulose fabrics by polypyrrole/silver nanocomposite films. *J. Colloid Interface Sci.*, *393,* 130–137.

[22] Sadki, S., Schottland, P., Brodie, N. & Sabouraud, G. (2000). The mechanisms of pyrrole electropolymerization. *Chem. Soc. Rev.*, *29*, 283–293.

[23] Heinze, J. (1990). Electronically conducting polymers. *Topics Curr. Chem.*, *152*, 1–47.

[24] Pron, A., Kucharski, Z., Budrowski, C., Zagórska, M., Krichene, S., Suwalski, J., Dehe, G. & Lefrant, S. (1985). Mossbauerspectroscopy studies of selected conducting polypyrroles. *J. Chem. Phys.*, *83*, 5923–5927.

[25] Armes, S. P. (1987). Optimum reaction conditions for the polymerization of pyrrole by iron(III) chloride in aqueous solution. *Synth. Met.*, *20*, 365–371.

[26] Omastová, M. & Mičušík, M. (2012). Polypyrrole coating of inorganic and organic materials by chemical oxidative polymerisation. *Chem. Pap.*, *66*, 392–414.

[27] Fan, J., Wan, M., Zhu, D., Chang, B., Pan, Z. & Xie, S. (1999). Synthesis and properties of carbon nanotubepolypyrrole composites. *Synthetic Met.*, *102*, 1266–1267.

[28] Han, Y., Shen, M., Lin, X., Ding, B., Zhang, L., Tong, H. & Zhang X. (2012). Ternary phase interfacial polymerization of polypyrrole/MWCNT nanocomposites with core–shell structure. *Synt. Met.*, *162*, 753– 758.

[29] Saba, J., Magga, Y., He, D., Miomandre, F. & Bai, J. (2013). Continuous electrodeposition of polypyrrole on carbon nanotube–carbon fiber hybrids as a protective treatment against nanotube dispersion. *Carbon, 51,* 20–26.

[30] Chen, G. Z., Shaffer, M. S. P., Coleby, D., Dixon, G., Zhou, W. & Fray, D. (2000). Carbon nanotube and polypyrrole composites: coating and doping. *Adv. Mater.*, *12*, 522–526.

[31] Chang, B. H., Liu, Z. Q., Sun, L. F., Tang, D. S., Zhou, W. Y., Wang, G., Qian, L. X., Xie, S. S., Fen, J. H. & Wan, M. X. (2000). Conductivity and magnetic susceptibility of nanotube- polypyrrole nanocomposites. *J. Low Temp. Phys.*, *119*, 41–48.

[32] Wu, T. -M. & Lin, S. -H. (2006). Synthesis, Characterization, and electrical properties of polypyrrole/multiwalled carbon nanotube composites. *J. Polym. Sci. Pol. Chem.*, *44*, 6449–6457.

[33] Qu, B., Xu, Y. -T., Lin, S. -J., Zheng, Y. -F. & Dai, L. -Z. (2010). Fabrication of Pt nanoparticles decorated PPy–MWNTs composites and their electrocatalytic activity for methanol oxidation. *Synt. Met.*, *160*, 732–742.

[34] Kim, Y. -Y., Yun, J., Kim, H. -I. & Lee, Y. -S. (2012). Effect of oxyfluorination on electromagnetic interference shielding of polypyrrole-coated multi-walled carbon nanotubes. *J. Ind. Eng. Chem.*, *18*, 392–398.

[35] Zhang, X., Zhang, J., Wang, R., Zhu, T. & Liu, Z. (2004). Surfactant-directed polypyrrole/CNT nanocables: synthesis, characterization, and enhanced electrical properties. *Chem. Phys. Chem.*, *5*, 998-1002.

[36] Zhang, X., Zhang, J. & Liu, Z. (2005). Conducting polymer/carbon nanotube composite films made by in situ electropolymerization using an ionic surfactant as the supporting electrolyte. *Carbon*, *43*, 2186–2191.

[37] Wu, T. M. & Lin, S. H. (2005). Synthesis, characterization, and electrical properties of polypyrrole/multiwalled carbon nanotube composites. *J. Polym. Sci. Polym. Chem.* 2006, *44*, 6449–6457.

[38] Goldman, D. & Lellouche, J. -P. (2010). An easy method for the production of functional polypyrrole/MWCNT and polycarbazole/MWCNT composites using nucleophilic multi- walled carbon nanotubes. *Carbon*, *48*, 4170–4177.

[39] Goldman, D. & Lellouche, J. P. (2010). An easy method for the production of functional polypyrrole/MWCNT and polycarbazole/MWCNT composites using nucleophilic multiwalled carbon nanotubes. *Carbon*, *48*, 4170–4177.

[40] Pumera, M., Smid, B., Peng, X., Golberg, D., Tang, J. & Ichinose, I. (2007). Spontaneous coating of carbon nanotubes with an ultrathin polypyrrole layer. *Chem. Eur. J.*, *13*, 7644 – 7649.

[41] Mi, H., Zhang, X., Xu, Y. & Xiao, F. (2010). Synthesis, characterization and electrochemical behavior of polypyrrole/carbon nanotube composites using organometallic-functionalized carbon nanotubes. *Appl. Surf. Sci.*, *256*, 2284–2288.

[42] Rana, S. & Cho, J. W. (2011). Core–shell morphology and characterization of carbon nanotube nanowires click coupled with polypyrrole. *Nanotechnology*, *22*, 275609 (7).

[43] Bhatia, R., Sangeeth, C. S. S., Prasad, V. & Menon, R. (2011). Preparation and characterization of multiwall carbon nanotube/polypyrrole coaxial fibrils. *Physica B*, *406*, 1727–1732.

[44] Wu, T. -M., Yen, Sh. -J., Chen, E. -Ch. & Chiang, R. -K. (2008). Synthesis, characterization, and properties of monodispersed magnetite coated multi-walled carbon nanotube/polypyrrole nanocomposites synthesized by in-situ chemical oxidative polymerization. *J. Polym. Sci. Pol. Phys.*, *46*, 727–733.

[45] Qu, B., Xu, Y. -T., Lin, S. -J., Zheng, Y. -F. & Dai, L. -Z. (2010). Fabrication of Pt nanoparticles decorated PPy–MWNTs composites and their electrocatalytic activity for methanol oxidation. *Synt. Met.*, *160*, 732–742.

[46] Conte, M. (2010). Supercapacitors Technical Requirements for New Applications. *Fuell Cells.*, *10*, 806–818.

[47] Ferraris, J. P., Eissa, M. M., Brotherston, I. D. & Loveday, D. C. (1998). Performance Evaluation of Poly 3-(Phenylthiophene) Derivatives as Active Materials for Electrochemical Capacitor Applications. *Chem. Mater.*, *10*, 3528–3535.

[48] Prasad, K. R., Koga, K. & Miura, N. (2004). Electrochemical deposition of nanostructured indium oxide: high-performance electrode material for redox supercapacitors. *Chem. Mater.*, *16*, 1845–1847.

[49] Shakir, I., Shahid, M., Nadeem, M. & Kang, D. J. (2012). Tin oxide coating on molybdenum oxide nanowires for high performance supercapacitor devices. *Electrochim. Acta*, *72*, 134– 137.

[50] Dias, H. V. R., Fianchini, M. & Rajapakse, R. M. G. (2006). Greener method for high-quality polypyrrole. *Polymer*, *47*, 7349–7354.

[51] Machida, S., Miyata, S. & Techagumpuch, A. (1989). Chemical synthesis of highly electrically conductive polypyrrole. *Synthetic Met.*, *31*, 311.

[52] John, R. & Wallace, G. G. (1993). Doping-dedoping of polypyrrole: a study using current-measuring and resistance-measuring techniques. *J. Electroanal. Chem.*, *354*, 145–160.

[53] Hussain, A. M. P. & Kumar, A. (2006). Enhanced electrochemical stability of all-polymer redox supercapacitors with modified polypyrrole electrodes. *J. Power Sources*, *161*, 1486– 1492.

[54] Frackowiak, E. & Beguin, F. (2001). Carbon materials for the electrochemical storage of energy in capacitors. *Carbon*, *39*, 937–950.

[55] Fu, Q., Gao, B., Dou, H., Hao, L., Lu, X., Sun, K., Jiang, J. & Zhang, X. (2011). Novel non-covalent sulfonated multiwalled carbon nanotubes from p-toluenesulfonic acid/glucose doped polypyrrole for electrochemical capacitors. *Synt. Met.*, *161*, 373–378.

[56] Hermas, A. A., Al-Juaid, S. S., Al-Thabaiti, S. A., Qusti, A. H. & Abdel Salam, M. (2012). In situ electropolymerization of conducting polypyrrole/carbon nanotubes composites on stainless steel: Role of carbon nanotubes types. *Prog. Org. Coat.*, *75*, 404– 410.

[57] Hu, Y., Zhao, Y., Li, Y., Li, H., Shao, H. & Qu, L. (2012). Defective super-long carbon nanotubes and polypyrrole composite for high-performance supercapacitor electrodes. *Electrochim. Acta*, *66*, 279– 286.

[58] Hughes, M., Chen, G. Z., Shaffer, M. S. P., Fray, D. J. & Windle, A. H. (2004). Controlling the nanostructure of electrochemically grown nanoporous composites of carbon nanotubes and conducting polymers. *Compos. Sci. Technol.*, *64*, 2325–2331.

[59] Peng, Ch., Zhang, Sh., Jewell, D. & Chen, G. Z. (2008). Carbon nanotube and conducting polymer composites for supercapacitors. *Prog. Nat. Sci.*, *18*, 777–788.

[60] Snooka, G. A., Kao, P. & Best, A. S. (2011). Conducting-polymer-based supercapacitor devices and electrodes. *J. Power Sources*, *196*, 1–12.

[61] Wang, J., Xu, Y. L., Chen, X. & Sun, X. F. (2007). Capacitance properties of single wall carbon nanotube/polypyrrole composite films. *Compos. Sci. Technol.*, *67*, 2981–2985.

[62] Lin, X. & Xu, Y. (2008). Facile synthesis and electrochemical capacitance of composites of polypyrrole/multi-walled carbon nanotubes. *Electrochim. Acta*, *53*, 4990–4997.

[63] Wang, J., Xu, Y., Yan, F., Zhu, J., Wang, J. & Xiao, F. (2010). Capacitive characteristics of nanocomposites of conducting polypyrrole and functionalized carbon nanotubes: effects of in situ dopant and film thickness. *J. Solid State Electrochem.*, *14*, 1565–1575.

[64] Paul, S., Choi, K. S., Lee, D. J., Sudhagar, P. & Kang, Y. S. (2012). Factors affecting the performance of supercapacitors assembled with polypyrrole/multi-walled carbon nanotube composite electrodes. *Electrochim. Acta*, *78*, 649–655.

[65] Hughes, M., Shaffer, M. S. P., Renouf, A. C., Singh, Ch., Chen, G. Z., Fray, D. J. & Windle, A. H. (2002). Electrochemical capacitance of nanocomposites films formed by coating aligned arrays of carbon nanotubes with polypyrrole. *Adv. Mater.*, *14*, 382–385.

[66] McQuade, D. T., Pullen, A. E. & Swager, T. M. (2000). Conjugated polymer-based chemical sensors. *Chem. Rev.*, *100*, 2537–2574.

[67] Bai, H. & Shi, G. Q. (2007). Review: Gas sensors based on conducting polymers. *Sensors*, *7*, 267–307.

[68] Karin, P. K. (2002). Chemical gas sensors based on organic semiconductor polypyrrole. *Crit. Rev. Anal. Chem.*, *32*, 121–140.

[69] Ram, M. K. & Bhethanabotla, V. R., Eds. (2010). *Sensors for Chemical and Biological Applications*; CRC Press: Boca Raton, FL, USA.

[70] Chen, Y., Li, Y., Wang, H. & Yang, M. (2007). Gas sensitivity of a composite of multiwalled carbon nanotubes and polypyrrole prepared by vapor phase polymerization. *Carbon*, *45*, 357–363.

[71] An, K. H., Jeong, S. Y., Hwang, H. R. & Lee, Y. H. (2004). Enhanced sensitivity of a gas sensor incorporating single-walled carbon nanocomposites. *Adv. Mater.*, *16*, 1005–1009.

[72] Hieu, N. V., Dung, N. Q., Tam, P. D., Trung, T. & Chien, N. D. (2009). Thin film polypyrrole/SWCNTs nanocomposites based NH3 sensor operated at room temperature. *Sens. Actuators B*, *140*, 500–507.

[73] Wang, Z. M., Tang, X. C., Xiao, Y. H., Yu, X. J., Zhang, L., Jia, D. Z. & Chen, G. C. (2011). Polypyrrole coated carbon nanotubes: Preparation, characterization, and gas-sensing properties. *J. Inorg. Mater.*, *26*, 961–968.

[74] Huyen, D. N., Tung, N. T., Vinh, T. D. & Thien, N. D. (2012). Synergistic Effects in the Gas Sensitivity of Polypyrrole/Single Wall Carbon Nanotube Composites. *Sensors.*, *12*, 7965-7974.

[75] Jiri, J. & Mira, J. (2003). Conducting polymers in electronic chemical sensors. *Nat. Mater.*, *2*, 19–25.

[76] Feng, X., Irle, S., Witek, H., Morokuma, K., Vidic, R. & Borguet, E. (2005). Sensitivity of ammonia interaction with single-walled carbon nanotube bundles to the presence of defect sites and functionalities. *J. Am. Chem. Soc.*, *127*, 10533–10538.

[77] Jin, E., Bian, X., Lu, X. & Wang, C. (2012). Fabrication of multiwalled carbon nanotubes/polypyrrole/ Prussian blue ternary composite nanofibers and their application for enzymeless hydrogen peroxide detection. *J. Mater. Sci.*, *47*, 4326–4331.

[78] Tam, P. D. & Van Hieu, N. (2011). Conducting polymer film-based immunosensors using carbon nanotube/antibodies doped polypyrrole. *Appl. Surf. Sci.*, *257*, 9817– 9824.

[79] Singh, K., Solanki, P. R., Basu, T. & Malhotra, B. D. (2012). Polypyrrole/multiwalled carbon nanotubesbased biosensor for cholesterol estimation. *Polym. Adv. Technol.*, *23*, 1084–1091.

[80] Singh, K., Singh, B. P., Chauhan, R. & Basu, T. (2012). Fabrication of Amperometric Bienzymatic Glucose Biosensor Based on MWCNT Tube and Polypyrrole Multilayered Nanocomposite. *J. Appl. Polym. Sci.*, *125*, E235–E246.

[81] Shahrokhian, S., Kamalzadeh, Z. & Saberi, R. -S. (2011). Application of Glassy Carbon Electrode Modified with a Bilayer of Multiwalled Carbon Nanotube and Polypyrrole Doped with Nitrazine Yellow for Voltammetric Determination of Naltrexone. *Electroanalysis*, *23*, 2925 – 2934.

[82] Baughman, R. H., Cui, C., Zakhidov, A. A., Iqbal, Z., Barisci, J. N., Spinks, G. M., Wallace, G. G., Mazzoldi, A., De Rossi, D., Rinzler, A. G., Jaschinski, O., Roth, S. & Kertesz, M. (1999). Carbon nanotube actuators. *Science*, *284*, 1340–1344.

[83] Smela, E. (2003). Conjugated polymer actuators for biomedical applications. *Adv. Mater.*, *15,* 481–494.

[84] Biso, M., Ansaldo, A., Picardo, E. & Ricci, D. (2012). Increasing the maximum strain and efficiency of bucky gel actuators by pyrrole oxidative polymerization on carbon nanotubes dispersed in an ionic liquid. *Carbon*, *50*, 4506-4511.

[85] Spinks, G. M., Xi, B., Truong, V. -T. & Wallace, G. G. (2005). Actuation behaviour of layered composites of polyaniline, carbon nanotubes and polypyrrole. *Synt. Met.*, *151*, 85–91.

[86] Mukai, K., Yamato, K., Asaka, K., Hata, K. & Oike, H. (2012). Actuator of double layer film composed of carbon nanotubes and polypyrroles. *Sensor. Actuat. B-Chem.*, *161*, 1010– 1017.

[87] Hung, F. S., Hung, F. Y. & Chiang, C. M. (2011). Crystallization and annealing effects of sputtered tin alloy films on electromagnetic interference shielding. *Appl. Surf. Sci.*, *257*, 3733- 3738.

[88] Kim, B. R., Lee, H. K., Park, S. H. & Kim, H. K. (2011). Electromagnetic interference shielding characteristics and shielding effectiveness of polyaniline-coated films. *Thin Solid Films*, *519*, 3492-3496.

[89] Yun, J., Im, J. S. & Kim, H. -I. (2012). Effect of Oxygen Plasma Treatment of Carbon Nanotubes on Electromagnetic Interference Shielding of Polypyrrole-Coated Carbon Nanotubes. *J. Appl. Polym. Sci.*, *126*, E39–E47.

[90] Li, Y., Chen, Ch., Zhang, S., Ni, Y. & Huang, J. (2008). Electrical conductivity and electromagnetic interference shielding characteristics of multiwalled carbon nanotube filled polyacrylate composite films. *Appl. Surf. Sci.*, *254*, 5766–5771.

[91] Gergely, A., Pászti, Z., Hakkel, O., Drotár, E., Mihály, J. & Kálmán, E. (2012). Corrosion protection of cold-rolled steel with alkyd paint coatings composited with submicron-structure types polypyrrole-modified nano-size alumina and carbon nanotubes. *Mater. Sci. Eng. B-Adv.*, *177,* 1571– 1582.

[92] Lu, Y., Li, T., Zhao, X., Li, M., Cao, Y., Yang, H. & Duan Y. Y. (2010). Electrodeposited polypyrrole/carbon nanotubes composite films electrodes for neural interfaces. *Biomaterials*, *31*, 5169– 5181.

[93] Swathi, S. K., Jeevananda, T. & Ramamurthy P. C. (2010). Fabrication of device quality films of high loaded PPy/MWCNT nanocomposites using pulsed laser deposition. *Org. Electron.*, *11*, 1489–1499.

INSULATING-CONDUCTIVE COMPOSITES AND CATALYTIC MATERIALS

In: Applied Surface Chemistry of Nanomaterials
Editors: Mohamed M. Chehimi and Jean Pinson
ISBN: 978-1-62808-351-4

Chapter 11

SURFACE COATING OF SOFT MATERIALS WITH CONDUCTING POLYMER-METAL NANOCOMPOSITES

Syuji Fujii[1] and Yoshinobu Nakamura[1,2,3]
[1]Department of Applied Chemistry, Faculty of Engineering
Osaka Institute of Technology, Osaka, Japan
[2]Graduate School of Engineering
Osaka Institute of Technology, Osaka, Japan
[3]Nanomaterials Microdevices Research Center,
Osaka Institute of Technology, Osaka, Japan

Keywords: Conducting polymer; Polypyrrole (PPy); Noble metal; Nanocomposites; Coating; Soft matter; Catalyst; Suzuki-Miyaura coupling reaction; Heck reaction

INTRODUCTION

Air-stable conducting polymer-noble metal nanocomposites have significant potential to allow the design of functional devices [1], and also exhibit enhanced sensing and catalytic capabilities superior to those of pure conducting polymers [2].

Despite the promise of these new materials and their widespread study, the scope of commercial uses remains small and relatively few viable technologies have emerged from the laboratory proof-of-concept stage. Limitations of processability such as low mechanical strength, poor flexibility, poor solubility in solvent, infusibility with heat and high cost have prevented the nanocomposite materials from making significant commercial impact.

Under this situation, there are several attempts in order to solve the problems shown above: syntheses of nanocomposites in forms of colloidal particles [3], fibers [4], tubes [5] and films on the hard substrates [6], and hybridization with polymeric materials with high processability [7].

In the present work, a facile and new chemical approach developed by our research group to enable one-step nanoprecise coating of conducting polymer-noble metal nanocomposite on soft materials with good processability, whose shape can be easily designed and controlled, without disrupting their morphological hierarchies is reviewed. We describe one-step deposition of conducting polymer-noble metal nanocomposite onto polymer latex particles and polymer fibers by chemical oxidative polymerization using metal salts in aqueous media (Figure 1). Using this coating method, the shape of nanocomposites can be easily controlled and processability of the nanocomposites is improved thanks to the inherited chemical and mechanical properties of the soft substrates. Another benefit from the introduction of noble metal nanoparticles to material surfaces is to avoid the spread of the nanoparticles to the environment and thus to prevent agglomeration of the metal nanoparticles. Characterization methods with respect to morphology, colloidal stability, surface/bulk chemical composition and conductivity using scanning and transmission electron microscopies, laser diffraction particle size analysis, X-ray photoelectron spectroscopy, thermogravimetric analysis, elemental analysis and four-point probe technique are presented. The nanocomposite-coated materials as a recyclable catalyst for aqueous organic reactions are also described.

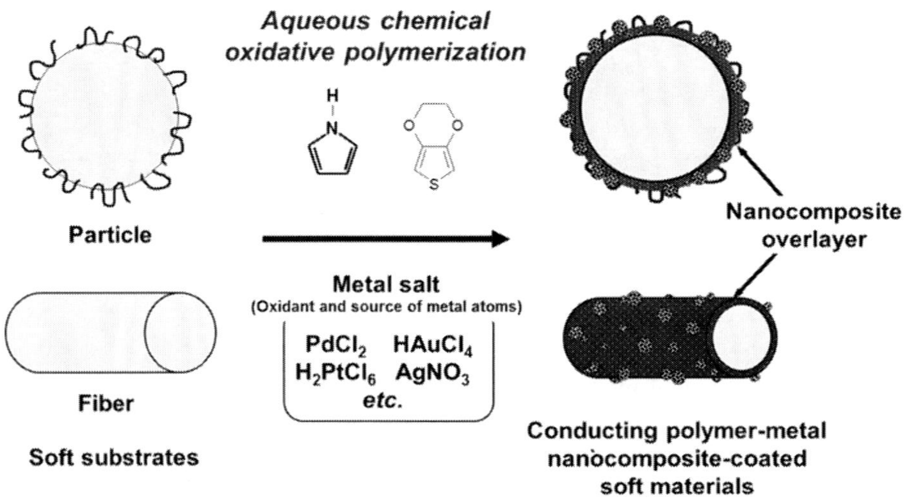

Figure 1. Scheme of conducting polymer-metal nanocomposite coating on soft substrates by one-step aqueous chemical oxidative polymerization.

This chapter is organized in the following four sections:

1. One-step synthesis of conducting polymer-noble metal nanocomposites by chemical oxidative polymerization
2. Conducting polymer-noble metal nanocomposite coating methodology
3. Characterization of conducting polymer-noble metal nanocomposite-coated materials
4. Organic coupling reactions using conducting polymer-metal nanocomposite-coated materials as a catalyst

1. One-Step Synthesis of Conducting Polymer-Noble Metal Nanocomposites by Chemical Oxidative Polymerization

Three main synthetic routes have been used to produce conducting polymer-noble metal nanocomposites: (1) synthesis of conducting polymer in the presence of metal nanoparticles [3a,8], (2) deposition of metal nanoparticles on a preformed conducting polymer matrix surface by redox reactions between conducting polymer and metal ions [9] and (3) a one-step chemical oxidative polymerization using metal salts as an oxidant [10].

Among them, chemical oxidative polymerization using metal salts is the simplest method for the synthesis of bulk conducting polymer-noble metal nanocomposites. It was demonstrated that chemical oxidative polymerization using metal salts such as hydrogen tetrachloroaurate (III), silver nitrate ($AgNO_3$), palladium(II) chloride ($PdCl_2$) and hydrogen hexachloroplatinate(IV), which act both as an oxidant and as a source of metal atoms, yielded well-dispersed metal nanoparticles in/on bulk conducting polymers. Selvan *et al.* [10a] polymerized pyrrole with tetrachloroauric acid as an oxidant in the presence of polystyrene (PS)-*b*-poly(2-vinylpyridine) copolymer micelles dispersed in toluene, which led to the fabrication of polypyrrole (PPy)-Au nanocomposites. Chen *et al.* [10b] have demonstrated a one-step facile and versatile synthetic route to PPy-Ag nanocomposites by chemical oxidative polymerization using $AgNO_3$ as an oxidant. Henry *et al.* [10c] suggested that $PdCl_2$ acts as an efficient oxidant for pyrrole to form PPy-Pd composite. More recently, Vasilyeva *et al.* [10d] described the synthesis of PPy-Pd nanocomposites via direct redox reaction between Pd(II) acetate and pyrrole in acetonitrile. Our research group has been studying on syntheses of conducting polymer-metal nanocomposites by chemical oxidative polymerization in *environmentally benign aqueous media*. This synthetic route using aqueous media is advantageous because the reaction takes place at room temperature and production on an industrial scale is much more likely compared to that using volatile organic solvents. We have succeeded in one-step synthesis of PPy-Au [11], PPy-Ag [12], PPy-Pd [13], PPy-Pt [14], polyaniline (PANI)-Ag [15], and poly(3,4-ethylenedioxythiophene) (PEDOT)-Pd [16] nanocomposites by aqueous chemical oxidative polymerization.

Although the chemical oxidative polymerization is simple and versatile method, the resulting nanocomposites are typically aggregated and therefore unsatisfactory for most applications.

2. Conducting Polymer-Noble Metal Nanocomposite Coating Methodology

In this section, conducting polymer-noble metal nanocomposite-coating methodologies are introduced. As substrates, colloidal polymer particles and polymeric fibers are used. Table 1 summarizes studies on conducting polymer-noble metal nanocomposite coating of soft materials by aqueous chemical oxidative polymerization. It describes kinds of nanocomposites, chemistries and shapes of substrates and nanocomposite loading percentages.

Table 1. Summary of studies on conducting polymer-metal nanocomposite-coated soft materials synthesized by aqueous chemical oxidative polymerization

Nanocomposite	Substrate	Substrate	Nanocomposite	Ref.
PPy-Pd	Particle	PS	2.5~53.0	13a,b
PPy-Au	Particle	PS	22	11
PPy-Pt	Particle	PS	10~50	14
PEDOT-Pd	Particle	PS	10~17	16
PPy-Pd	Fiber	Polyamide	< 3.5~5.2	13c

2.1. Conducting Polymer-Metal Nanocomposite Coating on Particles

Conducting polymer-metal nanocomposite coating is generally conducted via aqueous chemical oxidative polymerization in the presence of colloidally stable seed particles, namely chemical oxidative seeded polymerization [11, 13a, 13b, 14, 16].

The monomer (for example, pyrrole) is expected to distribute in the polymerization media and on the surface of the seed particles. The monomer can additionally be present within the seed particles for the polymer particles which can be swollen with the monomer. Nevertheless, polymerization should occur exclusively in aqueous solution and/or on the particle surface when hydrophobic particles such as polystyrene (PS) and poly(methyl methacrylate) (PMMA) are used as substrates, because the ionic metal oxidant is not expected to diffuse into the hydrophobic polymer particles and polymerization within the particles interior is therefore highly unlikely. The nanocomposite precipitates onto the particle surface if formed in the bulk aqueous phase or forms directly on the particle surface. The nanocomposite therefore accumulates on the seed particle surface, minimizing interfacial free energy.

One of keys of this method is to retain colloidal stability of the particles before, during and after the nanocomposite coating. In order to attain the colloidal stability, the help of colloidal stabilizer which gives steric stabilization is needed. Steric stabilization involves the use of a suitable polymeric stabilizer that form solvated layer on the surface of the particles. This outer layer of adsorbed polymer provides a steric barrier against the particle aggregation, thus resulting in a stable dispersion of nanocomposite-coated particles.

The dispersion stability of the synthesized particles can be assessed by optical microscopy and laser diffraction particle size distribution analysis. Optical microscopy studies can visually confirm the colloidal stability of the particles in wet state. Even with the nano/submicrometer-sized particles are used as seed particles, aggregations (generally > 1 μm) can be observed. It is advantageous that the size and shape of the aggregates can be discussed using the optical microscopy images. Laser diffraction particle size distribution analysis is an established technique for sizing micrometer-sized particles and can be used to characterize the dispersion state of the particles in aqueous media. The particle sizes and distributions obtained for dilute aqueous suspensions of the uncoated PS seed latex and the PPy-Pd nanocomposite-coated PS latexes are presented in Figure 2.

The particle size increases slightly with an increase of pyrrole monomer added in the polymerization system (2.5, 5.0 and 10 wt% of added pyrrole monomer based on the seed particles), while maintaining narrow particle size distributions, which suggests a high degree of dispersion of the composite latex particles.

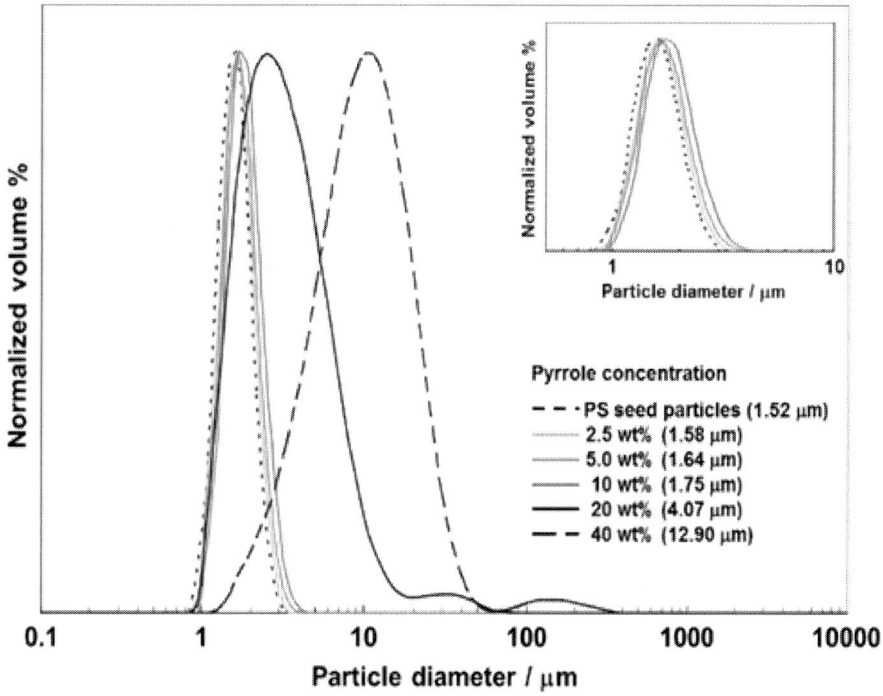

Figure 2. Laser diffraction particle size distribution curves obtained for uncoated PS seed particles and PPy-Pd nanocomposite-coated PS particles synthesized at various Py concentrations. Inset shows magnified laser diffraction particle size distribution curves obtained for colloidally stable PPy-Pd nanocomposite-coated PS particles (pyrrole: 2.5, 5.0 and 10 wt.% based on the seed particles). (Reproduced with permission from Ref. 13a).

On the other hand, for the 20 and 40 wt% pyrrole monomer systems, the PPy-Pd nanocomposite deposition on the PS particles causes appreciable flocculation, as indicated by the significant increase in the apparent particle diameter and diameter distribution. The coagulations obtained for the 20 and 40 wt% systems seem to be due to the excessive precipitation of the nanocomposite nuclei from the aqueous medium, which are then completely absorbed by the PS particles and caused the nanocomposite overlayer to interfere with the steric stabilization mechanism conferred by the steric colloidal stabilizer. The nanocomposite coating has been achieved for the polymer particles with diameters ranging from near one micrometer up to 84 μm, and PPy-Au [11], PPy-Pd [12a,12b], PPy-Pt [14], PEDOT-Pd [16] nanocomposites can coat the seed particles, successfully.

2.2. Conducting Polymer-Metal Nanocomposite Coating on Fibers

Conducting polymer-metal nanocomposite coating on the fibrous polymer substrates can be conducted in the similar way with the case of colloidal particles [13c]. Figure 3 shows digital camera and optical microscopy images of polyamide fibers before and after PPy-Pd nanocomposite coating.

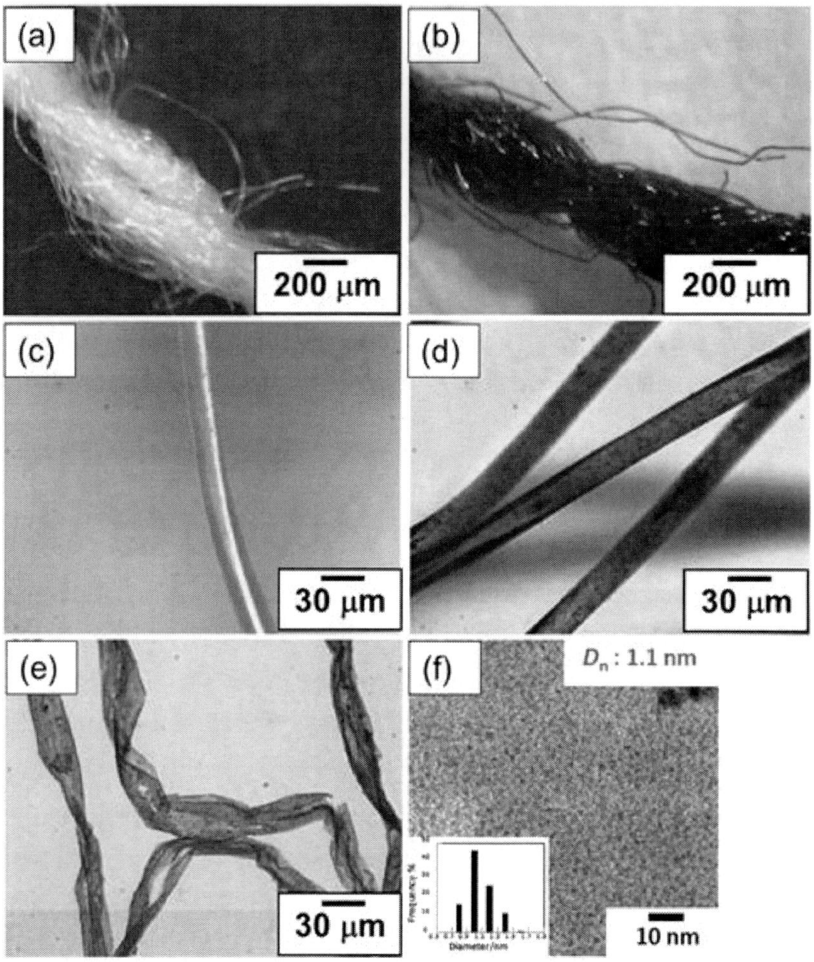

Figure 3. Digital camera (a,b) and optical microscopy (c,d) images of polyamide fibers before (a,c) and after (b,d) PPy-Pd nanocomposite coating. Optical (e) and TEM (f) images of PPy-Pd nanocomposite-coated polyamide fibers after extraction of polyamide component. The inset of (f) shows the particle size distribution of Pd nanoparticles.

These images indicate no signs of appreciable destruction of fibrous morphologies. Good control over the nanocomposite loading can be attained by simply varying the weight ratio of the substrate and pyrrole monomer.

Nanocomposite coating can be easily scaled up: nanocomposite coating can be conducted on woven cloth which consists of fibers, such as a T-shirt. Figure 4 shows digital camera images of a T-shirt before and after PPy-Pd nanocomposite coating by chemical oxidative polymerization in aqueous media at room temperature. The color of the T-shirt changed from white to black after the chemical oxidative polymerization, which indicates the nanocomposite coating. The color of the T-shirt was not bleached even after five times sonication in aqueous media, which indicated that the fibers were firmly coated with the PPy-Pd nanocomposite.

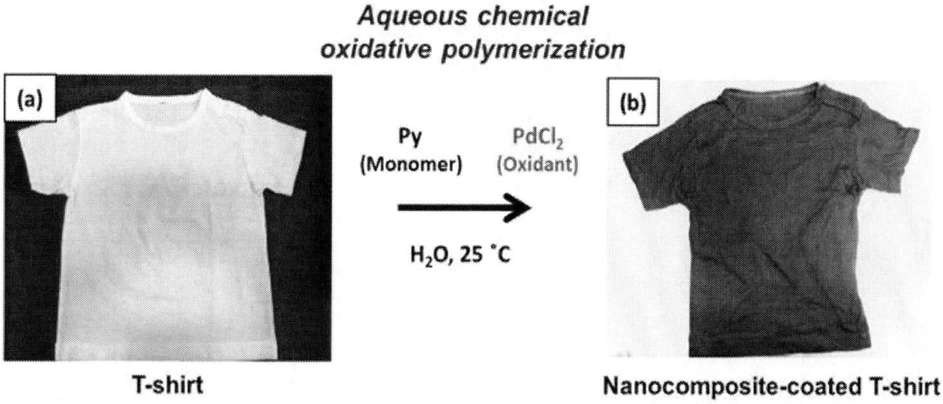

Figure 4. Digital camera images of T-shirt (Cotton 100 %) before (a) and after (b) PPy-Pd nanocomposite coating. (Reproduced with permission from Ref. 13c).

Substrate to be coated is not limitted to polyamide, but also include diacetate [13c], cotton [7a,13c], silk [13c], viscose [13c], wool [13c], cellulose [7b] and poly(ethylene terephthalate) [7d]. The deposition of nanocomposite on the inorganic fibers such as silica and quartz is also possible.

3. Characterization of Conducting Polymer-Noble Metal Nanocomposite-Coated Materials

Characterization of the conducting polymer-noble metal nanocomposite-coated materials is crucial for optimization of polymerization conditions and understanding the synthesis-structure-property relationships. In this section, characterization methods of chemical composition, morphology and conductivity are described.

3.1. Chemical Composition

Fourier transform infrared spectroscopy is a fundamental method to check hybridization of conducting polymer-metal nanocomposite and polymeric substrates. Unfortunately, however, it is somewhat difficult to determine the nanocomposite loading amount quantitatively.

Elemental microanalysis is a powerful method to determine conducting polymer-metal and conducting polymer loading amounts on the composite materials. For example, the percentage mass of the PPy-Pd nanocomposite loading on the composite materials can be determined by comparing the nitrogen content to that of the PPy-Pd composite bulk powder (N = 6.36%). (Here, the substrates contain no nitrogen). The percentage mass of the PPy component loading can also be calculated by comparing the nitrogen content to that of PPy homopolymer bulk powder, assuming that the PPy component in the PPy-Pd nanocomposite and the PPy homopolymer have the same chemical structure. The weight ratios of the PPy and Pd components in the nanocomposite-coated materials (particles and fibers) were calculated to be around 39:61, which is in good agreement with the theoretical value. These results indicate that the polymeric substrates do not interfere with the chemical oxidative polymerization of pyrrole using metal salt.

In order to confirm the presence of a metal component in the composite materials, X-ray diffraction measurement (XRD) studies can be utilized. In the nanocomposite-coated materials, clear peaks due to metal nanoparticles are observed, on the other hand, amorphous polymer substrates shows only broad peaks. From XRD pattern, metal crystalline size can be also estimated using a Scherrer equation.

X-ray photoelectron spectroscopy (XPS) studies offer surface chemical informations (~ 10 nm) of the composite materials. Figure 5 shows XPS survey spectra of the PPy-Pd nanocomposite-coated PS particles, the PPy-Pd nanocomposite bulk powder and the PS seed particles. Signals due to Pd, in addition to those due to C, N, O and Cl, are clearly apparent for the PPy-Pd bulk powder and the PPy-Pd nanocomposite-coated PS particles. The O 1s signal should arise from the partial over-oxidation of the PPy backbone. A signal due to Cl 2p indicates that the cationic PPy chains are doped with chloride anions (from the $PdCl_2$ oxidant and NaCl).

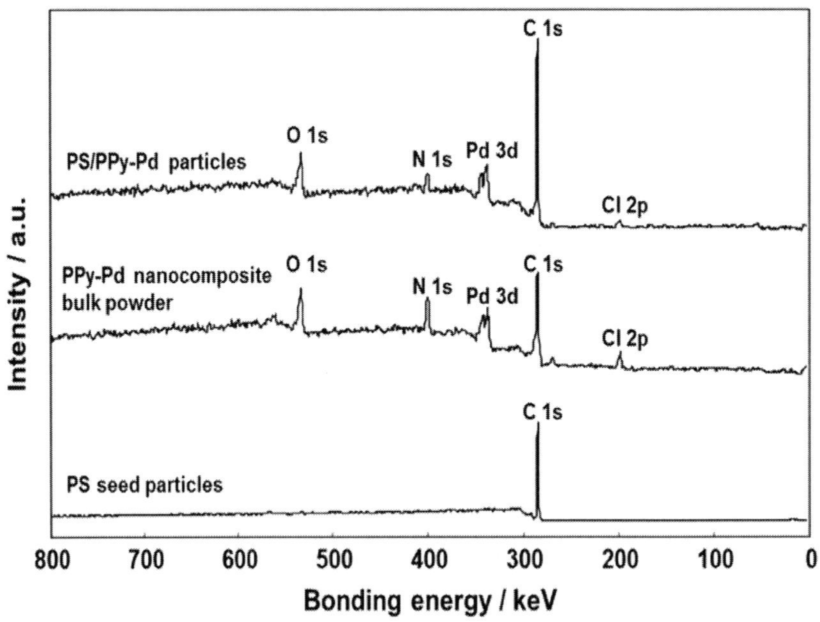

Figure 5. XPS survey spectra obtained for PPy-Pd nanocomposite-coated PS particles (pyrrole 10 wt% system), PPy-Pd nanocomposite bulk powder and uncoated PS seed particles. (Reproduced with permission from Ref. 13a)

From the Cl/N atomic ratios of the PPy-Pd bulk powder and composite particles, doping level can be estimated. High resolution Pd 3d XPS spectrum can confirm the reduction of Pd(II) to Pd(0) occurred. From the XPS results, it has been confirmed that the PPy-Pd nanocomposite coated the polymeric substrates.

3.2. Morphology

Optical microscopy, scanning electron microscopy (SEM) and transmission electron microscopy (TEM) are utilized in order to characterize shape and both surface and inner morphologies of the nanocomposite-coated materials.

SEM studies offer informations on surface nanomorphology of the materials. It can be confirmed that the smooth surface of the original substrates changed to rough surface with nuclei ranging between 50 and 250 nm after PPy-Pd nanocomposite coatings, which indicates surface coating of the substrates (Figure 6) [13]. The nuclei on the surface seem to be due to the precipitation of PPy-Pd nanocomposite nuclei from the aqueous medium, which are then adsorbed by the surfaces. Chehimi and coworkers reported ~100 nm PPy spheres and Ag triangular sheets coat substrates in the case of PPy-Ag nanocomposite system [7b]. The core-shell nature of the nanocomposite-coated materials can be readily verified by solvent extraction of the substrate core, followed by morphological examination of the insoluble nanocomposite residues. Extraction of core component from the composite materials using the solvents, which dissolve substrate core but not nanocomposite shell, resulted in insoluble black residues.

Examination of the black PPy-Pd nanocomposite residues by optical microscope revealed 'capsule' and 'tube' morphologies, with diameters corresponding to those of the original coated particle and fiber substrates, respectively (Figure 3). The solvent extraction experiment confirms that the composite materials do possess a core-shell morphology. The structure of nanocomposite shell can be investigated using a TEM in detail (Figures 3 and 6). The TEM images of PPy-Pd nanocomposite shell revealed that there are two Pd size distributions (approximately 6 nm and 1 nm). The 6-nm Pd nanoparticles formed aggregates (50~250 nm), which were observed in bump-like projections on a substrate surface. Heterogeneous character of the continuous shell was composed of a more transparent host material (presumably the PPy polymer) with incorporated 1 nm-sized Pd nanoparticles that have a higher absorbance for TEM electrons (dark elements dispersed in the PPy matrix.). An interesting characteristic of the conducting polymer capsules and tubes is that the transport rates of small molecules into the tube core are affected by the oxidation state of the conductive polymer [17], a feature with potential application in many molecular uptake and release scenarios.

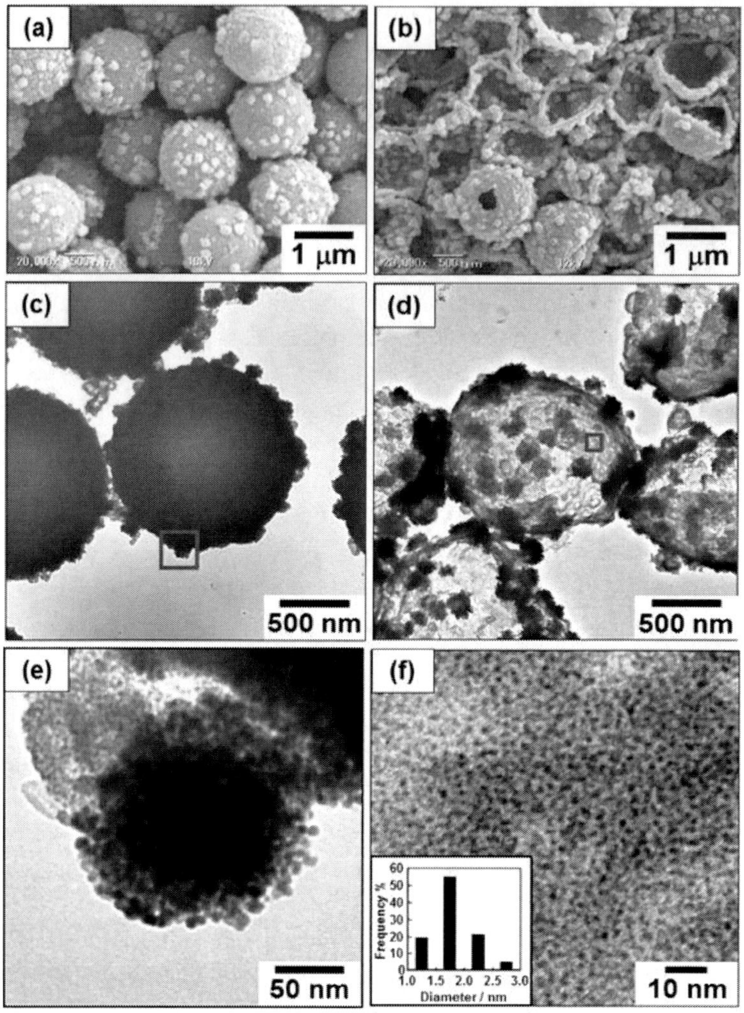

Figure 6. (a,b) SEM and (c,d,e,f) TEM images of PPy-Pd nanocomposite-coated PS particles synthesized at Py concentration of 10 wt% based on PS seed particles before (a,c) and after (b,d) extraction of PS seed particles. Parts 5e and 5f are magnified images of the areas indicated in (c) and (d), respectively. The inset of (f) shows the particle size distribution of Pd nanoparticles. (Reproduced with permission from Ref. 13a).

3.3. Conductivity

Electrical conductivities are usually conducted on pressed pellets of the dried nanocomposite-coated materials using the four-point probe technique. For example, a conductivity of PPy-Pd nanocomposite bulk powder has been measured to be 3.0×10^{1} S cm^{-1}, and PPy-Pd nanocomposite-coated materials show conductivities ranging between 3.0×10^{-1} and 2.4×10^{-14} S cm^{-1} [13]. The observed conductivity is invariably at least an order of magnitude lower than the conductivity of the corresponding nanocomposite bulk powder. This is not really surprising considering the presence of electrically insulating polymeric substrate and the increased number of resistive interparticle/fiber contacts at the nano/submicrometer level. Although the conductivities of the nanocomposite-coated materials are relatively low, it is worth emphasizing that pressed pellets prepared from a heterogeneous admixture of pristine polymeric substrates and nanocomposite bulk powder at the same weight ratio with nanocomposite-coated materials has even lower electrical conductivity, which is below the lower limit for the four-point-probe set-up ($<10^{-14}$ S cm^{-1}). More efficient electrical conduction occurs in the nanocomposite-coated materials, because the electrons can flow with lower resistance between adjacent fibers via the conductive pathway provided by the nanocomposite overlayers, without interference from the underlying electrically insulating fibrous cores. This result indicates that the composite materials have lower percolation threshold comparing with the heterogeneous mixture of the polymeric materials with nanocomposite bulk powder.

4. Organic Coupling Reactions Using Conducting Polymer-Metal Nanocomposite-Coated Materials as a Catalyst

The application of metal nanoparticles in catalysis has become an important frontier of research in recent years [18]. Metal nanoparticles offer high catalytic efficiency due to their high surface-to-volume ratio. The development of immobilized metal nanoparticle catalysts and the use of aqueous media have been of great interest in recent organic chemistry [19]. The simple recovery of catalysts by filtration and their reuses results in enhancing the economical evaluation of the reaction. At the same time, there is a prospect that the environmental pollution caused by residual metals in the waste fluid will be decreased. Our nanocomposite-coated materials can work as an efficient recyclable catalyst in environmentally benign aqueous media [20, 13a-13c].

Using the PPy-Pd nanocomposite-coated PS particles, the Suzuki-Miyaura coupling reaction of bromobenzene with 4-methylphenylboronic acid proceeded efficiently to give 4-methylbiphenyl in 99% yield (Figure 7).

After the first reaction, the composite particles can be recovered and successively subjected to the second through the fifth run of the coupling reaction under the same conditions to afford 4-methylbiphenyl in 99%, 99%, 98%, and 94% yield, respectively. Importantly, inductively coupled plasma atomic emission spectroscopy analyses confirmed that reaction media contained barely detectable levels of Pd and that the Pd loading in the particles did not change even after the fifth run, which indicated that no/little Pd nanoparticles detached from the PPy-Pd nanocomposite-coated particles. The catalyst can be also recycled four times without any loss of activity for Heck reaction of iodobenzene with acrylic acid.

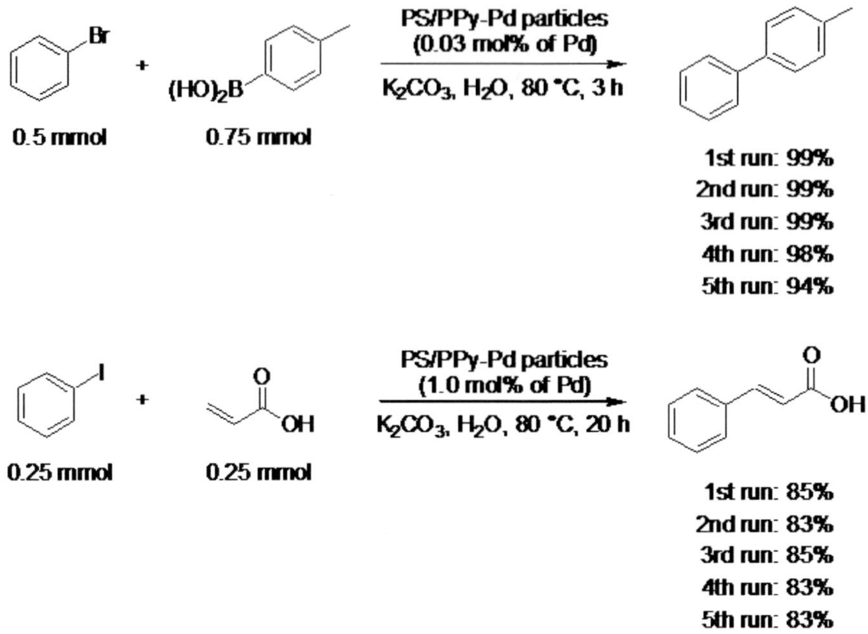

Figure 7. Aqueous cross coupling reactions catalyzed with PPy-Pd nanocomposite-coated polymer particles. (Reproduced with permission from Ref. 20)

The nanocomposite-coated fibers also work as a catalyst for C-C bond formation. More importantly, the fiber-supported metal nanoparticle catalyst can be easily recovered using tweezers (even with fingers!) from the reaction media, which can overcome a disadvantage in that either centrifugation or gravity-assisted sedimentation over several hours is required to collect and reuse the particle-supported catalyst. The advantages of a nanocomposite-coated material-based catalytic system are high activity, air and temperature stability and ease of separation.

CONCLUSION

The purpose of this chapter is to present recent developments in our ongoing studies on one-step and one-pot conducting polymer-noble metal nanocomposite-coating on soft materials. This chapter provides synthesis, characterization and application of the nanocomposite-coated materials. This coating method can be also applied to hard substrates including inorganic materials. The coating nanoengineering developed in this study provides a pathway to various practical hybrid materials with desired functions thanks to the rich varieties of natural/synthetic materials as substrates, metals and conducting polymers.

ACKNOWLEDGMENTS

Prof. Kensuke Akamatsu, Dr. Takaaki Tsuruoka and Prof. Hidemi Nawafune of Konan University are thanked for the TEM studies. Dr. Atsushi Ohtaka and Dr. Niklaas J. Buurma are thanked for their assistance in organic reaction experiments. Mr. Soichiro Matsuzawa, Mr. Hiroyuki Hamasaki, Ms. Mizuho Kodama and Mr. Hiroaki Takeoka are thanked for their experimental supports.

The works conducted by S.F. shown in this chapter were partially supported by a Grant-in-Aid for Scientific Research on Innovative Areas "Molecular Soft-Interface Science" and and "New Polymeric Materials Based on Elemental-Blocks" from the Ministry of Education, Culture, Sports, Science, Japan Science and Technology Agency, Japan and Adaptable and Seamless Technology transfer Program (A-STEP), Japan (AS232Z01455D) and Tanaka Kikinzoku Group Research Fund (MMS prize). One of the reviewers is also thanked for fruitful comments on the manuscript.

REFERENCES

[1] Gangopadhyay, R.; De, A. Conducting polymer nanocomposites: a brief overview, *Chem. Mater.*, 2000, *12*, 608-622.

[2] Tian Z. Q.; Lian Y. Z.; Wang J. Q.; Wang S. J.; Li W. H. J. Electrochemical and XPS studies on the generation of silver clusters in polyaniline films, *Electroanal. Chem.*, 1991, *308*, 357-363.; (b) Drelinkiewicz, A.; Hasik, M.; Kloc, M. Pd/polyaniline as the catalysts for 2-ethylanthraquinone hydrogenation. The effect of palladium dispersion, *Catal. Lett.*, 2000, *64*, 41-47.; (c) Kitani A.; Akashi T.; Sugimoto, K.; Ito, S. Electrocatalytic oxidation of methanol on platinum modified polyaniline electrodes, *Synth. Met.*, 2001, *121*, 1301-1302.; (d) Radford, P. T.; Creager, S. E. Dual-stream flow injection method for amplified electrochemical detection of ferrocene derivatives, *Anal. Chim. Acta.*, 2001, *449*, 199-209.

[3] Marinakos, S. M.; Novak, J. P.; Brouseau III, L. C.; House, A. B.; Edeki, E. M.; Feldhaus, J. C.; Feldheim, D. L. Gold particles as templates for the synthesis of hollow polymer capsules. Control of capsule dimensions and guest encapsulation, *J. Am. Chem. Soc.*, 1999, *121*, 8518-8522.; (b) Mangeney, C.; Bousalem, S.; Connan, C.; Vaulay, M.-J.; Bernard, S.; Chehimi, M. M. Latex and hollow particles of reactive polypyrrole: preparation, properties, and decoration by gold nanospheres, *Langmuir*, 2006, *22*, 10163-10169.; (c) Hamasaki, H.; Fukui, N.; Fujii, S.; Yusa, S.; Nakamura, Y. Sterically stabilized polypyrrole-palladium nanocomposite particles synthesized by aqueous chemical oxidative dispersion polymerization, *Colloid Polym. Sci.*, 2013, *291*, 223-230.; (d) Fujii, S.; Hamasaki, H.; Abe, H.; Yamanaka, S.; Ohtaka, A.; Nakamura, E.; Nakamura, Y. One-step synthesis of magnetic iron-conducting polymer-palladium ternary nanocomposite microspheres and their use as recyclable catalyst, *J. Mater. Chem. A.*, 2013, *1*, 4427-4430.

[4] Gallon, B. J.; Kojima, R. W.; Kaner, R. B.; Diaconescu, P. L. Palladium nanoparticles supported on polyaniline nanofibers as a semi-heterogeneous catalyst in water, *Angew. Chem.*, 2007, *119*, 7389-7392.; (b) Hamasaki, H.; Kanadani, H.; Nakamura, Y.; Fujii, S. One step synthesis of conducting polymer-palladium nanocomposite fibers by aqueous chemical oxidative polymerization, *Chem. Lett.*, 2012, *41(9)*, 982-983.

[5] Qiu, L.; Peng, Y.; Liu, B.; Lin, B.; Peng, Y.; Malik, M. J.; Yan, F. Polypyrrole nanotubesupported gold nanoparticles: An efficient electrocatalyst for oxygen reduction

and catalytic reduction of 4-nitrophenol, *Appl. Catal. A General*, 2012, *413-414*, 230-237.

[6] Chen, G.; Wang, Z.; Yang, T.; Huang, D.; Xia, D. Electrocatalytic hydrogenation of 4-chlorophenol on the glassy carbon electrode modified by composite polypyrrole/palladium film, *J. Phys. Chem. B*, 2006, *110*, 4863-4868.; (b) Rakesh, K. P.; Lakshminarayanan, V. Enhanced electrocatalytic activity of Pd-dispersed 3,4-polyethylenedioxythiophene film in hydrogen evolution and ethanol electro-oxidation reactions, *J. Phys. Chem. C*, 2010, *114*, 8507-8514.

[7] Babu, K. F.; Dhandapani, P.; Maruthamuthu, S.; Kulandainathan, M. A. One pot synthesis of polypyrrole silver nanocomposite on cotton fabrics for multifunctional property, *Carbohydrate Polym.*, 2012, *90*, 1557-1563.; (b) Attia, M. F.; Azib, T.; Salmi, Z.; Singh, A.; Decorse, P.; Battaglini, N.; Lecoq, H.; Omastová, M.; Higazy, A. A.; Elshafei, A. M.; Hashem, M. M.; Chehimi, M. M. One-step UV-induced modification of cellulose fabrics by polypyrrole/silver nanocomposite films, *J. Colloid Interface Sci.*, 2013, *393*, 130-137.; (c) Ijeri, V. S.; Nair, J. R.; Gerbaldi, C.; Bongiovanni, R. M.; Peenazi, N. Metallopolymer capacitor in "one pot" by self-directed UV-assisted process, *ACS Appl. Mater. Interfaces*, 2010, *2*, 3192-3200.; (d) Singh, A.; Salmi, Z.; Joshi, N.; Jha, P.; Kumar, A.; Decorse, P.; Lecoq, H.; Truong-Lau, S.; Chehimi, M. M.; Aswal, D. K.; Gupta, S. K. Photo-induced synthesis of polypyrrole-silver nanocomposite films on *N*-(3-trimethoxysilypropyl)pyrrole-modified biaxially oriented polyethylene terephthalate flexible substrates., *RSC Adv.*, 2013, 3, 5506-5523.

[8] Martins, C. R.; Almeida, Y. M.; Nascimento, G. C.; Azevedo, W. M. Metal nanoparticles incorporation during the photopolymerization of polypyrrole, *J. Mater. Sci.*, 2006, *41*, 7413-7418.

[9] Kang, E. T.; Ting, Y. P.; Neoh, K. G.; Tan, K. L. Spontaneous and sustained gold reduction by polyaniline in acid solution, *Polymer*, 1993, *34*, 4994-4996.

[10] Selvan, S. T.; Spatz, J. P.; Klok, H.-A.; Möller, M. Gold-polypyrrole core-shell particles in diblock copolymer micelles, *Adv. Mater.*, 1998, *10*, 132-134.; (b) Chen, A.; Kamata, K.; Nakagawa, M.; Iyoda, T.; Wang, H.; Li, X. Formation process of silver-polypyrrole coaxial nanocables synthesized by redox reaction between $AgNO_3$ and pyrrole in the presence of poly(vinylpyrrolidone), *J. Phys. Chem. B*, 2005, *109*, 18283-18288.; (c) Freund, M. S.; Henry, M. C.; Hsueh, C. C.; Timko, B. P. Reaction of pyrrole and chlorauric acid — A new route to composite colloids, *J. Electrochem. Soc.*, 2001, *148*, D155-D162.; (d) Vasilyeva, S. V.; Vorotyntsev, M. A.; Bezverkhyy, I.; Lesniewska, E.; Heintz, O.; Chassagnon, R. Synthesis and characterization of palladium nanoparticle/polypyrrole composites, *J. Phys. Chem. C.*, 2008, *112*, 19878-19885.

[11] Fujii, S.; Mouri, M.; Aichi, A.; Akamatsu, K.; Nawafune, H.; Nakamura, Y. Synthesis of polypyrrole-gold nanocomposite-coated polystyrene latex particles by simultaneous organic-inorganic deposition method, *J. Adhes. Soc. Jpn.*, 2008, *44*, 12-18.

[12] Fujii, S.; Aichi, A.; Akamatsu, K.; Nawafune, H.; Nakamura, Y. One-step synthesis of polypyrrole-coated silver nanocomposite particles and their application as a coloured particulate emulsifier, *J. Mater. Chem.*, 2007, *17*, 3777-3779.

[13] Fujii, S.; Matsuzawa, S.; Nakamura, Y.; Ohtaka, A.; Teratani, T.; Akamatsu, K.; Tsuruoka, T.; Nawafune, H. Synthesis and characterization of polypyrrole-palladium nanocomposite-coated latex particles and their use as a catalyst for Suzuki coupling reaction in aqueous media, *Langmuir*, 2010, *26*, 6230-6239.; (b) Fujii, S.; Matsuzawa,

S.; Hamasaki, H.; Nakamura, Y.; Bouleghlimat, A.; Buurma, N. J. Polypyrrole-palladium nanocomposite coating of micrometer-sized polymer particles toward a recyclable catalyst, *Langmuir*, 2012, *28(5)*, 2436-2447.; (c) Fujii, S.; Kodama, M.; Matsuzawa, S.; Hamasaki, H.; Ohtaka, A.; Nakamura, Y. (Editor, Abbass Hashim) "Conducting Polymer-Metal Nanocomposite Coating on Fibers" in "Advances in Nanocomposite Technology", 2011, Chapter 2, 47-72. ISBN 978-953-308-55-0, InTech.

[14] Fujii, S.; Takeoka, H.; Hamasaki, H.; Nakamura, Y. Synthesis and characterization of polypyrrole-platinum nanocomposite-coated polymer particle. *manuscript in preparation.*

[15] Fujii, S.; Nishimura, Y.; Aichi, A.; Matsuzawa, S.; Nakamura, Y.; Akamatsu, K.; Nawafune, H. Facile one-step route to polyaniline-silver nanocomposite particles and their application as a colored particulate emulsifier, *Synth. Met.*, 2010, *160*, 1433-1437.

[16] Hamasaki, H.; Maekawa, Y.; Matsuzawa, S.; Nakamura, Y.; Fujii, S. Synthesis of poly(3,4-ethylenedioxythiophene)-palladium nanocomposite-coated polymer particles by chemical oxidative seeded dispersion polymerization, *Chem. Lett.*, 2012, *41*, 1658-1659.

[17] Abidian, M. R.; Kim, D.-H.; Martin, D. C. Conducting-polymer nanotubes for controlled drug release. *Adv. Mater.*, 2006, *18*, 405-409.

[18] Ogasawara, S.; Kato, S. Palladium nanoparticles captured in microporous polymers: a tailor-made catalyst for heterogeneous carbon cross-coupling reactions, *J. Am. Chem. Soc.*, 2010, *132*, 4608-4613.; (b) Astruc, D. Palladium nanoparticles as efficient green homogeneous and heterogeneous carbon-carbon coupling pre-catalysts: a unifying view, *Inorg, Chem.*, 2007, *46*, 1884-1894.

[19] Lamblin, M.; Nassar-Hardy, L.; Hierso, J.-C.; Fouquet, E.; Felpin, F.-X. Recyclable heterogeneous palladium catalysts in pure water: sustainable developments in Suzuki, Heck, Sonogashira and Tsuji-Trost reactions, *Adv. Synth. Catal.*, 2010, *352*, 33-79.

[20] Ohtaka, A.; Kono, Y.; Teratani, T.; Fujii, S.; Matsuzawa, S.; Nakamura, Y.; Nomura, R. Polypyrrole-palladium nanocomposite-coated latex particles as a heterogeneous catalyst in water, *Catal. Lett.*, 2011, *141*, 1097-1103.

In: Applied Surface Chemistry of Nanomaterials
Editors: Mohamed M. Chehimi and Jean Pinson
ISBN: 978-1-62808-351-4

Chapter 12

CARBON FIBER AND NANO/MICRO HYBRID STRUCTURES AND THEIR INTERFACE IN ADVANCED COMPOSITES

Delong He[1], Jing Zhang [1], Johan Saba[1, 2], Jinkai Yuan[1], Fabien Miomandre[2] and Jinbo Bai[1]*

[1]Laboratoire de Mécanique des Sols, Structures et Matériaux, Ecole Centrale Paris, Châtenay-Malabry, France

[2]Laboratoire PPSM, Ecole Normale Supérieure de Cachan, Cachan Cedex, France

Keywords: Carbon fiber, PAN, Pitch, Structure, Fabrication, Properties, Polymer matrix composite, Surface treatment, Oxidation, Non-oxidation, Interfacial shear strength, Interface, Interphase, Nanomaterials, Carbon nanotubes, Nano/micro hybrid, Synthesis, Chemical vapor deposition, Catalyst, Multiscale polymer composites, Interlaminar shear strength, Multifunctionality

1. INTRODUCTION

Carbon fiber reinforced polymers (CFRPs) have been widely used in the fields of aeronautics & aerospace, automotive, entertainment and industrial construction, due to their light weight, high strength and modulus and high chemical stability under a wide range of harsh environmental conditions. This kind of composites usually consists of continuous carbon fibers (CFs) as the reinforcement, and a thermosetting or thermoplastic polymer as the matrix. Especially, advanced CFRP composites take a fast increase in the proportion of materials used in the aerospace and aeronautic vehicles, and are expected to replace the currently used metal alloys, with the improvement of their properties.

The CF structure and their interfacial properties are two key factors which influence the properties of the composites. The former is greatly associated with the primary materials and

* Corresponding author: Dr. Jinbo BAI, e-mail address: jinbo.bai@ecp.fr.

the fabrication process parameters used such as the oxidation and graphitization degrees; the latter is essential to efficiently bear the load transfer between fiber and matrix. The CF interfacial properties can be greatly modified by surface treatments, polymer sizing and nanomaterials grafting. It is worth underlining that the grafting of carbon nanotubes (CNTs) on CFs is a promising method to build high performance hierarchical CFRP composites. Therefore, a cost-efficient production of these multi-scale composites is highly desired to meet the requirements of commercialization and human sustainable development.

In this chapter, we will briefly introduce in the first part some basic knowledge about CF properties, structure and fabrication process, as well as their application in composites. Then, we will present the modification of CF surface, and their influence on composites properties. Lastly, we will present the nano/micrometer hybrid structures formed by in-situ grafting processes, and their improved reinforcement in the composites.

2. Carbon Fibers and Polymer Composites

2.1. Carbon Fiber Structures and Fabrication Process

In spite of their short history, carbon fibers have already gotten a great success in commercial application, due to their specific properties, e.g. light weight, high tensile modulus and strength, high thermal and chemical stability, high thermal conductivity, low electrical resistance, and excellent creep resistance, etc. In fact, the intrinsic properties of CFs are closely related to the sp2 hybridization of carbon. A carbon fiber contains a lot of crystallites [1] having various graphitization degrees. That is, the carbon layer spacing distance and organization are different from those of graphite. The proportion of these crystallites compared to amorphous part could vary from 0 to 100 %, correspondingly resulting in varied fiber properties. The high modulus (a measure of the stiffness of an elastic material and is a quantity used to characterize materials. It is defined as the ratio of the stress along an axis over the strain along that axis in the range of stress in which Hooke's law holds) of CFs stems from the fact that the carbon layers tend to be parallel to the fiber axis. The higher is the graphitic degree of CFs, the higher the thermal conductivity and the lower the electrical resistivity along the fiber direction.

The commercially available CFs are generally classified into two categories: general-purpose (GP) and high-performance (HP). The GP-type CF is characterized by mainly containing amorphous and isotropic structure, low tensile strength and modulus, and low cost. In contrast, the HP one is characterized by high strength (the strength of a material is its ability to withstand an applied stress without failure) and modulus [2]. According to their tensile strength and modulus, CFs can also be classified into high modulus (HM), intermediate modulus (IM), high strength (HT) type, etc.

The CF structure is directly determined by their precursor material and the fabrication processing which usually includes the fiber formation, stabilization and carbonization [3]. Nowadays, the commercially available CFs are mainly pitch-based and polyacrylonitrile-based (PAN-based) ones. The pitch-based fibers have high carbon yield, and are easily graphitized, therefore they can attain high moduli, and high thermal and electrical conductivities. From their part, the PAN-based fibers tend to develop high tensile strength

CFs. Because of this, up to now the PAN-based CFs have a larger market share than the pitch-based ones. However, a great progress has been made recently in the production of high strength pitch-based CFs. With the development of cost-efficient techniques, their market share is expected to become bigger and bigger for the abundance and low price of pitch.

2.1.1. PAN-Based Carbon Fibers

PAN $(C_3H_3N)_n$ is a linear polar polymer with a glass transition temperature of ~ 80 °C and a melting temperature of ~ 320 °C. Because of its highly polar nature, PAN decomposes before it melts. Wet spinning is therefore widely used to produce the PAN precursor fibers. The spinning solution consists normally of 10-30 wt. % PAN or (PAN copolymer) in a polar solvent [4]. Then, the PAN precursor fibers are generally stabilized by oxidizing in air at 180-300 °C under controlled tension and speed [2]. The stabilization serves to

1) convert thermoplastic PAN into nonthermoplastic cyclic compound to eliminate the relaxation and chain scission, and
2) increase the carbon yield during the subsequent carbonization step.

The PAN fibers are carbonized at 500 -1500 °C in an inert atmosphere. During this step, most of the noncarbon elements within the fiber (~ 50-60 wt %) are volatilized in the form of gases which include methane, hydrogen, hydrogen cyanide, water, carbon monoxide, carbon dioxide, ammonia, etc.[4] In general, the degree of preferred crystallite orientation and, thus, the modulus of the carbon fiber increase with the increase of heat treatment temperature [4]. Graphitization is carried out at 1500-3000 °C under an inert atmosphere (N_2 or Ar). In this step, very little gas is evolved, but the crystallite size and their preferred orientation are increased. In other words, the fiber becomes more graphitic [2].

The microstructure of PAN-based CFs can be generally described by the "Skin-Core" model proposed by Johnson [5]. The fiber structure is fibrillar in nature, moreover it is heterogeneous in the center and skin regions. In the skin region, the carbon layers are parallel to the fiber surface, and there exist needle-shaped microvoids between crystallites, whereas, in the core region the carbon layers exhibit a random texture (called turbostratic texture). For Type II PAN-Based CFs (heat-treatment temperature higher than 1500 °C), no skin structure is observed and the crystallites are smaller than those of Type I (heat-treatment temperature higher than 2000 °C) [5]. A high proportion of carbon layers are not parallel to the fiber surface, and the exposed carbon layer edges enhance the bonding between the CFs and the matrix when the fibers are used in the composite [2].

2.1.2. Pitch-Based Carbon Fibers

Petroleum pitches are another type of precursor which is widely used for the production of carbon fibers. They could be isotropic or anisotropic. This latter is called mesophase which is a liquid crystalline material consisting of aromatic oligomers where aromatic planes are stacked approximately parallel to each other [6]. The mesophase pitches have a high degree of aromaticity and a high carbon proportion (~ 88 – 93 wt. %) [2]. The pitch-based precursor fibers are commonly made using melt spinning. The fabrication process involves extruding the melted pitch into a gaseous atmosphere (e.g. N_2) through nozzles directed downward, so that the extruded fibers are cooled and solidified [2]. The spinning temperature is very important because it determines directly the pitch viscosity which in turn controls the microstructures of the resulting fibers. Then, as-spun pitch-based fibers are stabilized by oxidation in air at 250 – 300 °C, in order to prevent them from softening and deforming under heating. Afterwards, the pitch-based fibers are usually carbonized at 700-2000 °C under inert atmosphere (e.g. N_2 or Ar). In this step, the non-carbon elements are eliminated in the form of gases, mainly including methane, hydrogen, etc. [2] The graphitization is optionally carried out at 2500–3000 °C under inert atmosphere. The yield of the pitch-based CF process is ~70-80 %, which is much higher than that of the PAN-based process (~ 50 %). In addition, the shrinkage in diameter of the pitch-based CFs is smaller than the PAN-based one.

Compared with PAN-based CFs, the mesophase pitch-based CFs have merely perfect axial orientation but very different transverse microstructures [7]. This is mainly due to the high graphitizability of the mesophase. In fact, their microstructure and axial molecular orientation are created as early as the liquid crystalline precursor flows through a capillary, and is extended for fiber formation [4]. They can be influenced by the flow state during extrusion. In addition, the changing of CF textural structures can also be achieved by modifying mesophase pitch precursor composition. Cho et al. [8] demonstrated that the incorporation of multi-walled carbon nanotubes into mesophase pitch resulted in a variation of the final CF structure from radial organization to random state. It was believed that the addition of nanotubes to mesophase pitch would nucleate a larger number of smaller-size domains.

2.2. Carbon Fiber Reinforced Polymer Composites

2.2.1. Composite Manufacturing

The light weight, high stiffness and strength, high thermal conductivity and low electrical resistivity make CFs excellent reinforcements to polymer matrix. The polymer matrix-based composites play a more and more important role in the fields of aeronautics, automotive and renew energies, with the increasing concern on global environment issues [9]. In these composites, CFs could be in the form of short-length powders, continuous bundles, 2D woven fabrics or 3D felts. The different combinations between CFs and matrix can generate a huge variety of composites exhibiting varied properties. According to the requirements of a specific application domain, the corresponding composites with appropriate structures and properties are designed and produced. For example, polymer matrix-based composites reinforced by continuous CFs are widely used in the components of aeronautic vehicles, automobiles, high pressure tanks, golf shafts, etc. Whereas short CF reinforced concretes are preferred in civil construction.

According to the polymer type, the polymer matrix-based composites are usually divided into two categories: thermosetting and thermoplastic. The widely used thermosetting polymers include epoxy, phenolic [10, 11], bismaleimide (BMI) [12], unsaturated polyester (PE) [13] and polyimide (PI). The thermoplastic polymers include Polyamide (PA), Polycarbonate (PC), Polyphenylene sulphide (PPS), Polyether ether ketone (PEEK), Polyether imide (PEI), and thermoplastic polyimide (PI). Up to now, the thermosetting polymers, especially epoxy, are the most widely used matrix materials for producing CF reinforced composites, due to their high stiffness and strength, excellent corrosion resistance and good thermal stability. However, the curing time of a thermosetting resin is long. Recently, thermoplastic polymers are increasingly used as the matrices in CF reinforced composites, thanks to their high ductility, high processing speed, their possibility to be reshaped and their recycling potential. The semi-crystalline thermoplastic polymers, like PEEK or PPS show good mechanical properties, and could withstand high temperatures. However, their processing temperature is relatively high, typically ranging from 300 to 400 °C.

The manufacturing methods used for CF reinforced composites with thermosetting or thermoplastic matrix are very different. For short CF reinforced composites, the CF powders are firstly mixed with a liquid resin (thermosetting), or a molten or dissolved thermoplastic polymer solution. Then, they can be formed by a variety of molding methods [2]:

1) injection molding, extrusion, calendaring, and thermoforming in the case of thermoplastic;
2) compressing molding, or matched die molding in the case of thermosetting.

The forming methods for continuous fiber composites are quite different. The commonly used ones are as follows:

1) Resin transfer molding (RTM) for complicated structures;
2) Filament winding for cylindrical or related structures;
3) Pultrusion for unidirectional fiber composites with constant cross section; and
4) Compression of prepregs by hot pressing or vacuum bag.

Here, a prepreg consists of the CFs which have been pre-impregnated with a polymer resin. It is a semi-product which is ready to be used in the manufacturing process. The CF reinforcement can be in unidirectional form or in fabric form. The prepregs are widely used to fabricate laminated composites which are largely used in the components for aircrafts, automobiles, trains, ships, etc. The curing of the prepregs to form a thermosetting-matrix composite is carried out under heating and pressure. The critical parameters, including heat up rate, curing time, pressure, vacuum and cooling rate, are specifically chosen according to prepreg laminate number, polymer types and the properties desired. For thermoplastic-based composites, the processing is rapid, whereas the cooling rate after lamination must be well controlled.

2.2.2. Composite Properties

Many composite properties are strongly dependent on the arrangement and distribution of fibers: the fiber architecture. This expression encompasses intrinsic features of the fibers, such as their diameter and length, as well the volume fraction of fibers and their alignment and packing arrangement [9]. Table 1 lists the mechanical properties of epoxy-matrix based composites with unidirectional CFs [14]. A comparison is given with the properties of the composites with standard grade and high modulus grade CFs. The CF volume ratio in the composites tested was 60 %, and that of CF fabrics was 50 %. The composites with high modulus CFs exhibit higher tensile modulus in 0° direction, but lower tensile strength and strain, compared to the composites with standard CFs. Evidently, the difference of tensile strength between 0° and 90 ° is as high as 30 times, and that of tensile modulus ~ 20 times.

Table 1. Mechanical properties of epoxy-matrix based composites with unidirectional CFs(standard grade and high modulus grade) and CF fabric (standard grade and high modulus grade) [14], cured at 120°C

Properties	Units	Std CF Fabric	HMCF Fabric	Std CF UD	HMCF UD
Young's Modulus 0°	GPa	70	85	135	175
Young's Modulus 90°		70	85	10	8
In-plane Shear Modulus		5	5	5	5
Ultimate Tensile Strength 0°	MPa	600	350	1500	1000
Ultimate Tensile Strength 90°		600	350	50	40
Ultimate Comp. Strength 0°	MPa	570	150	1200	850
Ultimate Comp. Strength 90°		570	150	250	200
Ultimate In-plane Shear Strength	MPa	90	35	70	60
Ultimate Tensile Strain 0°	%	0.85	0.40	1.05	0.55
Ultimate Tensile Strain 90°		0.85	0.40	0.50	0.50
Ultimate Comp. Strain 0°	%	0.80	0.15	0.85	0.45
Ultimate Comp. Strain 90°		0.80	0.15	2.50	2.50
Ult. In-plane shear strain	%	1.80	0.70	1.40	1.20

Note: Std CF Fabric=Standard carbon fiber fabric, HMCF Fabric=High modulus carbon fiber fabric, UD = unidirectional.

Fibers @ 0° (UD), 0/90° (fabric) to loading axis, Dry, Room Temperature, Vf = 60% (UD), 50% (fabric).

The anisotropy of the composite properties can be decreased by varying CF layer laying directions. For example, the mechanical properties of an epoxy composite with CF layers in 0°, 30°, 60°, 90°, 120° and 150°are quasi-isotropic, as reported in reference [2]. In addition, the use of CF fabric can also increase the homogeneity of the continuous CF composites, as shown in Table 1. However, the realization of the isotropic composite property is with the compromise of the longitudinal properties of CFs.

Under real service conditions, the CFRPs suffer from fatigue. In the fiber direction, the CFRPs are much stiffer that metals. Their endurance limit reaches about 60–80% of the fracture stress, whereas the metal endurance limit is only about 30% of the fracture stress [15]. However, a stress system that develops only a small working strain in the main fiber direction may easily generate strains normal to the fibers or at the fiber/resin interface that may be high enough to cause any deterioration. Moreover, in the composites, the damage accumulation occurs in a general way rather than in a localized one, and the failure does not always occur by the propagation of a single macroscopic crack [16]. The micro-structural mechanisms of damage accumulation, include fiber breakage, matrix cracking, fiber debonding, transverse ply cracking, and delamination [16, 17]. These mechanisms occur independently or interactively depending on material variables and testing conditions.

For commonly-used laminar structure composites, delamination is one of the most prevalent life-limiting crack growth modes, as it may cause severe reduction of in-plane strength and stiffness, leading to catastrophic failures of the whole structure. Delamination may be introduced by external loading as static bending, compression or tension, cyclic fatigue or by impacts of low-to-high energies during manufacturing or in service [18]. In addition, it could also be affected by the interaction between the applied stress system and the environmental factors, such as the temperature, moisture, radiation and/or contact with various types of chemicals [19]. A number of methods have been proposed in order to improve the delamination resistance of the laminar composites, e.g. three-dimensional weaving [20], stitching [21], etc. However, in these methods, the interlaminar properties are enhanced at the cost of in-plane mechanical properties. In contrast, the introduction of nanomaterials in the polymer matrix or the use of multi-scale hybrid reinforcements could be a promising way to improve the composite interlaminar properties without deteriorating their in-plane properties [22, 23].

The high thermal conductivity and electrical conductivity of CFs, especially the high modulus pitch-based fibers, make these fibers highly effective for increasing the thermal and electrical conductivities of polymers [2]. Due to their anisotropic structures, the unidirectional CF polymer-matrix composites exhibit high electrical and thermal conduction in fiber direction. But, in transverse and through-thickness directions, the transport properties are mainly dominated by the polymer matrix. The in-plane conductivities can be improved by using CF fabrics [24]. The addition of conductive fillers (like micrometer CFs or CNTs) [25] and three-dimensional weaving [26] methods are used to improve the through-thickness electrical and thermal conductivities. In particular, the addition of CNTs is expected to improve both the interlaminar shear strength and the electrical and thermal conductivities of the composites, and to develop multifunctional CF polymer-matrix composites [27, 28].

3. Carbon Fiber Surface Treatment

The interface in fiber-reinforced composites is a surface formed by a common boundary of fiber and matrix for the transfer of loads. However, such boundary is rarely devoid of chemical interaction, and therefore we can also define a region, called the interphase, being the volume of material affected by the interaction at the interface [29]. Drzal et al. [30] gave a following definition of the interphase: "the interfacial region is the region beginning at the point in the fiber at which the properties differ from those of the bulk filler and ending at the point in the matrix at which the properties become equal to those of the bulk matrix. The chemistry, polymer chain mobility, degree of cure, crystallinity are likely to be altered in that region." The interphase will be formed from the parts of long chain polymers of the matrix, which are perturbed by interaction with the reinforcement. The nature of such interactions in the interphase varies considerably from one reinforcement/matrix system to the next [29].[9] The interfacial regions play a vital role in the mechanical properties of the resulting composite materials. Good interfacial adhesion provides composites with structural integrity and efficient load transfer from fiber to matrix [31, 32].

The pristine surface of carbon fiber is non-polar and made of highly crystallized graphitic basal planes with inert structures [33].The interfacial bonding strength between CFs and the resin matrix is thus low [34]. In order to improve the adhesion between the CFs and the matrix in a composite, different surface treatments and sizing of CFs are often performed after their carbonization. The increase of surface adhesion is achieved by numerous ways, such as increasing the wettability, increasing the mechanical interlocking, or introducing surface functional groups [33, 34]. Generally, the surface treatments can be divided into oxidative and non-oxidative treatments. Moreover, CFs are always coated with a thin film of sizing agent after being electrolytic and wash scrubbed, to improve their handle ability. The sizing agents usually present as solutions or emulsions of polymeric components and assistants [35-37].

3.1. Oxidative Surface Treatments

There are various methods of oxidative treatments, including dry oxidation in the presence of gases, plasma etching and wet oxidation [38].

Dry oxidative treatments are normally performed with air, oxygen and CO_2 at low or elevated temperatures. Herrick et al. [39] treated rayon-based CF in air at 500 °C for 16 h to slightly improve the interlaminar shear strength (ILSS). ILSS was improved by 45 % when the temperature was raised to 600 °C, although a serious weight loss was accompanied. Scola et al. [40] treated the fibers for 60 s in N_2 containing 0.1-1.8 % O_2 at 1000-1500 °C to improve their bonding characteristics in a resin. There was no significant degradation of mechanical properties of the fibers. Dai et al. studied the effect of heat treatment on carbon fiber surface properties and fibers/epoxy interfacial adhesion [41]. T300B CFs were heated in a vacuum drying chamber at 150, 180 and 200 °C for several hours using controlled processing cycles. It demonstrated that the content of activated carbon atoms (conjunction with oxygen, nitrogen and hydroxyl) on the treated CF surface and the polar surface energy decreased with increasing the heat treatment temperature. Compared with the untreated fibers, the wettability studied by dynamic contact angle test between CF and E51 epoxy resin became worse. The results of micro-droplet tests demonstrated that the interfacial shear strength (IFSS) of T300B/epoxy reduced after the heat treatment process. This was attributed to the decrement of the amount of reactive functional groups in the interfacial region.

Plasma treatment has become a popular method for improving the fiber-matrix adhesion in recent years [42-52]. Plasma is a partially or fully ionized gas containing electrons, radicals, ions and neutral atoms or molecules. The principle of a plasma treatment is the formation of active species in a gas induced by a suitable energy transfer. Typical gases used to create a plasma include air, oxygen, ammonia, nitrogen and argon. Erden et al. [52] used continuous atmospheric plasma oxidation (APO) to introduce oxygen functionalities on the surface of CFs in order to improve the interfacial adhesion between CFs and polyamide-12 (PA-12). After the APO treatment, CFs became more hydrophilic due to the introduction of polar oxygen-containing groups on the fiber surface, which also resulted in an increase of fiber surface energy. And the fiber tensile strength remained unaffected. The IFSS between CFs and PA-12 increased from 40 to 83 MPa with up to 4 min of APO treatment. This can be attributed to the increase of surface oxygen content from 7 at.% to 16 at.%, which yielded more hydrogen bonds between fibers and PA-12 matrix.

A number of liquid-phase oxidizing agents (e.g. nitric acid, acidic potassium permanganate, acidic potassium dichromate, hydrogen peroxide and ammonium bicarbonate) have also been used to treat CF surface. These liquid-phase treatments do not cause excessive pitting and hence degradation of the fiber strength [38].

Anodic oxidation is most widely used for treatment of commercial CFs as it is fast, uniform and suited to mass production [53-56]. CFs act as an anode in a suitable electrolyte bath. A potential is applied to the fiber to liberate oxygen on the surface. Typical electrolytes include nitric acid, sulfuric acid, sodium chloride, potassium nitrate, sodium hydroxide, ammonium hydroxide and so on.

3.2. Non-Oxidative Surface Treatments

Non-oxidative methods, including the deposition of an active form of carbon, plasma polymerization and grafting of polymers onto the fiber surface [57] have been used for the CF surface treatments. Whiskerization involves the growth of thin and high strength single crystals, such as silicon carbide (SiC), silicon nitride (Si_3N_4) and titanium dioxide (TiO_2) at right angles to the fiber surface [58]. Many polymerizable organic vapors are used for plasma polymerization process, such as polyamide, polyimide, organosilanes, propylene, and styrene monomers. Plasma polymerization is demonstrated to increase the polar component of surface free energy of carbon fibers [59, 60].

3.3. Sizing of Carbon Fibers

Sizing of CFs refers to the coating of organic materials applied to fiber surface during the manufacture [61-63]. Sizing is reputed to protect the fiber surface from damage during subsequent textile processing, aid in handling, provide a chemical link between the fiber surface and the matrix and thus to improve the fiber-matrix adhesion [64].

Sizing can be achieved by deposition of polymers from solutions onto the fiber surface. The fibers pass through a sizing bath filled with organic solutions. The choice of sizing materials depends on the fiber type and the matrix resin. They must be compatible with the matrix resin, which allows the resin to penetrate into the fiber bundle and interact with the fiber surface. Typical sizing materials include epoxy, urethane, polyester and others. The sizing amount is 0.5-1.5 wt.% of the fiber and the sizing layer is hundreds of nanometers thick [63].

The sizing agents can change the CF surface properties, as well as the wettability and chemical reactions with epoxy matrices [65-67]. Numerous studies also proved that the sizing agent can significantly impact the interfacial adhesion between CF and the resin matrix [67-69]. Zhang et al. demonstrated that molecular weights of the epoxy sizing agents had influence on the values of IFSS and aging resistance properties of the CF/epoxy composites [70]. The emulsifier content in sizing agent was also proved to affect the IFSS and ILSS of the CF/epoxy [71]. Moreover, multiscale simulation method was used to analyze cross-linked network structures within the interphase region, formed by fiber sizing agent, epoxy and hardener, in-between CF and matrix [72]. The results indicated that mutual diffusions and chemical reactions will take place under processing conditions and therefore determine the crosslink density and properties of the interphase region. Drzal et al. [30] have studied the effect of sizing on the adhesion of CF to epoxy matrix. They found that the sizing layer created a brittle interface layer between the fiber and matrix which improved the IFSS. Dai et al. [68] have investigated the influence of sizing on the CF/matrix interfacial adhesion by comparing sized and desized T300B and T700SC CFs. They found that the desized CFs presented less concentration of activated carbon atoms (conjunction with oxygen and nitrogen) and lower polar surface energy, but higher dispersive surface energy and IFSS. The sizing agent on T300B and T700SC fiber surface was negative for the interfacial bonding. This is contrary to the general principles. Desizing reduced the acid parameter of CF surface which promoted bonding strength at the fiber/epoxy interface. The IFSS of T300B/epoxy increased from 64 MPa to 88 MPa after desizing, with an improvement of 38 %. This was attributed to the increment of work of adhesion. The IFSS of desized T700SC/epoxy (89 MPa) was 9 % greater than that of T700SC/epoxy (82 MPa). The thicker sizing might result in weak layer in the interface region. They concluded that IFSS for CF/epoxy systems depended not only on the chemical bonding but also on the physical and adhesive interactions.

4. Nano/Micrometer Hybrid Structures

4.1. Carbon Nanotube Introduction

Nanomaterials, defined as having at least one dimension between 1 and 100 nm, are largely used as reinforcements in polymer-matrix nanocomposites [73] for various applications from electronics to aerospace, due to their distinguished structures and properties differing from micrometer and bulk materials [74]. In particular, nanomaterials have a substantially high surface-area-to-volume ratio. Whatever in form of nanoparticles, nanowires, nanotubes, or nanosheets, their large surface area can create huge interfacial zones in polymer matrix, with the properties quite different from those of the matrix.

Carbon nanotubes (CNTs) are one-dimensional carbon materials consisting of single or multiple graphene layers wrapped into seamless cylinders or concentric tubes [75]. According to the layer number, they are generally classified into single-walled carbon nanotubes (SWCNTs) and multi-walled carbon nanotubes (MWCNTs). CNTs have light weight, high aspect ratio, outstanding mechanical properties (e.g. tensile modulus ~ 1 TPa, and tensile strength ~50 GPa) [76, 77], and excellent thermal conductivity (3000 W $m^{-1}K^{-1}$) [78]. Concerning the electrical properties, SWCNTs can behave as metals or semiconductors depending on their chirality [79]. Both SWCNTs and MWCNTs show ballistic transport characteristics [80]. They have high carrier mobility (~10,000 $cm^2 V^{-1} s^{-1}$), and can transport a high electrical current density (~4 $\times 10^9$ A cm^{-2}) [81]. These excellent properties make CNTs ideal candidates as multifunctional reinforcements in composites.

As introduced in the precedent part, CFRP laminate composites exhibit relatively poor properties in the in-plane transverse direction and in the through-thickness direction, being dominated by polymer matrix and CF/matrix interface. The adhesion between CFs and polymer matrix might be generated by chemical bonding, van der Waals bonding, and mechanical interlocking. A good wettability is therefore the prerequisite condition to have an intimate contact between CFs and polymer matrix. Usually, the graphitic CFs have physically smooth and chemically inert surface [82]. Certain surface treatments and coating involving the modification of the fiber surface chemistry are carried out in order to improve the interfacial adhesion between CFs and the polymer matrix [83]. An optimum of adhesion is necessary to ensure an efficient stress transfer between matrix and fibers (0 ° direction), and to guarantee that a maximum stress level can be maintained across the interface and from fiber to matrix without disruption (90 ° direction) [83]. Besides, the insulative characteristics of polymers greatly limit the electrical and thermal transport properties of composites. Therefore, the addition of nanomaterials like CNTs into composites can be a promising strategy to improve their mechanical properties, as well as electrical and thermal conductivities in the transverse and in the through-thickness directions [84].

4.2. Incorporation of CNTs into CFRP Composites

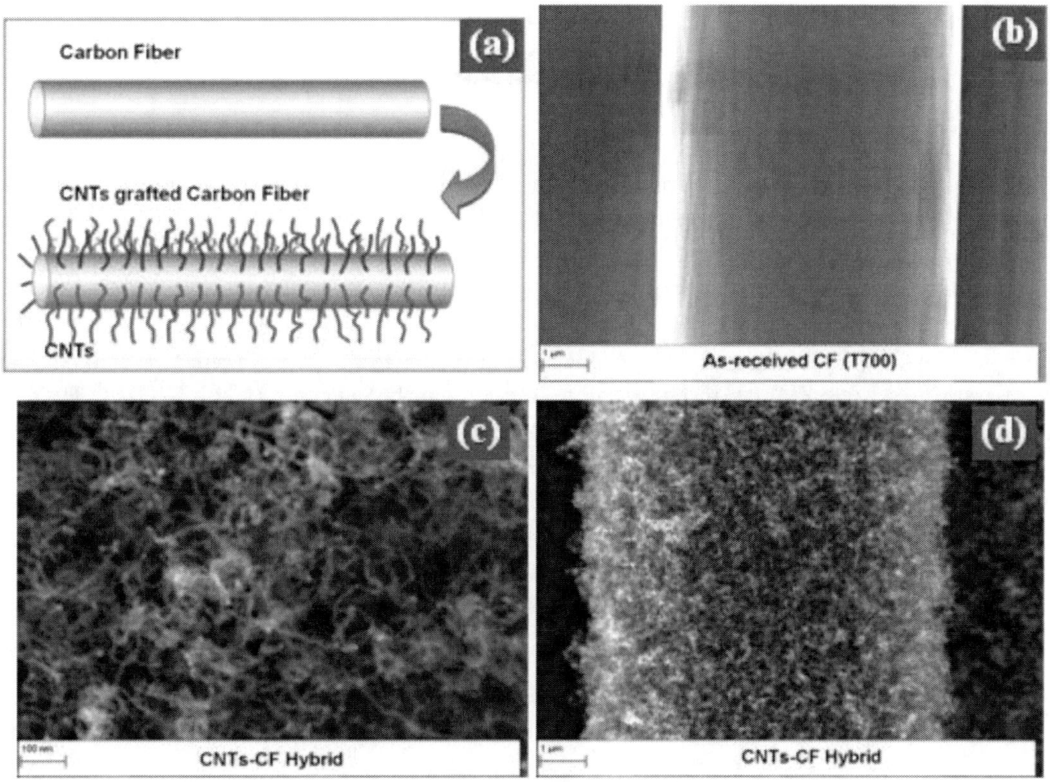

Figure 1. (a) Schematic of CF before and after CNT grafting; (b) SEM images of as-received CF (type T700) and (c-d) SEM images of CF after grafting CNTs by CVD.

A number of methods dealing with the incorporation of CNTs into CFRP composites have been reported in literature, such as dispersing CNTs in a polymer matrix [85], attaching CNTs on the fiber/fabric surface by electrophoresis [86] or chemical ways [87], inserting CNTs on the prepreg surface by transfer-printing vertically aligned CNT forests [88], and grafting on the fiber/fabric surface by CVD to develop multiscale hybrid structures [89, 90]. The first two methods concern the modification of the polymer matrix properties, and the last one relies on the modification of both the CF/polymer matrix interface and the matrix surrounding fiber surface. The main problems concerning mixing CNTs with polymer matrix are the greatly increased matrix viscosity upon adding CNTs, and the agglomeration of CNTs during dispersion. These problems will influence the followed preparation of the composites and thus their properties. The deposition of CNTs on a CF surface by electrophoresis is easy to be scaled-up in industrial applications, however, this technique is limited by the difficulty to obtain uniform coating, and the weak bonding between the fiber and the CNTs.[23] In contrast, the direct grafting of CNTs on CFs by CVD is an efficient way to incorporate high content of CNTs in the polymer matrix. As shown in Figure 1, the grafted CNTs are individually distributed on the CF surface, which is highly desired to avoid the CNT agglomeration and to obtain good dispersion in composites. Furthermore, the technique could be easily integrated into the industrial production processes used for CF surface treatment and the prepreg or composite preparation, once the CNT growth on CF tows or clothes is conducted in a continuous way.

4.3. CVD Production of CNTs/Carbon Fiber Hybrids

In general, a CVD process for CNT synthesis consists in the thermal decomposition of carbon sources (hydrocarbons or other carbon-containing species) on the surface of nanosized catalyst particles supported on certain substrates. It involves homogeneous gas phase reactions, and heterogeneous chemical reactions on or near the vicinity of a heated substrate surface, leading to the deposition of powders or films [91]. The three components required in the system are carbon source, heat source and catalyst (or catalyst precursors). As it is well known, nowadays CNTs can be produced using a wide range of metal catalysts and hydrocarbon sources [92, 93]. The carbon sources can be gaseous (CH_4, C_2H_2, C_2H_4 etc.), liquid (alcohol, benzene, toluene, xylene, etc.) or solid (camphor, naphthalene, etc.). The catalyst particles (Fe, Ni, Co, etc.) can be either previously deposited on the substrates or in situ produced from catalyst precursors. Then, the catalyst particles are heated to sufficiently high temperature (400-1200 °C) to decompose the carbon sources under an inert atmosphere (Ar, N_2, etc.) with the gas (H_2, NH_3, etc.). A variety of materials could serve as substrates for nanotube growth, including Al_2O_3, SiO_2, SiC, MgO, zeolite, graphite etc.

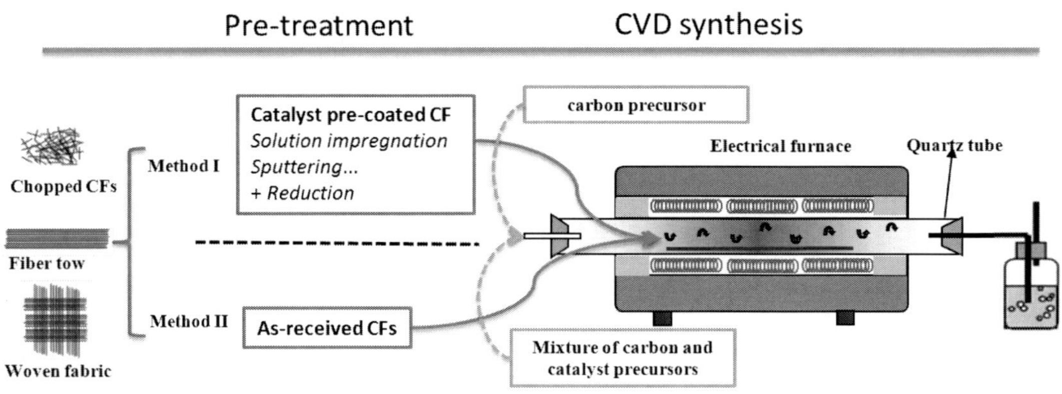

Figure 2. Schematic of the CVD system used for growing CNTs on carbon fibers.

Figure 2 schematically shows the CVD system used for in situ growing CNTs on CFs. It consists of three main parts: vapor precursor supply system, CVD reactor and effluent gas handling system. Compared to other CNT synthesis methods, the main advantage of CVD is the selective growth of CNTs on CFs at moderately high temperatures and in a controlled way. This is very important, because the high temperature processes are likely to degrade the intrinsic properties of CFs [94]. As shown in Figure 2, there are generally two different ways to prepare the catalyst particles which are required in the followed CNT growth processing. One is to deposit in advance the catalyst particles on the CF surface, the catalyst pre-coated CFs being then introduced into the CVD reactor for growing CNTs. We note it as the *catalyst pre-coating CVD method*. The other method is the in-situ deposition of the catalyst particles on CFs inside the CVD reactor during the CNT nucleation and growth. This is achieved by simultaneously injecting a mixture of carbon source and catalyst precursor. We note it as the *floating catalytic CVD*.

By using the catalyst pre-coating CVD method, Thostenson et al. [89] first reported the direct growth of CNTs on CFs to create multiscale CNTs/CF hybrid composites. They showed increased interfacial shear strength (IFSS) between CF and epoxy matrix after grafting CNTs on CFs. Later on, more and more attention was paid to the development of this kind of nano-micrometer hybrid structures. As shown in Table 2, more detailed studies were conducted on the influence of the processing parameters, such as catalyst, hydrocarbon source [95], temperature [96], and growth time. The general process used consists of the following steps:

1) removing any polymer sizing by heat-treatment or acid washing,
2) depositing catalyst layer onto CF surface,
3) reducing the catalyst layer and converting it into nanoparticles, and
4) growing CNTs by thermally decomposing carbon precursors in the CVD reactor.

It can be seen that, in these processes, the catalyst coating is a time-consuming but important step, because it has significant impact on the following processes: CNT nucleation, growth, and their distribution on CFs, as well as on CNT structure. A thin catalyst layer can be deposited by a variety of different techniques, like thermal (magnetron) sputtering, metal salt solution impregnation, polymer–metal complexes [97], thermal evaporation [98], etc. However, it is often very difficult to obtain uniformly distributed metal particles with homogeneous diameter. Besides, it is worth mentioning that for catalyst coating, the as-received CFs have to be unsized by heat treatment, or by harsh chemical treatment with strong acids or bases. Moreover, a time as long as 1~2 hours is often required to reduce the pre-coated metal oxides or salts or the deposited thin metal layer into nanometric particles. And the chemical reactions are commonly conducted at a relatively high temperature (more than 700 °C) for 30-60 min. All these processes could create the surface flaws on CFs [99], and increase the metal diffusion into the fiber, resulting in a deterioration of the fiber properties.

No time-consuming and delicate catalyst pre-coating process is one of the most distinguished features which make the floating catalytic CVD differ form the above discussed method. The as-received CFs (sized or unsized) can be directly introduced into the CVD reactor for CNT grafting. Therefore, this is a highly efficient method to in-situ develop CNTs/CF hybrid structures. Additionally, the reduced handling steps can greatly decrease the damage of CFs generated during any pre-treatment procedure. These are significantly important for potentially industrial large-scale production, and their application in advanced structural components of aircrafts and automobiles. As mentioned earlier, in the floating catalytic CVD, a mixture of carbon source and catalyst precursor is simultaneously injected into the CVD reactor. The commonly used catalyst precursors are typically organometallic compounds which contain transition metal, e.g. ferrocene (FeCp2), cobaltocene (FeCp2), nickelocene (NiCp2)... (where, Cp is cyclopentadienyl group, C_5H_5). The carbon sources commonly used are liquid hydrocarbons such as xylene and toluene, or their mixture with acetylene, ethylene, or other carbon organic substances. The CVD conditions reported in the literature are listed in Table 3.

It can be seen from the Table 3 that the CNT synthesis temperatures used are still a bit high, commonly above 750 °C. But, the stay time of CFs inside the CVD reactor is greatly decreased, compared to the catalyst-precoating CVD method. It could be as short as several minutes. This is highly desired for large-scale production. Another point which should be underlined is that the nucleation of catalyst particles on the surface of CFs is much more difficult than that on ceramic fibers and particles such as alumina [117-121], silicon carbide [90, 122], aluminum silicate or quartz fibers [123]. This is believed to be mainly due to the higher diffusion rate of transition metals into CFs [124]. Thus, the CNT growth rate is lower on CFs than on the ceramic substrates. Therefore, it requires finding optimized CVD conditions for each kind of CFs in order to get an optimum CNT growth rate and density. In addition, some research studies on introducing a buffer layer on CF surface have been conducted in order to improve the CNT growth efficiency [124, 125]. The deposited buffer layers like amorphous Si or SiO_2 prevent the catalyst diffusion, and promote the growth of vertically aligned CNTs with a higher growth rate.

All the above mentioned studies are done in the closed CVD reactors with static CFs or CF fabrics. Up to now, no study on continuously CNTs grafting on moving carbon fibers has been reported. However, this is highly desired for the large-scale application of the CNTs/CFs based hierarchical composites in aerospace and automotive. To achieve the continuous production of CNTs/CFs, more efforts are still needed on understanding the fundamental aspects of the CVD process as well as on large-scale procedure optimization.

4.4. Enhanced Properties of the CNTs/CF Hybrids-Based Composites

In-situ growth of CNTs on CFs by CVD provides a promising way to highly efficiently develop the CNTs/CF hybrid structures. These nano/micrometer hybrids are expected to replace the currently used commercial CFs to develop multi-scale hierarchical composites, in which the reinforcement elements from macroscale to nanoscale are schematically shown in Figure 3. The introduction of CNTs on the CF surface greatly modifies the CF interfacial properties, surface area and the properties of the interphase zone [133] between CF and matrix. The polymer matrix behaviors are also greatly varied by the CNT incorporation and the generated huge interfacial and interphase zones. Based on these reinforcing mechanisms from nanometer scale to macroscopic scale, the perpendicularly grafted CNTs on CFs are expected to increase the interfacial adhesion between the CF and the matrix, and also to improve the composite performance, especially in transverse direction and also in through-thickness direction.

Table 2. CNT growth on catalyst-precoated carbon fibers or carbon fiber fabrics by CVD

Product	Carbon fiber pretreatment			CVD condition				Authors & Published year
	Unsizing	Catalyst deposition	Reduction	Catalyst	T	Carbon source	time	
CNT/CFs Or C cloth	vacuum heat-treatment (700 °C)	magnetron sputtering	660 °C	stainless steel	660 °C	acetylene	30 min	Thostenson et al.,[89] 2002 Jo et al.[100], 2005
CNTs/CFs		ferrofluide/acetone	400 °C/20 min	Fe	550-1000 °C	methane	10–120 min	Zhu et al.,[96] 2003
CNTs/CFs		solution impregnation $Ni(NO_3)_2$/acetone	300 °C/ H_2	Ni	500 °C	methane, *n*-hexane, cyclopentadiene, benzene	-	Otsuka et al.,[101] 2004
CNTs/CFs	vacuum heat-treatment (380°C/2 h)	electro-deposition $NiCl_2.6H_2O$	H_2/20min	Ni	Hot-filament CVD	methane	15 min	Makris,et al.,[102], Riccardis,et al.,[103] 2005
CNTs/Fabric		solution impregnation $Ni(NO_3)_2$/acetone	400 °C/H_2 /60 min	Ni	530 °C	methane	60 min	Tzeng et al.,[104] 2006
CNTs/C cloth	unsized	powder spreading	no	Ni+S powder	1050–1200 °C	natural gas	2 h	Gong et al.,[105] 2007
CNTs/CFs		sputtering	H_2	Ni or Fe	750-890 °C	acetylene Or carbon monoxide	2.5-30 min	Susi et al.,[95] 2008
CNTs/CFs	acid oxidation + base wash	incipient wetness $Fe(NO_3)_3 \cdot {}_9H_2O$/ethanol		Fe	750 °C	acetylene	60 min	Qian et al.,[106,107] 2008

Table 2. (Continued)

Product	Carbon fiber pretreatment			CVD condition				Authors, published year
	Unsizing	Catalyst deposition	Reduction	catalyst	T	Carbon source	time	
CNTs/Fabric		thermal evaporation		Fe	820 °C	methane+acetylen e	30 min	Kepple et al.[98] 2008
CNTs/CFs mesopitch	nitric acid	solution impregnation $Co(NO_3)_2$,$Fe(NO_3)_2$,$Ni(NO_3)_3$	750 °C/4h	Co,Fe,Ni	750 °C	acetylene	30 min	Zhao et al.,[108] 2008
CNTs/CF preform		solution impregnation $NiNO_3$	450-550 °C/ H_2	Ni	750 °C	carbon monoxide	45 min	Chen et al.,[109] 2009
CNTs/CFs	acetone	polymer-metal complex			750 °C	methane+ ethylene	30 min	Shafranska et al.[97], 2009
CNTs/CFs or fabric	nitric acid (90 °C/2 h)	solution impregnation $Ni(NO_3)_2.6H_2O$	400 °C / H_2 / 60 min	Ni	700 °C	methane	60 min	Hung et al.,[110] 2009
CNTs/CFs	acid cleaning	dip coating $NiCl_2 \cdot 2H_2O$	400 °C/ H_2/30-60 min	Ni	700 °C	acetylene		Sharma et al.,[111, 112] 2011
CNTs/ Activated CF		solution impregnation $Ni(NO_3)_2.6H_2O$	H_2	Ni	750 °C	acetylene	40 min	Wang et al.,[113] 2011
CNTs/CF or fabric	heat-treatment (450 °C)	electroless dip coating acidic bath	500 °C/H2/20 mi n	Ni	550 °C	acetylene	5 -25 min	Agnihotri et al.,[114] 2011
CNTs/Fabric		solution impregnation Nickel nitrate or Iron-cobalt salts	600 °C /H_2/60-90min	Ni Fe-Co	600-750 °C	ethylene	varied	Lomov et al.,[115] 2011
CNTs/Fabric	coating Al_2O_3 layer ($Al(NO_3)_3 \cdot 9H_2O$)	incipient impregnation $Fe(NO_3)_3 \cdot 9H_2O$		Fe	750 °C	acetylene	30 min	An et al.,[116] 2011

Table 3. CNT growth on carbon fibers or woven fabrics by floating catalytic CVD

Products	Pretreatment	CVD condition Temperature /Carbon source / Carrier gas/ time	Authors &Published year
CNTs/CFs	Coating SiO_2 layer (Tetraethylorthosilicate/toluene, 850 °C)	850 °C/Ferrocene-toluene/Ar/1-10 min	Delmas et al.[125] 2012
CNTs/CFs	no	780°C/Ferrocene-xylene/H_2-Ar/5min	Lachman et al.[126] 2012
CNTs/CFs	Ferrocene-acetone drop (700 °C)	750 °C/Aetylene/H_2-N_2/30 min	An et al.[127] 2012
CNTs/CFs	Heat-treatment (800 °C/Ar/1 h) Or Coating SiO_2 layer	850 °C /Ferrocene-(xylene:ethanol:ethylene diamine) /Ar /30-120 min	Lv et al.[128] 2011
CNTs/CFs	Coating amorphous Si	850 °C /ferrocene–camphor/ N_2 /1-5 min	de Resende,et al.[124], 2010
CNTs/CFs	$MgSO_4$ in alcohol	900 °C/Iron Phthalocyanine powder /Ar-H_2 /15 min Or 800 °C/Ferrocene-xylene /Ar-H_2/30min	Sager et al.[129] 2009
CNTs/CF	no	750-800°C/Ferrocene-xylene /Ar-H_2/5-30min	Zhang et al.[94] 2009
CNTs/CFs (tow, cloth &felt)	no	750 °C/Ferrocene-toluene/Ar/15-90min	Mathur et al.[130] 2008
CNTs/CFs	no	800 °C-1000 °C/Ferrocene-xylene& Xylene-H_2S/H_2-Ar/0.5-10 min	Sonoyama et al.[131] 2006
CNTs/CFs	no	700-900 °C/Ferrocene-xylene/Ar /-	Zhao et al.[132] 2005

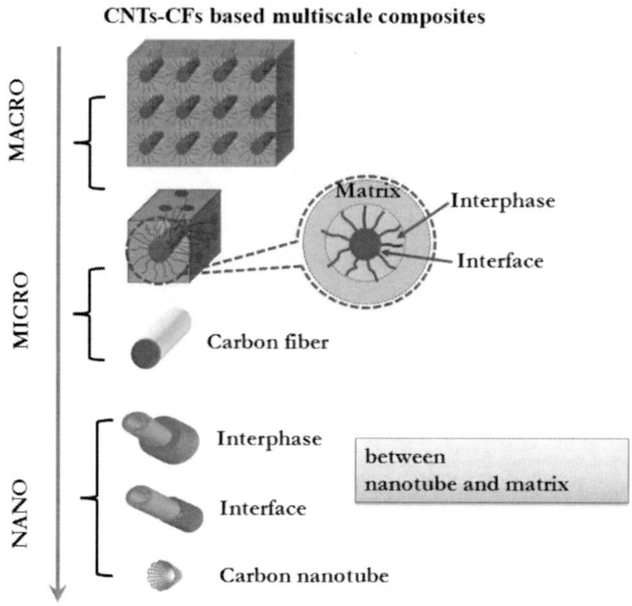

Figure 3. Schematic of the multiscale reinforcement elements of the composites based on the CNTs/CFs hybrids. The introduction of CNTs on the CF surface modifies CF surface properties, surface area and the properties of the interphase zone between CF and matrix.

1. Effect on the Interfacial Shear Strength (IFSS)

The fiber-matrix adhesion plays a crucial role in the CFRP composites. It ensures that a maximum stress level can be maintained across the fiber interface and from fiber to matrix without disruption [83]. It is usually quantified using interfacial shear strength (IFSS). The methods widely used for measuring the fiber-matrix adhesion are single fiber pull-out or push-out [106] and single-fiber fragmentation [129]. In particular, the single fiber fragmentation technique is a convenient and reproducible method which has been widely used to evaluate the fiber-matrix adhesion in the fiber reinforced composites. Figure 4 shows two kinds of the breaks in epoxy matrix after single fiber fragmentation test. In general, the desized CF weakly adheres to the polymer matrix. When the tensile stain in the fiber exceeds its failure strain, the fiber fractures, and debonds from the matrix, as shown in Fig 4a-b. The grafting of CNTs on CF surface could generally improve the fiber-matrix adhesion. The beak of CF is accompanied with the matrix fracture during the fragmentation test. And the CF could break into shorter length, as shown in Figure 4c-d.

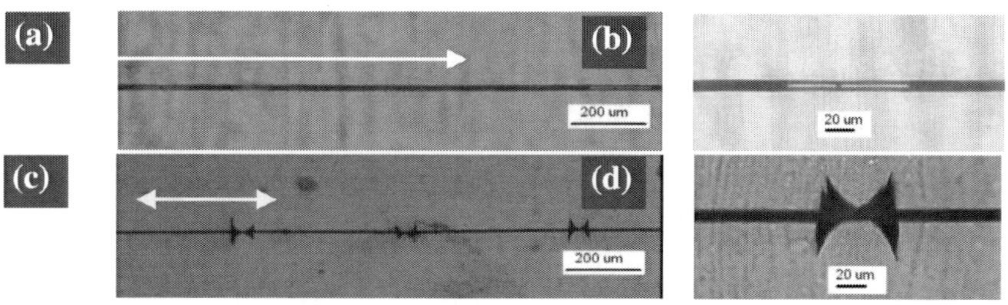

Figure 4. Optical microscopy images of the typical breaks in epoxy matrix after single fiber fragmentation test: (a-b) desized carbon fiber, and (c-d) CNTs grafted carbon fiber.

Table 4 summarizes some of recently reported research works concerning the interfacial reinforcement of the CNTs-grafted CFs in a polymer matrix. It can be seen that the grafting of CNTs on carbon fibers contributes to an improvement of the CF-polymer matrix interfacial connection.

Table 4. Summary of reported IFSS improvement of the CFs grafted with CNTs by CVD

Reinforcement-Polymer matrix	CNTs/CF preparation condition	Test type	IFSS improvement	Authors & Published year
CNTs/CF-Epoxy	Solution impregnation/CVD/750°C	Single fiber pull-out	60%	Qian et al.[106] 2008
CNTs/CF-Epoxy	Solution impregnation/CVD/800-900°C	Single fiber fragmentation	11% to 71%	Sager et al.[129] 2009
CNTs/CFs-Polyester	Solution impregnation/CVD/550°C	Single fiber pull-out	88 %	Agnihotri et al.[114] 2011
CNTs/CF-Epoxy	SiO_2 coating/Floating CCVD/850°C	Single fiber fragmentation	175%	Lv, et al. [128] 2011
CNTs/CF-Epoxy	Solution impregnation/CVD/750°C	Single fiber pull-out	110%	An et al. [116] 2011
CNTs/CF-Epoxy	Floating CCVD/780°C	Single fiber fragmentation	~90 %	Lachman et al.[126] 2012
CNTs/CF-Epoxy	Floating CCVD/750°C	Single fiber pull-out	~94%	An et al.[127] 2012

Thostenson et al. [89] first studied in 2002 the interfacial adhesion between the CVD prepared CNTs/CFs hybrids and the epoxy matrix. The single-fiber composite fragmentation tests demonstrated that the CNTs grafted CFs showed a ~15 % improvement in the IFSS, compared to the unsized CF. Afterwards, the CNTs/CF hybrids prepared by different CVD processes were used to study the influence of CNT grafting on the interface between the CFs and the polymer matrix. The improvement of the IFSS, as high as 175 %, was obtained in each case (see Table 4). The improved fiber-matrix interface after CNT grafting can be attributed to the increased surface area, and the mechanical interlocking between CNTs and matrix.

The quality of the interface between CNTs/CFs and matrix depends greatly on the CNT structure, density, length, and organization states on CFs, as well as on the CVD procedure used. Sager et al.[129] studied the influence of CNT organization state (random or aligned) on the fiber-epoxy interface. The IFSS improvements of 71 % and 11 % were obtained for CFs with random CNTs and radially aligned ones, respectively. However, it is worth mentioning that these two kinds of CNTs/CFs used were prepared using two different CVD conditions, as given in Table 4. Lv et al. [128] studied the variation of the IFSS between CNTs/CF and epoxy with the CNT orientation and length. They reported that MWCNTs with a perpendicular alignment and long length showed a higher IFSS in epoxy composites than random ones. They attributed this to a better wettability and a large contact interface between the hybrids and the resin in the case of well-aligned MWCNTs. The IFSS increased when the length of aligned MWCNTs on CFs increased from 16.6–47.2 μm, but it decreased slightly when the MWCNT length continued to increase from 63.5 and 108.6 μm. They considered that this slight decrease was due to the split MWCNT clusters resulting in asymmetrical stress distribution. Also, it is worth noting that the difference between the CF treatments (with or without SiO_2 coating) used for growing random and aligned CNTs.

Despite the improvement of IFSS of the CNTs/CFs-matrix, severe chemical and thermal treatments as well as high temperature CVD synthesis can degrade the intrinsic CF properties. Compared to the as-received sized commercial carbon fibers, a 55 % decrease in the fiber strength, but without impact on the fiber modulus, was observed after CNT grafting by Qian et al. [106] Then, Sager et al. [129] also reported that CVD processing generated a significant reduction of the fiber tensile strength, by an average of 37 % in the case of the radially aligned MWCNTs and by 30 % in the case of randomly oriented MWCNTs. Similarly, a reduction in the average fiber tensile modulus by 9 % and 13 % was observed in the radially aligned and randomly oriented MWCNT coated fibers, respectively. The significant decreases in the CF tensile properties mentioned above might be due to the severe damage to the CFs during the chemical pretreatments and the CVD synthesis. Conversely, the smaller decreases of the fiber properties were mentioned in the research of Lv, et al. [128] and An et al. [127]. Moreover, slight increases in the fiber tensile strength (~ 9 %) and in the tensile modulus (4 %) were also reported after grafting CNTs on CFs in the case of Hung, et al [110]. A summary of the influence of CNT grafting on the CF properties is given in Table 5. It can be seen that the pretreatments and CVD synthesis conditions have significant effects on the CNTs/CF properties. However, an optimum operation process could improve in both the fiber interfacial shear strength and the fiber tensile properties.

2. Effect in the Bulk Composite Properties

Due to the availability of CNTs-grafted CFs, only a few studies have been published on the bulk mechanical properties of the composites based on the CNTs/CFs. However, their improved fiber-matrix interface is expected to improve in turn the composite damage tolerance performance, especially in the in-plane transverse and in the through-thickness-directions, when the CNTs/CFs are used as reinforcements in polymer matrix. Indeed, Mathur et al. [130] studied the mechanical properties of the phenolic composites based on the CNTs grafted unidirectional (UD) CF tows, bi-directional (2D) CF cloth and three dimensional (3D) CF felt by CVD. It was found that the flexural strength was improved by 20 % for UD, 75 % for 2D and 66 % for 3D hybrid composites as compared to the ones prepared by neat reinforcements (without CNT growth) under identical conditions. Additionally, the flexural modulus of these composites was also improved by 28 %, 54 % and 46 %, respectively. They also found that the mechanical properties of the hybrid composites increased with the amount of deposited CNTs. An improvement of 27 % was obtained by Bekyarova et al. [86] in the interlaminar shear strength (ILSS) of the epoxy composites based on the CFs coating with 0.25 wt.% CNTs by electrophoresis. An enhancement of the out-of-plane electrical conductivity of 2 fold and of ~30 % was obtained by the deposition of SWCNTs and MWCNTs, respectively. Kepple et al. [98] demonstrated a 50 % improvement of the fracture toughness (Mode I) of the epoxy composites based on the CNTs grafted woven CFs, as well as a ~ 5% increase in the flexural modulus.

Table 5. Summary of the effect of CVD grafting CNTs on CF tensile properties

Fiber type	CNT growing preparation condition	Fiber tensile strength (%)	Fiber tensile modulus (%)	Authors & Published year
Unsized C320.00A	Solution impregnation /CVD/750°C	-55	unaffected	Qian et al.[106] 2008
Thornel T650	Solution impregnation /CVD/800-900°C	-37(aligned NTs) -30(random NTs)	-9 (aligned NTs) -13% (random NTs)	Sager et al.[129] 2009
PAN-based graphitized, USSR	Solution impregnation /CVD/ 700 °C	9.2%	4.4%	Hung, et al [110] 2009
TORAY T300	SiO_2 coating /Floating CCVD/850°C	-7.8 to -33.5	-	Lv, et al. [128] 2011
Chinese Academy of Sciences, T300	Floating CCVD/750°C	-1 to –9	unaffected	An et al.[127] 2012

Besides, the enhanced thermal conductivity of the hybrid structures has also been reported. Recently, Naito et al. [134] measured the thermal conductivities of high tensile strength PAN-based (T1000GB) and high modulus pitch-based (K13D) CFs with CNTs grown on them by CVD using a thermal diffusivity meter. The results showed that the thermal conductivities of both types of fibers were improved by grafting CNTs. The improvement for the CNTs-grafted T1000GB and K13D fibers were 47 % and 30 %, respectively, compared to their original conductivities (12.6 ± 1.0 W $m^{-1}K^{-1}$ for T1000GB and 745.5 ± 16.0 W $m^{-1}K^{-1}$ for K13D).

Similarly, Veedu et al. [90] showed remarkable improvements in the interlaminar fracture toughness, hardness, delamination resistance, in-plane mechanical properties, damping, thermoplastic behavior, and thermal and electrical conductivities of the composites of SiC fabrics when introducing vertically aligned MWCNTs.

The above-mentioned primary but promising results demonstrate that the CNTs/CFs have great application potential as multifunctional reinforcements in advanced composite materials. The in-situ grafting of CNTs on CF tows, woven fabrics or felts can greatly improve the composite performance and damage tolerance, especially in through-thickness direction. This is mainly attributed to:

1) the improved matrix resistance to crack initiation and propagation,
2) the mechanisms of crack-tip bridging and pullout of CNTs at nanoscale resulting in the improved delamination and impact damage resistance,
3) the improved fiber/matrix interface properties to efficiently transfer the stress from the matrix to the fibers, and
4) the created CNT networks inside the insulative polymer matrix to increase the electrical and thermal conductions.

The multi-scale CNTs/CFs-polymer composites take advantage of the excellent properties of the components at different scales. Their attractive properties make them potentially interesting for large applications in structure materials and multifunctional composites which are widely used in aerospace, automotive and sporting goods industries.

CONCLUSION AND PERSPECTIVES

The interfacial properties play an important role in the load transfer from the fibers to the polymer matrix, and thus show a significant influence on the mechanical behaviors of the long CF reinforced composites. Due to their inert properties, a number of surface treatment methods have been developed to modify physically or (and) chemically the CF surface properties, which improve the interfacial connection between CFs and the matrix. Compared with the conventionally used surface treatments, the introduction of nanomaterials such as CNTs in the fiber/polymer composites shows a number of great advantages due to their large specific surface area and its own excellent mechanical and multifunctional properties. The CNTs can be introduced by direct mixing with polymer matrix or by deposition on CF surface before the resin impregnation process. However, all these methods are limited by the difficulty of efficiently dispersing CNTs in the polymer matrix, and also by the weak CNT concentration that can be introduced into the composites. In contrast, the developing of multiscale CNTs/CF hybrid structures becomes highly desired in order to benefit from the excellent CNT properties and to ameliorate the composite mechanical properties, while at the same time improving their electrical and thermal conductivities. The nano/micro CNTs/CF hybrids could be synthesized in a controlled way by CVD method, which is simple and easy to be scaled-up. Moreover the CNT geometry, organization and distribution can be controlled by varying the CVD parameters like the temperature, carbon sources, catalyst precursors, etc. Thus, an important amount of CNTs (mass and voume ratio) could be easily introduced into the composite system. Indeed, the extensive studies on single fiber composites have reported that these hybrid structures show improved interfacial adhesion strength with polymer matrix, compared with the CFs without grafted CNTs. However, only a few investigations on the macroscopic composites have been reported. This is mainly due to the present availability of high-quality CNTs/CF hybrid structures. Therefore, the reinforcing mechanisms of CNTs in the composite properties under various loading conditions are still unclear up to now, for example, their influence on the delamination resistance of laminated composites.

The CNTs/CF hybrid structures show great application potential in fiber reinforced high performance composites used in a number of fields such as aeronautics, aerospace and automotive. In order to produce large-scale hybrid structures, it is necessary to investigate in detail the CVD techniques which allow continuously grafting homogeneous CNTs on carbon fibers. Meanwhile, a certainly high production rate is desired in order to integrate the CVD equipment into an already available CF industrial production line. The presence of large amount of CNTs on the CF surface makes great variation of CF surface geometry and surface area. Consequently, the penetration of polymer resin into the carbon fiber layers will be different from the commercially available fibers with relatively smooth surface. Therefore, a well adapted composite preparation procedure is also required to develop high quality CNTs/CF-polymer composites. It is also significant to conduct a detailed performance investigation of the CNTs/CF reinforced composites to understand the CNT reinforcing mechanisms from nanoscale, to micro and macroscale, in order to develop high performance CNTs/CF-polymer composites with multiple functionalities.

REFERENCES

[1] Larsen, J. V.; Smith, T. G., Carbon Fiber Structure. *NOLTR 71-166, Naval Ordnance Laboratory, White Oak, Silver Spring, MD* October 1971.

[2] Chung, D. D. L., Carbon Fiber Composites. Butterworth-Heinemann, London, 1994.

[3] Bunsell, A. R., *Fibre reinforcements for composite materials*. Elsevier Science Publishers: Amsterdam, 1988; p 90.

[4] Edie, D. D., *Carbon* 1998, *36*, 345.

[5] Johnson, D. J., *J. Phys. D-Appl. Phys.* 1987, *20*, 286.

[6] Mochida, I.; Korai, Y.; Ku, C.-H.; Watanabe, F.; Sakai, Y., *Carbon* 2000, *38*, 305.

[7] Edie, D. D.; Fox, N. K.; Barnett, B. C.; Fain, C. C., *Carbon* 1986, *24*, 477.

[8] Cho, T.; Lee, Y. S.; Rao, R.; Rao, A. M.; Edie, D. D.; Ogale, A. A., *Carbon* 2003, *41*, 1419.

[9] Hull, D.; Clyne, T. W., *An Introduction to Composite Materials*. Cambridge University Press: 1996.

[10] Choi, M. H.; Jeon, B. H.; Chung, I. J., *Polymer* 2000, *41*, 3243.

[11] Kim, S. S.; Park, D. C.; Lee, D. G., *Compos. Struct.* 2004, *66*, 359.

[12] Bao, L. R.; Yee, A. F., *Polymer* 2002, *43*, 3987.

[13] Gamstedt, E. K.; Skrifvars, M.; Jacobsen, T. K.; Pyrz, R., *Compos. Pt. A-Appl. Sci. Manuf.* 2002, *33*, 1239.

[14] Carbon fiber/epoxy composite product data, Performance Composites Limited. http://www.performance-composites.com/carbonfibre/mechanicalproperties_2.asp.

[15] Kawai, M.; Koizumi, M., *Compos. Pt. A-Appl. Sci. Manuf.* 2007, *38*, 2342.

[16] Harris, B., *Engineering composite materials* The Institute of materials, London, 1999; p 130-131.

[17] Wen, J.; Xia, Z.; Choy, F., *Compos. Pt. B-Eng.* 2011, *42*, 77.

[18] Kim, J. K.; Sham, M. L., *Compos. Sci. Technol.* 2000, *60*, 745.

[19] Selzer, R.; Friedrich, K., *Compos. Pt. A-Appl. Sci. Manuf.* 1997, *28*, 595.

[20] Gerlach, R.; Siviour, C. R.; Wiegand, J.; Petrinic, N., *Compos. Sci. Technol.* 2012, *72*, 397.

[21] Dransfield, K.; Baillie, C.; Mai, Y. W., *Compos. Sci. Technol.* 1994, *50*, 305.

[22] Shafi-Ullah Khan; Kim, J.-K., *Int'l J. Aeronautical &Space Sci.* 2011, *12*, 19.

[23] Lubineau, G.; Rahaman, A., *Carbon* 2012, *50*, 2377.

[24] Gaier, J. R.; YoderVandenberg, Y.; Berkebile, S.; Stueben, H.; Balagadde, F., *Carbon* 2003, *41*, 2187.

[25] Han, S. J.; Chung, D. D. L., *Compos. Sci. Technol.* 2011, *71*, 1944.

[26] Schuster, J.; Heider, D.; Sharp, K.; Glowania, M., *Compos. Sci. Technol.* 2008, *68*, 2085.

[27] Lonjon, A.; Demont, P.; Dantras, E.; Lacabanne, C., *J. Non-Cryst. Solids* 2012, *358*, 1859.

[28] Kim, S. K.; Kim, J. T.; Kim, H. C.; Rhee, K. Y.; Kathi, J., *J. Macromol. Sci. Part B-Phys.* 2012, *51*, 358.
[29] Jesson, D. A.; Watts, J. F., *Polym. Rev.* 2012, *52*, 321.
[30] Drzal, L. T.; Rich, M. J.; Koenig, M. F.; Lloyd, P. F., *J. Adhes.* 1983, *16*, 133.
[31] Piggott, M. R., *Carbon* 1989, *27*, 657.
[32] Piggott, M. R., *Polym.Compos.* 1987, *8* (5), 291.
[33] Guo, H.; Huang, Y. D.; Meng, L. H.; Liu, L.; Fan, D. P.; Liu, D. X., *Mater. Lett.* 2009, *63*, 1531.
[34] Matrab, T.; Nguyen, M. N.; Mahouche, S.; Lang, P.; Badre, C.; Turmine, M.; Girard, G.; Bai, J.; Chehimi , M. M., *J. Adhes.* 2008, 84, 684
[35] Ho, K. K. C.; Beamson, G.; Shia, G.; Polyakova, N. V.; Bismarck, A., *J. Fluorine Chem.* 2007, *128*, 1359.
[36] Paipetis, A.; Galiotis, C., *Compos. Pt. A-Appl. Sci. Manuf.* 1996, *27*, 755.
[37] Chen, W.; Yu, Y.; Li, P.; Wang, C.; Zhou, T.; Yang, X., *Compos. Sci. Technol.* 2007, *67*, 2261.
[38] Kim, J. K.; Mai, Y. W., *Engineered Interfaces in Fiber Reinforced Composites*. Elsevier Science Ltd: Oxford, 1998; p 186-189.
[39] Herrick, J. W.; Gruber, P. E.; Mansur, F. T., Surface treatments for fibrous carbon reinforcements. AFML-TR-66-178, Part I. Air Force Materials Laboratory, 1966.
[40] Scola, D. A.; Basche, M. Treatment of carbon fibers. United States Patent 3720536; 1970.
[41] Dai, Z.; Zhang, B.; Shi, F.; Li, M.; Zhang, Z.; Gu, Y., *Appl. Surf. Sci.* 2011, *257*, 8457.
[42] Morra, M.; Occhiello, E.; Garbassi, F.; Nicolais, L., *Compos. Sci. Technol.* 1991, *42*, 361.
[43] Commerçon, P.; Wightman, J. P., *J. Adhes.* 1992, *38*, 55.
[44] Yuan, L. Y.; Chen, C. S.; Shyu, S. S.; Lai, J. Y., *Compos. Sci. Technol.* 1992, *45*, 1.
[45] Yuan, L. Y.; Shyu, S. S.; Lai, J. Y., *Compos. Sci. Technol.* 1992, *45*, 9.
[46] Allred, R. E.; Schimpf, W. C., *J. Adhes. Sci.Technol.*1994, *8*, 383.
[47] Dilsiz, N.; Ebert, E.; Weisweiler, W.; Akovali, G., *J. Colloid Interface Sci.* 1995, *170*, 241.
[48] Vaidyanathan, N. P.; Kabadi, V. N.; Vaidyanathan, R.; Sadler, R. L., *J. Adhes.* 1995, *48*, 1.
[49] Chand, N.; Schulz, E.; Hinrichsen, G., *J. Mater. Sci. Lett.* 1996, *15*, 1374.
[50] Bogoeva-Gaceva, G.; Mäder, E.; Haüssler, L.; Dekanski, A., *Compos. Pt. A-Appl. Sci. Manuf.* 1997, *28*, 445.
[51] Pittman Jr, C. U.; Jiang, W.; He, G.-R.; Gardner, S. D., *Carbon* 1998, *36*, 25.
[52] Erden, S.; Kingsley, K. C. H.; Lamoriniere, S.; Lee, A. F.; Yildiz, H.; Bismarck, A., *Plasma Chem. Plasma Process.* 2010, *30*, 471.
[53] Yumitori, S.; Nakanishi, Y., *Compos. Pt. A-Appl. Sci. Manuf.* 1996, *27* (11), 1059.
[54] Yumitori, S.; Nakanishi, Y., *Compos. Pt. A-Appl. Sci. Manuf.* 1996, *27*, 1051.
[55] Fukunaga, A.; Ueda, S., *Compos. Sci. Technol.* 2000, *60,* 249.
[56] Liu, X.; Yang, C.; Lu, Y., *Appl. Surf. Sci.* 2012, *258*, 4268.
[57] Morgan, P., *Carbon Fibers and Their Composites*. CRC Press, New York, 2005, p347-370.
[58] Goan, J. C.; Prosen, S. P., Interfacial bonding in graphite fiber-resin composites. In *Interfaces in Composites*, Philadelphia: ASTM, 1969, pp 3-26.

[59] Donnet, J. B.; Guilpain, G., *Composites* 1991, *22*, 59.
[60] Dagli, G.; Sung, N.-H., *Polym. Compos.* 1989, *10*, 109.
[61] Fitzer, E.; Weiss, R., *Carbon* 1987, *25*, 455.
[62] Yumitori, S.; Wang, D.; Jones, F. R., *Composites* 1994, *25*, 698.
[63] Broyles, N. S.; Chan, R.; Davis, R. M.; Lesko, J. J.; Riffle, J. S., *Polymer* 1998, *39*, 2607.
[64] Sugiura, N.; Maki, N. Sizing agent for carbon fiber, method for sizing carbon fiber by said sizing agent, sized carbon fiber and knitted or woven fabric using said carbon fiber. United States Patent 7135516: Mitsubishi Rayon Co., Ltd. (Tokyo, JP); 2006.
[65] Xu, B.; Wang, X.; Lu, Y., *Appl. Surf. Sci.*2006, *253,* 2695.
[66] Tran, L. Q. N.; Fuentes, C. A.; Dupont-Gillain, C.; Van Vuure, A. W.; Verpoest, I., *Colloids Surf. A: Phys.Eng. Aspects* 2011, *377,* 251.
[67] Dilsiz, N.; Wightman, J. P., *Colloids Surf. A: Phys.Eng. Aspects* 2000, *164*, 325.
[68] Dai, Z.; Shi, F.; Zhang, B.; Li, M.; Zhang, Z., *Appl. Surf. Sci.* 2011, *257*, 6980.
[69] Ramanathan, T.; Bismarck, A.; Schulz, E.; Subramanian, K., *Compos. Sci. Technol.* 2001, *61*, 599.
[70] Zhang, R. L.; Huang, Y. D.; Su, D.; Liu, L.; Tang, Y. R., *Mater. Des.* 2012, *34*, 649.
[71] Zhang, R.; Huang, Y.; Liu, L.; Tang, Y.; Su, D.; Xu, L., *Mater. Des.* 2012, *33*, 367.
[72] Liu, H.; Li, M.; Lu, Z.-Y.; Zhang, Z.-G.; Sun, C.-C.; Cui, T., *Macromolecules* 2011, *44*, 8650.
[73] Coleman, J.N.; Khan, U.; Gun'ko, Y.K., *Adv. Mater.* 2006, 18, 689.
[74] Hussain, F.; Hojjati, M.; Okamoto, M.; Gorga, R. E., *J. Compos Mater.* 2006, *40*, 1511.
[75] Harris, P. J. F., *Carbon Nanotubes and Related Structures*. Cambridge University Press: 1999.
[76] Wong, E. W.; Sheehan, P. E.; Lieber, C. M., *Science* 1997, *277*, 1971.
[77] Yu, M. F.; Lourie, O.; Dyer, M. J.; Moloni, K.; Kelly, T. F.; Ruoff, R. S., *Science* 2000, *287*, 637.
[78] Kim, P.; Shi, L.; Majumdar, A.; McEuen, P. L., *Phys. Rev. Lett.* 2001, *87*, 215502.
[79] Wilder, J. W. G.; Venema, L. C.; Rinzler, A. G.; Smalley, R. E.; Dekker, C., *Nature* 1998, *391*, 59.
[80] Frank, S.; Poncharal, P.; Wang, Z. L.; de Heer, W. A., *Science* 1998, *280* , 1744.
[81] Hong, S.; Myung, S., *Nat. Nanotechnol.* 2007, *2*, 207.
[82] Li, J.; Cheng, X. H., *Mater. Chem. Phys.* 2008, *108*, 67.
[83] Drzal, L. T.; Madhukar, M., *J.Mater. Sci.* 1993, *28*, 569.
[84] Coleman, J. N.; Curran, S.; Dalton, A. B.; Davey, A. P.; McCarthy, B.; Blau, W.; Barklie, R.C., *Phys. Rev. B* 1998, *58*, R7492.
[85] Yokozeki, T.; Iwahori, Y.; Ishiwata, S.; Enomoto, K., *Compos. Pt. A-Appl. Sci. Manuf.* 2007, *38*, 2121.
[86] Bekyarova, E.; Thostenson, E.T.; Yu, A.; Kim, H.; Gao, J.; Tang, J.; Hahn, H.T.; Chou, T.W.; Itkis, M.E.; Haddon, R.C., *Langmuir* 2007, *23*, 3970.
[87] Laachachi, A.; Vivet, A.; Nouet, G.; Ben Doudou, B.; Poilâne, C.; Chen, J.; Bai, J.; Ayachi, M. H., *Mater. Lett.* 2008, 62, 394.
[88] Garcia, E. J.; Wardle, B. L.; Hart, A.J., *Compos. Pt. A-Appl. Sci. Manuf.* 2008, *39*, 1065.
[89] Thostenson, E.T.; Li, W.Z.; Wang, D.Z.; Ren, Z.F.; Chou, T. W., *J. Appl. Phys.* 2002, *91*, 6034.

[90] Veedu, V.P.; Cao, A.Y.; Li, X.S.; Ma, K.G.; Soldano, C.; Kar, S.; Ajayan, P.M.; Ghasemi-Nejhad, M.N., *Nat. Mater.* 2006, *5*, 457.
[91] Choy, K. L., *Prog. Mater. Sci.* 2003, *48*, 57.
[92] Karthikeyan, S.; Mahalingam, P.; Karthik, M., *E-J. Chem.* 2009, *6*, 1.
[93] Teo, K. B. K. a. S., C. and Chhowalla, M. and Milne, W.I., Catalytic synthesis of carbon nanotubes and nanofibers. In *Encyclopedia of Nanoscience and Nanotechnology*, Nalwa, H. S., Ed. American Scientific Publishers: Stevenson Ranch, CA, 2003; pp 665-686.
[94] Zhang, Q. H.; Liu, J. W.; Sager, R.; Dai, L. M.; Baur, J., *Compos. Sci. Technol.* 2009, *69*, 594.
[95] Susi, T.; Nasibulin, A. G.; Jiang, H.; Kauppinen, E. I., *J. Nanomater.* 2008, *2008,* 7.
[96] Zhu, S.; Su, C. H.; Lehoczky, S. L.; Muntele, I.; Ila, D., *Diam. Relat. Mat.* 2003, *12*, 1825.
[97] Shafranska, O.; Voronov, A.; Kohut, A.; Wu, X. F.; Akhatov, I. S., *Carbon* 2009, *47*, 3137.
[98] Kepple, K. L.; Sanborn, G. P.; Lacasse, P. A.; Gruenberg, K. M.; Ready, W. J., *Carbon* 2008, *46*, 2026.
[99] Kim, K. J.; Yu, W. R.; Youk, J. H.; Lee, J., *ACS Appl. Mater. Interfaces* 2012, *4*, 2250.
[100] Jo, S. H.; Huang, J. Y.; Chen, S.; Xiong, G. Y.; Wang, D. Z.; Ren, Z. F., *J. Vac. Sci. Technol. B* 2005, *23*, 2363.
[101] Otsuka, K.; Abe, Y.; Kanai, N.; Kobayashi, Y.; Takenaka, S.; Tanabe, E., *Carbon* 2004, *42*, 727.
[102] Makris, T. D.; Giorgi, R.; Lisi, N.; Pilloni, L.; Salernitano, E.; De Riccardis, M. F.; Carbone, D., Fuller. Nanotub. Carbon Nanostruct. 2005, *13*, 383.
[103] Riccardis, M. F. D.; Carbone, D.; Makris, T. D.; Giorgi, R.; Lisi, N.; Salernitano, E., *Carbon* 2006, *44*, 671.
[104] Tzeng, S.-S.; Hung, K.-H.; Ko, T.-H., *Carbon* 2006, *44*, 859.
[105] Gong, Q. J.; Li, H. J.; Wang, X.; Fu, Q. G.; Wang, Z. W.; Li, K. Z., *Compos. Sci. Technol.* 2007, *67*, 2986.
[106] Qian, H.; Bismarck, A.; Greenhalgh, E. S.; Kalinka, G.; Shaffer, M. S. P., *Chem. Mater.* 2008, *20*, 1862.
[107] Qian, H.; Bismarck, A.; Greenhalgh, E. S.; Shaffer, M. S. P., *Compos. Pt. A-Appl. Sci. Manuf.* 2010*, 41*, 1107.
[108] Zhao, J. O.; Liu, L.; Guo, Q. G.; Shi, J. L.; Zhai, G. T.; Song, J. R.; Liu, Z. J., *Carbon* 2008, *46*, 380.
[109] Chen, J.; Xiong, X.; Xiao, P., *Mater. Chem. Phys.* 2009, *116* , 57.
[110] Hung, K. H.; Kuo, W. S.; Ko, T. H.; Tzeng, S. S.; Yan, C. F., *Compos. Pt. A-Appl. Sci. Manuf.* 2009, *40*, 1299.
[111] Sharma, S. P.; Lakkad, S. C., *Compos. Pt. A-Appl. Sci. Manuf.* 2011, *42*, 8.
[112] Sharma, S. P.; Lakkad, S. C., *Surf. Coat. Technol.* 2009, *203*, 1329.
[113] Wang, L. P.; Huang, Z. C.; Zhang, M. Y.; Huang, Q. Z., Fabrication of ACF/CNT composites by CVD method. In *Advances in Composites, Pts 1 and 2*, Bu, J. L.; Jiang, Z. Y.; Jiao, S., Eds. Trans Tech Publications Ltd: Stafa-Zurich, 2011, 150-151, 452.
[114] Agnihotri, P.; Basu, S.; Kar, K. K., *Carbon* 2011, *49* , 3098.
[115] Lomov, S. V.; Gorbatikh, L.; Kotanjac, Å. e.; Koissin, V.; Houlle, M.; Rochez, O.; Karahan, M.; Mezzo, L.; Verpoest, I., *Compos. Sci. Technol.* 2011, *71*, 315.

[116] An, F.; Lu, C.; Guo, J.; He, S.; Lu, H.; Yang, Y., *Appl. Surf. Sci.* 2011, *258*, 1069.
[117] Garcia, E. J.; Wardle, B. L.; John Hart, A.; Yamamoto, N., *Compos. Sci. Technol.* 2008, *68*, 2034.
[118] Yamamoto, N.; John Hart, A.; Garcia, E. J.; Wicks, S. S.; Duong, H. M.; Slocum, A. H.; Wardle, B. L., *Carbon* 2009, *47*, 551.
[119] He, D.; Bozlar, M.; Genestoux, M.; Bai, J., *Carbon* 2010, *48*, 1159.
[120] He, D.; Li, H.; Bai, J., *Carbon* 2011, *49*, 5359.
[121] He, D.; Li, H.; Li, W.; Haghi-Ashtiani, P.; Lejay, P.; Bai, J., *Carbon* 2011, *49*, 2273.
[122] Ci, L. J.; Zhao, Z. G.; Bai, J. B., *Carbon* 2005, *43*, 883.
[123] Zhang, Q.; Qian, W.; Xiang, R.; Yang, Z.; Luo, G.; Wang, Y.; Wei, F., *Mater. Chem. Phys.* 2008, *107*, 317.
[124] de Resende, V. G.; Antunes, E. F.; de Oliveira Lobo, A.; Oliveira, D. A. n. L.; Trava-Airoldi, V. J.; Corat, E. J., *Carbon* 2010, *48*, 3655.
[125] Delmas, M.; Pinault, M.; Patel, S.; Porterat, D.; Reynaud, C.; Mayne-L'Hermite, M., *Nanotechnology* 2012,*23*, 8.
[126] Lachman, N.; Carey, B. J.; Hashim, D. P.; Ajayan, P. M.; Wagner, H. D., *Compos. Sci. Technol.* 2012, *72*, 1711.
[127] An, F.; Lu, C.; Li, Y.; Guo, J.; Lu, X.; Lu, H.; He, S.; Yang, Y., *Mater. Des.* 2012, *33*, 197.
[128] Lv, P.; Feng, Y.-y.; Zhang, P.; Chen, H.-m.; Zhao, N.; Feng, W., *Carbon* 2011, *49*, 4665.
[129] Sager, R. J.; Klein, P. J.; Lagoudas, D. C.; Zhang, Q.; Liu, J.; Dai, L.; Baur, J. W., *Compos. Sci. Technol.* 2009, *69*, 898.
[130] Mathur, R. B.; Chatterjee, S.; Singh, B. P., *Compos. Sci. Technol.* 2008, *68*, 1608.
[131] Sonoyama, N.; Ohshita, M.; Nijubu, A.; Nishikawa, H.; Yanase, H.; Hayashi, J.-i.; Chiba, T., *Carbon* 2006, *44*, 1754.
[132] Zhao, Z.-G.; Ci, L.-J.; Cheng, H.-M.; Bai, J.-B., *Carbon* 2005, *43*, 663.
[133] Jones, F. R., *J. Adhes. Sci. Technol.* 2010, *24*, 171.
[134] Naito, K.; Yang, J.-M.; Xu, Y.; Kagawa, Y., *Carbon* 2010, *48*, 1849.

INDEX

A

acetone, 44, 113, 333, 334, 335
acetonitrile, 44, 100, 127, 264, 265, 305
acetophenone, 25, 105
ACF, 345
acidic, 10, 73, 75, 96, 100, 102, 108, 114, 115, 119, 128, 213, 248, 289, 326, 334
acrylate, 122, 129, 148
acrylic acid, 107, 116, 313
activated carbon, 326, 327
actuation, 291, 292
actuators, ix, 107, 157, 291, 292, 298
acylation, 11, 75
additives, 154, 208, 220, 221, 225, 226, 258
adhesion, 22, 24, 69, 165, 325, 326, 327, 328, 336
adhesive interaction, 328
adjustment, 7
adsorption, viii, 4, 48, 49, 59, 76, 78, 79, 80, 82, 98, 101, 108, 109, 114, 146, 150, 195, 196, 197, 200, 217, 219, 221, 224, 225, 226, 232, 242, 243, 245, 290
aerosols, 214
aerospace, 146, 319, 328, 332, 340
AFM, 34, 36, 43, 47, 49, 122, 248, 262
aggregation, 4, 6, 7, 10, 25, 81, 109, 248, 249, 306
aggregation process, 6
AIBN, 14, 67, 69, 122
Air Force, 342
alcohols, 17
aldehydes, 71
alkaline media, 289
alkenes, 20
alkylation, 60, 64, 65, 66, 75, 76, 104, 105
alters, 153
aluminum oxide, 205, 211, 216, 258
amidation, 62
amine(s), 16, 17, 18, 22, 41, 60, 62, 63, 64, 66, 69, 73, 78, 82, 99, 101, 102, 113, 115, 213, 220, 226, 232, 243, 244, 245, 250
amine group, 99, 244, 250
amino, 10, 14, 15, 16, 17, 44, 63, 70, 71, 77, 99, 102, 104, 106, 109, 111, 242, 244, 245, 276
amino acid(s), 14, 16, 70, 71, 99
amino groups, 10, 15, 16, 244
ammonia, 14, 21, 104, 109, 273, 297, 321, 326
ammonium, 70, 81, 109, 113, 115, 117, 257, 326, 327
amplitude, 36, 46, 47, 177
anchoring, 10, 67, 72, 109, 118, 125, 126, 246
aniline, 67, 101, 109, 122, 123, 160, 280, 293, 294
anisotropic particles, ix, 207, 208, 209, 211, 215, 220, 221, 224, 226, 227, 229
anisotropy, 146, 206, 207, 212, 217, 225, 226, 324
annealing, 4, 6, 13, 15, 58, 65, 66, 68, 77, 83, 111, 213, 229, 298
annihilation, 190
anodization, 211
antibody, 104, 115, 117, 227
anticancer drug, 115
antigen, 25, 116
aqueous solutions, 58, 77, 108, 214, 260
aqueous suspension, 80, 84, 307
arginine, 115
argon, 15, 34, 326
aromatic compounds, 59, 77, 79, 82
aromatic rings, 10, 21, 79
asbestos, 114
atmosphere, 7, 9, 13, 14, 15, 121, 122, 163, 214, 219, 220, 226, 228, 229, 321, 330
atmospheric pressure, 14
atomic force, 34, 36, 248
atomic force microscope, 34, 36
atoms, ix, 14, 57, 58, 59, 65, 66, 99, 173, 174, 176, 179, 180, 181, 183, 186, 187, 190, 192, 195, 199, 209, 241, 250, 262, 305, 326, 327

ATRP, 20, 63, 73, 118, 119, 121, 122, 124, 125, 127, 128, 129, 247, 249, 250
attachment, 25, 41, 42, 43, 44, 45, 59, 65, 71, 96, 98, 104, 113, 116, 117, 126, 150, 244, 246, 247
Au nanoparticles, 100, 242, 247, 249
automobiles, 322, 323, 331
azodiisobutyronitrile, 69
Azomethine ylide, 70, 99

B

bacteria, 22, 24
bacteriophage, 218
band gap, 33, 59, 60, 66, 73, 95, 110, 290
barriers, 59
basal layer, 146
base, 47, 73, 108, 269, 322, 333
basicity, 222, 224
batteries, 273, 287
behaviors, 197, 243, 332, 340
bending, 8, 177, 178, 179, 292, 324
benign, x, 305, 313
benzene, 16, 19, 24, 75, 77, 122, 148, 214, 247, 293, 330, 333
benzoyl peroxide, 21, 60, 66, 67, 68, 69
bias, 34, 38, 176, 180, 183, 184, 185, 186, 187, 190, 191, 192
bicarbonate, 326
bile, 81
binding energy, 101, 230
biocompatibility, 3, 33, 63, 83, 227, 242, 257
biodegradability, 115, 227
biodegradation, 98, 115
biological systems, 3, 217
biomedical applications, 3, 63, 80, 83, 103, 130, 227, 231, 298
biomolecules, 4, 14, 16, 49
biopolymer, 228
biopolymers, 10
biosafety, 242
biosensors, 34, 49, 63, 71, 82
biotechnological applications, 241
biotin, 15, 17, 63, 227, 276
bisphenol, 148
blood, 115, 290
BMI, 322
Boltzmann constant, 39
bonding, 12, 43, 44, 47, 76, 77, 79, 80, 101, 105, 116, 145, 146, 148, 198, 248, 267, 321, 326, 328, 330, 343
bonds, 4, 12, 13, 14, 16, 17, 20, 21, 23, 47, 60, 63, 83, 98, 104, 105, 119, 125, 126, 246, 250, 283
bottom-up, 34, 49, 207, 208
breakdown, 157, 184
bromination, 14, 17, 98
bromine, 14, 73, 98, 164
building blocks, 230
bulk materials, 206, 241, 328
butadiene, 74, 99
butyl methacrylate, 20, 129, 244

C

calibration, 272
cancer, 3, 18, 44, 78, 80, 104, 115, 116, 117, 130
cancer cells, 78, 80, 115, 116, 117
candidates, 79, 206, 217, 225, 227, 288, 328
capillary, 7, 244, 322
caprolactam, 151
capsule, 311, 315
Carbenes, 99
carbides, 6
carbohydrate, 23, 77
carbon atoms, 14, 57, 58, 59, 60, 61, 65, 73, 76, 83, 102, 146
carbon dioxide, 50, 68, 321
carbon materials, 121, 328
carbon monoxide, 321, 333, 334
carbon nanotubes, v, vi, vii, viii, ix, 4, 8, 22, 39, 57, 67, 70, 83, 95, 100, 101, 107, 109, 111, 113, 114, 115, 117, 119, 122, 123, 130, 142, 145, 147, 214, 217, 250, 251, 279, 281, 284, 295, 296, 297, 298, 299, 320, 322, 328, 343, 344
carbonization, 320, 321, 325
carbonyl groups, 8, 10
carboxyl, 4, 6, 8, 10, 16, 18, 60, 61, 62, 63, 71, 248, 282
carboxylic acid, 9, 10, 16, 17, 60, 61, 62, 63, 65, 68, 70, 73, 75, 82, 119, 120, 151, 152, 226, 227, 232, 245
carboxylic acids, 61, 68, 70, 120, 226, 232
carboxylic groups, 70, 98, 107
carboxymethyl cellulose, 155
carcinoma, 18
casting, 273, 290
catalysis, 99, 232, 249, 250, 313
catalyst, x, 58, 72, 76, 127, 208, 214, 286, 303, 304, 313, 314, 315, 316, 317, 319. 330, 331, 332, 333, 334, 340
catalytic activity, 71, 76, 124, 250
catalytic hydrogenation, 123
catalytic system, 314
catastrophic failure, 324
cathode materials, 293
C-C, 8, 20, 21, 128, 241, 283, 314
cell surface, 22, 78

cellulose, 109, 125, 129, 155, 294, 309, 316
cellulose xanthate, 109
ceramic, 6, 332
chain propagation, 21
chain scission, 321
chain transfer, 119
chains conformation, 249
challenges, 25, 173, 242
chelates, 115
chemical bonds, 157
chemical deposition, 211
chemical etching, 228
chemical interaction, 282, 325
chemical pretreatments, 338
chemical properties, 60, 81, 241, 279, 281
chemical reactions, x, 17, 20, 96, 98, 327, 330, 331
chemical reactivity, 16, 59, 68
chemical stability, 3, 33, 34, 44, 47, 319, 320
chemical structures, viii
chemical vapor deposition, 5, 35, 57, 95, 211
chemicals, 6, 10, 325
chemisorption, 5
China, 139
chirality, 110, 111, 145, 328
chitosan, 63, 111, 115
chloride anion, 310
chlorination, 6, 14, 66, 98
chlorine, 14, 16, 65, 66
chloroform, 14, 16, 120
cholesterol, 246, 290, 298
chromatograms, 245
chromatography, ix, 245
classes, 222
cleaning, 6, 38, 44, 334
cleavage, 17, 80, 101, 102
clinical application, 247
closure, 73
clusters, 61, 217, 315, 338
CMC, 155, 156, 198
C-N, 104
CNT dispersion, 153, 154, 162, 165
CNT networks, 340
CNT surface chemistry, 152
CO2, 7, 98, 326
coatings, 3, 130, 229, 281, 292, 298, 311
cobalt, ix, 206, 208, 214, 218, 224, 225, 230, 231, 334
color, 4, 45, 309
commercial, 82, 221, 228, 242, 281, 292, 303, 320, 326, 332, 338
compatibility, 158, 242
complementary DNA, 44, 45, 46, 47, 77
complexity, 189
composite, 322, 323, 338, 341
composite mechanical properties, 340
composition, x, 4, 57, 61, 111, 211, 218, 224, 232, 242, 243, 250, 276, 304, 309, 322
compounds, 7, 23, 60, 66, 68, 69, 76, 78, 79, 83, 102, 104, 106, 195, 218, 219, 221, 222, 250, 332
compression, 324
condensation, 14, 17, 60, 62, 70
conductance, ix, 62, 66, 102, 110, 114, 155, 162, 173, 174, 175, 176, 177, 178, 179, 180, 181, 182, 183, 184, 185, 186, 187, 188, 189, 190, 191, 192, 193, 194, 195, 196, 197, 198, 199, 200, 260
conducting polymer composites, 296
conduction, 74, 160, 166, 176, 192, 193, 195, 280, 313, 325
conductivity, x, 15, 33, 58, 79, 95, 97, 112, 145, 153, 154, 155, 157, 162, 165, 166, 167, 173, 174, 257, 258, 261, 262, 273, 279, 281, 285, 288, 289, 290, 298, 304, 309, 312, 320, 322, 325, 328, 339
conductor, ix, 173, 174, 190, 195, 206, 209
conductors, 173, 174, 280
configuration, 39, 41, 60, 145, 177, 185, 186, 191, 194
confinement, 207
conjugation, viii, 18, 22, 59, 60, 71, 77, 83
connectivity, 166
construction, 130, 319, 322
contamination, 6
COOH, 8, 9, 12, 16, 21, 98, 101, 103, 104, 105, 110, 114, 115, 116, 117, 120, 124, 128, 130, 151, 152, 243, 282, 284, 289, 290, 292
cooling, 37, 58, 323
coordination, 209, 226
copolymer, 69, 81, 82, 111, 119, 120, 125, 126, 213, 216, 305, 316, 321
copolymerization, 124
copolymers, 71, 81, 107, 119, 120
copper, 22, 24, 112, 113, 121, 123, 179, 185, 186, 187, 188, 197, 198, 199
corrosion, 281, 292, 322
cost, 4, 6, 146, 190, 211, 288, 303, 320, 321, 325
cotton, 100, 125, 309, 316
Coulomb interaction, 108
counterbalance, 17
covalent bond, 5, 6, 9, 47, 67, 83, 118, 125, 150, 151
covalent bonding, 47, 83, 118
Covalent modification, 25, 109, 110
covering, 192, 273
CPP, 150, 151
creep, 320
crossing over, 249
crown, 101
crystal structure, 79, 230

crystalline, 6, 35, 57, 73, 148, 210, 226, 260, 285, 310, 321, 322, 323
crystallinity, 325
crystallites, 4, 6, 15, 320, 321
crystals, 26
CTA, 45
CuAAC, 123, 124
cubic system, 219
curing process, 243
Curtius rearrangement, 62, 63
CVD, 5, 35, 50, 57, 66, 67, 95, 96, 112, 205, 228, 329, 330, 331, 332, 333, 334, 335, 337, 338, 339, 340, 345
cyanide, 15, 40
cycles, 10, 44, 46, 47, 106, 175, 191, 265, 270, 271, 288, 289, 326
cycling, 191, 288
Cycloaddition, 70, 71, 72, 73, 74, 99, 123, 124, 127
cyclodextrins, 198
cyclohexanone, 25
cyclopentadiene, 74, 333
Cyclopropanation, 73
cysteine, 49, 246
cytochrome, viii, 4, 34, 44, 245
cytoplasm, 116
cytoskeleton, 217
cytotoxicity, 115, 116

D

damping, 339
decomposition, 21, 66, 68, 69, 119, 125, 128, 129, 213, 221, 229, 293
deconvolution, 283
decoration, 74, 286, 315
defect site, 59, 75, 105, 297
defects, 39, 58, 59, 77, 83, 102, 105, 109, 164, 174, 195
deformation, 80, 98
degradation, 47, 69, 98, 102, 326
dehydration, 225
denaturation, 46, 47
dendritic cell, 116
deoxyribonucleic acid, 71
deposition rate, 181, 184
deposits, 267
derivatives, 15, 16, 17, 19, 24, 25, 57, 60, 62, 64, 66, 71, 72, 73, 75, 76, 77, 78, 79, 80, 99, 102, 111, 294, 315
desorption, 290
destruction, 10, 66, 308
detectable, 227, 313
detection, viii, ix, 34, 43, 44, 45, 46, 75, 78, 113, 114, 195, 246, 247, 248, 250, 272, 273, 290, 293, 298, 315
detonation, 5, 6, 7, 8, 12, 15, 25
deviation, ix, 224, 288
DFT, 101
diamines, 14
diamonds, vii, 15
Diazonium salt, 34, 100, 122, 123
dielectric constant, 39
dielectric permittivity, 146, 157, 158, 159, 161, 162, 165, 167
Dielectric permittivity, 160
dielectrics, vii, 167
Diels-Alder reaction, 21, 74
dienes, 21, 74
diffraction, x, 304, 306, 307
diffusion, 40, 41, 45, 48, 78, 175, 180, 181, 199, 227, 269, 289, 290, 331, 332
diffusivity, 339
diisocyanates, 62
dimethacrylate, 244, 245, 246, 247
dimethylformamide, 66
dimethylsulfoxide, 15
dipoles, 49, 60, 83
direct observation, 115
discordance, 166
discrimination, 47
diseases, 3
disorder, 163
dispersion, 6, 17, 62, 65, 70, 77, 78, 79, 82, 83, 99, 101, 105, 106, 107, 108, 109, 110, 111, 115, 123, 128, 146, 148, 149, 153, 154, 155, 156, 157, 159, 161, 162, 165, 167, 242, 273, 281, 282, 290, 294, 306, 307, 315, 317, 329
displacement, 15, 16, 178, 194, 231
dissociation, 65, 128
distribution, 6, 16, 38, 73, 166, 197, 198, 208, 218, 242, 250, 261, 283, 306, 307, 308, 312, 323, 331, 338, 340
diversity, 208
DMF, 101, 108, 119, 262, 263
DNA, viii, 16, 25, 34, 44, 45, 46, 47, 71, 76, 77, 78, 80, 106, 109, 115, 155, 217, 218, 272
DNase, 80
DOI, 134, 135
domain structure, 194
dopamine, 19, 24, 43, 63, 114, 195, 196, 246
dopants, 289
doping, 33, 39, 41, 42, 66, 79, 80, 258, 267, 272, 280, 288, 290, 295, 311
double bonds, 65, 73, 75, 119, 164
double helix, 80

drawing, 181, 183, 186, 285
drug delivery, vii, 3, 4, 25, 63, 126, 129
drug release, 115, 317
drugs, 5, 18, 25, 76, 80, 246
drying, 7, 326
DSM, 163
ductility, 323
DWNT, 99
dyes, 23, 24, 78

E

elaboration, 206, 207, 208, 211, 218, 225, 230, 232, 248, 251
electric field, 43, 157
electrical conductivity, 60, 65, 73, 74, 79, 82, 83, 146, 153, 155, 162, 167, 261, 280, 281, 284, 298. 313, 325, 338
electrical properties, 66, 73, 153, 290, 293, 295, 328
electrical resistance, 320
electrocatalyst, 315
electrochemical behavior, 295
electrochemical deposition, 175, 186, 265, 285
electrochemical impedance, 45, 46
electrochemistry, viii, 33, 47, 49, 50, 67, 73, 98, 100, 103, 187, 190, 198, 257, 263, 273
electrode surface, 47, 181, 247, 257, 269
electrodeposition, 181, 211, 212, 213, 214, 227, 228, 231, 268, 269, 274, 281, 294
electrodes, 33, 34, 40, 41, 43, 44, 47, 112, 113, 114, 130, 157, 176, 177, 178, 179, 180, 183, 184, 186, 190, 191, 193, 195, 196, 197, 247, 265, 266, 273, 274, 296, 297, 299, 315
electroless deposition, 213, 217, 218
electrolysis, 293
electrolyte, 15, 33, 44, 181, 183, 186, 191, 196, 269, 273, 275, 284, 288, 289, 295, 326
electromagnetic, ix, 152, 178, 292, 295, 298
electron, x, 6, 20, 24, 34, 42, 43, 44, 47, 48, 49, 58, 59, 61, 65, 66, 68, 72, 73, 74, 77, 79, 99, 103, 105, 106, 114, 115, 148, 155, 163, 166, 173, 174, 177, 184, 195, 206, 214, 243, 280, 281, 282, 284, 290, 304
electron diffraction, 214
electron microscopy, 24, 61
electronic structure, 111, 167, 176
electrons, vii, 15, 49, 64, 68, 100, 104, 159, 166, 173, 174, 190, 193, 195, 290, 311, 313, 326
electrophoresis, 329, 338
electroreduction, 199, 290
electrospinning, 262, 263
ellipsometry, vii
elongation, 176
emission, 17, 43, 78, 279, 313
emulsions, 326
encapsulation, 315
endothelial cells, 228
endurance, 191, 324
energy, ix, 6, 15, 33, 58, 59, 68, 71, 74, 96, 101, 108, 157, 161, 176, 177, 226, 230, 272, 273, 279, 288, 292, 296, 326, 328
energy density, 108, 161, 288
energy transfer, 71, 326
engineering, viii, 74, 146, 151, 153, 156, 162, 165, 167
entanglements, 146
entrapment, 47, 49
environment, 12, 15, 26, 115, 174, 178, 192, 280, 304, 322
environmental conditions, 319
environmental factors, 325
environmental stimuli, 250
environments, 33, 47, 174
enzyme, 16, 34, 114, 228
EPC, 216
epitaxial growth, 57, 231
epithelial cells, 115
Epoxide, 64, 65
epoxy groups, 64
ESR, 282
ester, 10, 16, 77, 78, 117, 148, 158, 221
estrogen, 115
etching, 34, 35, 36, 37, 38, 39, 40, 49, 175, 181, 182, 183, 184, 186, 326
ethanol, 35, 113, 282, 316, 333, 335
etherification, 65
ethers, 63, 101
ethyl acetate, 76
ethylene, 101, 105, 113, 115, 124, 129, 222, 228, 243, 244, 245, 246, 247, 262, 332, 334, 335
ethylene glycol, 105, 115, 129, 222, 243, 247
ethylene oxide, 262
eukaryotic, 23, 217
eukaryotic cell, 23, 217
evaporation, 58, 180
evolution, 50, 103, 189, 316
excitation, 16, 17, 80
exclusion, 25, 101
excretion, 115
experimental condition, 25, 40, 41, 98, 176, 212, 214, 217, 263, 266, 268
exploitation, 60, 95
explosives, 6
exposure, 19, 45, 47, 65, 100, 210, 273, 290, 291
extraction, 247, 308, 311, 312
extrusion, 322, 323

F

fabrication, x, 5, 34, 39, 41, 67, 73, 173, 175, 176, 178, 179, 181, 185, 186, 195, 200, 248, 249, 259, 295, 296, 298, 299, 305, 320, 321, 345
Fermi level, 60, 74
ferromagnetic, ix, 68, 194, 206, 207, 208, 209, 210, 211, 212, 213, 214, 217, 218, 221, 222, 224, 225, 226, 227, 286
ferromagnetism, ix
filament, 177, 333
fillers, 152, 154, 157, 158, 325
film thickness, 297
films, vii, ix, 14, 33, 34, 35, 43, 68, 69, 82, 99, 112, 154, 155, 156, 180, 195, 246, 247, 258, 263, 266, 269, 272, 284, 292, 294, 295, 297, 298, 299, 303, 315, 316, 330
filtration, 100, 313
financial, 26, 233
financial support, 26, 233
fixation, 22
flame, 210
flaws, 331
flexibility, 26, 58, 147, 152, 303
flocculation, 307
fluctuations, 184, 185, 197, 200
fluid, 213, 313
fluorescence, 4, 5, 16, 26, 78, 104, 106, 115
Fluorination, 65, 66, 98
fluorine, 13, 14, 65, 66, 98, 163, 283
fluorine atoms, 14, 66
folic acid, 116
food, 113
force, 45, 48, 80, 109, 146, 164, 166, 265
Ford, 142, 143
formula, 192, 193, 222
fouling, 33
fracture stress, 324
fracture toughness, 339
fractures, 336
France, 3, 26, 57, 95, 131, 145, 173, 205, 241, 257, 319
free energy, 306, 327
free radical copolymerization, 243
free radicals, 68, 104
Friedel-Crafts acylation, 75
fruits, 113
FTIR, 8, 9, 10, 11, 12, 13, 14, 19, 20, 282
FTIR spectroscopy, 10
fuel cell, vii, 50, 287
fullerene, vii, 15, 23, 83, 145, 214
Functionalization, v, 16, 17, 20, 21, 41, 58, 59, 60, 63, 67, 68, 70, 72, 73, 74, 76, 83, 97, 205, 225

G

gas sensors, 113, 273, 297
GCE, 49
gel, 247, 298
gelation, 77
gemini surfactants, 108
gene therapy, 3
geometry, 59, 101, 166, 211, 340, 341
Germany, 33, 239
glass transition temperature, 321
glasses, vii
glucose, 43, 75, 78, 199, 228, 290, 296
glucose oxidase, 78, 228, 290
glue, 77
glycans, 23
glycerol, 120, 123, 124, 224
glycol, 70, 109, 112, 116, 117, 127, 283
glycoproteins, 25
GNP, ix, 242, 247, 248
Gold, vii, 208, 209, 231, 241, 242, 243, 246, 249, 250, 251, 315, 316
gold nanoparticles, vii, ix, 21, 67, 70, 71, 73, 74, 77, 99, 106, 242, 243, 245, 246, 247, 248, 249, 250, 315
grain boundaries, 68
grapheme oxide, viii
graphene, v, 57, 58, 59, 60, 61, 65, 69, 71, 74, 76, 79, 228
graphene sheet, 58, 59, 60, 61, 63, 64, 66, 68, 73, 74, 75, 76, 77, 78, 79, 80, 81, 82, 83, 95, 145, 228, 279
graphite, 6, 15, 57, 58, 59, 60, 61, 62, 63, 66, 67, 68, 69, 71, 73, 75, 76, 77, 78, 79, 80, 82, 83, 84, 95, 106, 145, 320, 330, 343
gravity, 314
grids, 288
growth, ix, 33, 38, 41, 58, 73, 112, 122, 127, 186, 187, 191, 207, 208, 209, 210, 211, 214, 217, 218, 219, 220, 222, 225, 226, 227, 228, 232, 247, 258, 264, 266, 269, 293, 324, 327, 330, 331, 332, 333, 335, 338
growth mechanism, 266, 269
growth modes, 324
growth rate, 332
growth time, 228, 331
guanine, 272

H

hafnium, 112
halogen, 16, 59, 65, 294

halogenation, 13,14, 60, 65,66, 98
hardener, 327
hardness, 44, 339
harmonization, 10
healing, vii, 77
heat transfer, 109
heating rate, 229
height, 37, 38, 39, 43, 45, 268
helical conformation, 111
helicity, 96
helium, 177
heme, 47, 48
heterogeneity, 61
hexane, 333
high aspect ratio, viii, 147, 157, 328
high strength, 95, 279, 319, 320, 321, 327
histogram, 176, 177, 180, 181, 188, 189, 190, 197, 198, 261
homogeneity, 150, 324
homolytic, 103
hormones, 25
host, 146, 165, 167, 212, 215, 227, 311
hot pressing, 323
House, 315
HRTEM, 122, 209, 282, 285
human, 25, 113, 117, 281, 292, 320
human body, 292
Hunter, 85, 277
hybrid, x, 62, 69, 73, 106, 113, 114, 120, 123, 124, 126, 214, 217, 230, 231, 232, 243, 249, 288, 293, 314, 319, 320, 325, 329, 331, 332, 338, 339, 340
hybridization, viii, 34, 44, 45, 46, 47, 76, 156, 303, 309, 320
hydrazine, 66, 105
hydrocarbons, 330, 332
hydrofluoric acid, 66
hydrogels, 80, 241
hydrogen, 7, 9, 12, 13, 14, 33, 34, 38, 39, 50, 59, 60, 62, 65, 66, 75, 76, 77, 79, 80, 116, 148, 149, 189, 198, 219, 248, 267, 273, 274, 282, 298, 305, 316, 321, 322, 326
hydrogen atoms, 14, 59, 60, 65
hydrogen bonds, 148, 282, 326
hydrogen cyanide, 321
hydrogen gas, 12, 273, 274
hydrogen peroxide, 75, 298, 326
hydrogenation, 12, 13, 20, 60, 65, 224, 241, 315, 316
hydrolysis, 10, 20, 76, 222
hydrophilicity, 63, 283
hydrophobicity, 58
hydroxide, 327
hydroxyl, 4, 6, 8, 9, 10, 11, 14, 17, 18, 19, 22, 24, 33, 60, 61, 62, 63, 65, 73, 82, 98, 153, 245, 262, 266, 267, 268, 269, 271, 326
hydroxyl groups, 4, 8, 10, 17, 19, 24, 60, 61, 62, 63
hysteresis, 212

I

illumination, 17, 35
immobilization, 4, 5, 10, 16, 25, 44, 45, 69, 113, 122, 124, 244, 245, 246, 248, 249, 272
immune response, 116
immunity, 178
impregnation, 331, 333, 334, 337, 339, 340
improvements, 157, 290, 337, 339
impurities, 6, 8, 98, 164
in vitro, 115, 116
in vivo, 18, 25, 115, 116
incompatibility, 281, 284
indium, 114, 211, 247, 290, 296
induction, 37, 205
industries, 340
infrared spectroscopy, 8, 309
inhomogeneity, 4
initiation, 244, 280, 339
injections, 18
injury, 269
insulation, 181
insulin, 4
integrity, 325
interface, vi,viii, ix, x, 20, 30, 33, 39, 69, 120, 130, 131, 137, 138, 139, 152, 208, 242, 243, 244, 250, 258, 259, 267, 268, 269, 292, 294, 315, 316, 319, 324, 325, 327, 328, 329, 336, 337, 338, 340, 342
interfacial adhesion, 65, 148, 152, 325, 326, 327, 328, 332, 337, 340
interfacial bonding, 325, 328
interference, 152, 292, 295, 298, 313
intermolecular interactions, 148
interphase, 293, 325, 327, 332, 336
Interphase, 319
ionic conduction, 184, 186
ions, 35, 38, 75, 183, 190, 196, 209, 213, 225, 249, 266, 268, 288, 326
IRI, 3
iron, vii, ix, 100, 206, 214, 219, 225, 227, 257, 260, 281, 284, 294, 315, 334, 335
irradiation, 4, 6, 11, 15, 21, 63, 65, 66, 69, 72, 73, 80, 98, 99, 116, 230
Islam, 137
islands, 79, 194
Isocyanate, 18, 61
isolation, 67, 105

J

Japan, 52, 303, 315

K

ketones, 10
kinetics, 47, 66, 103, 105, 221
KOH, 6

L

labeling, 115
lactic acid, 120, 129
lactose, 114
laminar, 324
lamination, 323
laser ablation, 5, 95
lead, 12, 13, 14, 43, 57, 98, 103, 146, 160, 219, 221, 226, 232, 258, 261, 275, 289
leakage, 157, 161, 184
ligand, 23, 72, 78, 113, 221, 222, 225, 226
light, 80, 106, 241, 258, 319, 320, 322, 328
lipids, 98
liquid crystals, 113
liquid phase, 14, 57
liquid-phase processes, ix, 207, 208, 225
liquids, 16, 33, 66, 67, 77, 108
lithium, 10, 66, 81, 104, 105, 121
lithography, 180
liver, 18
low temperatures, 176
lung cancer, 19
Luo, 51, 53, 85, 138, 143, 201, 202, 233, 234, 235, 345
lysine, 49, 117
lysozyme, 4, 106

M

macromolecules, 147, 148, 152, 162, 167
macroradicals, 69, 119
magnesium, 122
magnetic composites, 225, 226, 232
magnetic field, 194, 210, 212, 217, 220, 227
magnetic moment, 209
magnetic particles, 206, 211, 222, 230
magnetic properties, 207, 214, 215, 217, 219, 221, 224, 229
magnetism, 68, 206
magnetization, 121, 174, 206, 207, 212, 218, 224, 225, 226
magnetizations, 227
magnitude, 66, 79, 81, 110, 114, 147, 154, 155, 165, 175, 212, 260, 280, 312
manipulation, 60, 151, 174
manufacturing, 146, 155, 323, 325
mapping, 68
market share, 321
MAS, 213
masking, 35
mass, 6, 128, 129, 153, 244, 275, 284, 288, 289, 309, 326, 340
material surface, 304
materials science, 58
matrix(s), 69, 124, 147, 148, 149, 150, 152, 153, 157, 158, 160, 162, 164, 212, 213, 214, 226, 227, 229, 242, 289, 311, 319, 320, 321, 322, 323, 324, 325, 326, 327, 328, 329, 331, 332, 336, 337, 338, 339, 340
matter, 10, 96, 303
measurements, 39, 40, 46, 66, 79, 102, 167, 174, 181, 189, 198, 249, 284, 310
mechanical properties, x, 69, 82, 146, 147, 148, 154, 156, 173, 232, 241, 242, 279, 281, 304, 323, 324, 325, 326, 328, 329, 338, 339
media, x, 4, 7, 83, 119, 197, 198, 199, 217, 225, 248, 249, 304, 305, 306, 309, 313, 314, 316
medical, x, 3, 227, 292
medicine, vii, 281
melt(s), 162, 164, 165, 167, 321
melting, 321
melting temperature, 321
membranes, 211, 218
memory, 288
metal complexes, 331
metal ion(s), 183, 186, 242, 250, 305
metal nanoparticles, ix, 73, 208, 243, 304, 305, 310, 313
metal oxides, 208, 211, 288, 331
metal salts, 243, 304, 305
metals, vii, 100, 102, 103, 176, 187, 189, 192, 193, 208, 212, 214, 222, 246, 294, 313, 314, 324, 328
meter, 339
methacrylic acid, 20, 246, 247
methanol, 113, 261, 286, 290, 295, 296, 315
methodology, 304
methyl methacrylate, 71, 125, 129
methylene blue, 249
mice, 18, 19, 25, 115
microcrystalline, 69
microelectronics, 43
microemulsion, 161

micrometer, 147, 228, 306, 307, 317, 320, 325, 328, 331, 332
microscope, vii, 34, 116, 163, 180, 185, 311
microscopy, 46, 78, 79, 116, 175, 179, 248, 306, 311, 336
microspheres, 315
microstructure(s), 160, 207, 321, 322
migration, 157
miniaturization, 46, 47, 173
Ministry of Education, 293, 315
MIP, 246, 247
mixing, 62, 71, 101, 124, 160, 162, 163, 164, 165, 242, 329, 340
MMA, 122, 129
MMS, 315
model system, 188
models, 25, 102
modern airplanes, x
modifications, viii, x, 43, 47, 60, 96, 100, 106, 115, 154
modulus, 145, 146, 147, 148, 149, 150, 151, 319, 320, 321, 323, 324, 325, 328, 338, 339
moisture, 325
molecular dynamics, 149, 228
molecular orientation, 322
molecular oxygen, 71
molecular weight, 109, 148, 155, 246, 258, 327
molecules, viii, 3, 4, 10, 14, 16, 17, 25, 41, 42, 44, 45, 59, 60, 62, 63, 67, 69, 71, 76, 79, 83, 96, 109, 113, 114, 146, 154, 163, 195, 196, 200, 227, 243, 244, 245, 246, 249, 273, 290, 311, 326
molybdenum, 296
monohydrogen, 258
monolayer, 41, 49, 63, 67, 70, 79, 80
monomers, 21, 121, 151, 198, 221, 242, 243, 244, 257, 259, 260, 262, 263, 265, 267, 268, 269, 273, 276, 284, 327
morphology, x, 36, 158, 165, 211, 219, 224, 225, 226, 227, 229, 232, 243, 249, 257, 261, 263, 265, 266, 281, 289, 290, 295, 304, 309, 311
mosaic, 218
Moses, 31
motif, 23, 78
mucin, 117
Multifunctionality, 319
multiples, 176, 189, 195
multiwalled carbon nanotubes, 115, 295, 296, 297, 298
muscles, 157
myoglobin, 245

N

Na^+, 104
NaCl, 44, 266, 291, 310
Nano/micro hybrids, x
nanobelts, 225
nanocomposites, v, vi, viii, x, 3, 77, 104, 117, 119, 120, 124, 130, 128, 145, 146, 148, 158, 230, 241, 243, 248, 250, 279, 281, 289, 293, 294, 295, 297, 299, 303, 304, 305, 307, 328
nanodevices, 59
nanoelectronics, 58, 59, 83, 96, 279
nanofibers, 210, 261, 262, 263, 265, 267, 271, 272, 273, 274, 298, 315, 344
nanohorns, 70
nanomaterials, iii, ix, 3, 4, 9, 57, 115, 169, 206, 211, 222, 230, 237, 319, 320, 325, 328, 329, 340
nanomedicine, vii, 70
nanometer(s), viii, 146, 173, 175, 211, 243, 262, 279, 332
nanometer scale, viii, 173, 332
nanoobjects, vii, viii, ix, x
nanoprecise coating, x, 304
nanoribbons, 58, 61, 67, 82
nanorods, 70, 72, 207, 218, 219, 220, 224, 225, 226, 280
nanostructured materials, 258
nanostructures, v, vi, viii, 4, 33, 34, 35, 38, 39, 43, 49, 50, 70, 206, 207, 208, 209, 210, 217, 218, 235, 249, 255, 258, 260, 261, 262, 265, 266, 269, 270, 275, 276
nanotechnology, vii, 200
nanotube, vii, 96, 97, 99, 107, 109, 115, 120, 122, 128, 129, 145, 146, 148, 150, 151, 153, 156, 158, 159, 164, 165, 166, 205, 216, 250, 293, 294, 295, 296, 297, 298, 330
naphthalene, 77, 330
NAS, 126, 129
natural advantages, 147
natural gas, 333
neutral, 33, 80, 81, 100, 107, 197, 258, 280, 289, 326
NH2, 15, 19, 21, 101, 105, 112, 113, 116, 123, 130, 243, 244, 272
NHS, 16, 20, 62, 77, 117
nickel, ix, 27, 34, 35, 37, 38, 206, 213, 214, 334
niobium, 49
NIR, 116
Nitrenes, 72, 125
nitrides, 35
nitrobenzene, 44
nitrogen, 4, 8, 12, 15, 35, 44, 66, 69, 72, 74, 226, 250, 273, 310, 326, 328
nitrogen dioxide, 273

nitrogen gas, 66, 69, 74
nitroxide, 63, 119
NMP, 119, 120, 128
NMR, 213
N-N, 102
Nobel Prize, 57, 280
noble metals, 179, 224, 293
Non-covalent modification, 106
non-polar, 325
norepinephrine, 83
nucleating agent, 222
nucleation, 208, 221, 222, 223, 231, 232, 261, 331, 332
nuclei, 18, 221, 307, 311
nucleic acid, 47, 63
nucleophiles, 65
nucleus, 116, 208

O

ODN nitrogenous nucleobases, 198
oil, 100
oleic acid, 220
oligomers, 21, 267, 280, 321
one dimension, 328
opportunities, 58, 76
optical microscopy, 306, 308
optical properties, 242, 272
optimization, 309, 332
optoelectronic properties, 80
optoelectronics, 110
organ(s), 25, 150, 218, 219, 220, 221, 225, 229, 231, 261, 295, 332
organic polymers, 62, 211, 241, 257, 294
organic solvents, 16, 17, 60, 77, 82, 104, 128, 229, 248, 257, 305
oxidation, ix, 9, 10, 12, 14, 34, 40, 41, 58, 60, 61, 68, 71, 72, 73, 77, 98, 103, 104, 106, 110, 113, 118, 119, 120, 208, 214, 225, 227, 229, 230, 232, 247, 257, 258, 259, 261, 262, 267, 268, 269, 271, 272, 280, 286, 294, 295, 296, 310, 311, 315, 316, 319, 320, 322, 326, 333
oxidation rate, 268
oxide nanoparticles, 206
oxygen, 4, 10, 12, 13, 15, 33, 38, 39, 49, 50, 58, 59, 60, 61, 68, 80, 83, 98, 262, 283, 290, 315, 326, 327
oxygen plasma, 39
ozone, 9, 10, 262

P

PAA, 124, 125, 153, 212
paclitaxel, 18
pairing, 104
palladium, 305, 315, 316, 317
PAN, 319, 320, 321, 322, 339
parallel, 106, 209, 212, 216, 224, 320, 321
Pareto, 30
pathways, 115, 118, 230, 241, 242, 243
PCA, 77
peptide(s), 114, 116, 218, 245, 246
perchlorate, 268, 269, 271
percolation, 153, 155, 157, 158, 159, 160, 162, 165, 166, 313
percolation theory, 158
periodicity, 36, 37
permit, vii, viii, ix, x, 105, 106, 107, 108, 109, 110, 111, 114, 117, 118, 120, 127, 232
permittivity, 157, 158, 159, 160, 162, 165, 166, 167, 222
peroxide, 67, 68, 69
Persistent Organic Pollutants (POPs), 114
perylene, 76, 79
petroleum, 321
pH, 7, 33, 39, 40, 44, 45, 46, 47, 48, 49, 78, 100, 102, 104, 108, 110, 114, 115, 124, 212, 215, 216, 247, 248, 250, 258, 261, 266, 267, 268, 269, 271, 272, 276, 291
pharmaceutical, 114, 290
pharmacology, 3
phenol, 61, 71, 124
phenylalanine, 73
pheochromocytoma, 114
Philadelphia, 343
phosphate, 33, 39, 40, 44, 45, 46, 48, 49, 265, 267, 291
phosphates, 222
phosphorus, 75
photodynamic therapy (PDT), 80
photoelectron spectroscopy, x, 248, 304
photographs, 17
photoluminescence, 95
photolysis, 69
photopolymerization, 247, 316
physical interaction, 162, 248
physical properties, 33, 206
physicochemical properties, 222, 230
physics, 58, 280
pi-stacking (π-stacking), 76
pitch, 319, 320, 321, 322, 325, 339
plasma membrane, 115
plastics, 294

platelets, 226
platform, 49, 50, 78, 80, 83, 112, 124, 246
platinum, 220, 266, 280, 315, 317
PMMA, 122, 124, 127, 129, 306
point defects, 4
polar, 10, 16, 60, 62, 75, 77, 82, 104, 108, 321, 326, 327, 328
polar groups, 10, 82
polarity, 48
polarization, 157, 267, 268, 269
pollution, 313
poly(ethylene terephthalate), 309
poly(methyl methacrylate), 63, 306
poly(vinylpyrrolidone), 316
polyacrylamide, 69, 122
polyamide fiber, 308
polyamine(s), 101, 272
polycarbonate, 165, 206, 211, 216
polycondensation, 120, 128
polycyclic aromatic hydrocarbon, 77
polydimethylsiloxane, 161
polydispersity, 260, 261
polyether, 107
polyimide, 148, 322, 327
polymer, v, vi, vii, 110, 117, 120, 128, 129, 130, 133, 138, 142, 145, 147, 152, 157, 168, 247, 251, 252, 253, 277, 279, 296, 303, 305, 306, 308, 309, 313, 316, 317, 319, 320, 322, 337, 341, 343
polymer chain(s), 111, 117, 119, 120, 121, 128, 130, 146, 148, 150, 151, 165, 242, 248, 250, 280, 283, 284, 325
polymer composites, ix, 62, 146, 147, 148, 152, 155, 156, 157, 158, 161, 162, 166, 167, 243, 319, 340, 341
polymer films, 243
polymer matrix, 82, 147, 148, 150, 152, 153, 156, 157, 161, 165, 226, 242, 280, 305, 322, 325, 328, 329, 332, 336, 337, 338, 340
polymer molecule, 150
polymer nanocomposites, viii, 58, 117, 118, 119, 121, 125, 128, 130, 146, 315
polymer structure, 129, 260
polymer systems, 243
polymer-based composites, 206
polymeric films, 250
polymeric materials, 303, 313
polymeric matrices, 246
polymeric medium, 155
polymerization process, 118, 122, 260, 327
polymerization time, 282
polymer-noble, x, 303, 304, 305, 309, 314
Polymers wrapping, 107
polypropylene, 65, 150
polysaccharide, 110, 228
polysaccharides, 228
polystyrene, 69, 71, 72, 73, 82, 121, 123, 160, 161, 248, 305, 306, 316
polystyrene latex, 316
polyurethane, 113
polyvinyl alcohol, 69
porosity, 211, 213
porous materials, 62
porphyrins, 71, 99, 104, 106, 113
positron emission tomography, 18
potassium, 15, 58, 67, 76, 104, 264, 265, 326, 327
precedent, 328
precipitation, 21, 58, 190, 225, 307, 311
principles, 328
probability, 105
probe, x, 15, 44, 45, 46, 47, 155, 304, 312
project, 293
proline, 25
propagation, 324, 339
propane, 243
propylene, 148, 285, 327
protection, 292, 298
protein structure, 78, 217
proteins, 4, 16, 25, 47, 49, 78, 98, 108, 109, 115
protons, 258, 267, 271, 280
pure water, 110, 183, 184, 317
purification, 10, 63, 273, 282
purity, 59, 111
PVA, 69
PVP, 124
pyrene, 77, 106, 111
pyrimidine, 80
pyrolysis, 16, 214
pyrolytic graphite, 57

Q

quantification, 102
quantization, 180, 181, 182, 199
Quantized conductance, ix
quantum dot(s), vii, 3, 4, 22
quantum phenomena, 173
quartz, 11, 13, 306, 309, 332
quercetin, 113
quinone, 61

R

radiation, vii, 211, 281, 325
radiation therapy, vii
radical, 66, 104, 118

radical formation, 69
radical polymerization, 20, 69, 71, 116, 119, 243, 244, 246, 247, 249
radical reactions, viii
radicals, 10, 59, 60, 66, 67, 68, 69, 83, 98, 100, 103, 104, 118, 119, 128, 230, 243, 262, 267, 268, 269, 271, 326
Radicals, 68, 119
RAFT, 119, 120, 122, 125, 126, 128
Raman spectra, 163
Raman spectroscopy, 65, 102, 163
ramp, 37
reactant, 127
reaction mechanism, 68, 103, 232, 245, 269
reaction medium, 230
reaction order, 103
reaction temperature, 104, 222
reaction time, 62, 221
reactions, viii, x, 8, 9, 10, 11, 12, 17, 18, 20, 21, 49, 59, 60, 62, 63, 64, 66, 70, 74, 99, 101, 109, 153, 154, 241, 250, 258, 273, 288, 304, 305, 314, 316, 317, 327, 330
reactive groups, 117
reactive oxygen, 61
reactive sites, 24, 65, 68, 70, 73
reactivity, 6, 21, 59, 60, 62, 63, 68, 74, 121, 151
reading, x
reagents, 14, 16, 66, 68, 98, 146
reality, 60, 61
receptors, 113
recognition, 4, 111, 114, 246
recovery, 78, 154, 313
recycling, 163, 323
reductive alkylation, 105
regenerate, 68
regression, 268
regression line, 268
rehabilitation, 292
reinforcement, 43, 64, 82, 147, 148, 150, 152, 167, 319, 320, 323, 325, 332, 336, 337
relaxation, 157, 165, 321
relaxation times, 157
repair, 77
repulsion, 61
requirements, 26, 247, 292, 320, 322
researchers, 57, 161
residues, 10, 107, 311
resistance, 46, 66, 95, 112, 114, 153, 155, 156, 173, 175, 183, 184, 186, 227, 232, 296, 313, 320, 322, 325, 327, 339, 340
resolution, 79, 98, 122, 163, 175, 206, 229, 230, 283, 311
response, 39, 45, 108, 115, 192, 246, 247, 248, 272, 273, 274, 292
response time, 272, 273
restoration, 58, 68, 83
RIE, 34, 35, 36, 39
rings, 60, 64, 65, 70, 71, 72, 73, 148, 149, 282
RNA, 25, 116
rods, 206, 224, 230
room temperature, 14, 37, 39, 41, 44, 64, 67, 71, 74, 75, 79, 121, 145, 158, 160, 163, 165, 166, 176, 187, 189, 209, 212, 213, 216, 224, 225, 226, 230, 262, 297, 305, 309
root, 35, 40
root-mean-square, 35
roughness, 35, 177, 192
routes, 15, 130, 208, 258, 259, 262, 305

S

safety, 4, 25, 282
salt concentration, 68
salts, viii, 20, 21, 23, 34, 41, 43, 60, 66, 67, 68, 71, 76, 100, 101, 102, 103, 104, 110, 114, 118, 121, 123, 128, 230, 305, 331, 334
sapphire, 209, 210
saturation, 113, 121, 207, 218, 224, 225, 227
scanning electron microscopy, 311
scanning tunneling microscope, vii, 175
scattering, 114, 164, 173, 174, 195, 249
scholarship, 130
science, 173, 243
scope, 8, 84, 242, 303
sedimentation, 314
seed, 35, 207, 208, 306, 307, 310, 312
seeding, 35
segregation, 58
selectivity, 41, 68, 104, 109, 114, 247, 290
self-assembly, 34, 79, 80, 126, 195, 207, 247, 281
semiconductor, 59, 175, 209, 290, 297
semiconductors, vii, 33, 100, 294, 328
sensing, ix, 3, 34, 43, 44, 45, 47, 49, 128, 246, 248, 249, 262, 290, 297, 303
sensitivity, 26, 41, 47, 113, 114, 195, 247, 251, 272, 273, 274, 290, 291, 297
sensors, ix, 34, 41, 47, 49, 50, 78, 110, 112, 113, 114, 140, 195, 241, 246, 272, 273, 274, 278, 279, 281, 290, 293, 297
serum, 20, 290
serum albumin, 20
shape, 37, 47, 57, 146, 192, 199, 206, 207, 211, 212, 221, 224, 225, 226, 229, 230, 232, 242, 258, 260, 269, 288, 289, 304, 306, 311

shear, 147, 162, 164, 166, 319, 324, 325, 326, 331, 336, 338
shear strength, 147, 319, 325, 326, 331, 336, 338
shortage, 164, 166
showing, 39, 48, 68, 110, 146, 151, 156, 158, 218, 245, 249, 261
Si_3N_4, 327
side chain, 16, 82, 83, 107, 148, 150
signals, 13, 45
signs, 308
silane, 18, 63, 75
Silanization, 63
silica, 64, 205, 206, 212, 213, 216, 226, 244, 309
silicon, 36, 41, 47, 58, 73, 79, 173, 178, 179, 226, 327, 332
silk, 309
silver, vii, 69, 163, 179, 190, 196, 294, 305, 315, 316, 317
Silver, vii, 341
simulation(s), 100, 111, 149, 226, 228, 327
single crystals, 4, 327
SiO_2, 3, 6, 35, 49, 112, 179, 213, 226, 330, 332, 335, 337, 338, 339
skeleton, 105, 146, 282
skin, 116, 155, 321
Slovakia, 279
sodium, 6, 7, 21, 44, 58, 63, 64, 65, 67, 69, 76, 77, 81, 104, 107, 109, 112, 113, 155, 222, 225, 227, 245, 283, 327
sodium hydroxide, 65, 69, 222, 327
solar cells, 71
sol-gel, 211, 213
solid phase, 75, 246
solid state, 198
solubility, 16, 21, 60, 64, 80, 128, 257, 262, 303
solvents, 16, 21, 62, 70, 74, 75, 82, 104, 108, 109, 229, 262, 311
species, viii, 14, 59, 60, 61, 63, 70, 75, 83, 96, 97, 101, 102, 104, 108, 110, 121, 125, 151, 208, 213, 217, 218, 222, 224, 225, 226, 227, 230, 242, 258, 276, 326, 330
specific adsorption, 199, 244
specific surface, 58, 76, 157, 166, 245, 340
spectroscopy, ix, 24, 71, 246, 249, 313
spin, 103, 111, 193
spleen, 18
Spring, 341
stability, x, 14, 16, 19, 22, 24, 34, 35, 47, 58, 65, 82, 83, 98, 125, 177, 183, 195, 209, 246, 257, 275, 281, 288, 289, 296, 304, 306, 314
stabilization, 306, 307, 320, 321
stabilizers, 259
standard deviation, 224
state(s), 60, 109, 148, 149, 153, 155, 162, 167, 174, 191, 195, 226, 230, 242, 244, 258, 280, 306, 311, 322, 337
steel, 95, 163, 292, 296, 298, 333
stem cells, 115
stimulus, 292
STM, vii, 34, 42, 43, 79, 175, 176, 177, 180, 181, 185, 187, 188, 189, 190, 194, 199
STM tip, 43, 175, 176, 187
stock, 221
stoichiometry, 61
storage, ix, 16, 58, 113, 157, 272, 273, 279, 288, 289, 296
stress, 147, 150, 152, 250, 292, 320, 324, 325, 328, 336, 338, 340
Stress-strain curves, 150, 151, 152
stretching, 8, 13, 164, 176, 177, 179, 181, 282
strong interaction, 157, 165, 282
structural changes, 12
structural defects, 80
styrene, 67, 69, 119, 121, 122, 129, 327
substitution, 14, 64, 65, 75, 111
substitution reaction, 14, 64, 75
success rate, 190
sucrose, 7
sulfate, 44, 81, 109
sulfur, 21, 195
sulfuric acid, 36, 101, 327
Sun, 30, 31, 84, 85, 86, 87, 88, 89, 90, 92, 136, 137, 139, 142, 169, 233, 236, 237, 238, 277, 295, 296, 297, 343
superparamagnetic, 68, 121
surface area, 39, 44, 49, 157, 243, 272, 273, 279, 328, 332, 336, 337, 341
surface chemistry, ix, 4, 5, 8, 15, 25, 96, 151, 152, 167, 243, 250, 328
surface energy, 209, 326, 328
surface engineering, 146
surface layer, 230
surface modification, viii, ix, x, 5, 96, 97, 100, 101, 102, 103, 104, 105, 111, 112, 123, 158, 208, 226, 245
surface properties, 62, 63, 82, 326, 327, 336, 340
surface reactions, 7, 109, 117
surface structure, 4, 108
surface tension, 214
surface treatment, 4, 152, 319, 320, 325, 327, 328, 330, 340, 342
surfactant, 67, 73, 81, 104, 107, 157, 197, 213, 216, 224, 226, 258, 283, 285, 295
surfactants, viii, 6, 59, 76, 77, 80, 81, 82, 84, 101, 104, 107, 108, 157, 197, 213, 226, 280, 281, 283, 294

susceptibility, 295
suspensions, 4
sustainable development, 317, 320
swelling, 248, 249
SWNTs, 110, 111, 113, 145, 148, 150, 151, 152, 155, 156
symmetry, 209, 211
synergistic effect, 290
synthetic methods, ix, 211, 259

T

tanks, 322
target, 25, 46, 230, 246, 247, 273
TCC, 45
techniques, viii, 4, 5, 34, 35, 58, 61, 73, 83, 95, 115, 118, 146, 178, 186, 213, 221, 250, 296, 321, 331, 340
technologies, 303
technology, vii, 34, 35, 173, 247, 262
TEM, vii, 17, 18, 22, 24, 150, 151, 158, 160, 161, 163, 164, 165, 167, 206, 212, 216, 223, 224, 226, 228, 229, 231, 248, 249, 259, 262, 284, 285, 286, 287, 308, 311, 312, 314
temperature, 6, 7, 12, 13, 14, 15, 34, 35, 39, 44, 46, 58, 74, 109, 120, 162, 163, 175, 176, 177, 178, 209, 213, 214, 221, 224, 225, 228, 229, 230, 231, 248, 249, 314, 321, 323, 325, 326, 330, 331, 338, 340
tensile strength, 95, 146, 149, 150, 320, 323, 326, 328, 338, 339
tension(s), 214, 321, 324
TEOS, 213
terminals, 115
testing, 324
tetraethoxysilane, 213
tetrahydrofuran, 16, 219
texture, 214, 321
TGA, 45, 47
therapeutic agents, 3
therapeutics, 3
therapy, 83, 115, 227
thermal activation, 72
thermal decomposition, 12, 60, 68, 69, 104, 330
thermal evaporation, 331, 334
thermal properties, viii, 58, 145
thermal stability, 145, 323
thermal treatment, 12, 15, 60, 73, 83, 154, 213, 214, 338
thermogravimetric analysis, x, 304
thermogravimetry, 107
thermolysis, 214, 221
thin films, 22
thinning, 98, 190
tin, 114, 211, 247, 290, 298
tin oxide, 114, 247, 290
tissue, 3, 78, 269, 292
tissue engineering, 78
titanium, 327
tobacco, 218
toluene, 16, 109, 111, 113, 120, 214, 244, 249, 305, 330, 332, 335
top-down, 34, 35, 49, 207, 208
topology, 242, 243, 250
toxicity, 4, 114
tracks, 127
transducer, 42, 44, 176, 180, 273, 274
transferrin, 16
transformation(s), 15, 18, 21, 64, 106
transistor, 41, 60
transistors, vii, 79, 81, 112
transition metal, 65, 250, 332
transition metal ions, 250
transmission, vii, ix, x, 68, 116, 150, 174, 192, 206, 246, 304, 311
transmission electron microscopy, 68, 116, 150, 206, 311
transport, 71, 81, 112, 116, 174, 184, 193, 206, 244, 284, 311, 325, 328, 329
transportation, 113
treatment, 6, 7, 9, 10, 11, 12, 13, 14, 34, 60, 61, 63, 65, 66, 69, 75, 76, 77, 79, 80, 96, 98, 105, 120, 155, 156, 213, 230, 283, 294, 319, 321, 326, 331, 333, 334, 335
tricyclic antidepressant, 247
trifluoroacetate, 105
trifluoroacetic acid, 125
trypsin, 16
tryptophan, 107
tumor(s), 18, 19, 115, 116
tumor growth, 19
tumorigenesis, 18
tunneling, 47, 48, 79, 160, 162, 166, 175, 177, 178, 184
tyramine, 114

U

ultrasound, 69
ultraviolet irradiation, 62
unique features, 33
United States (USA), 294, 297, 342, 343
urea, 46
urethane, 327
uric acid, 43, 114
urinary tract, 18

urine, 25
USSR, 339
UV, 11, 17, 19, 20, 24, 69, 73, 80, 95, 100, 102, 125, 129, 213, 244, 247, 249, 294, 316
UV irradiation, 11, 20, 69, 73
UV light, 102, 125

V

vacancies, 59
valence, 74, 122, 176, 192, 193, 230, 280
valve, 292
vapor, ix, 14, 205, 207, 208, 209, 210, 211, 226, 232, 297, 319, 330
vapor-liquid-solid, 208
vapor-solid (VS), 208
variables, 324
variations, 39, 46, 249
varieties, 206, 212, 314
vehicles, 3, 115, 288, 319, 322
versatility, 218, 242, 288
vessels, 115
vibration, 8, 177, 282
vinylidene fluoride, 154
viruses, 217, 218
viscose, 309
viscosity, 109, 322, 329
visualization, 122
VLS, 206, 207, 208, 209

W

Washington, 91
waste, 313
water vapor, 11, 14
weakness, viii, 156
weight gain, 14
weight loss, 326
weight ratio, 308, 310, 313
wettability, 269, 325, 326, 327, 328, 338
wetting, 146
wires, ix, 34, 36, 37, 43, 174, 179, 181, 198, 199, 206, 209, 212, 217, 224, 225, 226
workers, 13, 124, 127, 189, 192, 194, 243, 244, 246

X

X-ray diffraction (XRD), 227, 230, 310
X-ray photoelectron spectroscopy (XPS), vii, 14, 98, 99, 102, 107, 124, 229, 230, 250, 283, 310, 311, 315

Y

yang, 33
yield, 6, 15, 16, 18, 21, 58, 75, 99, 119, 124, 148, 213, 214, 224, 226, 261, 272, 284, 313, 320, 321, 322

Z

zinc, 211
zirconia, 6